Endocannabinoid Regulation of Monoamines in Psychiatric and Neurological Disorders

Elisabeth J. Van Bockstaele
Editor

Endocannabinoid Regulation of Monoamines in Psychiatric and Neurological Disorders

Springer

Editor
Elisabeth J. Van Bockstaele
Department of Neuroscience
Thomas Jefferson University
Philadelphia, Pennsylvania
USA

ISBN 978-1-4614-7939-0 ISBN 978-1-4614-7940-6 (eBook)
DOI 10.1007/978-1-4614-7940-6
Springer New York Heidelberg Dordrecht London

Library of Congress Control Number: 2013943280

Printed on acid-free paper

Springer is part of Springer Science+Business Media (www.springer.com)

Contents

Contributors

Mary E. Abood, PhD Center for Substance Abuse Research, Department of Anatomy and Cell Biology, Temple University School of Medicine, Philadelphia, PA, USA

María Álvaro-Bartolomé, MSc Laboratori de Neurofarmacologia, Institut Universitari d'Investigació en Ciències de la Salut (IUNICS), Universitat de les Illes Balears (UIB), and Redes Temáticas de Investigación Cooperativa en Salud–Red de Trastornos Adictivos (RETICS-RTA), Palma de Mallorca, Spain

Luís F. Callado, MD, PhD Departamento de Farmacología e Instituto Biocruces, Universidad del País Vasco/Euskal Herriko Unibertsitatea (UPV/EHU), Centro de Investigación Biomédica en Red de Salud Mental (CIBERSAM), Leioa, Bizkaia, Spain

Ana Franky Carvalho Life and Health Science Research Institute (ICVS), School of Health Sciences, University of Minho, Braga, Portugal

ICVS/3B's – PT Government Associate Laboratory, Braga/Guimarães, Portugal

Linda Console-Bram, PhD Center for Substance Abuse Research, Department of Anatomy and Cell Biology, Temple University School of Medicine, Philadelphia, PA, USA

Silvain S. Dang Department of Psychology, The University of British Columbia, Vancouver, Canada

Lieve Desbonnet Biosciences Institute, University College Cork, Cork, Ireland

Samuel DeWitt School of Behavioral and Brain Sciences, University of Texas at Dallas, Dallas, TX, USA

Ahmet Dogrul Department of Medical Pharmacology, Gulhane Military Academy of Medicine, Ankara, Turkey

Susana Esteban, PhD Laboratori de Neurofarmacologia, Institut Universitari d'Investigació en Ciències de la Salut (IUNICS), Universitat de les Illes Balears (UIB), and Redes Temáticas de Investigación Cooperativa en Salud–Red de Trastornos Adictivos (RETICS-RTA), Palma de Mallorca, Spain

Francesca M. Filbey School of Behavioral and Brain Sciences, University of Texas at Dallas, Dallas, TX, USA

Zdeněk Fišar, PhD Department of Psychiatry, First Faculty of Medicine, Charles University in Prague and General University Hospital in Prague, Prague, Czech Republic

Jesús A. García-Sevilla, MD, PhD Laboratori de Neurofarmacologia, Institut Universitari d'Investigació en Ciències de la Salut (IUNICS), Universitat de les Illes Balears (UIB), and Redes Temáticas de Investigación Cooperativa en Salud–Red de Trastornos Adictivos (RETICS-RTA), Palma de Mallorca, Spain

Boris Gorzalka Department of Psychology, The University of British Columbia, Vancouver, Canada

Sven Kroener School of Behavioral and Brain Sciences, University of Texas at Dallas, Dallas, TX, USA

Ken Mackie, MD Department of Psychological & Brain Sciences, Bloomington, IN, USA

Gabriella Gobbi, MD, PhD Neurobiological Psychiatry Unit, Department of Psychiatry, McGill University, Montreal, Canada

Andrea Giuffrida Department of Pharmacology, UT Health Science Center San Antonio, San Antonio, TX, USA

Samir Haj-Dahmane, PhD Research Institute on Addictions, University at Buffalo, State University of New York, Buffalo, NY, USA

Cecilia J. Hillard, PhD Neuroscience Research Center, Medical College of Wisconsin, Milwaukee, WI, US

Jahan Marcu Center for Substance Abuse Research, Department of Anatomy and Cell Biology, Temple University School of Medicine, Philadelphia, PA, USA

Julien Matricon Department of Pharmacology, UT Health Science Center San Antonio, San Antonio, TX, USA

J. Javier Meana, MD, PhD Departamento de Farmacología e Instituto Biocruces, Universidad del País Vasco/Euskal Herriko Unibertsitatea (UPV/EHU), Centro de Investigación Biomédica en Red de Salud Mental (CIBERSAM), Leioa, Bizkaia, Spain

Colm M.P. O'Tuathaigh School of Medicine, University College Cork, Cork, Ireland

Melik Seyrek Department of Medical Pharmacology, Gulhane Military Academy of Medicine, Ankara, Turkey

Roh-Yu Shen, PhD Research Institute on Addictions, University at Buffalo, State University of New York, Buffalo, NY, USA

Ahmet Ulugol Department of Medical Pharmacology, Faculty of Medicine, Trakya University, Edirne, Turkey

Leyre Urigüen, BSc., PhD Departamento de Farmacología e Instituto Biocruces, Universidad del País Vasco/Euskal Herriko Unibertsitatea (UPV/EHU), Centro de Investigación Biomédica en Red de Salud Mental (CIBERSAM), Leioa, Bizkaia, Spain

Elisabeth J. Van Bockstaele Department of Neuroscience, Farber Institute for Neurosciences, Thomas Jefferson University, Philadelphia, PA, USA

John L. Waddington Molecular & Cellular Therapeutics, Royal College of Surgeons in Ireland, Dublin, Ireland

Bulent Yalcin Department of Anatomy, Gulhane Military Academy of Medicine, Ankara, Turkey

Chapter 1
Endocannabinoids and Monoamines: Modulating the Modulators

Elisabeth J. Van Bockstaele

Abstract The past decade has seen a tremendous growth in knowledge related to cannabinoid receptor signaling in brain. In addition, the impact and consequences of cannabinoid modulation of monoaminergic circuits are steadily emerging demonstrating a significant interaction between these two systems in a variety of psychiatric (affective disorders) and neurological disorders (neurodegeneration, pain). Areas to be covered in the accompanying chapters include an overview of the endocannabinoid system, a summary of current cannabinoid receptor nomenclature, and pharmacological principles as well as electrophysiological, biochemical, and behavioral evidence for cannabinoid modulation of dopaminergic, noradrenergic, and serotonergic circuitry.

As the most commonly used illicit drug, cannabis poses a serious risk for psychopathology (Ferdinand et al. 2005). Frequent use doubles the risk for depression and anxiety, and significantly decreases multiple indices of psychosocial functioning (Lundqvist 1995a, b, 2005, 2010; Lundqvist et al. 2001; Patton et al. 2002; Anglin et al. 2012). Although exposure to delta-9-tetrahydrocannabinol (THC), the primary psychoactive component of cannabis, is associated with a number of adverse psychological effects, recent evidence suggests that administration of synthetic cannabinoid receptor agonists/antagonists may hold some therapeutic potential. Furthermore, altering endogenous cannabinoid signaling by manipulating the metabolism and uptake of endogenous cannabinoids may provide clinical benefits. Therefore, elucidating neural targets of cannabinoids has significant public health relevance.

The past 2 decades have seen a tremendous growth in knowledge related to cannabinoid modulation of monoaminergic circuits and their interactions in a variety of psychiatric and neurological disorders. Despite increasing evidence from preclinical data suggesting that therapeutic use of cannabinoid-based drugs may outweigh any potential risks in certain serious medical conditions, the debate surrounding its

E. J. Van Bockstaele (✉)
Department of Neuroscience, Farber Institute for Neurosciences, Thomas Jefferson University, 900 Walnut Street, Suite 417, Philadelphia, 19107 PA, USA
e-mail: ejv101@jefferson.edu
e-mail: elisabeth.vanbockstaele@jefferson.edu
e-mail: evanbockstaele@gmail.com

E. J. Van Bockstaele (ed.), *Endocannabinoid Regulation of Monoamines in Psychiatric and Neurological Disorders,* DOI 10.1007/978-1-4614-7940-6_1,

widespread utility continues as regulatory concerns preclude a smooth transition of promising preclinical studies into clinical trial testing. This may persist in the near future as state and federal governments debate over regulation of medicinal applications of cannabis. Applications for medicinal cannabinoids that are already under investigation include the treatment of nausea, anorexia, neurodegeneration, inflammation, excitotoxicity, and pain. The appetitive and antiemetic properties of cannabinoids have led to the approval of their use in chemotherapy and AIDS patients. There is growing evidence for therapeutic cannabinoid effects on inflammatory and excitotoxic cellular processes that are linked to epilepsy, Parkinson's disease, amyotrophic lateral sclerosis, spasticity, and central nervous system (CNS) injury. The chapters, herein, review and discuss current insights into the brain endocannabinoid system, cannabinoid receptor signaling on synaptic plasticity, and potential therapeutic applications with a particular focus on endocannabinoid modulation of dopaminergic, noradrenergic, and serotonergic circuitry.

In the CNS, the endocannabinoid system (ECS) is involved in a variety of physiological functions because of abundant expression of its receptors and ligands (Herkenham et al. 1991; Mackie 2008). Endocannabinoids, anandamide (arachidonoyl ethanolamide) and 2-arachidonoylglycerol (2-AG), are arachidonic acid derivatives that exist as precursor lipids in the plasma membrane and are synthesized by the action of specific lipases under certain physiological or pathological conditions (Piomelli et al. 1998; Piomelli 2003, 2005; Basavarajappa 2007). Anandamide and 2-AG have been implicated in the control of emotional reactivity, motivated behaviors, and energy homeostasis primarily by actions on brain cannabinoid (CB) type 1 receptors (CB1r) (Martin 1986; Mechoulam, Parker et al. 2002). As one of the most abundant G protein coupled receptors (GPCRs) in the mammalian brain, CB1 receptors have been implicated in the regulation of learning and memory, food intake, pain, and mood. A second cannabinoid receptor, CB2r, expressed primarily in cells of the immune and hematopoietic systems has been reported in brain. Using a radiolabeled cannabinoid receptor agonist, Herkenham et al. (1991) mapped cannabinoid receptor binding sites throughout the rat brain. Within the brainstem, cannabinoid receptors are sparsely expressed in comparison to regions highly enriched in cannabinoid receptors such as the hippocampus, basal ganglia, cortex, and cerebellum. However, regions with low to moderate cannabinoid receptor binding were noted in noradrenergic brainstem nuclei such as the locus coeruleus (LC) and nucleus of the solitary tract (Herkenham et al. 1991). Interestingly, there have been reports of a lack of correlation between the density of CB1 receptors and the efficiency of receptor coupling (Breivogel et al. 1997).This may explain, in part, why functionally important responses can be manifested in areas with sparse CB1r labeling such as the brainstem and hypothalamus (Jamshidi and Taylor 2001; Rademacher et al. 2003). Mechanisms including receptor dimerization (Mackie 2005) or changes in signal amplification have been suggested. Along these lines, amplification of CB1 signaling has been reported by involvement of a protein kinase A-dependent phosphorylation of DARPP-32, achieved via modulation of dopamine D2 and adenosine A2A transmission (Andersson et al. 2005).

In Chap. 2, Kenneth Mackie describes the components of the endocannabinoid signaling system as well as an important functional role for endocannabinoids—their role in modulating diverse forms of synaptic plasticity. The notion that endocannabinoids can inhibit synaptic transmission, coupled with the observation that endocannabinoids are often produced under conditions of intense neuronal activity, underscores the importance of cannabinoid-induced modulation of synaptic transmission in a surprisingly diverse number of ways.

In Chap. 3, a summary of basic pharmacological definitions, principles, and mechanisms underlying cannabinoid receptor activation and current receptor nomenclature for classifying a target as a cannabinoid receptor is provided by Marcu et al. The authors consider and discuss a large number of emerging reports indicating that the resulting effects of endo-, phyto-, and synthetic cannabinoid interactions cannot be definitively explained based on a two-cannabinoid receptor theory. Therefore, the authors review the actions of endocannabinoids not restricted to the CB1r and CB2r, including additional GPCRs, ion channels, ion channel receptors (i.e., transient receptor potential cation channel; TRP) and nuclear receptors (peroxisome proliferator-activated receptor).

While amino acid transmitter systems (Kano et al. 2009) represent an important target of the ECS and exogenous cannabinoid-based drugs, interactions with monoaminergic circuitry has revealed important consequences for global effects on behavior. Accumulating evidence indicates a significant role of the cannabinoid system in the regulation of basal ganglia function, particularly with respect to reward, psychomotor function, and motor control. Dysfunction in the ECS is likely to impact dopamine- and basal ganglia related neurospsychiatric disorders, including drug addiction, psychosis, Parkinson's disease and Huntington's disease. The distribution of components of the ECS within basal ganglia networks suggest that the motivational and motor effects of cannabinoid-based ligands are modulated, in part, by dopamine transmission. In Chap. 4, De Witt et al. summarize a role for direct and indirect mechanisms underlying cannabinoid modulation of dopaminergic transmission. Specifically, the authors review existing evidence of cannabinoid modulation of excitatory and inhibitory networks in the reward system with the net effect of regulating the overall doapminergic 'reward tone'. Then, they discuss emerging evidence that cannabis exerts its addictive properties through effects of the ECS on the brain reward neurocircuitry. Specifically, the authors describe evidence for cue-elicited craving for marijuana, and, importantly, how in the absence of cannabis itself, cannabis-associated cues trigger activation in the reward pathway implicated in the neuropathology of addiction.

The use of synthetic cannabinoid receptor agonists/antagonists or compounds targeting endocannabinoid synthesis/metabolism in brain has received widespread attention as these approaches may hold some therapeutic potential for neurological and psychiatric disorders and has stimulated investigations into manipulating endogenous cannabinoids for potential clinical benefit. In Chap. 5, Matricon and Giuffrida discuss interactions between cannabinoids, dopamine and glutamate in the basal ganglia and review how targeting the cannabinoid receptor system might constitute an integrated pharmacotherapeutic approach in addressing the pathophysiology of disorders characterized by dopamine dysfunction, such as Parkinson's disease and schizophrenia. Specifically, the authors summarize evidence supporting direct and indirect cannabinoid receptor agonists as promising antiparkinsonian, antidyskinetic, and antipsychotic-like properties in animal models but highlight the lack of large-scale clinical studies to translate these preclinical findings into new therapies.

As the ECS plays a role in the regulation of mood, accumulating evidence supports changes in the ECS by chronic treatment with antidepressants, including serotonin and/or norepinephrine reuptake inhibitors as well as monoamine oxidase inhibitors. In Chap. 6, Fisar reviews preclinical and clinical data supporting a critical role for monoamine neurotransmission in the neurochemistry of mood disorders. He discusses the pathophysiology of mood disorders from a perspective of dysfunction in energy metabolism of neurons, modulation of inflammatory pathways, changes in activities of transcription factors, neurotrophic factors and other components involved in neuroplasticity and apoptosis. The chapter continues from a perspective of neuromodulation of synapses by cannabinoids summarizing evidence showing that cannabinoids have the capacity to produce increased hippocampal neurogenesis and that this is positively correlated with its antidepressant effects.

The ability of cannabinoid agonists to enhance norepinephrine release plays a critical role in the mood altering properties and cognitive effects of cannabis-based compounds. One of the most significant behavioral signs associated with cannabinoid administration relates to impairment in attention, vigilance, and cognitive processing (Casswell and Marks 1973; Chait 1992). Long-term cannabis use results in impairment of attention that worsens with increasing years of regular use (Solowij et al. 2002). Studies examining the effects of cannabinoids on attention (Hillyard and Kutas 1983; Naatanen 1990) have shown that chronic cannabis use affects information processing (Kempel et al. 2003) where users are unable to effectively focus (Solowij et al. 1995). One neurochemical target at which cannabinoids may interact to have global effects on behavior is brain noradrenergic circuitry. Moreover, the noradrenergic system continues to be an important target in the development of new therapies for affective disorders because of its critical role in the modulation of emotional state and regulation of arousal and stress responses (Heninger and Charney 1988; Charney et al. 1989; Ballenger 2000; Carrasco and Van de Kar 2003). In Chap. 7, Carvalho and Van Bockstaele discuss anatomical, biochemical, and behavioral evidence for cannabinoid modulation of noradrenergic circuits and review the role of norepinephrine in cannabinoid-induced behaviors, specifically aversion. The authors summarize studies showing that brain noradrenergic transmitter and receptors are significantly impacted by cannabinoids. They review how, under basal conditions, exposure to a synthetic CB1r agonist increases anxiety-like behaviors that correlate with increases in multiple indices of brain noradrenergic activity. Interestingly, a different consequence to the regulation of norepinephrine by cannabinoids is observed under conditions of stress. Specifically, stress-induced increases in cortical NE levels are significantly attenuated by prior treatment with a CB1r agonist suggesting complex actions of cannabinoids on noradrenergic circuitry that vary under basal vs stress conditions. These findings indicate that, with respect to monoamine release, CB1r modulation is complex and can involve either stimulation or inhibition of neurotransmitter released depending on neuronal state (Mackie 2005; Kano et al. 2009). This is consistent with studies showing that modulation of neurotransmitter release by cannabinoid receptor agonists can be different depending on neuronal firing rate (Roloff and Thayer 2009) and is likely to be dynamically regulated by stress. In Chap. 8, Gorzalka and Dang discuss evidence supporting the hypothalamic-pituitary-adrenal (HPA) axis as a major area of interaction between the endocannabinoid and the noradrenergic systems in mediating stress responses. By defining and paralleling the endocannabinoid and noradrenergic systems as 'gatekeepers', the authors discuss the role of norepinephrine in mediating physiological responses to stress by mobilizing the HPA axis and the ECS as preventing maladaptive HPA hyperactivation during chronic stress. Furthermore, the authors expand on the sexual dimorphism in both systems, and implications for psychiatric disorders, specifically depression.

Considerable evidence has accumulated to support the hypothesis that the ECS is altered by stress exposure and modulates stress responses through effects on synaptic activity. These data have important implications for therapeutic treatment of disorders in which hyperactive HPA axis activity contributes to disease. The CB1r is present in stress responsive circuits (frontal cortex, amygdala, and hypothalamus) that are essential to the expression of anxiety (Herkenham et al. 1990; Roloff and Thayer 2009; Oropeza 2005 #11). Acute restraint stress has been shown to increase the synthesis of endogenous endocannabinoids in limbic forebrain areas (Patel et al. 2005). In addition, release of endocannabinoids has been shown to mediate opioid-independent stress-induced analgesia by actions in the periaqueductal gray. Complex interactions exist between the cannabinoid system and stress responsivity. Low doses of cannabinoid agonists administered in familiar, nonstressful environments, typically result in positive responses such as enhanced euphoria and a reduction in anxiety (Hollister 1986). However, dysphoric reactions are commonly manifested as panic, anxiety, and paranoia and occur in response to high doses of consumption or when the drug is administered in environments that are stressful (Greggand Campbell 1976; Gregg et al. 1976). In Chap. 9, Hillard reviews how glucocorticoids mobilize endocannabinoids and how endocannabinoid-CB1r signaling serves as a primary regulator of synaptic plasticity via changes in presynaptic release, specifically subserving short-term, activity-driven changes in synaptic strength as well as other forms of presynaptic plasticity. She further discusses preclinical models that have suggested that therapeutic agents such as fatty acid amide hydroylxase (FAAH) inhibitors should be examined in humans for treatment of anxiety and depressive disorders that are characterized by excessive or prolonged HPA axis activation. This is based on studies showing that FAAH inhibition inhibits stress-induced increases in circulating glucocorticoids, reduces anxiety in adverse environments (Patel and Hillard 2006), and decreases immobility in rats in the forced swim assay (Gobbi et al. 2005) (also see Chap. 13).

In Chap. 10, Urigüen and García-Sevilla highlight findings from numerous experimental studies on the role of endocannabinoids and CB1rs in the modulation of brain monoaminergic systems: i.e., neuronal (spontaneous firing rate) activity and synthesis and release of the corresponding neurotransmitter. The authors also discuss the effects of cannabinoid drugs on the activity of presynaptic monoaminergic receptors (autoreceptors and heteroreceptors) that regulate the synthesis and release of classic neurotransmitters and participate in the mechanisms of action of antidepressant drugs. Finally, the authors discuss the possible relevance of the ECS and CB1rs in the pathophysiology and treatment of major depression and schizophrenia, with a special focus on evidence from postmortem human brain studies.

Haj-Dahmane and Shen, in Chap. 11, review the current understanding of the cellular mechanisms by which the ECS modulates the function of the serotonergic system and how stress mediators regulate endocannabinoid signaling in the dorsal raphe nucleus (DRN). In the projection areas, endocannabinoids modulate serotonin transmission by suppressing serotonin release and regulating the expression and function of serotonin receptors (i.e., 5-HT_{1A} and 5-HT_{2A}). At the level of the DRN, endocannabinoid signaling controls the excitability of 5-HT neurons primarily by modulating the strength of glutamatergic and GABAergic inputs impinging on DRN 5-hydroxytryptamine (5-HT) neurons. The authors then highlight the discovery that DRN 5-HT can synthesize and release endocannabinoids in an activity-dependent "phasic" mode, which represents an additional mechanism that enables 5-HT neurons to fine-tune their electrical activity and control central 5-HT transmission. Finally, the authors discuss the implications of endocannabinoid signaling in the DRN as a key modulator and integrator mediating the homeostatic response to stress explaining that a dysfunction of endocannabinoid signaling in the 5-HT system could contribute to stress-related mood disorders. In Chap. 12, Gobbi discusses how CB1r agonists, antagonists, and FAAH inhibitors modulate the firing activity of 5-HT neurons located in the DRN. While the CB1 receptor agonist WIN 55,212-2 produces a bell-shaped curve, increasing 5-HT firing at low doses (0.1–0.3 mg/kg) and decreasing firing at higher doses (>0.3 mg/kg), the FAAH inhibitor URB597 produces a sigma-shaped curve, with a plateau at the highest doses tested (0.3 mg/kg). THC produces a mixed response on 5-HT firing activity with 26 % of neurons showing an increase, 33 % showing a decrease, and 42 % showing no response. However, after 4 days, intraperitoneal (i.p.) injections of THC (1 mg/kg) produced a significant elevation of firing. These findings indicate that CB1r agonists and FAAH inhibitors interact with the 5-HT system and that these effects are related to emotional behaviors. Dogrul, in Chap. 13, reviews novel strategies for the development of novel therapeutics for pain such as, using peripherally restricted CB1 agonists, CB2 agonists or combining low doses of analgesic drugs from different pharmacological groups with the goal of developing additive or synergistic combinations with enhanced pain relief and reduced CNS effects.

O'Tuathaigh et al., in Chap. 14, review a large number of experimental studies examining the relationship between cannabis use, psychosis, and the influence of moderating environmental and genetic background factors. The authors explain that, although a minority of cannabis users develop subclinical symptoms or a clinical psychotic disorder, potential amplification of cannabis risk when interacting with genetic and other environmental risk factors may contribute to progression of the disorder. Focusing on clinical and preclinical studies, the authors elaborate on genetic data that provide convergent evidence for the notion of an interaction between cannabis and individual genetic vulnerability, with a focus on genes encoding proteins implicated in DA signaling.

In summary, improving treatments for increasingly prevalent and devastating psychiatric and neurological illnesses is needed. Although challenges exist with medicinal cannabis, the potential for the development of compounds designed to modulate endocannabinoid levels or the use of synthetic cannabinoids with well-defined pharmacological properties may provide significant clinical benefit for psychiatric and neurological disorders. The potential for establishing cannabinoid-monoaminergic interactions as a novel target in the development of improved treatment strategies for psychiatric disorders is exemplified by the effectiveness of the CB1r agonist, nabilone, in the management of symptoms of post-traumatic stress disorder (Fraser 2009). Taken with recent evidence that the endocannabinoid and noradrenergic systems interact in stress-related memory consolidation (Hill and McEwen 2009; Campolongo et al. 2009), targeting interactions between these two systems may represent a novel approach for the treatment of stress-induced anxiety disorders. The potential for establishing cannabinoid-monoaminergic interactions as a novel target in the development of improved treatment strategies for neurological disorders is also promising but will require large-scale clinical studies to determine whether promising preclinical findings translate into new therapies.

References

Andersson M, Usiello A et al (2005) Cannabinoid action depends on phosphorylation of dopamine- and cAMP-regulated phosphoprotein of 32 kDa at the protein kinase A site in striatal projection neurons. J Neurosci 25(37):8432–8438

Anglin DM, Corcoran CM et al (2012) Early cannabis use and schizotypal personality disorder symptoms from adolescence to middle adulthood. Schizophr Res 137(1–3):45–49

Ballenger JC (2000) Anxiety and depression: optimizing treatments. Prim Care Companion J Clin Psychiatry 2(3):71–79

Basavarajappa BS (2007) Neuropharmacology of the endocannabinoid signaling system-molecular mechanisms, biological actions and synaptic plasticity. Curr Neuropharmacol 5(2):81–97

Breivogel CS, Sim LJ et al (1997) Regional differences in cannabinoid receptor/G-protein coupling in rat brain. J Pharmacol Exp Ther 282(3):1632–1642

Campolongo P, Roozendaal B et al (2009) Endocannabinoids in the rat basolateral amygdala enhance memory consolidation and enable glucocorticoid modulation of memory. Proc Natl Acad Sci U S A 106(12):4888–4893

Carrasco GA, Van de Kar LD (2003) Neuroendocrine pharmacology of stress. Eur J Pharmacol 463(1–3):235–272

Casswell S, Marks D (1973) Cannabis induced impairment of performance of a divided attention task. Nature 241(5384):60–61

Chait LD, Pierri J (1992) Effects of smoked marijuana on human performance: a critical review. In: Murphy L, Bartke A (eds) Marijuana/Cannabinoids: Neurobilogy and Neurophysiology CRC Press, Boca Raton, 387–423

Charney DS, Woods SW et al (1989) Noradrenergic function in generalized anxiety disorder: effects of yohimbine in healthy subjects and patients with generalized anxiety disorder. Psychiatry Res 27(2):173–182

Ferdinand RF, Ende J et al (2005) Cannabis–psychosis pathway independent of other types of psychopathology. Schizophr Res 79(2–3):289–295

Fraser GA (2009) The use of a synthetic cannabinoid in the management of treatment-resistant nightmares in posttraumatic stress disorder (PTSD). CNS Neurosci Ther 15(1):84–88

Gobbi G, Bambico FR et al (2005) Antidepressant-like activity and modulation of brain monoaminergic transmission by blockade of anandamide hydrolysis. Proc Natl Acad Sci U S A 102(51):18620–18625

Gregg JM, Campbell RL et al (1976) Cardiovascular effects of cannabinol during oral surgery. Anesth Analgesia 55(2):203–213

Gregg JM, Small EW et al (1976) Emotional response to intravenous delta9tetrahydrocannabinol during oral surgery. J Oral Surgery 34(4):301–313

Heninger GR, Charney DS (1988) Monoamine receptor systems and anxiety disorders. Psychiatr Clin North Am 11(2):309–326

Herkenham M, Lynn AB et al (1990) Cannabinoid receptor localization in brain. Proc Natl Acad Sci U S A 87(5):1932–1936

Herkenham M, Lynn AB et al (1991) Characterization and localization of cannabinoid receptors in rat brain: a quantitative in vitro autoradiographic study. J Neurosci 11(2):563–583

Hill MN, McEwen BS (2009) Involvement of the endocannabinoid system in the neurobehavioral effects of stress and glucocorticoids. Prog Neuropsychopharmacol Biol Psychiatry 34(5):791–797

Hillyard SA, Kutas M (1983) Electrophysiology of cognitive processing. Annu Rev Psychol 34:33–61

Hollister LE (1986) Health aspects of cannabis. Pharmacological Rev 38(1):1–20

Jamshidi N, Taylor DA (2001) Anandamide administration into the ventromedial hypothalamus stimulates appetite in rats. British J Pharmacol 134(6):1151–1154

Kano M, Ohno-Shosaku T et al (2009) Endocannabinoid-mediated control of synaptic transmission. Physiol Rev 89(1):309–380

Kempel P, Lampe K et al (2003) Auditory-evoked potentials and selective attention: different ways of information processing in cannabis users and controls. Neuropsycho Biol 48(2):95–101

Lundqvist T (1995a) Chronic cannabis use and the sense of coherence. Life Sci 56(23–24):2145–2150

Lundqvist T (1995b) Specific thought patterns in chronic cannabis smokers observed during treatment. Life Sci 56(23–24):2141–2144

Lundqvist T (2005) Cognitive consequences of cannabis use: comparison with abuse of stimulants and heroin with regard to attention, memory and executive functions. Pharmacol Biochem Behav 81(2):319–330

Lundqvist T (2010) Imaging cognitive deficits in drug abuse. Curr Top Behav Neurosci 3:247–275

Lundqvist T, Jonsson S et al (2001). Frontal lobe dysfunction in long-term cannabis users. Neurotoxicol Teratol 23(5):437–443

Mackie K (2005) Cannabinoid receptor homo- and heterodimerization. Life Sci 77(14):1667–1673

Mackie K (2008) Cannabinoid receptors: where they are and what they do. J Neuroendocrinol 20(Suppl 1):10–14

Martin BR (1986) Cellular effects of cannabinoids. Pharmacol Rev 38(1):45–74

Mechoulam R, Parker LA et al (2002) Cannabidiol: an overview of some pharmacological aspects. J Clin Pharmacol 42(11 Suppl) 11–19

Naatanen R (1990) The role of attention in auditory information processing revealed by event-related brain potentials. Behav Brain Sci 13:201–288

Oropeza VC, Page ME, Van Bockstaele EJ (2005) Systemic administration of WIN 55,212-2 increases norepinephrine release in the rat frontal cortex. Brain Res. 1046(1–2):45–54

Patel S, Hillard CJ (2006) Pharmacological evaluation of cannabinoid receptor ligands in a mouse model of anxiety: further evidence for an anxiolytic role for endogenous cannabinoid signaling. J Pharmacol Exp Ther 318(1):304–311

Patel S, Roelke CT et al (2005) Inhibition of restraint stress-induced neural and behavioral activation by endogenous cannabinoid signaling. Eur J Neurosci 21(4):1057–1069

Patton GC, Coffey C et al (2002) Cannabis use and mental health in young people: cohort study. BMJ 325(7374):1195–1198

Piomelli D (2003) The molecular logic of endocannabinoid signaling. Nat Rev Neurosci 4(11):873–884

Piomelli D (2005) The endocannabinoid system: a drug discovery perspective. Curr Opin Investig Drugs 6(7):672–679

Piomelli D, Beltramo M et al (1998) Endogenous cannabinoid signaling. Neurobiol Dis 5(6): 462–473

Rademacher DJ, Patel S et al (2003) Microinjection of a cannabinoid receptor antagonist into the NTS increases baroreflex duration in dogs. American journal of physiology. Heart Circ Physiol 284(5):1570–1576

Roloff AM, Thayer SA (2009) Modulation of excitatory synaptic transmission by Delta 9-tetrahydrocannabinol switches from agonist to antagonist depending on firing rate. Mol Pharmacol 75(4):892–900

Solowij N, Michie PT et al (1995) Differential impairments of selective attention due to frequency and duration of cannabis use. Biol Psychiatry 37(10):731–739

Solowij N, Stephens R et al (2002) Does marijuana use cause long-term cognitive deficits? JAMA 287(20):2653–2654

Chapter 2
Endocannabinoid-Mediated Synaptic Plasticity

Ken Mackie

Abstract Endocannabinoids are ubiquitous lipid signaling molecules that mimic some of the actions of phytocannabinoids such as delta-9-tetrahydrocannabinol. Endocannabinoids are a component of the endocannabinoid signaling system, which comprises the endocannabinoids, the enzymes that synthesize and degrade endocannabinoids, and cannabinoid receptors. Within the central nervous system (CNS), endocannabinoids serve as modulators of both long-term and short-term synaptic plasticity. This review will briefly review the signaling of cannabinoid-1 (CB_1) and cannabinoid-2 (CB_2) receptors and then explore some of the roles endocannabinoids play in mediating diverse forms of synaptic plasticity, with an emphasis on recent findings.

Endocannabinoids (eCBs) are small lipid signaling molecules, so named because they often engage the same receptors as the well-known phytocannabinoid, delta-9-tetrahydrocannabinol (THC). Work over the past 20 years firmly establishes that eCBs participate in signaling in many parts of the body, and especially in the nervous system (Katona and Freund 2012). This review will consider an important functional role for eCBs—their role in modulating diverse forms of synaptic plasticity. However, first, the components of the eCB signaling system will be considered, with an emphasis on those components that are most relevant for synaptic plasticity, and then their regulation.

All well-characterized eCBs are arachidonic acid derivatives. One of their key features is that they exist as precursor lipids in the cell membrane and are liberated by the action of specific lipases under certain physiological or pathological conditions. The two most studied eCBs are anandamide (arachidonoyl ethanolamide) and 2-arachidonoyl glycerol (2-AG). The precursors of anandamide are the N-arachidonoyl phosphatidyl ethanolamines (NAPEs). Anandamide can be produced from NAPEs by several different pathways (Ahn et al., 2008). The precursor of 2-AG is chiefly phosphatidyl bisphosphate (PIP2). 2-AG is primarily produced from PIP2 by the sequential action of a phospholipase C (PLC) and one of two diacyl glycerol lipases (Tanimura et al. 2010). The completely different routes of anandamide and 2-AG synthesis suggest that they are produced under different physiological

K. Mackie (✉)
Department of Psychological & Brain Sciences, 1101 E 10th St., Bloomington, IN 47401, USA
e-mail: kmackie@indiana.edu

E. J. Van Bockstaele (ed.), *Endocannabinoid Regulation of Monoamines in Psychiatric and Neurological Disorders*, DOI 10.1007/978-1-4614-7940-6_2,

conditions. In general, this is what has been found (Hohmann et al. 2005; Liu et al. 2008; Puente et al. 2011).

Similar to eCB synthesis, eCB degradation largely occurs via different pathways. Most anandamide is degraded by fatty acid amino hydrolase (FAAH) (McKinney and Cravatt 2005; Ahn et al. 2009). In contrast, 2-AG can be degraded by several serine hydrolases (monoacyl glycerol lipase, alpha beta hydrolase domain-containing 6, alpha beta hydrolase domain-containing 12, and FAAH) (Blankman et al. 2007). Thus, anandamide and 2-AG breakdown will be differentially regulated and inhibition of the respective pathways can be a useful tool to identify the eCB involved in a specific form of eCB-mediated synaptic plasticity.

For the purposes of this review, we will primarily consider the cannabinoid-1 (CB_1) and cannabinoid-2 (CB_2) receptors. (It is important to note that eCBs can interact with a wide variety of receptors and other molecules including other G protein-coupled receptors (GPCRs), transcription factors, and ion channels (Zygmunt et al. 1999; Fu et al. 2003; Oz 2006; McHugh et al. 2010). However, with a few notable exceptions (Melis et al. 2008; Mazzola et al. 2009; Chavez et al. 2010), these do not yet have an established role in eCB-mediated synaptic plasticity). The CB_1 and CB_2 receptors were both cloned a little more than 20 years ago. CB_1 receptors are highly expressed in the central nervous system (CNS) but are present throughout the body and play roles in processes as diverse as metabolism, reproduction, and immune regulation (Nagarkatti et al. 2009; Talwar and Potluri 2011; Ward and Raffa 2011). CB_2 receptors are less highly expressed, which has made identification of the cell types expressing them somewhat more problematic (Atwood and Mackie 2010). It is well accepted that CB_2 receptors are expressed in several types of immune cells, particularly cells of macrophage lineage, including microglia. The extent of their expression in neurons and other glia is more contentious (Atwood and Mackie 2010). A striking feature of CB_2 receptors is their high inducibility. For example, in the experimental allergic encephalitis (EAE) model of multiple sclerosis, CB_2 mRNA levels can increase more than 100 fold (Maresz et al. 2005). The low level of CB_2 receptor expression under basal conditions and the lack of suitably sensitive antibodies has, at the level of anatomy, led to much confusion (reviewed in Atwood and Mackie 2010). Thus, the most conclusive evidence for a role of CB_2 in the CNS outside of microglia comes from functional and molecular studies (e.g., Xi et al. 2011; den Boon et al. 2012; Zarruk et al. 2012). However, these studies lack anatomical precision, and often it is hard to conclusively determine which cell type(s) are involved. Conclusive resolution of these issues will require cell type-specific deletion of CB_2 receptors.

Both CB_1 and CB_2 receptors are GPCRs. They primarily couple to Gi/Go G proteins, thus their dominant signaling pathways include inhibition of adenylyl cyclase, activation of mitogen-activated protein (MAP) kinases, inhibition of some voltage-dependent calcium channels, and activation of G protein-gated inwardly rectifying potassium (GIRK) channels (Howlett et al. 2002). Nonetheless, it is important to appreciate that both receptors can couple to alternative pathways. For example, CB_1 can stimulate adenylyl cyclase (Glass and Felder 1997; Felder et al. 1998) and both receptors can release calcium from intracellular stores (Sugiura et al. 1997; Lauckner et al. 2005; Shoemaker et al. 2005). In considering activation of CB_1 and CB_2 receptors by eCBs, two other properties of these ligands need to be considered. The first is efficacy. Efficacy is a measure of how completely a particular ligand can activate a receptor. There is good agreement that anandamide is a lower efficacy agonist than 2-AG (Mackie et al. 1993; Luk et al. 2004; Sugiura et al., 2006). The consequences of this depend on receptor number and the efficiency of the receptor's coupling to downstream signaling pathways. In general, low receptor density and/or poor coupling to downstream effectors will cause a low-efficacy agonist to have a diminished cellular response relative to a high-efficacy agonist (e.g., Luk et al. 2004). Under these conditions the low-efficacy agonist is considered to be a partial agonist. Conversely, under conditions where receptor density is high or effector coupling is strong, low- and high-efficacy agonists may have indistinguishable cellular responses. The second important property is functional selectivity. Functional selectivity refers to the ability of different agonists to differentially activate distinct signaling pathways, despite both activating the receptor (Kenakin and Miller 2010). Both CB_1 and CB_2 ligands can show functional selectivity; however, the functional selectivity of commonly encountered CB_2 agonists is particularly striking (Atwood et al. 2012a, b).

Most Gi/Go-coupled GPCRs also modulate ion channels. CB_1 receptors inhibit several voltage-dependent calcium channels, particularly N (Cav2.2) and P/Q (Cav2.1) channels (Mackie and Hille 1992; Mackie et al. 1995; Twitchell et al. 1997). In addition, CB_1 receptors activate GIRK channels (Mackie et al. 1995). CB_2 receptors likely modulate the same types of ion channels, although the strong functional selectivity of different CB_2 ligands means that only some ligands can do this. For example, 2-AG potently inhibits calcium channels, whereas anandamide does not (Atwood et al. 2012a, b).

The property of cannabinoid receptors to activate GIRK channels and inhibit calcium channels suggests that they will likely dampen neuronal excitability and inhibit synaptic transmission. A large number of studies support this contention (Roth 1978; Shen et al. 1996; Levenes et al. 1998; Misner and Sullivan 1999; Vaughan et al. 1999; Hajos et al. 2000; Takahashi and Linden 2000). Inhibition of synaptic transmission by CB_1 receptors is an example where ligand efficacy is important. For example, THC, a low-efficacy CB_1 agonist, has little effect on synaptic transmission in some model systems and can actually antagonize inhibition of synaptic transmission by 2-AG (Shen and Thayer 1999; Straiker and Mackie 2005). However, whether THC inhibits synaptic transmission depends on many factors, including the frequency of stimulation (Roloff and Thayer 2009; Hoffman and Lupica 2012).

The concept that eCBs can inhibit synaptic transmission, coupled with the observation that eCBs are often produced under conditions encountered during vigorous synaptic transmission, gave rise to a series of studies to determine if eCBs produced in this way could modulate synaptic transmission. Indeed, eCBs generated during intense neuronal activity can modulate synaptic transmission in a surprisingly diverse number of ways (Kano et al. 2009; Castillo et al. 2012).

The first type of synaptic plasticity demonstrated to be mediated by eCBs was depolarization-induced suppression of inhibition (DSI) (Ohno-Shosaku et al. 2001;

Wilson and Nicoll 2001). DSI is a phenomenon where intense depolarization (e.g., repeated action potentials or a 1–5 s step depolarization to 0 mV) of a postsynaptic neuron leads to a transient (tens of seconds) suppression of inhibitory transmission onto that neuron (Llano et al. 1991; Pitler and Alger 1992; Pitler and Alger 1994). An analogous phenomenon involving excitatory transmission mission is called depolarization-induced suppression of excitation (DSE) (e.g., Kreitzer and Regehr 2001). These phenomena have been extensively studied in both the hippocampus and cerebellum. Work from a number of investigators has arrived at the following canonical mechanism (however, note there is not complete agreement on these steps (Kano et al. 2009)): depolarization of the postsynaptic cell leads to an increase in intracellular calcium (can be entry through calcium channels and/or release from intracellular stores) that activates diacyl glycerol lipase (DAGL) alpha. DAGL then cleaves the acyl chain in the one position on diacyl glycerol (DAG), generating 2-AG. 2-AG then travels (possibly by diffusion or via an undefined carrier) to the presynaptic terminal, where it engages presynaptic CB_1 receptors, inhibiting calcium channels (and possibly also inhibiting the vesicular release machinery) to suppress synaptic transmission. DSI (and DSE) are terminated as 2-AG is degraded, either by MGL (Pan et al. 2009; Straiker et al. 2009) and/or cyclooxygenase-2 (COX-2) (Kim and Alger 2004; Straiker et al. 2011). The participation of COX-2 in terminating DSE may have important therapeutic implications. For example, if COX-2 is increased (e.g., following ischemic injury), the duration of DSE will be shortened and glutamate release increased, which may exacerbate excitotoxicity. In addition, COX-2 metabolites of 2-AG (e.g., prostaglandin E2 glycerol) can enhance excitatory synaptic transmission and long-term potentiation in the hippocampus (Sang et al. 2006; Yang et al. 2008).

A second form of transient modulation of synaptic transmission by eCBs is metabotropic suppression of inhibition (MSI) or excitation (MSE) (Maejima et al. 2001; Varma et al. 2001; Kim et al. 2002). This is a functionally distinct pathway from DSI/DSE. In MSI/MSE, activation of a postsynaptic Gq/11-linked GPCR stimulates PLCbeta, leading to the production of DAG. DAGL then cleaves the DAG to 2-AG, which then traverses the synapse to activate presynaptic CB_1 receptors, inhibiting synaptic transmission (Kano et al. 2009). In theory, any appropriately positioned, postsynaptic Gq/11-linked GPCR should be able to elicit MSI/MSE. However, the most commonly encountered receptors mediating MSI/MSE are the group I metabotropic glutamate receptors (i.e., mGluR1 and mGluR5) and the M1 and M3 muscarinic receptors.

Although DSI/DSE and MSI/MSE can occur independently of one another, they can also synergize. In this situation, a brief depolarization, combined with modest activation of the Gq/11-linked receptor, increases intracellular calcium. This increased intracellular calcium stimulates the activity of PLCbeta, leading to higher levels of DAG production (and possibly greater DAGL activity), which increases 2-AG production (Hashimotodani et al. 2005). In this way eCBs can serve as a coincidence detector between depolarization and activation of metabotropic receptors.

The preceding discussion has focused on CB_1 receptor-mediated forms of short-term synaptic plasticity. In these cases, CB_1 involvement has been firmly established by antagonism of the plasticity with a variety of CB_1 receptor antagonists or the absence of the plasticity in CB_1 receptor knockout mice. Thus, under normal conditions (i.e., an acute brain slice or cultured neurons), CB_2 receptors have not been observed to participate in these short-term forms of synaptic plasticity in the brain regions studied. However, these experiments left open the question if CB_2 receptors *can* participate in short-term forms of synaptic plasticity. We addressed this question by transfecting CB_2 receptors into hippocampal neurons cultured from CB_1 receptor knockout mice. Expression of CB_2 receptors into these neurons recovered 2-AG-mediated inhibition of synaptic transmission as well as DSE (Atwood et al. 2012a, b). Thus, CB_2 appears capable of supporting short-term forms of eCB-mediated synaptic plasticity, if it is appropriately expressed in neurons.

Apart from this example, there is additional evidence that CB_2 receptors can influence synaptic transmission or neuronal excitability. Activation of CB_2 receptors in layer 2/3 of the rodent prefrontal cortex increased activity of calcium-activated chloride currents, reducing spontaneous activity (den Boon et al. 2012). In addition, activation of CB_2 receptors decreased action potential (but not action potential-independent γ-aminobutyric acid (GABA) release in rat medial entorhinal cortex (Morgan et al. 2009)).

The above has focused on short-term synaptic plasticity. However, shortly after the description of the depolarization and metabotropic receptor forms of eCB-mediated synaptic plasticity discussed above, long-term depression (LTD) mediated by eCBs (eLTD) was described (Gerdeman et al. 2002; Robbe et al. 2002). This has been thoroughly studied in both excitatory (e.g., Gerdeman et al. 2002; Robbe et al. 2002; Peterfi et al. 2012) and inhibitory connections (e.g., Chevaleyre and Castillo 2003). eLTD occurs at some CB_1-expressing synapses following prolonged low-frequency stimulation (e.g., 1 Hz, 10 min (Robbe et al. 2002)). It often appears to require prolonged activation of postsynaptic group I mGluR receptors, leading to continued synthesis of eCBs (likely, 2-AG), sustained activation of CB_1 receptors, and persistent inhibition of neurotransmitter release (possibly mediated by RIM1alpha (Chevaleyre et al. 2007)). Like other forms of LTD, eLTD synaptic depression persists after the cessation of the inducing stimulus (in this case, CB_1 production). Other forms of eLTD that vary from this canonical pathway have been reported in hippocampus (CA1) from young (<P10) rats (Yasuda et al. 2008) and in cultured autaptic hippocampal neurons (Kellogg et al. 2009). The former form of eLTD is notable for likely involving activation of potassium channels (Yasuda et al. 2008); whereas the latter form involves CB_1 receptors signaling via Gi/o-independent G proteins (Kellogg et al. 2009).

The above paradigm for eLTP generally involves the direct action of a Gq/11-linked GPCR, followed by 2-AG production, prolonged stimulation of CB_1 receptors, and inhibition of neurotransmission that exceeds the duration of eCB production. A related form of long-term synaptic depression has been demonstrated following the activation of two (membrane) steroid hormone receptors, the glucocorticoid receptor and the α isoform of the estrogen receptor. In the case of the glucocorticoid receptor, activation of this receptor in hypothalamic parvocellular neurons leads to a long-lasting inhibition of glutamate release (Di et al. 2003; Evanson et al. 2010;

Tasker and Herman 2011). This leads to inhibition of corticotropin-releasing hormone (CRH)-secreting neurons and suppression of the hypothalamic–pituitary–adrenal (HPA) axis. In the case of estrogen receptor-mediated eLTD, activation of a membrane-associated α form of the estrogen receptor leads to long-term inhibition of CB_1-expressing inhibitory synapses onto CA1 pyramidal neurons. Notable aspects of this latter form of eLTD is that it (1) only occurs in female rats, (2) requires mGluR1 signaling, and (3) appears to involve anandamide, and not 2-AG (Huang and Woolley 2012). The glucocorticoid receptor-mediated form of eLTD requires a G protein (as it is blocked by inclusion of GDPβS in the recording pipette), but whether this is a metabotropic glutamate receptor has not been tested. With the identification of these two steroid hormone receptor–mediated forms of eLTD, it is interesting to speculate that similar forms of LTD may be evoked by mineralocorticoid or androgen receptors stimulating GPCR activation.

Another form of activity-dependent modulation of neuronal excitability is slow self-inhibition (SSI). This form of eCB-mediated modulation of neuronal excitability has been reported in neocortical low-threshold spiking interneurons (Bacci et al. 2004), in a population of cerebellar basket cells (Kreitzer et al. 2002), and in a fraction of cortical pyramidal neurons (Marinelli et al. 2009). The likely signaling pathway for this phenomenon is that repeated depolarization of the neuron increases intracellular calcium, which activates DAGL and increases 2-AG production. 2-AG then activates a potassium conductance (likely GIRK channels) (Marinelli et al. 2008). In contrast to the forms of synaptic plasticity discussed earlier, SSI involves cell autonomous 2-AG signaling.

The above forms of eCB-mediated modulation of synaptic transmission and neuronal excitability have considered exclusively the domain of inter- or intra-neuronal signaling, with no involvement of glial cells. There are two major ways that glial cells may influence neuronal excitability and synaptic transmission in an eCB-dependent fashion. One is that glial cells, particularly astrocytes and microglial cells, can produce prodigious amounts of eCBs (Walter et al. 2002; Walter et al. 2003; Stella 2004). The other is that the glial cells may be expressing the cannabinoid receptors and influencing synaptic plasticity in a paracrine fashion. Considerable evidence has emerged over the past 5 years that glial cells, particularly astrocytes, participate as active CB_1-expressing partners in some forms of eCB-mediated synaptic plasticity. This involvement was surprising to some in the field, as immunocytochemical studies showed high levels of CB_1 expression in some GABAergic neurons and intermediate levels in a subset of excitatory synapses. CB_1 expression in astrocytes, when noted. for example (Rodriguez et al., 2001), was only a small fraction of the levels observed in neuronal elements. However, density of CB_1 receptor expression does not necessarily correlate with "importance," as has been amply shown in prior studies (Azad et al. 2003; Marsicano et al. 2003; Domenici et al. 2006; Monory and Lutz 2009).

Several anatomical features of astrocytes are important when considering their role as potential mediators and modulators of eCB action (Ventura and Harris 1999). The first is that most central synapses are embedded in glial endfeet. This means that glial membranes are never far from the source of eCB production (primarily dendrites). The second is that the ramifications of a single astrocyte can extend over a considerable range, thus potentially transducing a local signal into one covering several hundred cubic microns. The third is that gap junction coupling between astrocytes will further increase the potential distance a signal can be transmitted. Thus, if CB_1 stimulation increases the concentration of a diffusible messenger (e.g., calcium) in one astrocyte, that messenger may affect a number of neighboring astrocytes, potentially influencing a volume of several thousand cubic microns.

That eCBs released from neurons can signal via astrocytic CB_1 receptors was first demonstrated about 5 years ago (Navarrete and Araque 2008). In these experiments, the investigators found that by depolarizing one neuron, eCBs were produced that activated neighboring astrocytic CB_1 receptors. These CB_1 receptors increased astrocytic intracellular calcium (interestingly, in a non-Gi/o-mediated fashion; this is a common feature of all forms of synaptic plasticity mediated by astrocytic CB_1 receptors and deserves additional study), causing release of glutamate from the astrocyte, which activated (presynaptic) mGluR1 receptors to increase glutamate release and synaptic efficacy (Navarrete and Araque 2010). Therefore, in classic DSE, glutamatergic transmission onto the depolarized cell is inhibited by activation of presynaptic CB_1 receptors, but when astrocytic CB_1 receptors are activated, glutamatergic transmission is enhanced. Thus, eCB production could either stimulate or inhibit glutamatergic transmission, leading to the question of what happens in vivo. These investigators found that whether eCB production enhanced or suppressed glutamatergic transmission had strong spatial dependency. If glutamatergic terminals were close to the site (<40 microns) (using the location of the neuron's soma) of eCB production, then inhibition dominated, whereas at more distant sites (between 60 and 100 microns), enhancement was seen (Navarrete and Araque 2008). Interestingly, if presynaptic CB_1 inhibition was prevented (by treatment of the slice with pertussis toxin), then activation of astrocytic CB_1 led to more pronounced synaptic enhancement (Navarrete and Araque 2008), suggesting that under basal conditions, production of eCBs in this system (CA1) leads to a suppression of proximal glutamatergic transmission and an enhancement of more distal glutamatergic transmission. This provides an additional mechanism for a network to "tune" synaptic strength, where neuronal activation strong enough to produce eCBs will decrease subsequent glutamatergic input onto those neurons, yet will strengthen glutamatergic inputs onto more distant, presumably less-stimulated, neurons.

A second form of astrocyte-mediated cannabinoid synaptic plasticity has been described more recently (Han et al. 2012). In this study, it was shown that activation of astrocyte CB_1 receptors by exogenous cannabinoids (systemically administered) led to LTD of CA3 to CA1 glutamatergic synapses. In addition to requiring astrocyte CB_1 receptors, this cannabinoid-dependent form of LTD (CB-LTD) also required activation of N-methyl d-aspartic acid (NMDA) receptors and was mediated by the loss of cell surface α-amino-3-hydroxy-5-methyl-4-isoxazole propionate (AMPA) receptors. Interestingly, the authors of this study were able to show a strong correlation between CB-LTD and THC-induced impairment of spatial working memory (Han et al. 2012), suggesting that the deleterious effects of THC on spatial working memory may be due to astrocyte-mediated synaptic depression at CA3>CA1 synapses.

The observation that CB_1 receptors on astrocytes can mediate opposing forms of synaptic plasticity (that is, LTD or long-term potentiation) deserves further consideration. In both cases the common mediator appears to be glutamate release from astrocytes. Drawing from the extensive literature of glutamate-mediating diverse forms of synaptic plasticity (Citri and Malenka 2008), it is conceivable that the discrepancy in the two studies discussed earlier may be due to the different time courses of synaptic stimulation (that is, brief and punctate in the first study and prolonged with the second study). The first form of astrocytic CB_1 stimulation may lead to brief, high local levels of glutamate, whereas the second will lead to more diffuse and prolonged elevations of extracellular glutamate. (This possibility is strengthened by the observation that blockade of extracellular glutamate uptake by threo-beta-benzyloxyaspartate (TBOA) induces a mechanistically similar form of LTD (Han et al. 2012).) There may also be experimental differences that explain the discrepancy. In the first study, experiments were conducted in slices prepared from hippocampus, which, while effectively preserving laminar neuronal connectivity, may disrupt the extensive network of interconnected astrocytes. In addition, the first study used minimal stimulation to investigate a small number of synaptic connections. The second study examined field potentials in anesthetized animals, so astrocyte connectivity would be maintained. (However, most of the experiments were conducted in anesthetized animals, which brings in the potential complicating factors of anesthesia.) Despite some uncertainties in interpretation, the above experiments establish that astrocytes can participate in cannabinoid-mediated synaptic plasticity and in cannabinoid-mediated behaviors. It is probable that future studies will more completely establish the precise mechanisms involved, which are likely varied depending on the form of stimulation and the precise behaviors involved.

A very recent study expanded the role of astrocytes in eCB-mediated synaptic plasticity to LTD in spike-timing-dependent plasticity (STDP). STDP is a phenomenon where repeated pairing of presynaptic stimulation with postsynaptic depolarization leads to persistent changes in synaptic strength (Feldman 2012). A key feature of most forms of STDP is that the order of stimulation is critically important, where the sign of plasticity (LTP vs. LTD) depends on whether presynaptic or postsynaptic stimulation occurs first. Some forms of STDP have been shown to involve CB_1 receptors and eCBs (e.g., Sjostrom et al. 2003; Tzounopoulos et al. 2007; Fino et al. 2010); however, a potential role for astrocytic CB_1 receptors was not investigated. A recent study examining STDP in the neocortex during the development of somatosensory barrel cortex found that STDP producing LTD at the glutamatergic synapse from layer IV to layer II/III required astrocytic CB_1 receptors (Min and Nevian 2012). This was established by using an LTD-producing STDP protocol where the postsynaptic neuron was depolarized 25 ms before afferent fibers were stimulated. The LTD produced under these circumstances was NMDA receptor dependent, was occluded by clamping astrocyte calcium levels, and was mimicked by inducing astrocytic calcium spikes together with afferent stimulation. Thus, the model in this case appears to be that prolonged production of eCBs during the STDP protocol coupled with glutamate release leads to activation of presynaptic NMDA receptors and persistent enhancement of glutamate release. It is important to note that this model requires compartmentalization of the astrocyte glutamate release (to the presynapse) and presynaptic glutamate release during the induction stage (simply increasing astrocyte calcium levels was insufficient to cause this form of LTD). However, the tight investiture of excitatory synapses by astrocyte processes and highly active glutamate uptake by astrocytes provides the necessary anatomical and functional substrates for compartmentalization of extracellular glutamate. It is interesting to note the similarities between this form to STDP LTD and CB LTD (Han et al. 2012) discussed previously.

The above brief overview of eCB-mediated synaptic plasticity highlights the rich repertoire of eCB signaling in the brain. Despite the many forms of eCB-mediated synaptic plasticity, several themes emerge: eCB plasticity is widespread and is involved in many different circuits, from the spinal cord to the cortex. Because CB_1-mediated synaptic plasticity occurs on both inhibitory and excitatory terminals, activation of these receptors and the various forms of synaptic plasticity that follow will be very state dependent and can be expected to have quite unpredictable effects at the network level, generally requiring experimentation to validate. The density of CB_1 receptors correlates poorly to their functional importance. An example of this is CB_1 receptors on astrocytes. While these receptors are very sparsely seen in immunocytochemical studies, synaptic plasticity elicited by these receptors appears to exert significant effects at the network and behavioral levels (Navarrete and Araque 2010; Han et al. 2012; Min and Nevian 2012). While in many cases CB_1-mediated synaptic plasticity is elicited by 2-AG, anandamide also frequently participates. The difficulties in demonstrating anandamide involvement primarily arise because of the diverse synthetic pathways for this eCB and the lack of specific inhibitors for its synthesis. Thus, more indirect approaches, such as inhibiting anandamide degradation by FAAH, must be used (keeping in mind the caveat that inhibition of anandamide degradation also affects other FAAH metabolites such as N-arachidonoyl glycerol and the acyl amides. In conclusion, in the 12 years since eCB-mediated synaptic plasticity was first demonstrated, it has been shown to involve an unexpectedly large number of types of synaptic plasticity as well as utilize numerous intracellular signaling pathways. It is likely that additional forms of eCB-mediated synaptic plasticity remain to be elucidated and our understanding of their roles in behavior will increase over the coming years.

References

Ahn K, McKinney MK, Cravatt BF (2008) Enzymatic pathways that regulate endocannabinoid signaling in the nervous system. Chemi Rev 108(5):1687–1707

Ahn K, Johnson DS, Cravatt BF (2009) Fatty acid amide hydrolase as a potential therapeutic target for the treatment of pain and CNS disorders. Expert Opin Drug Discov 4(7):763–784

Atwood BK, Mackie K (2010) CB2: a cannabinoid receptor with an identity crisis. Br J Pharmacol 160(3):467–479

Atwood BK, Straiker A, Mackie K (2012a) CB(2) cannabinoid receptors inhibit synaptic transmission when expressed in cultured autaptic neurons. Neuropharmacology 63(4):514–523

Atwood BK, Wager-Miller J, Haskins C, Straiker A, Mackie K (2012b) Functional selectivity in CB(2) cannabinoid receptor signaling and regulation: implications for the therapeutic potential of CB(2) ligands. Mol Pharmacol 81(2):250–263

Azad SC, Eder M, Marsicano G, Lutz B, Zieglgansberger W, Rammes G (2003) Activation of the cannabinoid receptor type 1 decreases glutamatergic and GABAergic synaptic transmission in the lateral amygdala of the mouse. Learn Mem 10(2):116–128

Bacci A, Huguenard JR, Prince DA (2004) Long-lasting self-inhibition of neocortical interneurons mediated by endocannabinoids. Nature 431(7006):312–316

Blankman JL, Simon GM, Cravatt BF (2007) A comprehensive profile of brain enzymes that hydrolyze the endocannabinoid 2-arachidonoylglycerol. Chemi Biol 14(12):1347–1356

Castillo PE, Younts TJ, Chavez AE, Hashimotodani Y (2012) Endocannabinoid signaling and synaptic function. Neuron 76(1):70–81

Chavez AE, Chiu CQ, Castillo PE (2010) Trpv1 activation by endogenous anandamide triggers postsynaptic long-term depression in dentate gyrus. Nat Neurosci 13(12):1511–1518

Chevaleyre V, Castillo PE (2003) Heterosynaptic LTD of hippocampal GABAergic synapses: a novel role of endocannabinoids in regulating excitability. Neuron 38(3):461–472

Chevaleyre V, Heifets BD, Kaeser PS, Sudhof TC, Castillo PE (2007) Endocannabinoid-mediated long-term plasticity requires cAMP/PKA signaling and RIM1alpha. Neuron 54(5):801–812

Citri A, Malenka RC (2008) Synaptic plasticity: multiple forms, functions, and mechanisms. Neuropsychopharmacol 33(1):18–41

den Boon FS, Chameau P, Schaafsma-Zhao Q, van Aken W, Bari M, Oddi S, Kruse CG, Maccarrone M, Wadman WJ, Werkman TR (2012) Excitability of prefrontal cortical pyramidal neurons is modulated by activation of intracellular type-2 cannabinoid receptors. Proc Natl Acad Sci U S A 109(9):3534–3539.

Di S, Malcher-Lopes R, Halmos KC, Tasker JG (2003) Nongenomic glucocorticoid inhibition via endocannabinoid release in the hypothalamus: a fast feedback mechanism. J Neurosci 23(12):4850–4857

Domenici MR, Azad SC, Marsicano G, Schierloh A, Wotjak CT, Dodt HU, Zieglgansberger W, Lutz B, Rammes G (2006) Cannabinoid receptor type 1 located on presynaptic terminals of principal neurons in the forebrain controls glutamatergic synaptic transmission. J Neurosci 26(21):5794–5799

Evanson NK, Tasker JG, Hill MN, Hillard CJ, Herman JP (2010) Fast feedback inhibition of the HPA axis by glucocorticoids is mediated by endocannabinoid signaling. Endocrinology 151(10):4811–4819

Felder CC, Joyce KE, Briley EM, Glass M, Mackie KP, Fahey KJ, Cullinan GJ, Hunden DC, Johnson DW, Chaney MO, Koppel GA, Brownstein M (1998) Ly320135, a novel cannabinoid CB1 receptor antagonist, unmasks coupling of the CB1 receptor to stimulation of camp accumulation. J Pharmacol Exp Ther 284(1):291–297

Feldman DE (2012) The spike-timing dependence of plasticity. Neuron 75(4):556–571

Fino E, Paille V, Cui Y, Morera-Herreras T, Deniau JM, Venance L (2010) Distinct coincidence detectors govern the corticostriatal spike timing-dependent plasticity. J Physiol 588(Pt 16): 3045–3062

Fu J, Gaetani S, Oveisi F, Verme JLo, Serrano A, Rodriguez De Fonseca F, Rosengarth A, Luecke H, Giacomo BDi, Tarzia G, Piomelli D (2003) Oleylethanolamide regulates feeding and body weight through activation of the nuclear receptor PPAR-alpha. Nature 425(6953):90–93

Gerdeman GL, Ronesi J, Lovinger DM (2002) Postsynaptic endocannabinoid release is critical to long-term depression in the striatum. Nat Neurosci 5(5):446–451

Glass M, Felder CC (1997) Concurrent stimulation of cannabinoid CB1 and dopamine D2 receptors augments camp accumulation in striatal neurons: evidence for a Gs linkage to the CB1 receptor. J Neurosci 17(14):5327–5333

Hajos N, Katona I, Naiem SS, MacKie K, Ledent C, Mody I, Freund TF (2000) Cannabinoids inhibit hippocampal GABAergic transmission and network oscillations. Eur J Neurosci 12(9):3239–3249

Han J, Kesner P, Metna-Laurent M, Duan T, Xu L, Georges F, Koehl M, Abrous DN, Mendizabal-Zubiaga J, Grandes P, Liu Q, Bai G, Wang W, Xiong L, Ren W et al (2012) Acute cannabinoids impair working memory through astroglial CB1 receptor modulation of hippocampal LTD. Cell 148(5):1039–1050

Hashimotodani Y, Ohno-Shosaku T, Tsubokawa H, Ogata H, moto KE, Maejima T, Araishi K, Shin HS, Kano M (2005) Phospholipase Cbeta serves as a coincidence detector through its Ca2+ dependency for triggering retrograde endocannabinoid signal. Neuron 45(2):257–268

Hoffman AF, Lupica CR (2012) Synaptic targets of Δ9-tetrahydrocannabinol in the central nervous system. Cold Spring Harb Perspect Med. doi:10.1101/cshperspect.a012237

Hohmann AG, Suplita RL, Bolton NM, Neely MH, Fegley D, Mangieri R, Krey JF, Walker JM, Holmes PV, Crystal JD, Duranti A, Tontini A, Mor M, Tarzia G, Piomelli D (2005) An endocannabinoid mechanism for stress-induced analgesia. Nature 435(7045):1108–1112

Howlett AC, Barth F, Bonner TI, Cabral G, Casellas P, Devane WA, Felder CC, Herkenham M, Mackie K, Martin BR, Mechoulam R, Pertwee RG (2002) International Union of Pharmacology. XXVII. Classification of cannabinoid receptors. Pharmacol Rev 54(2):161–202

Huang GZ, Woolley CS (2012) Estradiol acutely suppresses inhibition in the hippocampus through a sex-specific endocannabinoid and mGluR-dependent mechanism. Neuron 74(5):801–808

Kano M, Ohno-Shosaku T, Hashimotodani Y, Uchigashima M, Watanabe M (2009) Endocannabinoid-mediated control of synaptic transmission. Physiol Rev 89(1):309–380

Katona I, Freund TF (2012) Multiple functions of endocannabinoid signaling in the brain. Ann Rev Neurosci 35:529–558

Kellogg R, Mackie K, Straiker A (2009) Cannabinoid CB1 receptor-dependent long-term depression in autaptic excitatory neurons. J Neurophysiol 102(2):1160–1171

Kenakin T, Miller LJ (2010) Seven transmembrane receptors as shapeshifting proteins: the impact of allosteric modulation and functional selectivity on new drug discovery. Pharmacol Rev 62(2):265–304

Kim J, Alger BE (2004) Inhibition of cyclooxygenase-2 potentiates retrograde endocannabinoid effects in hippocampus. Nat Neurosci 7(7):697–698

Kim J, Isokawa M, Ledent C, Alger BE (2002) Activation of muscarinic acetylcholine receptors enhances the release of endogenous cannabinoids in the hippocampus. J Neurosci 22(23):10182–10191

Kreitzer AC, Regehr WG (2001) Retrograde inhibition of presynaptic calcium influx by endogenous cannabinoids at excitatory synapses onto Purkinje cells. Neuron 29(3):717–727

Kreitzer AC, Carter AG, Regehr WG (2002) Inhibition of interneuron firing extends the spread of endocannabinoid signaling in the cerebellum. Neuron 34(5):787–796

Lauckner JE, Hille B, Mackie K (2005) The cannabinoid agonist WIN55,212-2 increases intracellular calcium via CB1 receptor coupling to Gq/11 G proteins. Proc Natl Acad Sci U S A 102(52):19144–19149

Levenes C, Daniel H, Soubrie P, Crepel F (1998) Cannabinoids decrease excitatory synaptic transmission and impair long-term depression in rat cerebellar Purkinje cells. J Physiol 510(3): 867–879

Liu J, Wang L, Harvey-White J, Huang BX, Kim HY, Luquet S, Palmiter RD, Krystal G, Rai R, Mahadevan A, Razdan RK, Kunos G (2008) Multiple pathways involved in the biosynthesis of anandamide. Neuropharmacology 54(1):1–7

Llano I, Leresche N, Marty A (1991) Calcium entry increases the sensitivity of cerebellar Purkinje cells to applied GABA and decreases inhibitory synaptic currents. Neuron 6(4):565–574

Luk T, Jin W, Zvonok A, Lu D, Lin XZ, Chavkin C, Makriyannis A, Mackie K (2004) Identification of a potent and highly efficacious, yet slowly desensitizing CB1 cannabinoid receptor agonist. Br J Pharmacol 142(3):495–500

Mackie K, Devane WA, Hille B (1993) Anandamide, an endogenous cannabinoid, inhibits calcium currents as a partial agonist in N18 neuroblastoma cells. Mol Pharmacol 44(3):498–503

Mackie K, Hille B (1992) Cannabinoids inhibit N-type calcium channels in neuroblastoma-glioma cells. Proc Natl Acad Sci U S A 89(9):3825–3829

Mackie KY, Westenbroek Lai R, Mitchell R (1995) Cannabinoids activate an inwardly rectifying potassium conductance and inhibit Q-type calcium currents in AtT20 cells transfected with rat brain cannabinoid receptor. J Neurosci 15(10):6552–6561

Maejima T, Hashimoto K, Yoshida T, Aiba A, Kano M (2001) Presynaptic inhibition caused by retrograde signal from metabotropic glutamate to cannabinoid receptors. Neuron 31(3): 463–475

Maresz K, Carrier EJ, Ponomarev ED, Hillard CJ, Dittel BN (2005) Modulation of the cannabinoid CB2 receptor in microglial cells in response to inflammatory stimuli. J Neurochem 95(2): 437–445

Marinelli S, Pacioni S, Bisogno T, Marzo VDi, Prince DA, Huguenard JR, Bacci A (2008) The endocannabinoid 2-arachidonoylglycerol is responsible for the slow self-inhibition in neocortical interneurons. J Neurosci 28(50):13532–13541

Marinelli S, Pacioni S, Cannich A, Marsicano G, Bacci A (2009) Self-modulation of neocortical pyramidal neurons by endocannabinoids. Nat Neurosci 12(12):1488–1490

Marsicano G, Goodenough S, Monory K, Hermann H, Eder M, Cannich A, Azad SC, Cascio MG, Gutierrez SO, Mvan der S, Lopez-Rodriguez ML, Casanova E, Schutz G, Zieglgansberger W, Marzo VDi et al (2003) CB1 cannabinoid receptors and on-demand defense against excitotoxicity. Science 302(5642):84–88

Mazzola C, Medalie J, Scherma M, Panlilio LV, Solinas M, Tanda G, Drago F, Cadet JL, Goldberg SR, Yasar S (2009) Fatty acid amide hydrolase (FAAH) inhibition enhances memory acquisition through activation of PPAR-alpha nuclear receptors. Learn Mem 16(5):332–337

McHugh D, Hu SS, Rimmerman N, Juknat A, Vogel Z, Walker JM, Bradshaw HB (2010) N-arachidonoyl glycine, an abundant endogenous lipid, potently drives directed cellular migration through GPR18, the putative abnormal cannabidiol receptor. BMC Neurosci 11:44

McKinney MK, Cravatt BF (2005) Structure and function of fatty acid amide hydrolase. Ann Rev Biochem 74:411–432

Melis M, Pillolla G, Luchicchi A, Muntoni AL, Yasar S, Goldberg SR, Pistis M (2008) Endogenous fatty acid ethanolamides suppress nicotine-induced activation of mesolimbic dopamine neurons through nuclear receptors. J Neurosci 28(51):13985–13994

Min R, Nevian T (2012) Astrocyte signaling controls spike timing-dependent depression at neocortical synapses. Nat Neurosci 15(5):746–53

Misner DL, Sullivan JM (1999) Mechanism of cannabinoid effects on long-term potentiation and depression in hippocampal CA1 neurons. J Neurosci 19(16):6795–6805

Monory K, Lutz B (2009) Genetic models of the endocannabinoid system. Curr Top Behav Neurosci 1:111–139

Morgan NH, Stanford IM, Woodhall GL (2009) Functional CB2 type cannabinoid receptors at CNS synapses. Neuropharmacology 57(4):356–368

Nagarkatti P, Pandey R, Rieder SA, Hegde VL, Nagarkatti M (2009) Cannabinoids as novel anti-inflammatory drugs. Future Med Chem 1(7):1333–1349

Navarrete M, Araque A (2008) Endocannabinoids mediate neuron-astrocyte communication. Neuron 57(6):883–893

Navarrete M, Araque A (2010) Endocannabinoids potentiate synaptic transmission through stimulation of astrocytes. Neuron 68(1):113–126

Ohno-Shosaku T, Maejima T, Kano M (2001) Endogenous cannabinoids mediate retrograde signals from depolarized postsynaptic neurons to presynaptic terminals. Neuron 29(3):729–738

Oz M (2006) Receptor-independent effects of endocannabinoids on ion channels. Curr Pharm Design 12(2):227–239

Pan B, Wang W, Long JZ, Sun D, Hillard CJ, Cravatt BF, Liu QS (2009) Blockade of 2-arachidonoylglycerol hydrolysis by selective monoacylglycerol lipase inhibitor 4-nitrophenyl 4-(dibenzo[d][1,3]dioxol-5-yl(hydroxy)methyl)piperidine-1-carboxylate (JZL184) enhances retrograde endocannabinoid signaling. J Pharmacol Exp Ther 331(2):591–597

Peterfi Z, Urban GM, Papp OI, Nemeth B, Monyer H, Szabo G, Erdelyi F, Mackie K, Freund TF, Hajos N, Katona I (2012) Endocannabinoid-mediated long-term depression of afferent

excitatory synapses in hippocampal pyramidal cells and GABAergic interneurons. J Neurosci 32(41):14448–14463
Pitler TA, Alger BE (1992) Postsynaptic spike firing reduces synaptic GABAA responses in hippocampal pyramidal cells. J Neurosci 12(10):4122–4132
Pitler TA, Alger BE (1994) Depolarization-induced suppression of GABAergic inhibition in rat hippocampal pyramidal cells: G protein involvement in a presynaptic mechanism. Neuron 13(6):1447–1455
Puente N, Cui Y, Lassalle O, Lafourcade M, Georges F, Venance L, Grandes P, Manzoni OJ (2011) Polymodal activation of the endocannabinoid system in the extended amygdala. Nat Neurosci 14(12):1542–1547
Robbe D, Kopf M, Remaury A, Bockaert J, Manzoni OJ (2002) Endogenous cannabinoids mediate long-term synaptic depression in the nucleus accumbens. Proc Natl Acad Sci U S A 99(12):8384–8388
Rodriguez JJ, Mackie K, Pickel VM (2001) Ultrastructural localization of the CB1 cannabinoid receptor in mu-opioid receptor patches of the rat caudate putamen nucleus. J Neurosci 21(3):823–833
Roloff AM, Thayer SA (2009) Modulation of excitatory synaptic transmission by Delta 9-tetrahydrocannabinol switches from agonist to antagonist depending on firing rate. Mol Pharmacol 75(4):892–900
Roth SH (1978) Stereospecific presynaptic inhibitory effect of delta9-tetrahydrocannabinol on cholinergic transmission in the myenteric plexus of the guinea pig. Can J Physiol Pharmacol 56(6):968–975
Sang N, Zhang J, Chen C (2006) PGE2 glycerol ester, a COX-2 oxidative metabolite of 2-arachidonoyl glycerol, modulates inhibitory synaptic transmission in mouse hippocampal neurons. J Physiol 572(Pt 3):735–745
Shen M, Piser TM, Seybold VS, Thayer SA (1996) Cannabinoid receptor agonists inhibit glutamatergic synaptic transmission in rat hippocampal cultures. J Neurosci 16(14):4322–4334
Shen M, Thayer SA (1999) Delta9-tetrahydrocannabinol acts as a partial agonist to modulate glutamatergic synaptic transmission between rat hippocampal neurons in culture. Mol Pharmacol 55(1):8–13
Shoemaker JL, Ruckle MB, Mayeux PR, Prather PL (2005) Agonist-directed trafficking of response by endocannabinoids acting at CB2 receptors. J Pharmacol Exp Ther 315(2):828–838
Sjostrom PJ, Turrigiano GG, Nelson SB (2003) Neocortical LTD via coincident activation of presynaptic NMDA and cannabinoid receptors. Neuron 39(4):641–654
Stella N (2004) Cannabinoid signaling in glial cells. Glia 48(4):267–277
Straiker A, Mackie K (2005) Depolarization-induced suppression of excitation in murine autaptic hippocampal neurones. J Physiol 569(Pt 2):501–517
Straiker A, Hu SS, Long JZ, Arnold A, Wager-Miller J, Cravatt BF, Mackie K (2009) Monoacylglycerol lipase limits the duration of endocannabinoid-mediated depolarization-induced suppression of excitation in autaptic hippocampal neurons. Mol Pharmacol 76(6): 1220–1227
Straiker A, Wager-Miller J, Hu SS, Blankman JL, Cravatt BF, Mackie K (2011) COX-2 and fatty acid amide hydrolase can regulate the time course of depolarization-induced suppression of excitation. Br J Pharmacol 164(6):1672–1683
Sugiura T, Kodaka T, Kondo S, Nakane S, Kondo H, Waku K, Ishima Y, Watanabe K, Yamamoto I (1997) Is the cannabinoid CB1 receptor a 2-arachidonoylglycerol receptor? Structural requirements for triggering a Ca2+ transient in NG108-15 cells. J Biochem 122(4):890–895
Sugiura T, Kishimoto S, Oka S, Gokoh M (2006) Biochemistry, pharmacology and physiology of 2-arachidonoylglycerol, an endogenous cannabinoid receptor ligand. Prog Lipid Res 45(5):405–446
Takahashi KA, Linden DJ (2000) Cannabinoid receptor modulation of synapses received by cerebellar Purkinje cells. J Neurophysiol 83(3):1167–80
Talwar R, Potluri VK (2011) Cannabinoid 1 (CB1) receptor–pharmacology, role in pain and recent developments in emerging CB1 agonists. CNS Neurol Disord Drug Targets 10(5):536–544

Tanimura A, Yamazaki M, Hashimotodani Y, Uchigashima M, Kawata S, Abe M, Kita Y, Hashimoto K, Shimizu T, Watanabe M, Sakimura K, Kano M (2010) The endocannabinoid 2-arachidonoylglycerol produced by diacylglycerol lipase alpha mediates retrograde suppression of synaptic transmission. Neuron 65(3):320–327

Tasker JG, Herman JP (2011) Mechanisms of rapid glucocorticoid feedback inhibition of the hypothalamic-pituitary-adrenal axis. Stress 14(4):398–406

Twitchell W, Brown S, Mackie K (1997) Cannabinoids inhibit N- and P/Q-type calcium channels in cultured rat hippocampal neurons. J Neurophysiol 78(1):43–50

Tzounopoulos T, Rubio ME, Keen JE, Trussell LO (2007) Coactivation of pre- and postsynaptic signaling mechanisms determines cell-specific spike-timing-dependent plasticity. Neuron 54(2):291–301

Varma N, Carlson GC, Ledent C, Alger BE (2001) Metabotropic glutamate receptors drive the endocannabinoid system in hippocampus. J Neurosci 21(24):188

Vaughan CW, McGregor IS, Christie MJ (1999) Cannabinoid receptor activation inhibits GABAergic neurotransmission in rostral ventromedial medulla neurons in vitro. Br J Pharmacol 127(4):935–940

Ventura R, Harris KM (1999) Three-dimensional relationships between hippocampal synapses and astrocytes. J Neurosci 19(16):6897–6906

Walter L, Franklin A, Witting A, Moller T, Stella N (2002) Astrocytes in culture produce anandamide and other acylethanolamides. J Biol Chem 277(23):20869–20876

Walter L, Franklin A, Witting A, Wade C, Xie Y, Kunos G, Mackie K, Stella N (2003) Nonpsychotropic cannabinoid receptors regulate microglial cell migration. J Neurosci 23(4):1398–1405

Ward SJ, Raffa RB (2011) Rimonabant redux and strategies to improve the future outlook of CB1 receptor neutral-antagonist/inverse-agonist therapies. Obesity 19(7):1325–1334

Wilson RI, Nicoll RA (2001) Endogenous cannabinoids mediate retrograde signalling at hippocampal synapses. Nature 410(6828):588–92

Xi ZX, Peng XQ, Li X, Song R, Zhang HY, Liu QR, Yang HJ, Bi GH, Li J, Gardner EL (2011) Brain cannabinoid CB(2) receptors modulate cocaine's actions in mice. Nat Neurosci 14(9):1160–1166

Yang H, Zhang J, Andreasson K, Chen C (2008) COX-2 oxidative metabolism of endocannabinoids augments hippocampal synaptic plasticity. Mol Cell Neurosci 37(4):682–695

Yasuda H, Huang Y, Tsumoto T (2008) Regulation of excitability and plasticity by endocannabinoids and PKA in developing hippocampus. Proc Natl Acad Sci U S A 105(8):3106–3111

Zarruk JG, Fernandez-Lopez D, Garcia-Yebenes I, Garcia-Gutierrez MS, Vivancos J, Nombela F, Torres M, Burguete MC, Manzanares J, Lizasoain I, Moro MA (2012) Cannabinoid type 2 receptor activation downregulates stroke-induced classic and alternative brain macrophage/microglial activation concomitant to neuroprotection. Stroke 43(1):211–219

Zygmunt PM, Petersson J, Andersson DA, Chuang H, Sorgard M, Marzo VDi, Julius D, Hogestatt ED (1999) Vanilloid receptors on sensory nerves mediate the vasodilator action of anandamide. Nature 400(6743):452–457

Chapter 3
Current Cannabinoid Receptor Nomenclature and Pharmacological Principles

Jahan Marcu, Linda Console-Bram and Mary E. Abood

Abstract The CB_1 and CB_2 cannabinoid receptors are members of the G protein-coupled receptor (GPCR) family that were isolated more than 20 years ago. CB_1 and CB_2 mediate the effects of Δ^9-tetrahydrocannabinol (Δ^9-THC), the principal psychoactive ingredient in marijuana and subsequently identified endogenous cannabinoids (endocannabinoids) anandamide and 2-arachidonyl glycerol. The discovery of additional sites of action for endocannabinoids as well as synthetic cannabinoid compounds suggests the existence of additional cannabinoid receptors. We review this evidence, as well as the current nomenclature for classifying a target as a cannabinoid receptor. We discuss basic pharmacological definitions and principles in order to place in context the mechanisms underlying cannabinoid receptor activation. Constitutive (agonist independent) activity and allosterism are observed with cannabinoid receptors. Allosteric modulation of cannabinoid receptors may usher in new classes of medicinal compounds capable of enhancing signals generated by endocannabinoids. Natural polymorphisms and alternative splice variants may also contribute to the pharmacological diversity of the cannabinoid receptors. Thus, each of the cannabinoid receptors is able to recognize multiple classes of compounds and produce an array of distinct downstream effects. However, many challenges await the field, including the classification of other GPCRs (i.e., GPR18 and GPR55) as bona fide cannabinoid receptors and developing strategies to target receptor conformations for harnessing specific pharmacological responses. The basic biology of the endocannabinoid system will continue to be revealed by ongoing investigations, and progress will partially depend upon the development of technologies that can assimilate current research trends and theories.

Abbreviations

abn-CBD	Abnormal cannabidiol
ACEA	Arachidonyl-2′-chlorethylamide
AEA	*N*-arachidonoyl ethanolamide, a.k.a. anandamide
2-AG	2-Arachidonoylglycerol
CB_1	Cannabinoid receptor 1

M. E. Abood (✉) · J. Marcu · L. Console-Bram
Center for Substance Abuse Research, Department of Anatomy and Cell Biology, Temple University School of Medicine, Philadelphia, PA 19140, USA
e-mail: mabood@temple.edu

E. J. Van Bockstaele (ed.), *Endocannabinoid Regulation of Monoamines in Psychiatric and Neurological Disorders*, DOI 10.1007/978-1-4614-7940-6_3,

CB_2	Cannabinoid receptor 2
CGRP	Calcitonin gene-related peptide
CHO	Chinese hamster ovary cells
CNR1(2) gene	Cannabinoid receptor 1(2) gene
EC_{50}	Half maximal effective concentration
ECS	Endocannabinoid system
GPCR	G protein-coupled receptor
GLYT2a	Glycine transporter 2a
GRK	G protein-coupled receptor kinase
GTPγS	Guanosine [gamma-thio] triphosphate
HEK293	Human embryonic kidney cells
HU210	11-Hydroxy-Δ^8-THC-dimethylheptyl
IUPHAR	International Union of Pharmacology
LPA	Lysophosphatidic acid
LPI	Lysophosphatidylinositol
MAPK	Mitogen-activated protein kinase
NADA	*N*-arachidonoyl dopamine
NAGly	*N*-arachidonyl glycine
NF-κB	Nuclear factor kappa-light-chain-enhancer of activated B cells
PEA	Palmitoylethanolamide
PSNCBAM-1	1-(4-Chlorophenyl)-3-[3-(6-pyrrolidin-1-ylpyridin-2-yl)phenyl] urea
Virodhamine	*O*-arachidonoyl-ethanolamine
PPAR	Peroxisome proliferator-activated receptor
PPARγ	Peroxisome proliferator-activated receptor gamma
RFLP	Restriction fragment length polymorphism
RT-PCR	Real-time polymerase chain reaction
SNPs	Small nucleotide polymorphisms
Δ^9-THC	(−)-Δ^9-Tetrahydrocannabinol
TLRs	Toll-like receptors
TRP	Transient receptor potential cation channel
TRPV	Transient receptor potential cation channel vanilloid

3.1 Introduction

The human body produces a vast amount of arachidonic acid derivatives, some of which have been identified as endocannabinoids; however, it was through studying (−)-Δ^9-tetrahydrocannabinol (Δ^9-THC) and related synthetic analogs that the first cannabinoid receptor (CB_1) was discovered (Devane et al. 1988) and subsequently cloned (Matsuda et al. 1990). Δ^9-THC is the primary psychoactive constituent in *Cannabis* (a.k.a. marijuana), hence the name "cannabinoid" receptor. To date, more than 80 cannabinoids have been isolated from *Cannabis sativa*, including Δ^9-THC, and many of these phytocannabinoids await screening for pharmacological activity (Turner et al. 1980; Ahmed et al. 2008; Radwan et al. 2008; Elsohly and

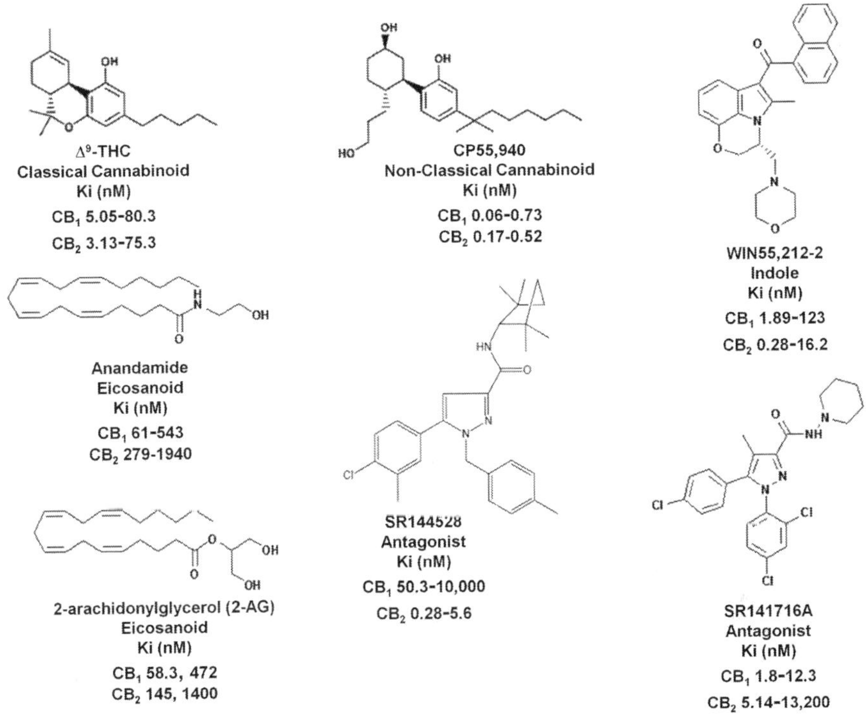

Fig. 3.1 The structures of prototypical cannabinoid compounds from each structural class with average K_i values for the displacement of a tritiated compound from rat, human, or mouse CB_1 and CB_2 receptors. The tritiated compound used is often [^{3}H]CP55940, but values are also derived from displacement of [^{3}H]SR141716A, [^{3}H]R-(+)-WIN55,212, or [^{3}H]HU-210. (For full reference see Pertwee et al. 2010)

Slade 2005). An arachidonic acid moiety, *N*-arachidonylethanolamide (AEA) was shown to activate CB_1, a member of the G protein-coupled receptor (GPCR) family, and named "anandamide" from the Sanskrit word for "bliss" (Devane et al. 1992). The identification of an endogenous ligand and the availability of novel ligands with cannabinoid receptor activity led to subsequent breakthroughs elucidating an "endocannabinoid system" (Di Marzo et al. 1998). A second cannabinoid receptor (CB_2) was isolated by a PCR-based strategy designed to isolate GPCRs in differentiated myeloid cells (Munro et al. 1993). The CB_2 receptor shares 44 % amino acid homology with CB_1, and a distinct yet similar binding profile, thus representing a receptor subtype. The most current nomenclature for cannabinoid receptors has been reported by a subcommittee of the International Union of Basic and Clinical Pharmacology (IUPHAR; Pertwee et al. 2010); a brief summary is presented here. A synopsis of basic pharmacological definitions and principles is also discussed with consideration for new developments in cannabinoid receptor pharmacology.

A range of pharmacological and genetic tools have been developed, and used to delineate "cannabinoid receptor"-mediated activity since the discovery of the first cannabinoid receptor. Five distinct classes of cannabinoid compounds have been identified (Fig. 3.1):

1. Classical cannabinoids (e.g., Δ^9-THC, 11-hydroxy-Δ^8-THC-dimethylheptyl (HU210)).
2. Nonclassical cannabinoids (e.g., CP55,940).
3. Indoles (e.g., WIN55,212).
4. Eicosanoids (e.g., the endogenous ligands; e.g., AEA, 2-arachidonoylglycerol).
5. Antagonists/inverse agonists (e.g., SR141716A and AM251 for CB_1, SR145528 and AM630 for CB_2; Devane et al. 1992; Eissenstat et al. 1995; Howlett 1995; Mechoulam et al. 1995; Xie et al. 1996; Rinaldi-Carmona et al. 1994, 1998).

In general, the agonists show little selectivity between the CB_1 and CB_2 receptors, while the antagonist compounds are highly selective (> 1,000-fold selective for CB_1 vs. CB_2 and vice versa with nanomolar affinity at the relevant receptor, see Fig. 3.1). The selectivity of these antagonists allows the discrimination of CB_1- vs. CB_2-mediated effects in vitro and in vivo. Despite a generalized nonselectivity with respect to agonists, there are some that exhibit selectivity for CB_1 vs. CB_2 receptors. One example is arachidonyl-2′-chlorethylamide (ACEA; Hillard et al. 1999b), which is highly selective for CB_1 (nanomolar affinity at CB_1 and > 1,000-fold selectivity for CB_1 vs. CB_2). HU-308, a Δ^9-THC analog, is a highly selective CB_2 agonist with nanomolar affinity at CB_2 and > 1,000-fold selectivity for CB_2 vs. CB_1 (Hanus et al. 1999). Several other compounds show > 100-fold selectivity and are generally classified as selective agonists (please see Pertwee et al. (2010) for more examples). However, these compounds are used at micromolar concentrations in vitro, and therefore the possibility exists that they may be acting at both receptors. Thus, additional controls should be performed to ensure the site of action of these compounds, i.e., experimentation in the presence of both CB_1 and CB_2 antagonists. Fortunately, in addition to the selective CB_1 and CB_2 antagonists that can be used to block agonist effects, there are also genetic tools available to the research community. CB_1 knockout mice have been generated in several laboratories; with both global (Zimmer et al. 1999; Ledent et al. 1999; Marsicano et al. 2002) and tissue-specific inactivation of CB_1, including select CNS neuronal populations (Marsicano et al. 2003), spinal cord-specific nociceptors (Agarwal et al. 2007), dorsal horn inhibitory interneurons (Pernia-Andrade et al. 2009), and liver-specific hepatocytes (Osei-Hyiaman et al. 2008). CB_2 knockout mice (global inactivation) have also been generated (Buckley et al. 2000).

Since the discovery of AEA, several other arachidonic acid derivatives are now considered to be endocannabinoids, having been shown to interact with the CB_1 and/or CB_2 cannabinoid receptors. 2-Arachidonoylglycerol (2-AG) was isolated from canine intestines in 1995 by Mechoulam and Fride (Mechoulam and Fride 1995) and demonstrated binding to both CB_1 and CB_2 receptors. Initially, both AEA and 2-AG were thought to bind with similar affinities to CB_1 and CB_2 (Pertwee 1999). However, other investigators found that 2-AG was more potent than AEA at eliciting increases in intracellular calcium (Sugiura et al. 2000) and this transient effect was also inhibited by CB_2 not CB_1 antagonists. It is now generally accepted that 2-AG acts as a full agonist, whereas AEA is a partial agonist, at both CB_1 and CB_2 receptors (Sugiura 2009; Gonsiorek et al. 2000). In addition, these endocannabinoids have recently demonstrated a differential role in memory and anxiety (Busquets-Garcia et al. 2011). Although both appear to be involved in anxiolytic responses, only AEA was reported to modulate memory consolidation. Anxiolytic responses evoked by 2-AG were mediated by CB_2, whereas CB_1 receptors mediated AEA anxiolytic effects. In addition, 2-AG antinociception has been reported to be mediated both CB_1 and CB_2 receptors (Guindon et al. 2011), whereas AEA-mediated antinociception is largely via CB_1 (Naidu et al. 2009). Collectively, these findings indicate that the biological effects of 2-AG and AEA are differentially modulated by the endocannabinoid system (ECS), perhaps reflective of their differing cannabinoid receptor potencies and/or a consequence of their regional levels at a particular instance due to the surrounding milieu.

Two additional endocannabinoids, homo-γ-linolenylethanolamide and docosatetraenylethanolamide, were isolated from brain (Hanus et al. 1993), and shown to compete for binding at CB_1 receptors; although these lipids have not been very well studied. Virodhamine, arachidonic acid, and ethanolamine joined by an ester linkage, has also been isolated and shown to act as a partial agonist at the CB_1 receptor and a full agonist at the CB_2 receptor (Porter et al. 2002). However, in another investigation, virodhamine was found to behave as a CB_1 receptor antagonist/inverse agonist (Steffens et al. 2005). *N*-arachidonoyl dopamine (NADA), is primarily a vanilloid receptor agonist, but has some activity at CB_1 receptors as well (Huang et al. 2002).

Another class of lipids have also been identified which have an effect on 2-AG-mediated events (Ben-Shabat et al. 1998). Although neither bind or activate CB_1 or CB_2 cannabinoid receptors, they significantly potentiate the apparent binding of 2-AG and its apparent capacity to inhibit adenylyl cyclase. Together these esters also significantly potentiate the behavioral effects produced by 2-AG. This enhancement of the biological activity of 2-AG by related, endogenous 2-acylglycerols, which alone show no significant activity in any of the tests employed, was termed an "entourage effect". The inactivation of 2-AG in neuronal cells is inhibited by 2-linoleoylglycerol, but not 2-palmitoylglycerol.

In addition, palmitoylethanolamide (PEA) has been suggested as a possible endogenous ligand at the CB_2 receptor (Facci et al. 1995). Subsequent studies showed no affinity for PEA at the CB_2 receptor (Showalter et al. 1996; Lambert et al. 1999; Griffin et al. 2000), and suggest that another GPCR may be responsible for PEA's actions (Franklin et al. 2003). Recently, PEA-induced calcium transients in sensory neurons was found to be a consequence of PPARα and TRPV1 channel activation, not CB receptor activation (Ambrosino et al. 2012). A metabolite of AEA, *N*-arachidonyl glycine (NAGly), present in bovine and rat brain as well as other tissues (Bradshaw et al. 2009), has been shown to suppress tonic inflammatory pain (Huang et al. 2001). This arachidonic acid-glycine conjugate, has poor affinity for CB_1 and CB_2 (Sheskin et al. 1997).

The actions of endocannabinoids are not restricted to the CB_1 and CB_2 receptors. Additional GPCRs as well as ion channels, ion channel receptors (i.e., transient receptor potential cation channel (TRP)) and nuclear receptors (peroxisome proliferator-activated receptor (PPAR)) have also been identified as sites of endocannabinoid interaction. Activation of transient receptor potential cation channel vanilloid (TRPV) receptors was demonstrated with both AEA (Zygmunt et al. 1999) and NADA (Huang et al. 2002). Activation via AEA was reported to induce vasodilation of isolated vascular preparations as a consequence of calcitonin gene-related peptide (CGRP; Zygmunt et al. 1999), whereas NADA activation of rat dorsal root ganglion and hippocampal TRPV receptors resulted in the release of substance P and CGRP (Huang et al. 2002; Hejazi et al. 2006). Evidence for phytocannabinoid interaction with TRP channels has also been demonstrated (De Petrocellis et al. 2011). NADA and AEA have also been shown to modulate calcium channels (White et al. 2001; Romano and Lograno 2006; Ross et al. 2009). The channels targeted by synthetic cannabinoids have recently been extensively reviewed (Pertwee et al. 2010).

Both 2-AG and AEA have been shown to mediate activities of PPARs (Lenman and Fowler 2007; Rockwell et al. 2006). Furthermore, findings from a recent study suggest that 2-AG activation of CB_1 receptors enables cross-talk between PPARγ and nuclear factor kappa-light-chain-enhancer of activated B cells (NF-κB; Du et al. 2011). Δ^9-THC, AEA, and NAGly have all been reported to potentiate the function of glycine receptors via allosteric interaction (Yevenes and Zeilhofer 2011; Hejazi et al. 2006). In addition, NAGly is a reversible, noncompetitive inhibitor of glycine transport via the glycine transporter 2a (GLYT2a; Wiles et al. 2006). The interaction between cannabinoids, both endogenous and synthetic, and Toll-like receptors (TLRs) has been the focus of much research, and has recently been reviewed (Downer 2011). Indeed, the relevance of the ECS in physiology is further complicated by demonstrations of endocannabinoid interaction with GABAergic/glutamatergic, biogenic amine, and opioid neurotransmission (Lopez-Moreno et al. 2008; Fisar 2012; Kirilly et al. 2012).

In addition to endocannabinoid activation of receptors other than CB_1 and CB_2, a synthetic cannabinoid compound also demonstrated interaction with noncannabinoid receptors. In particular, WIN55,212 as well as AEA elicited guanosine [gamma-thio] triphosphate (GTPγS) activity in brain membranes from CB_1 receptor knockout mice (Breivogel et al. 2001). These effects were not sensitive to inhibition by SR141716A. This same phenomenon has also been demonstrated in a second strain of CB_1 receptor knockout mice (Monory et al. 2002). The identity of this receptor remains unknown. That AEA produces the full range of behavioral effects (antinociception, catalepsy, and impaired locomotor activity) in CB_1 receptor knockout mice (Di Marzo et al. 2000) may be related to this receptor or may be due to AEA's ability to act at TRP channels (Zygmunt et al. 1999). Another putative cannabinoid receptor subtype which is responsive to WIN55,212 and CP55,940 and blocked by capsazepine has been found in the hippocampus (Hajos et al. 2001). These receptors are found on excitatory (pyramidal) axon terminals and have been shown to suppress glutamate release in CB_1 receptor knockout animals.

Cannabinoids including AEA elicit cardiovascular effects via peripherally located CB_1 receptors (Ishac et al. 1996; Jarai et al. 1999; Wagner et al. 1999). Abnormal cannabidiol (abn-CBD, a neurobehaviorally inactive synthetic cannabinoid), AEA, and a stable analog of AEA (methanandamide) caused hypotension and mesenteric vasodilation in wild-type mice as well as in mice lacking CB_1 receptors or both CB_1 and CB_2 receptors (Jarai et al. 1999). As a consequence of these findings, in addition to the lack of abn-CBD binding to CB_1 and CB_2 receptors observed in this study, the existence of an endothelial "abn-CBD receptor" has been suggested. In contrast to the studies described above with AEA-stimulated GTPγS activity, the cardiovascular and endothelial effects mediated by the "abn-CBD receptor" were SR141716A-sensitive. These effects were not due to activation of TRPV receptors as the TRPV receptor antagonist capsazepine did not inhibit these endothelium-dependent cardiovascular effects (Zygmunt et al. 1999). The cannabidiol analog and selective inhibitor of the "abn-CBD receptor", O-1918, inhibits the vasorelaxant effects of abn-CBD and AEA (Offertaler et al. 2003). Furthermore, a lack of abn-CBD binding to CB_1 and CB_2 was corroborated in this study, and the authors reported that O-1918 did not bind to either of these cannabinoid receptors. The putative "abn-CBD receptor" has also been characterized in immortalized and primary microglia (Walter et al. 2003; Kreutz et al. 2009; Franklin and Stella 2003). These studies provide evidence that the "abn-CBD receptor" is involved in microglial migration.

It is well accepted that cannabinoids play a role in immune function. The exact nature of this involvement has not been resolved. As previously mentioned, NAGly suppresses inflammatory pain independent of CB_1 and CB_2 (Huang et al. 2001). In 2006, a group of investigators suggested that NAGly is the endogenous ligand for GPR18 (Kohno et al. 2006), another candidate cannabinoid receptor. In a recent report, NAGly was found to induce apoptosis of proinflammatory macrophages, further supporting the role of NAGly in inflammation (Takenouchi et al. 2012). Attenuation of apoptosis following knock-down of GPR18 expression by siRNA supports a role for GPR18 in immune function. The finding that abn-CBD, AEA, and NAGly act as full agonists at GPR18 led to the suggestion that the "abn-CBD receptor" and the GPR18 receptor are one and the same (McHugh et al. 2010; McHugh et al. 2011). However, this premise requires further substantiation in light of a current study, which reported that NAGly is not an agonist at GPR18 (Lu et al. 2012).

The endocannabinoids 2-AG and AEA have also been reported to bind to the lysophosphatidylinositol (LPI)-sensitive receptor, GPR55 (Ryberg et al. 2007). The findings that Δ^9-THC, CBD, and the synthetic cannabinoid CP55940 also bind to GPR55 led this group to postulate that GPR55 is a novel cannabinoid receptor. However, GPR55 exhibits only 10–15 % homology to cloned CB_1 and CB_2 receptors (Baker et al. 2006). More importantly, the reported pharmacology of GPR55 is conflicting; studies from different laboratories have found widely discrepant results as summarized in several recent reviews (Ross 2009; Sharir and Abood 2010; Pertwee et al. 2010; Henstridge et al. 2011).

A large number of studies and emerging reports indicate that the resulting effects of endo-, phyto-, and synthetic cannabinoid interactions cannot be definitively explained based on the two-cannabinoid receptor theory. Activation of previously orphaned G protein receptors, GPR18, and GPR55, by endo-, phyto-, and synthetic cannabinoids suggest that these receptors may have a role in the wide ranging neuro-modulatory effects of the ECS (reviewed in Stella 2010; Pertwee et al. 2010).

3.2 Cannabinoid Receptor Nomenclature

Defining a cannabinoid receptor has increasingly become more complex. The IUPHAR committee on Receptor Nomenclature and Drug Classification, Subcommittee on Cannabinoid Receptors, has proposed a set of criteria for classifying a target as a cannabinoid receptor (Pertwee et al. 2010). This committee consists of a number of scientists who are actively involved in cannabinoid research, and who regularly review, new targets and new nomenclature for cannabinoid receptors. The current criteria are as follows:

1. It should be activated at its orthosteric site and with significant potency by an established CB_1/CB_2 receptor ligand.
2. It should be activated by at least one established endogenous CB_1/CB_2 receptor agonist at "physiologically relevant" concentrations.
3. If it is a GPCR, it should display significant amino acid sequence similarity with the CB_1 or the CB_2 receptor and hence be a member of the α group of Class A rhodopsin-type GPCRs.
4. It should not be a "well-established" non-CB_1, non-CB_2 receptor or channel, especially if there is already strong evidence that (a) it is activated endogenously by a non-CB_1, non-CB_2 receptor ligand with appropriate potency and relative intrinsic activity and (b) it is not activated endogenously by any endocannabinoid with appropriate potency and relative intrinsic activity.
5. It should be expressed by mammalian cells that are known to be exposed to concentrations of endogenously released endocannabinoid molecules capable of eliciting a response.

These IUPHAR criteria have been partially met with respect to candidate cannabinoid targets GPR18 and GPR55. AEA, an undisputed endocannabinoid and Δ^9-THC have been reported to act as full agonists at GPR18 (McHugh et al. 2011). Interestingly, NAGly, a AEA metabolite, was described by these investigators as more potent than AEA at GPR18. The site of GPR18 activation, orthosteric or allosteric, by Δ^9-THC, AEA, and NAGly has not been elucidated. Findings of AEA-mediated activation at GPR55 remain controversial. Previous studies from our laboratory (Kapur et al. 2009) indicated that GPR55 was not activated by AEA, whereas Ryberg at al. (2007) reported that AEA was equipotent at GPR55, CB_1, and CB_2. Recent studies from our laboratory demonstrated inhibition of GPR55 signaling by AEA and virodhamine (Sharir et al. 2012). However, once again, the location of activation, orthosteric vs. allosteric at this receptor by these endocannabinoids is yet unknown. In opposition to the third IUPHAR criteria, CB_1 and CB_2 belong to the Class A rhodopsin α-group, whereas both GPR18 and GPR55 are members of the Class A rhodopsin δ-group of GPCRs (Fredriksson et al. 2003). Nucleotide sequence homology between CB_1 and CB_2 is reported to be 44 %; 68 % within the residues of the transmembrane domain (Munro et al. 1993). Both GPR18 and GPR55 share little (less than 15 %) homology with CB_1 and CB_2 (Pertwee et al. 2010).

Since AEA also binds to TRPV channels, additional studies are needed to ensure that AEA has a greater potency and affinity at cannabinoid receptors as opposed to TRP receptors. The localization of receptors, along with endogenous cannabinoids (virodhamine and AEA as well as AEA's metabolite NAGly), and synthetic/degradative endocannabinoid enzymes, within the same peripheral and/or brain tissue lends support for GPR18 and GPR55 as cannabinoid receptors with respect to IUPHAR criteria number 5 (Bradshaw et al. 2009; Howlett et al. 2002; Porter et al. 2002; Di Marzo et al. 1994; Stella 2010).

Further research is required to fully characterize GPR18 and GPR55 prior to definitive classification as cannabinoid receptors. Such studies should include: competitive binding experiments of labeled agonists in transfected and nontransfected cell lines, displacement binding assays, modeling of binding pockets, point mutations of binding pocket domain(s), and development of knock-out mice. Data from these experiments, along with the development of high-potency synthetic agonists, and antagonists will provide the necessary insight into whether or not these two GPCRs should join the ranks of cannabinoid receptors.

3.3 Pharmacological Principles

Consequent to the interaction of endocannabinoids with a multitude of endogenous receptor systems, classification of orphaned GPCRs as cannabinoid receptors should proceed prudently. The breadth of knowledge gained from cannabinoid research brings to light the relevance of ligand concentration with respect to conclusions regarding cannabinoid involvement in biological events. A suggested "rule of thumb" in determining cannabinoid-mediated effects is that cannabinoid compounds generally ligate their receptor(s) in the nanomolar range (Stella 2010). Hence, the use of concentrations greater than 1 μM may produce off-target effects. For example, cannabinoid agonists were demonstrated to elicit increases in intracellular calcium and arachidonic acid release in both transfected and nontransfected Chinese hamster ovary cells (CHO) cells at a concentration of 10 μM (10–100-fold greater than the K_i at the CB_1 receptor; Felder et al. 1992). The increases in calcium and arachidonic acid release were not observed at concentrations close to the agonist's K_i values. Consequently, cannabinoids at these high concentrations elicited receptor and nonreceptor-mediated effects. In the field of cannabinoid pharmacology, the nature of the cannabinoid compound–receptor interaction has been upstaged by the biological effect that it imparts. In the quest for answers regarding the purpose of endocannabinoids, it is useful to review the theory of drug–receptor interaction.

The field of pharmacology has its roots in the desire to protect mankind from ailments. Chemicals were introduced into the body as a means of alleviating symptoms. Due to the intrinsic curiosity of man, the science of pharmacology expanded to include how the chemical interacts with biological systems to produce its effects (pharmacodynamics), and how the drug is handled by the body (pharmacokinetics). For a comprehensive review of basic pharmacology, the reader is referred to Katzung et al. (2009). Conceptually, the receptor, site of drug interaction with the body, was borne from experiments of Ehrlich and Langley in the nineteenth and twentieth century's. In the most classical sense, the receptor was considered a membrane-bound protein. With the identification of receptors, searches for the endogenous compound(s) which interact with the receptor began. Hence binding of a drug/endogenous compound with the receptor elicits a biological effect.

Traditionally, when an endogenous compound binds to a receptor causing activation of the receptor to yield a biological response, the compound is referred to as an agonist. Conversely, if a compound binds to a receptor, at the same site as the agonist, and inhibits the biological response it is an antagonist. Chemicals like therapeutic compounds are designed to either mimic or inhibit this biological response to avoid illness or improve symptomatology. The interaction of receptor and agonist or antagonist was presumed reversible and competitive with the biological effect being proportional to the number of receptors occupied. The agonist was thought to shift the conformation of the receptor to an active state, whereas the antagonist permitted the receptor to remain in an inactive state. As our knowledge of receptors and biological effectors became more expansive, the basic agonist/antagonist definition also expanded. For example, a new class of compounds for GPCRs, allosteric modulators has been identified. A compound can now be regarded as an agonist whether it binds to the same (*orthosteric*) or distinct (*allosteric*) site as the endogenous compound to elicit its effect (Fig. 3.2). Recently, allosteric modulators have been identified for cannabinoid receptors (Iliff et al. 2011).

Agonist binding to the orthosteric site initiates a conformational change concomitant with the dissociation of the G protein from the receptor and exchange of the bound GDP for GTP (Kenakin 2001). In the absence of bound endogenous agonist, GPCRs exist in an inactive conformation, whereas receptors are coupled to G proteins, bound by GDP. Orthosteric binding of receptor molecules is modeled as a competitive saturable process. However, interactions at an allosteric site may not be competitive. The receptor may undergo covalent modification such that the conformational state induced by the allosteric agonist shifts the equilibrium of receptor state. This alteration in receptor state may either enhance (positive allostery) or attenuate (negative allostery) the ability of the receptor to couple to its G protein(s), thereby affecting the continuation/magnitude of the biological response. Wang et al. (2009) provide a comprehensive discussion of allosterism (Wang et al. 2009). Accurate identification and characterization of novel drugs as allosteric modulators (positive vs. negative) is imperative in drug development (Conn et al. 2009).

The ability of an agonist to interact with its receptor to produce a certain level of response is related to the compound's intrinsic activity or efficacy. Some agonists yield a reduced level of response. These are known as partial agonists (Fig. 3.3). The maximal effect of a partial agonist is independent of the number of receptors occupied and receptor affinity. Rather the ability of the partial agonist to induce G protein receptor coupling (intrinsic activity/efficacy) is reduced, resulting in a

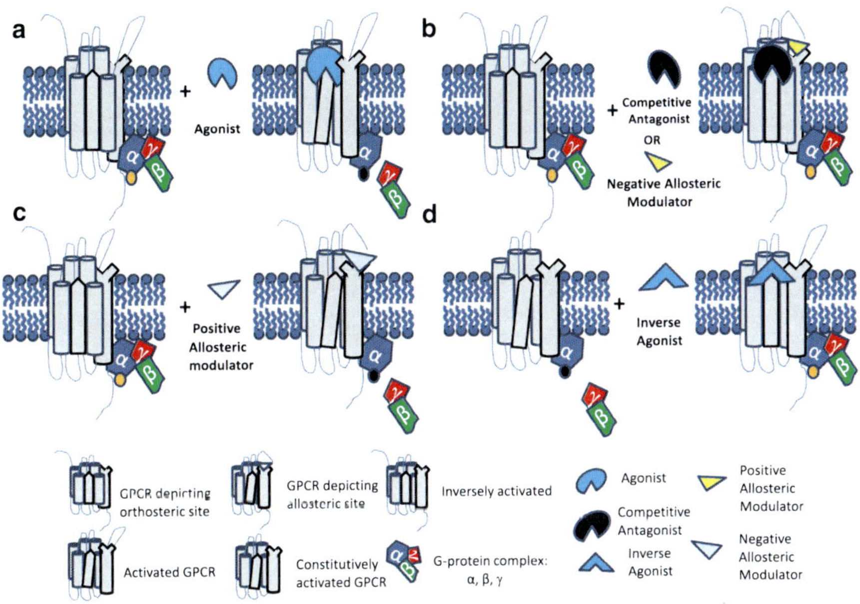

Fig. 3.2 Schematic of G protein-coupled receptor (GPCR). **a** Depiction of inactive receptor (*left*) and agonist-activated receptor (*right*). In the inactive state, the G protein–GDP-bound subunit complex is bound to the receptor protein (*left*); whereas upon binding of agonist at the orthosteric site, the receptor is activated and the G protein subunits βγ dissociate from the GTP-bound α subunit (*right*). **b** Illustration of binding of either competitive antagonist or negative allosteric modulator at orthosteric and allosteric sites, respectively. Note that the receptor is not activated by either of these ligands. **c** Binding of a positive allosteric modulator, at an allosteric site. Note that this ligand does not activate the receptor in and of itself. **d** Constitutive activity is demonstrated on the *left*. The receptor is in an activated state, bound to the α subunit of the GDP–G protein complex, in the absence of agonist. Presumably in this state the receptor conformation is different than the agonist–receptor conformation, indicated by a difference in receptor shape and shading. Binding of an inverse agonist, at the orthosteric site of a constitutively active receptor causes a "disactivation" of constitutive activity, a presumed conformational change, and recoupling of the G protein subunits with the receptor. Note that the position of the G protein complex is shown bound to the receptor in a slightly different location to illustrate a different G protein–receptor conformational state, accommodating decreased basal activity consequent to inverse agonist binding

Fig. 3.3 Concentration-response analysis with agonists illustrating full and partial agonists. Shown are exemplary curves of a full agonist (○), and a partial agonist (Δ)

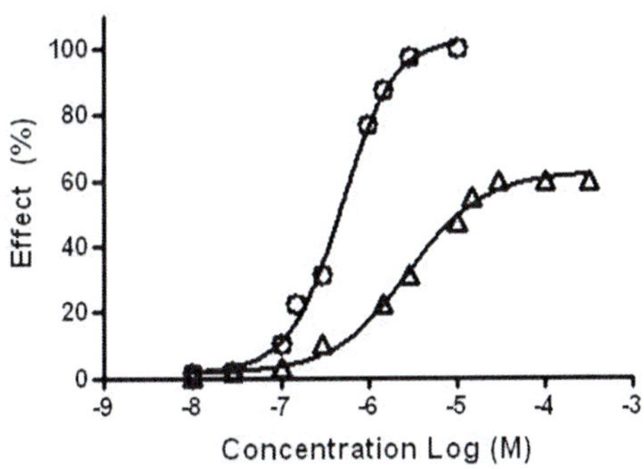

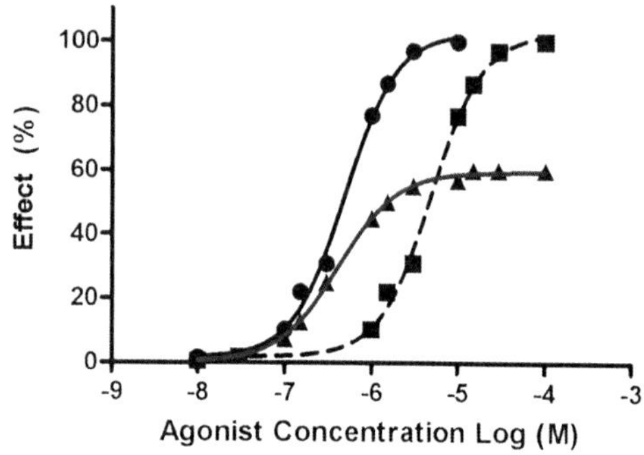

Fig. 3.4 Concentration-response analysis in the presence of competitive and noncompetitive antagonists. The concentration response curves are depicted for agonist alone (●), in the presence of a competitive antagonist (■), and in the presence of a noncompetitive antagonist (▲)

submaximal effect. In the presence of a full agonist, partial agonists can appear to be competitive antagonists as they compete with the agonist for the orthosteric site on the receptor. The response of a full agonist will be augmented in the presence of low concentrations of a partial agonist. However, as the concentration of partial agonist increases the effect of the full agonist is attenuated. Thus, the partial agonist appears as if it is a competitive antagonist with respect to the biologic response being measured. Antagonism of an agonist–receptor response can be either described as competitive or noncompetitive inhibition. If the compound is a neutral (silent) competitive antagonist, having no intrinsic activity/efficacy of its own, the response of the agonist in the presence of competitive antagonist will be attenuated. Similar to agonist binding, antagonists can either bind to the orthosteric site OR allosteric site. Orthosteric antagonist-binding results in either blockade or dimunition of endogenous agonist response whereas binding of an antagonist to an allosteric site results in an attenuated endogenous agonist response. Competitive antagonism presumes that competitive antagonist binding occurs at the receptor site where the endogenous agonist binds, the orthosteric site (Fig. 3.4). As the name implies, competition exists between agonist and antagonist for the same site on the receptor. In the presence of this type of antagonist, a greater concentration of agonist will be required to elicit the same magnitude of response observed with agonist alone, resulting in an increased EC_{50} value. Unlike competitive antagonism, noncompetitive antagonism renders the receptor unavailable for agonist binding. A modification of the receptor is induced, be it a covalent alteration or a conformational alteration, the orthosteric site is not readily available to the endogenous agonist. However, the effects of the noncompetitive antagonist are insurmountable. Hence, despite increasing concentrations of agonist, the maximal agonist-induced response will not be attained. The EC_{50} of the agonist often is not different in the presence of noncompetitive antagonist. This scenario can be mimicked by negative allosteric modulation. Whether a compound noncompetitively antagonizes a receptor-mediated response via binding at the orthosteric, or to an allosteric site, the antagonism produced yields a receptor unrecognizable to the endogenous agonist, and/or an alteration in the agonist-mediated response.

An additional type of antagonism, uncompetitive antagonism, has been described with respect to NMDA receptor activity (Lipton 2004). Uncompetitive antagonists require receptor activation by an agonist at the orthosteric site prior to being capable of binding to an allosteric site. Inhibition of the agonist-mediated biologic response by uncompetitive antagonists is greater in the presence of high levels of agonist in comparison to lower levels of agonists. Thus far, uncompetitive antagonism has not been described for cannabinoid receptors, although it poses an interesting pursuit.

The finding that antagonists exhibited "negative intrinsic activity" was first observed decades ago in a receptor recombinant system (Costa and Herz 1989). This negative intrinsic activity, the ability to yield an opposite effect compared with an agonist's "positive" intrinsic activity, has been coined inverse agonism. The concept of inverse agonism is extensively reviewed (Kenakin 2004; Greasley and Clapham 2006). Compounds which exhibit negative intrinsic activity, i.e., inhibitors of basal G protein activity, are called inverse agonists. With in vitro systems, inverse agonists often exhibit competitive antagonism, also referred to as functional antagonism.

To accommodate the observation of "negative intrinsic activity", the ternary complex model of receptors, a model accommodating GPCRs, was extended to include an additional receptor state. In this state, receptors spontaneously become activated thereby promoting uncoupling of the G protein receptor complex (Kenakin 2001). This spontaneous receptor activity is described as constitutive activity. A priori, constitutive activity assumes receptor activity in the absence of endogenous compound. Constitutive activity is described in detail in Sect. 3.4 of this chapter.

As discussed previously, allosterism, either positive or negative, affects the agonist-mediated biologic response. Therefore, exploitation of allosterism can aid in the development of novel therapeutic compounds or perhaps aide in enhancing the biologic effect of an existing therapeutic agent with low intrinsic activity. In addition, if a therapeutic agent exhibits off-target effects at a particular dose, an allosteric modulator could be used such that a lower concentration of the therapeutic compound could be used to achieve the same response without the unwanted off-target effects. Allosteric modulators could also be a valuable research tool to obtain more reliable results from binding assays investigating orthosteric sites and constitutive activity. For example, it has been proposed that an allosteric modulator could stabilize the ligand–receptor interaction during filtration (Hulme and Trevethick 2010). An allosteric modulator could reduce unwanted receptor–ligand dissociation during the wash steps, thereby increasing reproducibility of binding parameters obtained from radioligand assays.

The concept of biased agonism, differential signaling for different agonists at the same receptor, has been demonstrated with many GPCRs since its initial reporting two decades ago (Andresen 2011). This finding also led to expansion of the two receptor state theory. Biased agonism suggests that there are multiple active states of a receptor leading to activation of multiple signal transduction pathways (Kenakin 1995). Any given agonist could bind to a particular active state in a particular region of the body to preferentially activate a given pathway. It has been suggested that biased agonism may be conferred as a consequence of homo- and/or heterodimerization. This concept may explain the interactions of endocannabinoids with multiple receptor systems, and/or direct interaction of cannabinoid receptor homodimers or heterodimers with other receptor systems. Functional CB_1 and CB_2 heteromers have been demonstrated in transfected cell lines as well as in rat brain (Callen et al. 2012). Heterodimerization and multimerization of cannabinoid receptors are reviewed in detail elsewhere (Hudson et al. 2010; Mackie 2005).

3.4 Constitutive Activity

Constitutive (agonist independent) activity is observed with the overexpression of many GPCRs (Lefkowitz et al. 1993). Experimental evidence for constitutively active CB_1 receptors was first noted when SR141716A, initially described as a CB_1 antagonist, was found to have inverse agonist properties with respect to stimulation of mitogen-activated protein kinase (MAPK) activity (Bouaboula et al. 1997). Cannabinoid agonists also stimulate MAPK in transfected CHO cells expressing CB_1 (Bouaboula et al. 1995). However, basal levels of MAPK activity were higher in CB_1-transfected cells compared with untransfected cells, suggesting the presence of autoactivated CB_1 receptors. SR141716A not only antagonized the agonist effect on MAPK, but also reduced basal MAPK activity in CB_1-transfected but not untransfected cells. Similarly, basal cAMP levels were reduced and SR141716A raised basal cAMP levels in transfected cells. The EC_{50} for SR141716A was similar to its IC_{50}, suggesting that these effects are a result of direct binding to unoccupied (precoupled) CB_1 receptors and not due to the presence of endogenous ligands in the cultures. A significantly higher EC_{50} would be predicted if endogenous agonists were competing with SR141716A. Subsequent studies extended these findings to CB_1 receptor-activated GTPγS binding (Landsman et al. 1997) and inhibition of calcium conductance (Pan et al. 1998). In addition, CB_1 receptors can sequester G proteins, making them unavailable to couple to other receptors (Vasquez and Lewis 1999). SR141716A is also an inverse agonist when CB_1 receptors are coexpressed with G protein-coupled potassium channels in *Xenopus* oocytes (McAllister et al. 1999). A study in primary cultures of rat cerebellar granule neurons presented evidence for inverse agonism by SR141716A on nitric oxide synthase activity (Hillard et al. 1999a). Evidence for inverse agonism was also reported in the guinea pig small intestine (Coutts et al. 2000). Constitutive receptor activity, a priori, occurs in the absence of endogenous ligands as previously mentioned. In a recent review, the authors caution identification of cannabinoid receptor activation as constitutive activity unless endogenous ligands are known not to be present (Howlett et al. 2011).

Mutations (either naturally occurring or engineered) can also give rise to constitutively active GPCRs. Mutations that result in constitutive activity may provide clues to the key amino acids involved in receptor activation. Generally, constitutively active receptors are also constitutively phosphorylated and desensitized, providing support for a model where a single active-state conformation is the target for phosphorylation, internalization and desensitization (Leurs et al. 1998). However, a study on the angiotensin II receptor and a series of studies on the CB_1 receptor suggest that GPCRs may possess several transition states, each associated with a distinguishable conformation during receptor activation and regulation (Houston and Howlett 1998; Thomas et al. 2000; Jin et al. 1999; Roche et al. 1999; Hsieh et al. 1999).

Nie and Lewis found that the C-terminal domain contributes to constitutive activity of CB_1 (Nie and Lewis 2001). Truncation of the distal C-terminal tail of the CB_1 receptor (at residue 417 in rat CB_1) enhanced both the constitutive activity and the ability of the receptor to sequester G proteins. Conversely, mutation of a highly conserved aspartate residue in TMH2, D2.50 (164 in rat CB_1) abolished G protein sequestration and constitutive receptor activity without disrupting agonist-stimulated activity at Ca^{2+} channels. They concluded that the distal C-terminal tail acts to constrain the receptor from activating G proteins, whereas the aspartate (D2.50) in the second transmembrane domain stabilizes the receptor in both the inactive RG(GDP) state and the active R*G(GTP) state.

An interaction between F3.36/W6.48 has also been proposed to be key to the maintenance of the CB_1 inactive state (Singh et al. 2002). Previous modeling studies had suggested that a F3.36/W6.48 interaction requires a F3.36 trans $\chi 1$/W6.48 g+ $\chi 1$ rotameric state. SR141716A stabilizes this F3.36/W6.48 aromatic stacking interaction, while WIN55,212-2 favors a F3.36 g+ $\chi 1$/W6.48 trans $\chi 1$ state (Singh et al. 2002). McAllister et al. explored this hypothesis in a mutation study of mouse CB_1 (McAllister et al. 2004). The F3.36(201)A mutation showed statistically significant increases in ligand-independent stimulation of GTPγS binding vs. wild-type CB_1. Basal levels for the W6.48(357)A mutant were not statistically different from wild-type CB_1. F3.36(201)A demonstrated a limited activation profile in the presence of multiple agonists. In contrast, enhanced agonist activation was produced by W6.48(357)A. These results suggest that a F3.36(201)/W6.48(357)-specific contact is an important constraint for the CB_1-inactive state that may need to break during activation. Modeling studies suggested that the F3.36(201)/W6.48(357) contact can exist in the inactive state of CB_1 and be broken in the activated state via a χ_1 rotamer switch (F3.36(201) trans, W6.48(357) g+)$\rightarrow$(F3.36(201) g+, W6.48(357) trans) as previously proposed. The F3.36(201)–W6.48(357) interaction therefore may represent a "toggle switch" for activation of CB_1. Similar results were reported with mutation of F3.36(200) in the human CB_1 receptor (Shen et al. 2006).

Constitutive activity has also been shown with the CB_2 receptor (Bouaboula et al. 1999c). CB_2 receptors expressed in CHO cells also sequester G_i proteins; the CB_2 inverse agonist SR144528 inhibits basal G protein activity as well as switching off MAPK activation from receptor tyrosine kinases and the GPCR lysophosphatidic acid (LPA) receptor (Bouaboula et al. 1999a). When expressed in heterologous systems, CB_2 receptors are constitutively phosphorylated and internalized (Bouaboula et al. 1999b). Autophosphorylation as well as agonist-induced phosphorylation occur on S352 and involves a G protein-coupled receptor kinase (GRK; Bouaboula et al. 1999c). In transfected HEK293 cells, mutations of CB_2 at H316Y, which corresponds to a single nucleotide polymorphism, caused higher constitutive activity than the CB_2 wild-type receptor (Carrasquer et al. 2010). These data suggest that CB_2 polymorphic receptors may contribute to the etiology of certain diseases.

Table 3.1 Summary of mutations and polymorphisms in the endocannabinoid system. (For review see Hillard et al. 2012)

Mutation type	Description	Disease associations	References
CNR1 trinucleotide repeat in 3′ UTR	AAT repeat	Schizophrenia, substance abuse disorders, Parkinson's disease, inverse relation between number of repeats and working memory performance	Zhang et al. 2004; Comings et al. 1997; Ujike et al. 2002; Barrero et al. 2005; Ruiz-Contreras et al. 2013
CNR1 SNPs or haplotypes	rs6454674; rs806380; rs806377; rs1049353; rs806379; rs1535255; rs2023239; rs806368; rs806369; rs1049353; rs4707436; rs12720071; rs3505747	Substance abuse disorders, depression, anxiety and eating disorders, obesity, schizophrenia, attention deficit disorder	Hopfer et al. 2006; Zuo et al. 2009; Zhang et al. 2004; Juhasz et al. 2009; Lazary et al. 2009; Ho et al. 2011; Okahisha et al. 2011; Mutombo et al. 2012; Marcos et al. 2012
CB_2 SNPs	rs2502992, rs2501432	Low bone mineral density or osteoporosis associated in at least three distinct human populations	Huang et al. 2009; Karsak et al. 2005; Karsak et al. 2009; Yamada et al. 2007

CNR1 cannabinoid receptor 1 gene, single polymorphism, CB_2 cannabinoid receptor 2

3.5 Gene Structure, Polymorphisms and Species Diversity

The roles of specific amino acids within cannabinoid receptors have been studied in detail and researchers have identified many requirements essential for a compound to bind and/or activate these receptors. There is mounting evidence that different types of cannabinoids may require different amino acids for binding and activation (reviewed in Abood 2005; Shim 2010). There are a number of published examples (described below) demonstrating a potentially vast amount of CB_1 and CB_2 gene divergence in human populations, which can arise from splice variations, polymorphisms, and somatic mutations (Table 3.1).

The human CB_1 receptor has distinct splice variant forms. A PCR amplification product was isolated that lacked 167 bp of the coding region of the human CB_1 receptor (Shire et al. 1995). This alternative splice form (hCB_{1a}) is unusual in that it is generated from the mRNA-encoding hCB_1, and not from a separate exon (Shire et al. 1995). When expressed, the hCB_{1a} clone would translate to a receptor truncated by 61 amino acid residues with 28 amino acid residues different at the NH_2-terminal. A second splice variant of the coding region has been reported, in which a 99 base portion of the coding exon is spliced out of the human mRNA leading to an in-frame deletion of 33 amino acids (Ryberg et al. 2005). This hCB_{1b} cDNA was isolated while cloning the previously reported splice variant. Both the hCB_{1a} and hCB_{1b} variants show altered ligand binding and GTPγS activity compared with CB_1 when the cDNAs are expressed in HEK293 cells (Ryberg et al. 2005). Of the six endocannabinoids tested, only 2-AG showed significant affinity for hCB_{1b}; furthermore, 2-AG acted as an inverse agonist at both variants. AEA was able to activate the variants at concentrations $>10\ \mu M$. However, Δ^9-THC, CP55940, WIN55,212, HU210, and SR141716 exhibited good affinity and GTPγS activity with the variants. Interestingly, when these splice variants were expressed in autaptic hippocampal neuronal culture, a different pharmacological profile was observed (Straiker et al. 2012). Instead of 2-AG acting as an inverse agonist, they found 2-AG to be an efficacious agonist for both hCB_{1a} and hCB_{1b} receptors. CB_{1a} and CB_{1b} expression has been detected in many tissues by RT-PCR (Ryberg et al. 2005; Shire et al. 1995). It will be important to confirm that the CB_{1a} and CB_{1b} receptor proteins are indeed expressed as splice variants often arise due to incomplete splicing during library construction and RT-PCR techniques. The construction of antibodies selective to CB_1 or CB_{1a}/CB_{1b} peptides would be useful to detect these proteins. Neither splice variant is present in rat or mouse, because the splice consensus sequence is absent in these genes (Bonner 1996).

Previous studies have suggested the presence of three exons upstream of the coding region of the CB_1 receptor (Bonner 1996). The genomic structure of the human CB_1 receptor has been reported (Zhang et al. 2004). In this study, three exons upstream of the coding exon were identified (a total of four exons), with a variation in the first exon. Five distinct variant exonic structures were demonstrated.

The CB_1 receptors are highly conserved among vertebrate species and have also been found in some invertebrates (Murphy et al. 2001; McPartland and Glass 2003; Elphick and Egertova 2001). Shortly after the cloning of the rat cannabinoid receptor, isolation of a human CB_1 receptor cDNA was reported (Gerard et al. 1991). The human CB_1 receptor has one less amino acid in the N-terminus compared with the other mammalian species (472 amino acids vs. 473 amino acids). The rat and human receptors are highly conserved, 93 % identity at the nucleic acid level and 97 % at the amino acid level. Similarly, the mouse and rat clones have 95 % nucleic acid identity (100 % amino acid identity) and the mouse and human clones have 90 % nucleic acid identity (97 % amino acid identity; Abood et al. 1997; Chakrabarti et al. 1995; Ho and Zhao 1996).

A molecular phylogenetic analysis which included the CB_1 receptor showed that the sequence diversity in 62 mammalian species varied from 0.41 to 27 % (Murphy et al. 2001). In addition to mammals, the CB_1 receptor has been isolated from birds (Soderstrom et al. 2000), fish (Yamaguchi et al. 1996), amphibia (Soderstrom et al. 2000; Cottone et al. 2003), and an invertebrate, *Ciona intestitinalis* (Elphick et al. 2003). This deuterostomian invertebrate CB receptor contains 28 % amino acid identity with CB_1, and 24 % with CB_2 (Elphick et al. 2003). Since a CB receptor ortholog has not been found in *Drosophila melanogaster* or *Caenorhabditis elegans*, it has been suggested that the ancestor of vertebrate CB_1 and CB_2 receptors originated in a deuterostomian invertebrate (Elphick et al. 2003). Several human CB_1 receptor polymorphisms have also been identified. The initial polymorphism found was a restriction fragment length polymorphism (RFLP) in the intron preceding the coding exon of the receptor (Caenazzo et al. 1991). The CB_1 receptor gene is intronless in its coding region, but possesses an intron 5′ to the coding exon with three putative upstream exons (Bonner 1996; Zhang et al. 2004).

A positive association between a microsatellite polymorphism ($(AAT)_n$) in the CB_1 gene and IV drug abuse has been described (Comings et al. 1997). This polymorphism has subsequently been localized 3′ to the coding exon of the CB_1 receptor (Zhang et al. 2004). Although there are differences between populations, the CB_1 $(AAT)_n$ polymorphism has also been associated with schizophrenia (Ujike et al. 2002) as well as with depression in Parkinson's disease (Barrero et al. 2005), providing genetic evidence for a role of the cannabinoid system in these disorders. There is also a significant inversely proportional association to the number of AAT repeats and working memory performance (Ruiz-Contreras et al. 2013).

The first polymorphism in the coding exon described was a silent mutation in T453 (G to A), a conserved amino acid present in the C terminal region of the CB_1 and CB_2 receptors that was a common polymorphism in the German population (rs1049353; Gadzicki et al. 1999). While this mutation is silent, analysis of several human sequences present in the database reveals that CB1K5 (accession #AF107262), a full length sequence, contains five nucleotide changes, three of which result in amino acid differences. Coincidentally, two amino acid differences are in the third transmembrane domain, F200L and I216V. The third variant is in the fourth transmembrane domain, V246A. A report by the group that submitted the sequence to the database revealed that this was a somatic mutation in an epilepsy patient; i.e., DNA obtained from their blood was unaltered, but DNA from the hippocampus showed the mutation (Kathmann et al. 2000). The presence of a somatic mutation rather than a polymorphism is generally indicative of the disease process in cancers (e.g., mutant p53 or APC expression in tumors, but not normal tissues (Baker et al. 1989; Lamlum et al. 2000)). CB_1 receptor polymorphisms may affect responsiveness to cannabinoids.

Several polymorphisms were studied in control and drug-abusing individuals from European, African, and Japanese ethnicities and found association with a 5′ "TAG" haplotype that was highly associated with substance abuse in all three populations (Zhang et al. 2004). Analysis of mRNA levels from postmortem brain samples of individuals with the TAG haplotype showed reduced expression for individuals expressing this allele. In sum, the genomic studies implicate the CB_1 receptor in drug addiction and mental health (Ho et al. 2011; Hopfer et al. 2006; Juhasz et al. 2009; Lazary et al. 2009; Marcos et al. 2012; Mutombo et al. 2012; Okahisa et al. 2011; Ruiz-Contreras et al. 2013; Zuo et al. 2009).

Polymorphisms in the CB_2 receptor have been identified as well (Karsak et al. 2005; Sipe et al. 2005). Polymorphisms of the human CB_2 gene are linked to osteoporosis in several studies (Karsak et al. 2005, 2009; Yamada et al. 2007). Karsak et al. examined CB_1 and CB_2 receptor DNA in a sample of French postmenopausal patients and female controls. The authors report that certain changes in CB_2 receptor, but not the CB_1 receptor, were strongly associated with osteoporosis (Karsak et al. 2005). A second study replicated these findings in a group of pre- and postmenopausal Japanese women (Yamada et al. 2007). In contrast, a recent study has found only nominally significant correlations with CB_2 polymorphisms and osteoporosis in a Chinese population; the role of the CNR2 gene in the etiology of Chinese osteoporosis thus requires further study in larger samples (Huang et al. 2009).

A recent study examined the role of CB_2 DNA or genes on hand bone strength (Karsak et al. 2009). The authors analyzed radiographic images and DNA samples from a Chevashian population, an ethnically homogeneous population of people of Bulgaric ancestry that live along the Volga River. Several single nucleotide polymorphisms (SNPs) were significantly associated with certain bone phenotypes as previously reported (Karsak et al. 2005). Two of the associated SNPs were in adjacent nucleotides ("double SNP" rs2502992–rs2501432) within the coding region of CB_2 and result in a nonconservative missense variant (Gln63Arg). This variant is probably functionally relevant as demonstrated by a differentially endocannabinoid-induced inhibition of T lymphocyte proliferation (Sipe et al. 2005). A less functional form of the CB_2 receptor appears to lead to weak hand bone strength and is associated with osteoporosis.

In addition to the human CB_2 receptor, clones have been isolated from mouse (Shire et al. 1996; Valk et al. 1997), rat (Griffin et al. 2000; Brown et al. 2002; Liu et al. 2009), dog (Ndong et al. 2011), the puffer fish *Fugu rubripes* (Elphick 2002) as well as zebrafish (McPartland et al. 2007). There is also information in the GenBank database on additional species. The CB_2 receptor shows less homology between species than does CB_1; for instance, the human and mouse CB_2 receptors share 82 % amino acid identity (Shire et al. 1996), and the mouse and rat 93 % amino acid identity. The human, rat, and mouse sequences diverge at the C-terminus; the mouse sequence is 13 amino acids shorter, whereas the rat clone is 50 amino acids longer than the human CB_2 (Brown et al. 2002).

The first evidence for alternative splice forms of CB_2 was in the C-terminus of the rat CB_2 receptor (Griffin et al. 2000; Brown et al. 2002). That this may give rise to rat-specific pharmacology of the CB_2 receptor was suggested by differences in ligand recognition with a number of compounds at the rat CB_2 receptor compared with the human CB_2 receptor in transfected cells (Griffin et al. 2000). The clone described in these studies was amplified from genomic DNA rat CB_2; however, this isoform has subsequently been shown to be the major splice form of rat CB_2 (Liu et al. 2009). Now, variants of the human and mouse CB_2 receptors have been reported as well (Liu et al. 2009).

In summary, the CB_1 and CB_2 receptor genes have diverse regulatory regions that may provide extensive flexibility in gene regulation of the receptors during health maintenance and disease progression. Alterations in ECS gene transcription may contribute to the occurrence of neurodegenerative or mental diseases. For example, genetic differences (i.e., haplotype blocks) in the cannabinoid receptor 1 gene (CNR1) are associated with mental and depressive disorders (Hillard et al. 2012). Several comparable studies document significant associations of the CNR1 with a specific disease, which suggests that the occurrences of certain polymorphisms are not capricious but harbingers of disease. Cannabinoid receptors may represent an excellent candidate for developing personalized medicine, as health professionals may be able to screen the gene encoding a cannabinoid receptor (i.e., CNR1 or CNR2) of a patient, and determine the class of cannabinoids to administer. For instance, a hypothetical patient could possess a polymorphism or acquire a mutation that hinders efficient interactions at cannabinoid receptors by endocannabinoids. In this scenario, a drug is chosen based on the patient's genotype. A different class of cannabinoids may be able to replace the endogenous compounds or increase the efficiency of receptor–ligand interactions. This type of personalized approach would require either selecting a compound that has structurally distinct binding requirements from endogenous compounds (i.e., an indole such as WIN55,212) or a compound that will enhance endocannabinoid activity at the target (i.e., an allosteric modulator such as Org27569). A "clinical endocannabinoid deficiency syndrome" resulting from defects in the ECS (i.e., receptor mutations, alterations in endocannabinoid production) has already been proposed to underlie certain diseases including treatment resistant conditions (Russo 2008). To date, a mutation is yet to be identified in the human cannabinoid receptor that results in conclusive alteration of ligand–receptor interactions; however, molecular biologists have discovered amino acids residues important for selective ligand recognition and maintaining receptor–ligand interactions in vitro (Song and Bonner 1996; Kapur et al. 2008). The efficacy of future cannabis-based clinical trials could be enhanced by developing patient-screening methods for polymorphisms or mutations in genes associated with the endocannabinoid system.

3.6 Conclusions

The multifarious nature of cannabinoid receptors allows for a single GPCR to recognize diverse classes of compounds and produce an array of distinct downstream effects. Allosteric modulation of cannabinoid receptors may usher in new classes of medicinal compounds capable of enhancing signals generated by endocannabinoids. Natural polymorphisms and alternative splice variants may also contribute to the pharmacological diversity of the cannabinoid receptors. The cannabinoid research field continues to generate significant discoveries since the identification of the CB_1 receptor more than 20 years ago. However, many challenges await the field, including the classification of other GPCRs (i.e., GPR18 and GPR55) as bona fide cannabinoid receptors and developing strategies to target receptor conformations for harnessing specific pharmacological responses. The basic biology of the endocannabinoid system will continue to be revealed by ongoing investigations, and progress will partially depend upon the development of technologies that can assimilate current research trends and theories.

Acknowledgments Linda Console-Bram and Jahan Marcu contributed equally to this chapter. The authors would like to thank Dr. Haleli Sharir for her suggestions, and acknowledge funding from NIH/NIDA DA023204 (Abood, PI), T32DA07237 (Unterwald, PI), and DA13429 (Unterwald, PI).

References

Abood ME (2005) Molecular biology of cannabinoid receptors. Handb Exp Pharmacol (168):81–115

Abood ME, Ditto KA, Noel MA, Showalter VM, Tao Q (1997) Isolation and expression of mouse CB1 cannabinoid receptor gene: comparison of binding properties with those of native CB1 receptors in mouse brain and N18TG2 neuroblastoma cells. Biochem Pharmacol 53:207–214

Agarwal N, Pacher P, Tegeder I, Amaya F, Constantin CE, Brenner GJ, Rubino T, Michalski CW, Marsicano G, Monory K, Mackie K, Marian C, Batkai S, Parolaro D, Fischer MJ, Reeh P, Kunos G, Kress M, Lutz B, Woolf CJ, Kuner R (2007) Cannabinoids mediate analgesia largely via peripheral type 1 cannabinoid receptors in nociceptors. Nat Neurosci 10(7):870–879

Ahmed SA, Ross SA, Slade D, Radwan MM, Zulfiqar F, Matsumoto RR, Xu YT, Viard E, Speth RC, Karamyan VT, ElSohly MA (2008) Cannabinoid ester constituents from high-potency Cannabis sativa. J Nat Prod 71(4):536–542

Ambrosino P, Soldovieri M, Russo C, Taglialatela M (2012) Activation and desensitization of Trpv1 channels in sensory neurons by the peroxisome proliferator-activated receptor alpha agonist palmitoylethanolamide. Br J Pharmacol 168(6):1430–1444

Andresen BT (2011) A pharmacological primer of biased agonism. Endocr Metab Immune Disord Drug Targets 11(2):92–98

Baker SJ, Fearon ER, Nigro JM, Hamilton SR, Preisinger AC, Jessup JM, vanTuinen P, Ledbetter DH, Barker DF, Nakamura Y et al (1989) Chromosome 17 deletions and p53 gene mutations in colorectal carcinomas. Science 244(4901):217–221

Baker D, Pryce G, Davies WL, Hiley CR (2006) In silico patent searching reveals a new cannabinoid receptor. Trends Pharmacol Sci 27(1):1–4

Barrero FJ, Ampuero I, Morales B, Vives F, Dios Luna Del Castillo J de, Hoenicka J, Garcia Yebenes J (2005) Depression in Parkinson's disease is related to a genetic polymorphism of the cannabinoid receptor gene (CNR1). Pharmacogenomics J 5(2):135–141

Ben-Shabat S, Fride E, Sheskin T, Tamiri T, Rhee MH, Vogel Z, Bisogno T, Petrocellis LD, Marzo VD, Mechoulam R (1998) An entourage effect: inactive endogenous fatty acid glycerol esters enhance 2-arachidonoyl-glycerol cannabinoid activity. Eur J Pharmacol 353:23–31

Bonner T (1996) Molecular biology of cannabinoid receptors. J Neuroimmunol 69:15–23

Bouaboula M, Poinot-Chazel C, Bourrie B, Canat X, Calandra B, Rinaldi-Carmona M, Le Fur G, Casellas P (1995) Activation of mitogen-activated protein kinases by stimulation of the central cannabinoid receptor CB1. Biochem J 312(2):637–641

Bouaboula M, Perrachon S, Milligan L, Canat X, Rinaldi-Carmona M, Portier M, Barth F, Calandra B, Pecceu F, Lupker J, Maffrand J-P, LeFur G, Casellas P (1997) A selective inverse agonist for central cannabinoid receptor inhibits mitogen-activated protein kinase activation stimulated by insulin or insulin-like growth factor 1. J Biol Chem 272:22330–22339

Bouaboula M, Bianchini L, McKenzie F, Pouyssegur J, Casellas P (1999a) Cannabinoid receptor CB1 activates the Na+/H+ exchanger NHE-1 isoform via Gi-mediated mitogen activated protein kinase signaling transduction pathways. FEBS Lett 449(1):61–65

Bouaboula M, Desnoyer N, Carayon P, Combes T, Casellas P (1999b) Gi protein modulation induced by a selective inverse agonist for the peripheral cannabinoid receptor CB2: implication for intracellular signalization cross-regulation. Mol Pharmacol 55(3):473–480

Bouaboula M, Dussossoy D, Casellas P (1999c) Regulation of peripheral cannabinoid receptor CB2 phosphorylation by the inverse agonist SR 144528. Implications for receptor biological responses. J Biol Chem 274:20397–20405

Bradshaw HB, Rimmerman N, Hu SS, Benton VM, Stuart JM, Masuda K, Cravatt BF, O'Dell DK, Walker JM (2009) The endocannabinoid anandamide is a precursor for the signaling lipid N-arachidonoyl glycine by two distinct pathways. BMC Biochem 10:14

Breivogel CS, Griffin G, Di Marzo V, Martin BR (2001) Evidence for a new G protein-coupled cannabinoid receptor in mouse brain. Mol Pharmacol 60(1):155–163

Brown SM, Wager-Miller J, Mackie K (2002) Cloning and molecular characterization of the rat CB2 cannabinoid receptor. Biochim Biophys Acta 1576(3):255–264

Buckley NE, McCoy KL, Mezey E, Bonner T, Zimmer A, Felder CC, Glass M (2000) Immunomodulation by cannabinoids is absent in mice deficient for the cannabinoid CB(2) receptor. Eur J Pharmacol 396(2–3):141–149

Busquets-Garcia A, Puighermanal E, Pastor A, la Torre R de, Maldonado R, Ozaita A (2011) Differential role of anandamide and 2-arachidonoylglycerol in memory and anxiety-like responses. Biol Psychiatry 70(5):479–486

Caenazzo L, Hoehe M, Hsieh W, Berrettini W, Bonner T, Gershon E (1991) HindIII identifies a two allele DNA polymorphism of the human cannabinoid receptor gene (CNR). Nucleic Acids Res 19(17):4798.

Callen L, Moreno E, Barroso-Chinea P, Moreno-Delgado D, Cortes A, Mallol J, Casado V, Lanciego JL, Franco R, Lluis C, Canela EI, McCormick PJ (2012) Cannabinoid receptors CB1 and CB2 form functional heteromers in brain. J Biol Chem 287(25):20851–20865

Carrasquer A, Nebane NM, Williams WM, Song ZH (2010) Functional consequences of nonsynonymous single nucleotide polymorphisms in the CB2 cannabinoid receptor. Pharmacogenet Genomics 20(3):157–166

Chakrabarti A, Onaivi ES, Chaudhuri G (1995) Cloning and sequencing of a cDNA encoding the mouse brain-type cannabinoid receptor protein. DNA Seq 5:385–388

Comings D, Muhleman D, Gade R, Johnson P, Verde R, Saucier G, MacMurray J (1997) Cannabinoid receptor gene (CNR1): association with i.v. drug use. Mol Psychiatry 2(2):161–168

Conn PJ, Christopoulos A, Lindsley CW (2009) Allosteric modulators of GPCRs: a novel approach for the treatment of CNS disorders. Nat Rev Drug Discov 8(1):41–54

Costa T, Herz A (1989) Antagonists with negative intrinsic activity at delta opioid receptors coupled to GTP-binding proteins. Proc Natl Acad Sci U S A 86(19):7321–7325

Cottone E, Salio C, Conrath M, Franzoni MF (2003) Xenopus laevis CB1 cannabinoid receptor: molecular cloning and mRNA distribution in the central nervous system. J Comp Neurol 464(4):487–496

Coutts A, Brewster N, Ingram T, Razdan R, Pertwee R (2000) Comparison of novel cannabinoid partial agonists and SR141716A in the guinea-pig small intestine. Br J Pharmacol 129(4): 645–652

De Petrocellis L, Ligresti A, Moriello AS, Allara M, Bisogno T, Petrosino S, Stott CG, Di Marzo V (2011) Effects of cannabinoids and cannabinoid-enriched Cannabis extracts on TRP channels and endocannabinoid metabolic enzymes. Br J Pharmacol 163(7):1479–1494

Devane WA, Dysarz FA 3rd, Johnson MR, Melvin LS, Howlett AC (1988) Determination and characterization of a cannabinoid receptor in rat brain. Mol Pharmacol 34(5):605–613

Devane WA, Hanus L, Breuer A, Pertwee RG, Stevenson LA, Griffin G, Gibson D, Mandelbaum A, Etinger A, Mechoulam R (1992) Isolation and structure of a brain constituent that binds to the cannabinoid receptor. Science 258:1946–1949

Di Marzo V, Fontana A, Cadas H, Schinelli S, Cimino G, Schwartz JC, Piomelli D (1994) Formation and inactivation of endogenous cannabinoid anandamide in central neurons. Nature 372(6507):686–691

Di Marzo V, Melck D, Bisogno T, De Petrocellis L (1998) Endocannabinoids: endogenous cannabinoid receptor ligands with neuromodulatory action. Trends Neurosci 21(12):521–528

Di Marzo V, Breivogel CS, Tao Q, Bridgen DT, Razdan RK, Zimmer AM, Zimmer A, Martin BR (2000) Levels, metabolism, and pharmacological activity of anandamide in CB(1) cannabinoid receptor knockout mice: evidence for non-CB(1), non-CB(2) receptor-mediated actions of anandamide in mouse brain. J Neurochem 75(6):2434–2444

Downer EJ (2011) Cannabinoids and innate immunity: taking a toll on neuroinflammation. ScientificWorldJournal 11:855–865

Du H, Chen X, Zhang J, Chen C (2011) Inhibition of COX-2 expression by endocannabinoid 2-arachidonoylglycerol is mediated via PPAR-gamma. Br J Pharmacol 163(7):1533–1549

Eissenstat MA, Bell MR, D'Ambra TE, Alexander EJ, Daum SJ, Ackerman JH, Gruett MD, Kumar V, Estep KG, Olefirowicz EM et al (1995) Aminoalkylindoles: structure-activity relationships of novel cannabinoid mimetics. J Med Chem 38(16):3094–3105

Elphick MR (2002) Evolution of cannabinoid receptors in vertebrates: identification of a CB(2) gene in the puffer fish Fugu rubripes. Biol Bull 202(2):104–107

Elphick MR, Egertova M (2001) The neurobiology and evolution of cannabinoid signalling. Philos Trans R Soc Lond B Biol Sci 356(1407):381–408

Elphick MR, Satou Y, Satoh N (2003) The invertebrate ancestry of endocannabinoid signalling: an orthologue of vertebrate cannabinoid receptors in the urochordate Ciona intestinalis. Gene 302(1–2):95–101

Elsohly MA, Slade D (2005) Chemical constituents of marijuana: the complex mixture of natural cannabinoids. Life Sci 78(5):539–548

Facci L, Toso RD, Romanello S, Buriani A, Skaper SD, Leon A (1995) Mast cells express a peripheral cannabinoid receptor with differential sensitivity to anandamide and palmitoylethanolamide. Proc Natl Acad Sci U S A 92:3376–3380

Felder CC, Veluz JS, Williams HL, Briley EM, Matsuda LA (1992) Cannabinoid agonists stimulate both receptor- and non-receptor-mediated signal transduction pathways in cells transfected with and expressing cannabinoid receptor clones. Mol Pharmacol 42(5):838–845

Fisar Z (2012) Cannabinoids and monoamine neurotransmission with focus on monoamine oxidase. Prog Neuropsychopharmacol Biol Psychiatry 38(1):68–77

Franklin A, Stella N (2003) Arachidonylcyclopropylamide increases microglial cell migration through cannabinoid CB2 and abnormal-cannabidiol-sensitive receptors. Eur J Pharmacol 474(2–3):195–198

Franklin A, Parmentier-Batteur S, Walter L, Greenberg DA, Stella N (2003) Palmitoylethanolamide increases after focal cerebral ischemia and potentiates microglial cell motility. J Neurosci 23(21):7767–7775

Fredriksson R, Lagerstrom MC, Lundin LG, Schioth HB (2003) The G-protein-coupled receptors in the human genome form five main families. Phylogenetic analysis, paralogon groups, and fingerprints. Mol Pharmacol 63(6):1256–1272

Gadzicki D, Muller-Vahl K, Stuhrmann M (1999) A frequent polymorphism in the coding exon of the human cannabinoid receptor (CNR1) gene. Mol Cell Probes 13:321–323

Gerard CM, Mollereau C, Vassart G, Parmentier M (1991) Molecular cloning of a human cannabinoid receptor which is also expressed in testis. Biochem J 279:129–134

Gonsiorek W, Lunn C, Fan X, Narula S, Lundell D, Hipkin RW (2000) Endocannabinoid 2-arachidonyl glycerol is a full agonist through human type 2 cannabinoid receptor: antagonism by anandamide. Mol Pharmacol 57(5):1045–1050

Greasley PJ, Clapham JC (2006) Inverse agonism or neutral antagonism at G-protein coupled receptors: a medicinal chemistry challenge worth pursuing? Eur J Pharmacol 553(1–3):1–9

Griffin G, Tao Q, Abood M (2000) Cloning and pharmacological characterization of the Rat CB2 cannabinoid receptor. J Pharmacol Exp Ther 292(3):886–894

Guindon J, Guijarro A, Piomelli D, Hohmann AG (2011) Peripheral antinociceptive effects of inhibitors of monoacylglycerol lipase in a rat model of inflammatory pain. Br J Pharmacol 163(7):1464–1478

Hajos N, Ledent C, Freund TF (2001) Novel cannabinoid-sensitive receptor mediates inhibition of glutamatergic synaptic transmission in the hippocampus. Neuroscience 106(1):1–4

Hanus L, Gopher A, Almog S, Mechoulam R (1993) Two new unsaturated fatty acid ethanolamides in brain that bind to the cannabinoid receptor. J Med Chem 36:3032–3034

Hanus L, Breuer A, Tchilibon S, Shiloah S, Goldenberg D, Horowitz M, Pertwee RG, Ross RA, Mechoulam R, Fride E (1999) HU-308: a specific agonist for CB(2), a peripheral cannabinoid receptor. Proc Natl Acad Sci U S A 96(25):14228–14233

Hejazi N, Zhou C, Oz M, Sun H, Ye JH, Zhang L (2006) Delta9-tetrahydrocannabinol and endogenous cannabinoid anandamide directly potentiate the function of glycine receptors. Mol Pharmacol 69(3):991–997

Henstridge CM, Balenga NA, Kargl J, Andradas C, Brown AJ, Irving A, Sanchez C, Waldhoer M (2011) Minireview: recent developments in the physiology and pathology of the lysophosphatidylinositol-sensitive receptor GPR55. Mol Endocrinol 25(11):1835–1848

Hillard C, Muthian S, Kearn C (1999a) Effects of CB(1) cannabinoid receptor activation on cerebellar granule cell nitric oxide synthase activity. FEBS Lett 459(2):277–281

Hillard CJ, Manna S, Greenberg MJ, DiCamelli R, Ross RA, Stevenson LA, Murphy V, Pertwee RG, Campbell WB (1999b) Synthesis and characterization of potent and selective agonists of the neuronal cannabinoid receptor (CB1). J Pharmacol Exp Ther 289(3):1427–1433

Hillard CJ, Weinlander KM, Stuhr KL (2012) Contributions of endocannabinoid signaling to psychiatric disorders in humans: genetic and biochemical evidence. Neuroscience 204:207–229

Ho BY, Zhao J (1996) Determination of the cannabinoid receptors in mouse x rat hybridoma NG108-15 cells and rat GH4C1 cells. Neurosci Lett 212(2):123–126

Ho BC, Wassink TH, Ziebell S, Andreasen NC (2011) Cannabinoid receptor 1 gene polymorphisms and marijuana misuse interactions on white matter and cognitive deficits in schizophrenia. Schizophrenia Res 128(1–3):66–75

Hopfer CJ, Young SE, Purcell S, Crowley TJ, Stallings MC, Corley RP, Rhee SH, Smolen A, Krauter K, Hewitt JK, Ehringer MA (2006) Cannabis receptor haplotype associated with fewer cannabis dependence symptoms in adolescents. Am J Med Genet 141B(8):895–901

Houston DB, Howlett AC (1998) Differential receptor-G-protein coupling evoked by dissimilar cannabinoid receptor agonists. Cell Signal 10(9):667–674

Howlett AC (1995) Pharmacology of cannabinoid receptors. Ann Rev Pharmacol Toxicol 35: 607–634

Howlett AC, Barth F, Bonner TI, Cabral G, Casellas P, Devane WA, Felder CC, Herkenham M, Mackie K, Martin BR, Mechoulam R, Pertwee RG (2002) International union of pharmacology. XXVII. Classification of cannabinoid receptors. Pharmacol Rev 54(2):161–202

Howlett AC, Reggio PH, Childers SR, Hampson RE, Ulloa NM, Deutsch DG (2011) Endocannabinoid tone versus constitutive activity of cannabinoid receptors. Br J Pharmacol 163(7): 1329–1343

Hsieh C, Brown S, Derleth C, Mackie K (1999) Internalization and recycling of the CB1 cannabinoid receptor. J Neurochem 73(2):493–501

Huang QY, Li GH, Kung AW (2009) Multiple osteoporosis susceptibility genes on chromosome 1p36 in Chinese. Bone 44(5):984–988

Huang SM, Bisogno T, Petros TJ, Chang SY, Zavitsanos PA, Zipkin RE, Sivakumar R, Coop A, Maeda DY, De Petrocellis L, Burstein S, Di Marzo V, Walker JM (2001) Identification of a new class of molecules, the arachidonyl amino acids, and characterization of one member that inhibits pain. J Biol Chem 276(46):42639–42644

Huang SM, Bisogno T, Trevisani M, Al-Hayani A, De Petrocellis L, Fezza F, Tognetto M, Petros TJ, Krey JF, Chu CJ, Miller JD, Davies SN, Geppetti P, Walker JM, Di Marzo V (2002) An endogenous capsaicin-like substance with high potency at recombinant and native vanilloid VR1 receptors. Proc Natl Acad Sci U S A 99(12):8400–8405

Hudson BD, Hebert TE, Kelly ME (2010) Ligand- and heterodimer-directed signaling of the CB(1) cannabinoid receptor. Mol Pharmacol 77(1):1–9

Hulme EC, Trevethick MA (2010) Ligand binding assays at equilibrium: validation and interpretation. Br J Pharmacol 161(6):1219–1237

Iliff HA, Lynch DL, Kotsikorou E, Reggio PH (2011) Parameterization of Org27569: an allosteric modulator of the cannabinoid CB(1) G protein-coupled receptor. J Comput Chem 32(10):2110–2126

Ishac EJN, Jiang L, Lake KD, Varga K, Abood ME, Kunos G (1996) Inhibition of exocytotic noradrenaline release by presynaptic cannabinoid CB1 receptors on peripheral sympathetic nerves. Br J Pharmacol 118:2023–2028

Jarai Z, Wagner J, Varga K, Lake K, Compton D, Martin B, Zimmer A, Bonner T, Buckley N, Mezey E, Razdan R, Zimmer A, Kunos G (1999) Cannabinoid-induced mesenteric vasodilation through an endothelial site distinct from CB1 or CB2 receptors. Proc Natl Acad Sci U S A 96(24):14136–14141

Jin W, Brown S, Roche J, Hsieh C, Celver J, Kovoor A, Chavkin C, Mackie K (1999) Distinct domains of the CB1 cannabinoid receptor mediate desensitization and internalization. J Neurosci 19:3773–3780

Juhasz G, Chase D, Pegg E, Downey D, Toth ZG, Stones K, Platt H, Mekli K, Payton A, Elliott R, Anderson IM, Deakin JF (2009) CNR1 gene is associated with high neuroticism and low agreeableness and interacts with recent negative life events to predict current depressive symptoms. Neuropsychopharmacology 34(8):2019–2027

Kapur A, Samaniego P, Thakur GA, Makriyannis A, Abood ME (2008) Mapping the structural requirements in the CB1 cannabinoid receptor transmembrane helix II for signal transduction. J Pharmacol Exp Ther 325(1):341–348

Kapur A, Zhao P, Sharir H, Bai Y, Caron MG, Barak LS, Abood ME (2009) Atypical responsiveness of the orphan receptor GPR55 to cannabinoid ligands. J Biol Chem 284(43):29817–29827

Karsak M, Cohen-Solal M, Freudenberg J, Ostertag A, Morieux C, Kornak U, Essig J, Erxlebe E, Bab I, Kubisch C, Vernejoul MC de, Zimmer A (2005) Cannabinoid receptor type 2 gene is associated with human osteoporosis. Hum Mol Genet 14(22):3389–3396

Karsak M, Malkin I, Toliat MR, Kubisch C, Nurnberg P, Zimmer A, Livshits G (2009) The cannabinoid receptor type 2 (CNR2) gene is associated with hand bone strength phenotypes in an ethnically homogeneous family sample. Human Genet 126(5):629–636

Kathmann M, Haug K, Heils A, Nothen M, Schlicker E (2000) Exchange of three amino acids in the cannabinoid CB1 receptor (CNR1) of an epilepsy patient. In: 2000 symposium on the cannabinoids, Burlington, Vermont, 2000. International Cannabinoid Research Society

Katzung BG, Master SB, Trevor AJ (eds) (2009) *Basic and Clinical Pharmacology,* 11th edn. McGraw Hill, USA

Kenakin T (1995) Agonist-receptor efficacy. II. Agonist trafficking of receptor signals [see comments]. Trends Pharmacol Sci 16(7):232–238

Kenakin T (2001) Inverse, protean, and ligand-selective agonism: matters of receptor conformation. Faseb J 15(3):598–611

Kenakin T (2004) Efficacy as a vector: the relative prevalence and paucity of inverse agonism. Mol Pharmacol 65(1):2–11

Kirilly E, Hunyady L, Bagdy G (2012) Opposing local effects of endocannabinoids on the activity of noradrenergic neurons and release of noradrenaline: relevance for their role in depression and in the actions of CB(1) receptor antagonists. J Neural Transm 120(1):177–186

Kohno M, Hasegawa H, Inoue A, Muraoka M, Miyazaki T, Oka K, Yasukawa M (2006) Identification of N-arachidonylglycine as the endogenous ligand for orphan G-protein-coupled receptor GPR18. Biochem Biophys Res Commun 347(3):827–832

Kreutz S, Koch M, Bottger C, Ghadban C, Korf HW, Dehghani F (2009) 2-Arachidonoylglycerol elicits neuroprotective effects on excitotoxically lesioned dentate gyrus granule cells via abnormal-cannabidiol-sensitive receptors on microglial cells. Glia 57(3):286–294

Lambert D, DiPaolo F, Sonveaux P, Kanyonyo M, Govaerts S, Hermans E, Bueb J, Delzenne N, Tschirhart E (1999) Analogues and homologues of N-palmitoylethanolamide, a putative endogenous CB(2) cannabinoid, as potential ligands for the cannabinoid receptors. Biochim Biophys Acta 1440:266–274

Lamlum H, Papadopoulou A, Ilyas M, Rowan A, Gillet C, Hanby A, Talbot I, Bodmer W, Tomlinson I (2000) APC mutations are sufficient for the growth of early colorectal adenomas. Proc Natl Acad Sci U S A 97(5):2225–2228

Landsman RS, Burkey TH, Consroe P, Roeske WR, Yamamura HI (1997) SR141716A is an inverse agonist at the human cannabinoid CB1 receptor. Eur J Pharmacol 334:R1–R2

Lazary J, Lazary A, Gonda X, Benko A, Molnar E, Hunyady L, Juhasz G, Bagdy G (2009) Promoter variants of the cannabinoid receptor 1 gene (CNR1) in interaction with 5-HTTLPR affect the anxious phenotype. Am J Med Genet 150B(8):1118–1127

Ledent C, Valverde O, Cossu G, Petitet F, Aubert J, Beslot F, Bohme G, Imperato A, Pedrazzini T, Roques B, Vassart G, Fratta W, Parmentier M (1999) Unresponsiveness to cannabinoids and reduced addictive effects of opiates in CB1 receptor knockout mice. Science 283(5400):401–404

Lefkowitz RJ, Cotecchia S, Samama P, Costa T (1993) Constitutive activity of receptors coupled to guanine nucleotide regulatory proteins. Trends Pharmacol Sci 14:303–307

Lenman A, Fowler CJ (2007) Interaction of ligands for the peroxisome proliferator-activated receptor gamma with the endocannabinoid system. Br J Pharmacol 151(8):1343–1351

Leurs R, Smit M, Alewijnse A, Timmerman H (1998) Agonist-independent regulation of constitutively active G-protein-coupled receptors. Trends Biochem Sci 23(11):418–422

Lipton SA (2004) Failures and successes of NMDA receptor antagonists: molecular basis for the use of open-channel blockers like memantine in the treatment of acute and chronic neurologic insults. NeuroRx 1(1):101–110

Liu QR, Pan CH, Hishimoto A, Li CY, Xi ZX, Llorente-Berzal A, Viveros MP, Ishiguro H, Arinami T, Onaivi ES, Uhl GR (2009) Species differences in cannabinoid receptor 2 (CNR2 gene): identification of novel human and rodent CB2 isoforms, differential tissue expression and regulation by cannabinoid receptor ligands. Genes Brain Behav 8(5):519–530

Lopez-Moreno JA, Gonzalez-Cuevas G, Moreno G, Navarro M (2008) The pharmacology of the endocannabinoid system: functional and structural interactions with other neurotransmitter systems and their repercussions in behavioral addiction. Addict Biol 13(2):160–187

Lu VB, Puhl HL, Ikeda SR (2012) N-arachidonyl Glycine (NAGly) does not activate G protein-coupled receptor 18 (GPR18) signaling via canonical pathways. Mol Pharmacol 83(1):267–282

Mackie K (2005) Cannabinoid receptor homo- and heterodimerization. Life Sci 77(14):1667–1673

Marcos M, Pastor I, la Calle C de, Barrio-Real L, Laso FJ, Gonzalez-Sarmiento R (2012) Cannabinoid receptor 1 gene is associated with alcohol dependence. Alcohol Clin Exp Res 36(2): 267–271

Marsicano G, Wotjak CT, Azad SC, Bisogno T, Rammes G, Cascio MG, Hermann H, Tang J, Hofmann C, Zieglgansberger W, Di Marzo V, Lutz B (2002) The endogenous cannabinoid system controls extinction of aversive memories. Nature 418(6897):530–534

Marsicano G, Goodenough S, Monory K, Hermann H, Eder M, Cannich A, Azad SC, Cascio MG, Gutierrez SO, Stelt M van der, Lopez-Rodriguez ML, Casanova E, Schutz G, Zieglgansberger W, Di Marzo V, Behl C, Lutz B (2003) CB1 cannabinoid receptors and on-demand defense against excitotoxicity. Science 302(5642):84–88

Matsuda LA, Lolait SJ, Brownstein MJ, Young AC, Bonner TI (1990) Structure of a cannabinoid receptor and functional expression of the cloned cDNA. Nature 346:561–564

McAllister S, Griffin G, Satin L, Abood M (1999) Cannabinoid receptors can activate and inhibit G protein-coupled inwardly rectifying potassium channels in a Xenopus oocyte expression system. J Pharmacol Exp Ther 291(2):618–626

McAllister SD, Hurst DP, Barnett-Norris J, Lynch D, Reggio PH, Abood ME (2004) Structural mimicry in class A G protein-coupled receptor rotamer toggle switches: the importance of the F3.36(201)/W6.48(357) interaction in cannabinoid CB1 receptor activation. J Biol Chem 279(46):48024–48037

McHugh D, Hu SS, Rimmerman N, Juknat A, Vogel Z, Walker JM, Bradshaw HB (2010) N-arachidonoyl glycine, an abundant endogenous lipid, potently drives directed cellular migration through GPR18, the putative abnormal cannabidiol receptor. BMC Neurosci 11:44

McHugh D, Page J, Dunn E, Bradshaw HB (2011) Delta(9)-THC and N-arachidonyl glycine are full agonists at GPR18 and cause migration in the human endometrial cell line, HEC-1B. Br J Pharmacol 2011(19):1476–5381

McPartland JM, Glass M (2003) Functional mapping of cannabinoid receptor homologs in mammals, other vertebrates, and invertebrates. Gene 312:297–303

McPartland JM, Glass M, Matias I, Norris RW, Kilpatrick CW (2007) A shifted repertoire of endocannabinoid genes in the zebrafish (Danio rerio). Mol Genet Genomics 277(5):555–570

Mechoulam R, Fride E (1995) The unpaved road to the endogenous brain cannabinoid ligands, the anandamides. In Pertwee RG. *Cannabinoid receptors*. Academic Press, Boston MA USA, pp 233–258

Mechoulam R, Ben-Shabat S, Hanus L, Ligumsky M, Kaminski NE, Schatz AR, Gopher A, Almog S, Martin BR, Compton DR, Pertwee RG, Griffin G, Bayewitch M, Barg J, Vogel Z (1995) Identification of an endogenous 2-monoglyceride, present in canine gut, that binds to cannabinoid receptors. Biochem Pharmacol 50(1):83–90

Monory K, Tzavara ET, Lexime J, Ledent C, Parmentier M, Borsodi A, Hanoune J (2002) Novel, not adenylyl cyclase-coupled cannabinoid binding site in cerebellum of mice. Biochem Biophys Res Commun 292(1):231–235

Munro S, Thomas KL, Abu-Shaar M (1993) Molecular characterization of a peripheral receptor for cannabinoids. Nature 365:61–65

Murphy WJ, Eizirik E, Johnson WE, Zhang YP, Ryder OA, O'Brien SJ (2001) Molecular phylogenetics and the origins of placental mammals. Nature 409(6820):614–618

Mutombo PB, Yamasaki M, Nabika T, Shiwaku K (2012) Cannabinoid receptor 1 (CNR1) 4895 C/T genetic polymorphism was associated with obesity in Japanese men. J Atheroscler Thromb 19(8):779–785

Naidu PS, Booker L, Cravatt BF, Lichtman AH (2009) Synergy between enzyme inhibitors of fatty acid amide hydrolase and cyclooxygenase in visceral nociception. J Pharmacol Exp Ther 329(1):48–56

Ndong C, O'Donnell D, Ahmad S, Groblewski T (2011) Cloning and pharmacological characterization of the dog cannabinoid CB(2) receptor. Eur J Pharmacol 669(1–3):24–31

Nie J, Lewis DL (2001) The proximal and distal C-terminal tail domains of the CB1 cannabinoid receptor mediate G protein coupling. Neuroscience 107(1):161–167

Offertaler L, Mo FM, Batkai S, Liu J, Begg M, Razdan RK, Martin BR, Bukoski RD, Kunos G (2003) Selective ligands and cellular effectors of a G protein-coupled endothelial cannabinoid receptor. Mol Pharmacol 63(3):699–705

Okahisa Y, Kodama M, Takaki M, Inada T, Uchimura N, Yamada M, Iwata N, Iyo M, Sora I, Ozaki N, Ujike H (2011) Association study of two cannabinoid receptor genes, CNR1 and CNR2, with methamphetamine dependence. Curr Neuropharmacol 9(1):183–189

Osei-Hyiaman D, Liu J, Zhou L, Godlewski G, Harvey-White J, Jeong WI, Batkai S, Marsicano G, Lutz B, Buettner C, Kunos G (2008) Hepatic CB1 receptor is required for development of diet-induced steatosis, dyslipidemia, and insulin and leptin resistance in mice. J Clin Invest 118(9):3160–3169

Pan X, Ikeda S, Lewis D (1998) SR 141716A acts as an inverse agonist to increase neuronal voltage-dependent Ca2+ currents by reversal of tonic CB1 cannabinoid receptor activity. Mol Pharmacol 54:1064–1072

Pernia-Andrade AJ, Kato A, Witschi R, Nyilas R, Katona I, Freund TF, Watanabe M, Filitz J, Koppert W, Schuttler J, Ji G, Neugebauer V, Marsicano G, Lutz B, Vanegas H, Zeilhofer HU (2009) Spinal endocannabinoids and CB1 receptors mediate C-fiber-induced heterosynaptic pain sensitization. Science 325(5941):760–764

Pertwee RG (1999) Pharmacology of cannabinoid receptor ligands. Curr Med Chem 6(8):635–664

Pertwee RG, Howlett AC, Abood ME, Alexander SP, Di Marzo V, Elphick MR, Greasley PJ, Hansen HS, Kunos G, Mackie K, Mechoulam R, Ross RA (2010) International Union of Basic and Clinical Pharmacology. LXXIX. Cannabinoid receptors and their ligands: beyond CB and CB. Pharmacol Rev 62(4):588–631

Porter AC, Sauer JM, Knierman MD, Becker GW, Berna MJ, Bao J, Nomikos GG, Carter P, Bymaster FP, Leese AB, Felder CC (2002) Characterization of a novel endocannabinoid, virodhamine, with antagonist activity at the CB1 receptor. J Pharmacol Exp Ther 301(3):1020–1024

Radwan MM, Ross SA, Slade D, Ahmed SA, Zulfiqar F, Elsohly MA (2008) Isolation and characterization of new Cannabis constituents from a high potency variety. Planta Med 74(3): 267–272

Rinaldi-Carmona M, Barth F, Heaulme M, Shire D, Calandra B, Congy C, Martinez S, Maruani J, Neliat G, Caput D, Ferrar P, Soubrie P, Breliere JC, Fur GL (1994) SR141716A, a potent and selective antagonist of the brain cannabinoid receptor. FEBS Lett 350:240–244

Rinaldi-Carmona M, Duigou AL, Oustric D, Barth F, Bouaboula M, Carayon P, Casellas P, Fur GL (1998) Modulation of CB1 cannabinoid receptor functions after a long-term exposure to agonist or inverse agonist in the chinese hamster ovary cell expression system. J Pharmacol Exp Ther 287(3):1038–1047

Roche J, Bounds S, Brown S, Mackie K (1999) A mutation in the second transmembrane region of the CB1 receptor selectively disrupts G protein signaling and prevents receptor internalization. Mol Pharmacol 56(3):611–618

Rockwell CE, Snider NT, Thompson JT, Vanden Heuvel JP, Kaminski NE (2006) Interleukin-2 suppression by 2-arachidonyl glycerol is mediated through peroxisome proliferator-activated receptor gamma independently of cannabinoid receptors 1 and 2. Mol Pharmacol 70(1):101–111

Romano MR, Lograno MD (2006) Cannabinoid agonists induce relaxation in the bovine ophthalmic artery: evidences for CB1 receptors, nitric oxide and potassium channels. Br J Pharmacol 147(8):917–925

Ross RA (2009) The enigmatic pharmacology of GPR55. Trends Pharmacol Sci 30(3):156–163

Ross HR, Gilmore AJ, Connor M (2009) Inhibition of human recombinant T-type calcium channels by the endocannabinoid N-arachidonoyl dopamine. Br J Pharmacol 156(5):740–750

Ruiz-Contreras AE, Carrillo-Sanchez K, Gomez-Lopez N, Vadillo-Ortega F, Hernandez-Morales S, Carnevale-Cantoni A, Espejel-Nunez A, Mendez-Diaz M, Prospero-Garcia O (2013) Working memory performance in young adults is associated to the AATn polymorphism of the CNR1 gene. Behav Brain Res 236(1):62–66

Russo EB (2008) Clinical endocannabinoid deficiency (CECD): can this concept explain therapeutic benefits of cannabis in migraine, fibromyalgia, irritable bowel syndrome and other treatment-resistant conditions? Neuro Endocrinol Lett 29(2):192–200

Ryberg E, Vu HK, Larsson N, Groblewski T, Hjorth S, Elebring T, Sjogren S, Greasley PJ (2005) Identification and characterisation of a novel splice variant of the human CB1 receptor. FEBS Lett 579(1):259–264

Ryberg E, Larsson N, Sjogren S, Hjorth S, Hermansson NO, Leonova J, Elebring T, Nilsson K, Drmota T, Greasley PJ (2007) The orphan receptor GPR55 is a novel cannabinoid receptor. Br J Pharmacol 152(7):1092–1101

Sharir H, Abood ME (2010) Pharmacological characterization of GPR55, a putative cannabinoid receptor. Pharmacol Ther 126(3):301–313

Sharir H, Console-Bram L, Mundy C, Popoff SN, Kapur A, Abood ME (2012) The endocannabinoids anandamide and virodhamine modulate the activity of the candidate cannabinoid receptor GPR55. J Neuroimmune Pharmacol 7(4):856–865

Shen CP, Xiao JC, Armstrong H, Hagmann W, Fong TM (2006) F200A substitution in the third transmembrane helix of human cannabinoid CB1 receptor converts AM2233 from receptor agonist to inverse agonist. Eur J Pharmacol 531(1–3):41–46

Sheskin T, Hanus L, Slager J, Vogel Z, Mechoulam R (1997) Structural requirements for binding of anandamide-type compounds to the brain cannabinoid receptor. J Med Chem 40(5):659–667

Shim JY (2010) Understanding functional residues of the cannabinoid CB1. Curr Top Med Chem 10(8):779–798

Shire D, Carillon C, Kaghad M, Calandra B, Rinaldi-Carmona M, Fur GL, Caput D, Ferrara P (1995) An Amino-terminal variant of the central cannabinoid receptor resulting from alternative splicing. J Biol Chem 270(8):3726–3731

Shire D, Calandra B, Rinaldi-Carmona M, Oustric D, Pessegue B, Bonnin-Cabanne O, Fur GL, Caput D, Ferrara P (1996) Molecular cloning, expression and function of the murine CB2 peripheral cannabinoid receptor. Biochim Biophys Acta 1307(2):132–136

Showalter VM, Compton DR, Martin BR, Abood ME (1996) Evaluation of binding in a transfected cell line expressing a peripheral cannabinoid receptor (CB2): identification of cannabinoid receptor subtype selective ligands. J Pharmacol Exp Ther 278(3):989–999

Singh R, Hurst DP, Barnett-Norris J, Lynch DL, Reggio PH, Guarnieri F (2002) Activation of the cannabinoid CB1 receptor may involve a W6 48/F3 36 rotamer toggle switch. J Pept Res 60(6):357–370

Sipe JC, Arbour N, Gerber A, Beutler E (2005) Reduced endocannabinoid immune modulation by a common cannabinoid 2 (CB2) receptor gene polymorphism: possible risk for autoimmune disorders. J Leukoc Biol 78(1):231–238

Soderstrom K, Leid M, Moore FL, Murray TF (2000) Behavioral, pharmacological, and molecular characterization of an amphibian cannabinoid receptor. J Neurochem 75(1):413–423

Song ZH, Bonner TI (1996) A lysine residue of the cannabinoid receptor is critical for receptor recognition by several agonists but not WIN55212-2. Mol Pharmacol 49(5):891–896

Steffens M, Zentner J, Honegger J, Feuerstein TJ (2005) Binding affinity and agonist activity of putative endogenous cannabinoids at the human neocortical CB1 receptor. Biochem Pharmacol 69(1):169–178

Stella N (2010) Cannabinoid and cannabinoid-like receptors in microglia, astrocytes, and astrocytomas. Glia 58(9):1017–1030

Straiker A, Wager-Miller J, Hutchens J, Mackie K (2012) Differential signalling in human cannabinoid CB1 receptors and their splice variants in autaptic hippocampal neurones. Br J Pharmacol 165(8):2660–2671

Sugiura T (2009) Physiological roles of 2-arachidonoylglycerol, an endogenous cannabinoid receptor ligand. Biofactors 35(1):88–97

Sugiura T, Kondo S, Kishimoto S, Miyashita T, Nakane S, Kodaka T, Suhara Y, Takayama H, Waku K (2000) Evidence that 2-arachidonoylglycerol but not N-palmitoylethanolamine or anandamide is the physiological ligand for the cannabinoid CB2 receptor. Comparison of the agonistic activities of various cannabinoid receptor ligands in hl-60 cells. J Biol Chem 275(1):605–612

Takenouchi R, Inoue K, Kambe Y, Miyata A (2012) N-arachidonoyl glycine induces macrophage apoptosis via GPR18. Biochem Biophys Res Commun 418(2):366–371

Thomas W, Qian H, Chang C, Karnik S (2000) Agonist-induced phosphorylation of the angiotensin II (AT(1A)) receptor requires generation of a conformation that is distinct from the inositol phosphate-signaling state. J Biol Chem 275(4):2893–2900

Turner CE, Bouwsma OJ, Billets S, Elsohly MA (1980) Constituents of Cannabis sativa L. XVIII—Electron voltage selected ion monitoring study of cannabinoids. Biomed Mass Spectrom 7(6):247–256

Ujike H, Takaki M, Nakata K, Tanaka Y, Takeda T, Kodama M, Fujiwara Y, Sakai A, Kuroda S (2002) CNR1, central cannabinoid receptor gene, associated with susceptibility to hebephrenic schizophrenia. Mol Psychiatry 7(5):515–518

Valk PJM, Hol S, Vankan Y, Ihle JN, Askew D, Jenkins NA, Gilbert DJ, Copeland NG, deBoth NJ, Lowenberg B, Delwel R (1997) The genes encoding the peripheral cannabinoid receptor and a-L- fucosidase are located near a newly identified common virus integration site. J Virol 71(9):6796–6804

Vasquez C, Lewis DL (1999) The CB1 cannabinoid receptor can sequester G-proteins, making them unavailable to couple to other receptors. J Neurosci 19(21):9271–9280

Wagner J, Varga K, Jarai Z, Kunos G (1999) Mesenteric vasodilation mediated by endothelial anandamide receptors. Hypertension 33(1 Pt 2):429–434

Walter L, Franklin A, Witting A, Wade C, Xie Y, Kunos G, Mackie K, Stella N (2003) Nonpsychotropic cannabinoid receptors regulate microglial cell migration. J Neurosci 23(4):1398–1405

Wang L, Martin B, Brenneman R, Luttrell LM, Maudsley S (2009) Allosteric modulators of g protein-coupled receptors: future therapeutics for complex physiological disorders. J Pharmacol Exp Ther 331(2):340–348

White R, Ho WS, Bottrill FE, Ford WR, Hiley CR (2001) Mechanisms of anandamide-induced vasorelaxation in rat isolated coronary arteries. Br J Pharmacol 134(4):921–929

Wiles AL, Pearlman RJ, Rosvall M, Aubrey KR, Vandenberg RJ (2006) N-Arachidonyl-glycine inhibits the glycine transporter, GLYT2a. J Neurochem 99(3):781–786

Xie XQ, Melvin LS, Makryiannis A (1996) The comformational properties of the highly selective cannabinoid receptor ligand CP-55,940. J Biol Chem 271:10640–10647

Yamada Y, Ando F, Shimokata H (2007) Association of candidate gene polymorphisms with bone mineral density in community-dwelling Japanese women and men. Int J Mol Med 19(5): 791–801

Yamaguchi F, Macrae AD, Brenner S (1996) Molecular cloning of two cannabinoid type-1 receptor genes from the puffer fish Fugu rubripes. Genomics 35:603–605

Yevenes GE, Zeilhofer HU (2011) Molecular sites for the positive allosteric modulation of glycine receptors by endocannabinoids. PLoS One 6(8):e23886.

Zhang PW, Ishiguro H, Ohtsuki T, Hess J, Carillo F, Walther D, Onaivi ES, Arinami T, Uhl GR (2004) Human cannabinoid receptor 1: 5′ exons, candidate regulatory regions, polymorphisms, haplotypes and association with polysubstance abuse. Mol Psychiatry 9(10):916–931

Zimmer A, Zimmer A, Hohmann A, Herkenham M, Bonner T (1999) Increased mortality, hypoactivity, and hypoalgesia in cannabinoid CB1 receptor knockout mice. Proc Natl Acad Sci U S A 96:5780–5785

Zuo L, Kranzler HR, Luo X, Yang BZ, Weiss R, Brady K, Poling J, Farrer L, Gelernter J (2009) Interaction between two independent CNR1 variants increases risk for cocaine dependence in European Americans: a replication study in family-based sample and population-based sample. Neuropsychopharmacology 34(6):1504–1513

Zygmunt P, Petersson J, Andersson D, Chuang H, Sorgard M, DiMarzo V, Julius D, Hogestatt E (1999) Vanilloid receptors on sensory nerves mediate the vasodilator action of anandamide. Nature 400:452–457

Chapter 4
Cue-Elicited Craving for Cannabis Activates the Reward Neurocircuitry Associated with the Neuropathology of Addiction

Samuel J. DeWitt, Sven Kroener and Francesca M. Filbey

Abstract Craving or the intense desire for a rewarding object or experience is an important factor in the etiology of addiction. Based on the incentive sensitization theory, addiction stems from drug-induced sensitization in dopaminergic reward structures, which attribute incentive-related salience to drug-associated cues. In this way, after repeated coupling with the drug, the cue can trigger similar primary responses in the brain's reward neurocircuitry as the drug itself. There is growing evidence that cannabis exerts its addictive properties through effects of the endo-cannabinoid system on the brain reward neurocircuitry. Specifically, the ubiquitous cannabinoid 1 (CB1) receptors play a key role in modulating reward pathways. In the present chapter, we describe the evidence for cue-elicited craving for marijuana, and, more specifically, how in the absence of cannabis itself, cannabis-associated cues trigger activation in the reward pathway implicated in the neuropathology of addiction.

4.1 Cannabinoid 1 Receptors and Reward

Understanding the complex interactions of the endocannabinoid system with the brain's reward neurocircuitry is critical for our understanding of the addictive properties of cannabis and, consequently, the effects of cue-elicited craving. In this section, we briefly review some of the mechanisms of how cannabinoids alter neural plasticity, which leads to addiction and cue-induced craving.

F. M. Filbey (✉) · S. J. DeWitt · S. Kroener
School of Behavioral and Brain Sciences, University of Texas at Dallas,
800 West Campbell Rd, 75080 Dallas, TX, USA
e-mail: francesca.filbey@utdallas.edu

E. J. Van Bockstaele (ed.), *Endocannabinoid Regulation of Monoamines in Psychiatric and Neurological Disorders*, DOI 10.1007/978-1-4614-7940-6_4,

4.1.1 Cannabinoid Receptors and Physiological Effects of Cannabinoid Receptor Activation in the Central Nervous System

Endocannabinoids are lipid signaling molecules with potent actions at cannabinoid receptors. To date, two cannabinoid receptors (CB1 and CB2) have been cloned and characterized pharmacologically. The first cannabinoid receptor, designated cannabinoid 1 (CB1), was discovered (Devane et al. 1988) and subsequently cloned (Matsuda et al. 1990) on the basis of its responsiveness to (−)-Δ9-tetrahydrocannabinol (Δ9-THC) and its synthetic analogs. Δ9-THC is the primary psychoactive component in marijuana (from Cannabis sativa) (Elsohly and Slade 2005). The CB1 receptors are expressed at high density throughout the central nervous system (CNS) (Herkenham et al. 1991; Matsuda et al. 1993; Egertová and Elphick 2000) and at lower density in peripheral tissues and immune cells (Galigue et al. 1995). The central distribution pattern of CB1 receptors is heterogeneous and reflects their ability to impair cognition and memory and to alter the control of motor function. Thus, CB1 receptors have been found in the hippocampus, basal ganglia, striatum, substantia nigra, globus pallidus, amygdala, cerebellum, as well as in the neocortex (Herkenham et al. 1990; Glass et al. 1997; Tsou et al. 1999).

CB2 receptors were initially thought to be predominantly expressed in the peripheral tissues and immune cells (Munro et al. 1993). More recent studies have reported low-level distribution of CB2 receptors in microglia and neurons throughout much of the CNS (Onaivi et al. 2008; Brusco et al. 2008; Gong et al. 2006; Van Sickle et al. 2005; Lanciego et al. 2011). However, despite converging functional evidence for a role of CB2 receptors in CNS function (Schmidt et al. 2012; Aracil-Fernández et al. 2012; Morgan et al. 2009), the extent and distribution pattern of CB2 receptors remains a matter of debate (Atwood and Mackie 2010). CB2 receptors are an attractive therapeutic target for pain management and modulation of the immune system without the overt psychoactive effects of CB1 receptor activation. They may also be involved in the development of drug abuse and in neuropsychiatric disorders, including psychosis and depression (Onaivi et al. 2008).

Cannabinoid receptors belong to the superfamily of G-protein-coupled receptors, and signal through the Go/i family of G-proteins (Devane et al. 1988; Matsuda et al. 1990). Activation of both of these receptors inhibits adenylate cyclase and stimulates mitogen-activated protein kinases (Bouaboula et al. 1995). In addition, stimulation of CB1 receptors inhibits N- type and P/Q-type voltage-gated calcium channels (Mackie and Hille 1992) and activates G-protein-coupled inward rectifier potassium (GIRK) currents (Mackie et al. 1995).

In the CNS, endocannabinoids are important retrograde signaling molecules that modulate synaptic transmission and mediate several forms of short-term and long-term synaptic plasticity (for review see Freund et al. 2003; Kano et al. 2009). Brief postsynaptic membrane depolarization triggers the synthesis and release of endocannabinoids, which diffuse through the membrane to act retrogradely at presynaptic CB1 receptors to induce a short-term suppression of neurotransmitter release at both gamma-aminobutyric acid (GABA)ergic (Ohno-Shosaku et al. 2001; Wilson and Nicoll 2001) and glutamatergic (Kreitzer and Regehr 2001) synapses, resulting in so-called depolarization-induced suppression of inhibition (DSI) and excitation (DSE), respectively. These effects typically persist for a minute or less. However, retrograde endocannabinoid signaling has also been shown to mediate activity-dependent long-term depression (LTD) of glutamatergic (Gerdeman et al. 2002; Haj-Dahmane and Shen 2010) and GABAergic synaptic transmission (Chevaleyre and Castillo 2003; Marsicano et al. 2002; Heifets and Castillo 2009) throughout the brain.

Activation of presynaptic CB1 receptors can also inhibit the evoked release of a number of excitatory or inhibitory neurotransmitters, including acetylcholine, noradrenaline, dopamine (DA), 5-hydroxytryptamine, D-aspartate, and cholecystokinin (see Howlett et al. 2002 for review). Finally, in addition to the modulation of neurotransmitter release, in several brain areas endocannabinoids have been shown to modulate postsynaptic neuronal excitability directly. For example, CB1 receptor-induced activation of an inward rectifier potassium current hyperpolarizes inhibitory interneurons and suppresses their firing activity (Kreitzer and Regehr 2001; Bacci et al. 2004).

4.1.2 Endocannabinoid-Induced Plasticity in the Mesocorticolimbic Reward Pathway

Synaptic plasticity in the ventral tegmental area (VTA) and targets of the mesocorticolimbic DA innervation contribute to the development and maintenance of drug addiction. Specifically, plasticity in the nucleus accumbens (NAc), prefrontal cortex (PFC) and amygdala, as well as the connections between these structures participate in the formation of conditioned associations between drug reward and external and internal cues associated with drug intake. Sidhpura and Parsons (2011), Gardner (2005), and Lupica et al. (2004) offer extensive reviews on the effects of cannabis on CB1 receptors in reward structures.

In animal models, the direct injection of THC into the VTA or NAc leads to an overflow of DA in these structures (Diana et al. 1998; Fadda et al. 2006; French et al. 2006; Gessa et al. 1998, 2006; Melis et al. 2004; Tanda et al. 1997). Furthermore, potent cannabinoid agonists WIN,212-2,HU210 and CP55040 have been shown to increase neuronal firing rates including in brain slices containing VTA (French et al. 1997; Gessa et al. 1998; Wu and French 2000; Cheer et al. 2000). Findings show co-localization of immunoreactivity for CB1 receptors and tyrosine-hydroxylase, the rate-limiting enzyme for the production of catecholamines found in DA neurons in the VTA, suggesting the possibility of a direct postsynaptic effect of cannabinoids on DA neurons (Wenger et al. 2003). However, more direct evidence suggests that CB1 agonists act on presynaptic GABA receptors in the VTA to reduce GABA release and disinhibit DA neurons in the VTA (Lupica et al. 2004; Szabo et al. 2002). This mechanism is similar to the action of opiates, such as heroin

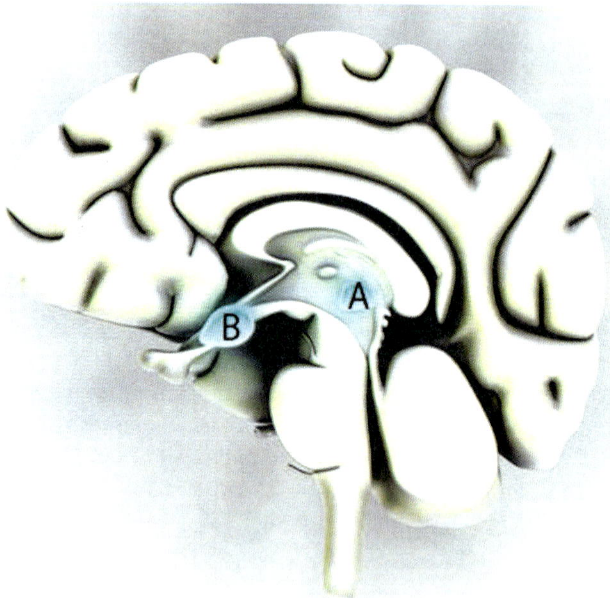

Fig. 4.1 Ventral tegmental area/nucleus accumbens loop (VTA/NAc). The VTA (**a**) and NAc (**b**) are proposed to moderate the tone of reward signaling (Gardner 2005) primarily via CB1 activation on GABAergic and glutamatergic neurons within, and projecting to both regions. This activation elicits short-term as well as long-term excitatory and inhibitory plasticity between the regions in a loop-like fashion, culminating in a moderation of reward signaling via dopamine (DA) release

and morphine, which also increase DA release in the NAc by reducing GABAergic inhibition of VTA DA neurons (Johnson and North 1992; Lupica et al. 2004).

In addition to its effect on GABAergic terminals in the VTA that control the baseline firing rate and bursting mode of VTA DA neurons, CB1 stimulation also has direct, albeit, opposing effects in the NAc. Specifically, it was shown that CB1 agonists inhibit both the release of GABA (most likely from intrinsic interneurons or axon collaterals of the GABAergic medium spiny neurons) onto NAc spiny projection neurons (Hoffman and Lupica 2001; Manzoni and Bockaert 2001), but also inhibit excitatory glutamatergic inputs, which most likely arise from prefrontal cortical areas (Robbe et al. 2001). How these effects modulate the activity of NAc projection neurons that project back to the VTA to inhibit the activity of DAergic VTA neurons, and, ultimately, the release of DA in the NAc and other components of the reward circuitry, is unclear. It is likely that the precise effects of CB1 activation on GABAergic and glutamatergic synaptic transmission within the NAc, and consequently the balance of inhibition and excitation in the NAc–VTA loop (Fig. 4.1) depend on the specific circumstances under which either of the systems predominates (Lupica et al. 2004). Finally, there is evidence that Δ9-THC increases extracellular DA levels via presynaptic stimulation of DA fibers in a number of reward-related structures including NAc and medial PFC (Chen et al. 1990a, b), therefore, contributing to the reward response in these regions.

Existing evidence seems to suggest that the net effect of cannabinoid modulation of excitation and inhibition in the reward system through CB1 activation in the VTA and NAc is to regulate the overall DAergic 'reward tone' (Gardner 2005). This cascade of events starts in the VTA; however, as outlined above, DA release in the NAc is further regulated by CB1 receptor modulation of GABA and glutamate release in the NAc, which may also further act to modulate DA tone via a feedback loop between the NAc and the VTA (Lupica et al. 2004). Because the mesocorticolimbic reward pathway is critical for the encoding of motivational salience, it can be inferred that cue-elicited craving mechanisms observed in behavioral studies are also associated with this reward neurocircuitry (Hyman et al. 2006). The next section describes these behavioral studies, which implement cannabis cue-elicited craving paradigms across a number of modalities.

4.2 Understanding Cue-Elicited Craving for Cannabis

In humans, the neurobiology of craving is well-established and has been more consistent compared to animal models (for review, Weiss 2005). The published human studies of cue-elicited craving for cannabis suggest that it is a reliable and valid phenomenon analogous to cue-elicited craving for other drugs of abuse (review in Filbey and DeWitt 2012). Thus, similar to response seen with cues for other drugs of abuse, cannabis-related cues presented in a variety of sensory modalities elicit increases in self-reported craving. Studies of cue-elicited craving for cannabis frequently utilize the 47-item Marijuana Craving Questionnaire (MCQ) (Heishman et al. 2001) of self-reported cannabis craving across four main domains related to cannabis use. The domains are: (a) compulsivity—an inability to control cannabis use (e.g., "*If I smoked a little marijuana right now, I would not be able to stop using it*"); (b) emotionality—use of cannabis in anticipation of relief from withdrawal or negative mood (e.g., "*I would feel more anxious if I smoked marijuana right now*"); (c) expectancy—anticipation of positive outcomes from smoking cannabis (e.g., "*Smoking marijuana would help me sleep better at night*"); and (d) purposefulness—intention and planning to use cannabis for positive outcomes (e.g., "*It would be great to smoke marijuana right now*"). Individuals rate their responses using a 1 (strongly disagree) to 7 (strongly agree) Likert scale. One of the many studies to utilize MCQ was in a group of 48 cannabis users (36 males, average days of cannabis use in the last month = 17.5) where audio presentations of imagery scripts were found to induce craving. The authors additionally noted that the magnitude of craving varied as a function of the amount of cannabis-related content in the script such that MCQ craving scores correlated positively with urge-intensity of the auditory scripts, with highest craving ratings seen immediately following the high-urge auditory script (Cronbach's alpha coefficient = 0.56, $p < 0.05$) (Singleton et al. 2002). It should be noted that, for this study, participants were not required to abstain from use prior to the experiment, thus increases in craving were not affected by abstinence, although may be confounded by time of last use.

Other modality of cannabis cue-exposure combine auditory scripts with tactile cues such as a used cannabis pipe or bong (Haughey et al. 2008; Schacht et al. 2009). In these studies, the auditory scripts instructed participants to focus on the tactile cue that they were holding and to imagine using it to smoke cannabis. In the large study of 105 daily cannabis users by Haughey et al. (2008), the effects of withdrawal were also measured along with subjective craving ratings. Cue-elicited craving was measured before and after a required 5-day abstinence period. The results showed an increase in craving as a result of both abstinence and cue exposure. This suggests that exposure to cannabis cues increases subjective craving beyond the effects induced by abstinence alone. A follow-up study by Schacht et al. 2009 used the same cue-elicited craving paradigm. For this study, the abstinence period prior to cue exposure and cannabis administration was 24 h. Analysis of the MCQ indicated that following the abstinence period, craving increased significantly after cue exposure compared to baseline scores. Both of these studies investigated differences in cue-elicited craving across different genotype groups for the CNR1 gene, which codes for CB1 receptors. Interestingly, Schacht et al. (2009) failed to replicate group differences in self-reported craving reported by Haughey et al. (2008). The researchers explain that this may be due to the shortened abstinence period used (1 day vs. 5 days). Cues delivered through the olfactory system have also been shown to elicit self-reported craving. This has been illustrated through cues such as a lit cannabis scent stick, which is often paired with auditory and tactile cues (McRae-Clark et al. 2011; Bordnick et al. 2009).

Virtual reality (VR) technology has also provided insight as to how contextual cues (i.e., environment conditioned with the use of cannabis) affect cue-elicited craving. For example, Bordnick et al. (2009) presented participants with a VR paradigm that incorporated audio, visual, and vibrotactile cues. The authors compared subjective craving during a scenario that included cannabis (i.e., a party where cannabis was present and being smoked, or a room with cannabis paraphernalia) with neutral scenarios (i.e., art gallery or nature). The results showed greater subjective craving during the cannabis cue VR environment compared to the neutral VR environments, suggesting contextual effects in cue-elicited craving for cannabis.

In addition to self-reported craving (i.e., MCQ), physiological measures have also provided evidence for cue-elicited craving for cannabis. For example, in an electroencephalogram (EEG) study that used cannabis-related pictures, event-related brain potentials (ERPs) during visual cue presentation induced an enhanced late positive complex (LPC) of the ERP signal in cannabis users relative to controls (Wolfling et al. 2008). Enhanced LPC is linked to neural networks of primary motivational systems, including limbic and striatal structures. This, in combination with greater skin conductance reflecting greater arousal in cannabis users during cue exposure, suggests increased motivational relevance for the cues compared to neutral stimuli (Lang and Davis 2006; Schupp et al. 2000, 2006; Amrhein et al. 2004). This phenomenon is shown to be strong in drug-dependent individuals (Carter and Tiffany 1999; Geier et al. 2000; Laberg and Ellertsen 1987). Studies have also shown that there is increased attentional bias towards cannabis cues compared with neutral cues that is consistent with what is seen with tobacco and alcohol. In a study where cannabis users were shown photos related to cannabis, cannabis users maintained their gaze on cues longer and had faster approach response times to cannabis-related stimuli than controls. The users also rated cannabis cues as being more pleasant on a valence rating scale (Field et al. 2006).

This phenomenon has also been shown to be present across the lifespan as studies of cue-elicited craving in adolescent populations have shown. For example, exposure to auditory, visual, and tactile cannabis cues resulted in increased skin conductance, heart rate, and higher self-reported craving using the MCQ in a group of 15 daily cannabis users (age range: 16–21 years) (Gray et al. 2008). In another study of 13 cannabis-dependent adolescents (age range: 14–17 years), visual and tactile cannabis cues were used in conjunction with EEG measures. Findings revealed increased self-reported craving and a heightened P300 ERP signal following handling of cannabis paraphernalia (Nickerson et al. 2011). Irregular P300 patterns (positive signal occurring 300 ms after target stimulation) have been consistently associated with motivational salience and arousal (Polich and Criado 2006) and a variety of drugs of abuse (Heroin: Franken et al. 2008; Franken et al. 2003; Lubman et al. 2007; Alcohol: Crego et al. 2012; Begleiter et al. 1984; Marijuana: Gallinat et al. 2012; Theunissen et al. 2012). Interestingly, subjective craving ratings for this study were not associated with increased P300 findings. This may be a limitation of the self-reported craving questionnaire utilized, which differed from the more validated MCQ.

Cannabis cue-elicited craving has also been shown to predict future cannabis-related problems for users (Cousijn et al. 2011) examined cannabis cue-elicited craving in 32 heavy cannabis users (age range: 18–25 years) compared to 39 non-using controls matched on age, gender, and estimated intelligence. Using an approach/avoidance cue paradigm, the participants were presented with cannabis cues (images of cannabis, paraphernalia, and individuals smoking cannabis) and neutral images (images of individuals and objects matched visually to cannabis cues). During the "approach" trials, the participants moved a joystick to increase the size of an image and during the "avoid" trials, the joystick was used to make images smaller. Differences between the trials were measured using reaction times (RTs), with a relatively faster approach RT compared to an avoid RT considered an "approach bias", and the reverse considered an "avoid bias". The results showed heavy cannabis users had a higher approach bias for cannabis cues than did non-using controls. There were no differences in approach bias between the two groups for neutral images. Importantly, the approach bias for cannabis cues seen in heavy users predicted cannabis-use at 6-month follow-up. Specifically, heavy users with stronger approach bias increased their weekly cannabis use, whereas users with lower approach bias (or even avoid biases) decreased their weekly use.

Taken together, the above findings suggest that cannabis cues, regardless of sensory modality, trigger subjective and physiological craving for cannabis in both adults and adolescents. A clear advantage across these studies is the consistent use of the measure of subjective craving (i.e., MCQ), which has increased the replicability of the studies (e.g., Haughey and Schacht studies). However, differences in study design limit our interpretation of the cannabis cue-elicited craving phenomenon. For example, the observed effects may have been diminished due to the wide array of cue types rather than consistent use of the participants' preferred/most frequently used modality. However, the consistent finding of increased craving speaks to the robustness of this cue-elicited phenomenon (Note: Haughey and Schacht studies asked participants to select from two cue options). The degree to which effects are driven by abstinence requirements for each study also remains to be determined. For example, the 24-h abstinence requirement in Schacht et al. 2009 resulted in increased craving following cue-exposure. As previously stated however, these results failed to replicate the group differences that Haughey et al. 2008 found when using a 5-day abstinence period. This highlights the challenge of finding a period of abstinence long enough to induce increased subjective craving, but short enough to avoid confounds of acute withdrawal symptoms. Regardless, these studies describe a robust effect of cues across multiple modalities on subjective craving and physiological responses which suggest that in cannabis users, conditioned cues have increased motivational salience and trigger craving. Not until recent studies using in vivo human neuroimaging techniques (e.g., Filbey et al. 2009), has the connection between behavioral and biological substrates of cue-elicited craving been linked. These emergent findings provide strong evidence for the role of cannabis cue-elicited craving in the pathology of addiction.

4.3 Evidence from Imaging Studies

As described earlier, cannabis cues lead to heightened subjective craving similar to the response to cannabis itself. Findings from the human brain imaging literature suggest that the regions within the reward network underlie the development and experience of craving for cannabis (Cousijn et al. 2012a, b; Filbey et al. 2009), showing similar responses as to other drugs such as cocaine, heroin, methamphetamine, alcohol, and tobacco (for review see Filbey et al. 2008; Goldstein and Volkow 2002; Hommer et al. 2011; Volkow et al. 2002, 2004).

The first brain imaging study to examine cue-elicited craving for cannabis was conducted on 38—3-day abstinent—regular cannabis users (age range: 18–50 years) who were exposed to tactile cannabis and neutral cues during a functional magnetic resonance imaging (fMRI) scan (Filbey et al. 2009). The comparisons of the blood oxygenated level dependent (BOLD) response during exposure to the cannabis cue (used cannabis pipe) vs. exposure to the neutral cue (pencil) showed greater activation in several structures in the reward pathway (Table 4.1). These areas include the VTA, anterior cingulate, insula, hippocampus, and amygdala, all of which play a role in reward processing and incentive motivation. Specifically, the anterior cingulate (ACC) has been implicated in decision-making processes surrounding reward and motivation. The insula has most recently been recognized for its role in interoceptive processes in response to cues. The amygdala plays a role

Table 4.1 Imaging studies of cannabis cue-elicited craving

Authors	Participants	Cannabis cue	Major regions of activation (*for cannabis cue*)
Filbey et al. 2009	• 38 regular cannabis users (18–25) • 72-h abstinent	• Tactile presentation of cannabis pipe and control cue (pencil)	• **VTA** • dorsal ACC cortex • **amygdala** • **OFC** • **NAc** • **insula**
Cousijn et al. 2012b	• 33 frequent cannabis users, 20 sporadic cannabis users and 21 controls (aged 18–25) • 24-h abstinent	• Visual presentation of cannabis-related cues, control cues (objects and people) and target cues (animals)	*Frequent users* • **posterior cingulate gyrus/precuneus** • medial frontal/orbitofrontal cortex[a] • superior frontal gyrus • **dorsal/ventral striatum**[a] • **ACC cortex**[a] *Sporadic Users* • **posterior cingulate gyrus/precuneus** • **medial frontal gyrus** *Controls* • **posterior cingulate gyrus/precuneus** • **DLPFC** *Frequent users > sporadic/ controls* • **VTA**
Cousijn et al. 2012	• 33 heavyCannabis users/36 controls(18–25) • 24-h abstinent	• Visual presentation of cannabis-related images and control images (with approach/avoid components)	*Users/controls* • ventral medial frontal gyrus • **posterior cingulate gyrus** *Users only* • **amygdala** • **insula** • **DLPFC** • **anterior cingulate**

Bold regions appear in more than one study
ACC anterior cingulate, *DLPFC* dorsolateral prefrontal cortex, *NAc* nucleus accumbens, *OFC* oribitofrontal cortex, *VTA* ventral tegmental area
[a] Only seen in users with high problem severity scores (CUDIT score)

in interpreting the emotional content of the cue, while the hippocampus is responsible for object recognition and memory processes related to the cues. Activation of orbitofrontal cortex (OFC) and nucleus accumbens (Nac) was also positively associated with problems related to marijuana use such that the greater the activation, the higher the score on a marijuana problem scale. The model presented in Fig. 4.2 integrates these findings with theories put forth in the literature on processes related to addiction (Koob and Volkow 2010). In this model of cue-elicited cannabis crav-

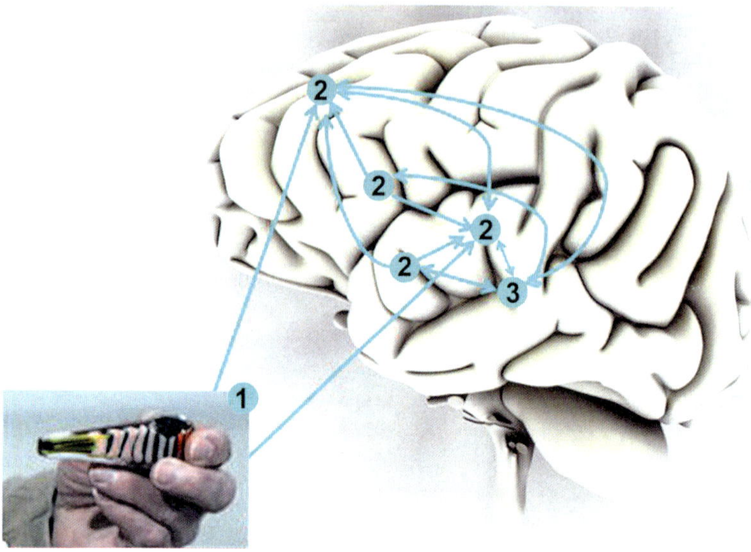

Fig. 4.2 A proposed model of cannabis cue-elicited craving. Based on neuroimaging findings (Filbey et al. 2009), it is proposed that in response to a conditioned stimulus (i.e., pipe), sensory information is *1* processed for salience in prefrontal areas (e.g., anterior cingulate (ACC) gyrus), as well as for emotional content in the amygdala. *2* signals between this pathway are modulated by projections from the insula that mediates interoceptive processes as well as projections from the hippocampus which underlie memories for contextual content. *3* these pathways form a loop with the ventral tegmental area (VTA), which gives rise to dopaminergic projections that are integral for the subjective feelings of craving

ing, the neurobiological response to cannabis cues triggers a cascade of events that underlies aspects of reward processing and attribution of salience. The model suggests multiple paths of convergence into the VTA that underlie the assessment of emotional context, memory integration, evaluation of valence, as well as internal representations of the stimuli.

The above findings have been replicated in two subsequent studies. The first of these studies examined how cue-elicited craving may be affected by severity of use. Using fMRI as in Filbey et al.'s (2009) study described earlier, age and gender matched young adults (age 18–25 years) were categorized into frequent cannabis users ($N=31$), sporadic cannabis users ($n=20$) and cannabis-naïve controls ($n=21$), of ages 18–25 (Cousijn et al. 2012a). Frequent cannabis use was defined either as using 10 or more times in the last month for at least the last 2 years with no treatment or having a history of treatment for cannabis use. Sporadic use was defined as using cannabis 1–50 times over the period of a lifetime and the control group was defined as being cannabis-naive. Cannabis-related problem severity was measured using the Cannabis Use Disorder Identification Test (CUDIT) (Adamson and Sellman 2003) and craving was measured using the MCQ. The event-related cue reactivity task implemented was an adaptation of one used to investigate nicotine craving (McClernon et al. 2005). Participants viewed cannabis images (photos of cannabis, people using cannabis, and cannabis-related paraphernalia), control images (neutral people and objects) and target images (photos of animals). The target images were utilized as a measure of attention such that participants were required to press a button when they saw an animal. Using a region of interest (ROI) analysis that focused on neural response in reward-related brain regions, results of this study showed an increased activation in VTA for frequent users compared to sporadic users and controls but not in orbitofrontal cortex (OFC), ACC, striatum, or amygdala (Table 4.1). Within the frequent users group, differences between high-problem and low-problem subgroups were observed such that the high-problem subgroup had greater neural response in bilateral OFC, bilateral ACC, bilateral NAc, and right caudate compared to the low- problem subgroup. These findings are in accord with findings by Filbey et al. (2009), which showed a positive correlation between cannabis problems and activation in OFC and NAc.

In a follow-up study by Cousijn et al., the researchers examined the neural basis of an approach bias (reported Cousijn et al. 2011; see Section 2) related to cue response, as well as possible changes over time. To that end, the researchers recruited 33 heavy cannabis users and thirty-six controls (age range: 18–25) matched on age, gender and estimated intelligence (Cousijn et al. 2012b). While the heavy cannabis users were defined similarly to their earlier imaging study described above (Cousijn et al. 2012a), the controls were not cannabis-naïve. In this study, controls were required to have used cannabis no more than 50 times in their life and none in the last year. Additionally, in this study, all participants were required to avoid alcohol and drugs 24 hours before the scan session. Participants completed a baseline scan and a 6-month follow-up over the phone. The researchers translated their behavioral paradigm to determine approach/avoidance bias (described in Section 2) (Cousijn et al. 2011) into a neuroimaging paradigm to determine the associated neural effects. Unlike the researchers' previous study involving the approach-avoidance paradigm, group differences in RTs were not found, with RTs for approach trials being shorter than avoid trials for both groups. Results showed activation of reward regions in the presence of the cannabis cue (Table 4.1). Specifically, heightened ventromedial prefrontal cortex (VMPFC) and posterior cingulate gyrus was seen in both groups during approach trials for cannabis as compared to avoid trials for cannabis. For users, decreased activation in DLPFC and in ACC was negatively associated with changes in problem severity at a 6 month follow up (as measured by the CUDIT) such that the weaker the approach bias activation in these regions, the larger the increase in problem severity. In the absence of a targeted intervention, which could explain such a decrease in problem severity the authors suggest that the combined regulatory aspects of DLPFC and evaluative aspects of ACC may explain why higher activation in these regions led to the observed decrease in problem severity associated with cannabis use.

Taken together, these neuroimaging studies provide support for the role of the reward regions in cannabis cue-elicited craving. Moreover, this activation appears to be related to frequency of use and cannabis-related problem severity. This suggests a clear neurobiological profile for cue-elicited craving rooted firmly in the reward and motivational salience neurocircuitry systems of the brain.

4.4 Summary and Conclusions

The evidence presented here characterizes cue-elicited craving for cannabis as a valid phenomenon, the mechanisms of which have a clear behavioral, molecular, and neural basis linked to the reward neurocircuitry implicated in addiction pathology. The endocannabinoid system mediates neural plasticity of reward structures involved in the development and maintenance of drug addiction. This plasticity influences the association between reward response and external cues. Cue-elicited craving paradigms have repeatedly been shown to elicit behavioral and physiological responses similar to what is seen with other drugs of abuse. Furthermore, self-reported craving after exposure to such cues has consistently been shown to increase. These increases in craving are shown to be predictive of cannabis problems associated with addiction pathology. Recent evidence from neuroimaging studies suggests that the neural response to cannabis cues is similar to those of other addictive stimuli (e.g., alcohol, tobacco).

Such evidence shows that in the absence of the drug itself, cannabis cues can elicit a similar reward response, suggesting that cue-elicited craving plays a vital role in the reward neurocircuitry related to addiction. Furthermore, neuroimaging findings provide a vital link between biology (molecular) and behavioral studies.

As discussed earlier, it is difficult to ascertain what the measured craving response, either self-reported or neurobiological is attributed to. Specifically, are these findings informing on the effects of withdrawal alone (e.g., 3-day abstinence in Filbey study)? And, if so, how would the craving response differ in an *ad libitum* sample? Stated differently, might there be differences due to the reinforcing properties of the cues, such as positive reinforcement in *ad libitum* states and negative reinforcement during withdrawal states? Future studies are needed to disentangle the nature of craving processes and how they may evolve during the course of drug use and protracted abstinence. In sum, while more work lies ahead in this field, the evidence suggests that the neural mechanisms in response to cannabis cues are analogous to those activated by the drug itself. These findings inform interventions targeted at avoiding cannabis-use relapse as the evidence suggests that the reward response associated with cannabis cues is a strong factor in relapse and continued abuse of the drug.

References

Adamson SJ, Sellman JD (2003) A prototype screening instrument for cannabis use disorder: the cannabis use disorders identification test (CUDIT) in an alcohol-dependent clinical sample. Drug Alcohol Rev 22(3):309–315

Amrhein C, Muhlberger A, Pauli P, Wiedemann G (2004) Modulation of event-related brain potentials during affective picture processing: a complement to startle reflex and skin conductance response? Int J Psychophysiol: official journal of the International Organization of Psychophysiology 54(3):231–240

Aracil-Fernandez A, Trigo JM, Garcia-Gutierrez MS, Ortega-Alvaro A, Ternianov A, Navarro D, Manzanares J (2012) Decreased cocaine motor sensitization and self- administration in mice overexpressing cannabinoid CB(2) receptors. Neuropsychopharmacology 37(7):1749–1763

Atwood BK, Mackie K (2010) CB2: a cannabinoid receptor with an identity crisis. Br J Pharmacol 160(3):467–479

Bacci A, Huguenard JR, Prince DA (2004) Long-lasting self-inhibition of neocortical interneurons mediated by endocannabinoids. Nature 431(7006):312–316

Begleiter H, Porjesz B, Bihari B, Kissin B (1984) Event-related brain potentials in boys at risk for alcoholism. Science, New York 225 (4669):1493–1496

Bordnick PS, Copp HL, Traylor A, Graap KM, Carter BL, Walton A, Ferrer M (2009) Reactivity to cannabis cues in virtual reality environments. J Psychoactive Drugs 41(2):105–112

Bouaboula M, Poinot-Chazel C, Bourrie B, Canat X, Calandra B, Rinaldi-Carmona M, Le Fur G, Casellas, P (1995) Activation of mitogen-activated protein kinases by stimulation of the central cannabinoid receptor CB1. Biochem J 312(2):637–641

Brusco A, Tagliaferro PA, Saez T, Onaivi ES (2008). Ultrastructural localization of neuronal brain CB2 cannabinoid receptors. Ann N Y Acad Sci 1139 450–457

Carter BL, Tiffany ST (1999) Cue-reactivity and the future of addiction research. Addiction. Abingdon, England 94(3):349–351

Cheer JF, Marsden CA, Kendall DA, Mason R, (2000) Lack of response suppression follows repeated ventral tegmental cannabinoid administration: an in vitro electrophysiological study. Neuroscience 99(4):661–667

Chen J, Paredes W, Lowinson JH, Gardner EL (1990a) Delta 9-tetrahydrocannabinol enhances presynaptic dopamine efflux in medial prefrontal cortex. Eur J Pharmacol 190(1–2):259–262

Chen JP, Paredes W, Li J, Smith D, Lowinson J, Gardner EL, (1990b) Delta 9-tetrahydrocannabinol produces naloxone-blockable enhancement of presynaptic basal dopamine efflux in nucleus accumbens of conscious, freely-moving rats as measured by intracerebral microdialysis. Psychopharmacology 102(2):156–162

Chevaleyre V, Castillo PE, (2003) Heterosynaptic LTD of hippocampal GABAergic synapses: a novel role of endocannabinoids in regulating excitability. Neuron 38(3):461–472

Cooper ZD, Haney M (2008) Cannabis reinforcement and dependence: role of the cannabinoid CB1 receptor. Addiction Biol 13(2):188–195

Cousijn J, Goudriaan AE, Wiers RW (2011) Reaching out towards cannabis: approach- bias in heavy cannabis users predicts changes in cannabis use. Addiction, England 106(9):1667–1674

Cousijn J, Goudriaan AE, Ridderinkhof KR, den Brink W van, Veltman DJ, Wiers RW (2012a) Approach-bias predicts development of cannabis problem severity in heavy cannabis users: results from a prospective FMRI study. PLoS One 7(9):e42394

Cousijn J, Goudriaan AE, Ridderinkhof KR, den Brink W van, Veltman DJ, Wiers RW. (2012b). Neural responses associated with cue-reactivity in frequent cannabis users. Addict Biol

CregoA,CadaveiraF,ParadaM,CorralM,Caamaño-IsornaF,RodràguezHolguànS(2012)Increased amplitude of P3 event-related potential in young binge drinkers. Alcohol , New York 46(5): 415–25

Devane WA, Dysarz FA 3rd, Johnson MR, Melvin LS, Howlett AC (1988) Determination and characterization of a cannabinoid receptor in rat brain. Mol Pharmacol 34(5):605–613

Diana M, Melis M, Gessa GL (1998) Increase in meso-prefrontal dopaminergic activity after stimulation of CB1 receptors by cannabinoids. Eur J Neurosci 10(9):2825–2830

Egertova M, Elphick MR (2000) Localisation of cannabinoid receptors in the rat brain using antibodies to the intracellular C-terminal tail of CB. J Comp Neurol 422(2):159–171

Elsohly MA, Slade D (2005). Chemical constituents of marijuana: the complex mixture of natural cannabinoids. Life Sci 78(5):539–548

Fadda P, Scherma M, Spano MS, Salis P, Melis V, Fattore L, Fratta W (2006) Cannabinoid self-administration increases dopamine release in the nucleus accumbens. Neuroreport 17(15):1629–1632

Field M, Eastwood B, Bradley BP, Mogg K (2006) Selective processing of cannabis cues in regular cannabis users. Drug Alcohol Depend 85(1):75–82

Filbey FM, DeWitt SJ (2012) Cannabis cue-elicited craving and the reward neurocircuitry. Prog Neuropsychopharmacol Biol Psychiatry 38(1):30–35

Filbey FM, Ray L, Smolen A, Claus ED, Audette A, Hutchison KE (2008) Differential neural response to alcohol priming and alcohol taste cues is associated with DRD4 VNTR and OPRM1 genotypes. Alcohol Clin Exp Res 32(7):1113–1123

Filbey FM, Schacht JP, Myers US, Chavez RS, Hutchison KE (2009) Marijuana craving in the brain. Proc Natl Acad Sci U S A 106(31):13016–13021

Franken IH, Stam CJ, Hendriks VM, den Brink W van (2003) Neurophysiological evidence for abnormal cognitive processing of drug cues in heroin dependence. Psychopharmacology 170(2):205–212

Franken IH, Dietvorst RC, Hesselmans M, Franzek EJ, Wetering BJ van de, Van Strien JW (2008) Cocaine craving is associated with electrophysiological brain responses to cocaine-related stimuli. Addict Biol 13(3–4):386–392

French ED, Dillon K, Wu X (1997) Cannabinoids excite dopamine neurons in the ventral tegmentum and substantia nigra. Neuroreport 8(3):649–652

Freund TF, Katona I, Piomelli D, (2003) Role of endogenous cannabinoids in synaptic signaling. Physiol Rev 83(3):1017–1066

Galiegue S, Mary S, Marchand J, Dussossoy D, Carriere D, Carayon P, Bouaboula M, Shire D, Le Fur G, Casellas P (1995) Expression of central and peripheral cannabinoid receptors in human immune tissues and leukocyte subpopulations. Eur J Biochem 232(1):54–61

Gallinat J, Rentzsch J, Roser P (2012) Neurophysiological effects of cannabinoids: implications for psychosis research. Curr Pharm Des 18(32):4938–49

Gardner EL (2005) Endocannabinoid signaling system and brain reward: emphasis on dopamine. Pharmacol Biochem Behav 81(2):263–284

Geier A, Mucha RF, Pauli P (2000) Appetitive nature of drug cues confirmed with physiological measures in a model using pictures of smoking. Psychopharmacology 150(3):283–291

Gerdeman GL, Ronesi J, Lovinger DM (2002) Postsynaptic endocannabinoid release is critical to long-term depression in the striatum. Nat Neurosci 5(5):446–451

Gessa GL, Melis M, Muntoni AL, Diana M (1998) Cannabinoids activate mesolimbic dopamine neurons by an action on cannabinoid CB1 receptors. Eur J Pharmacol 341(1):39–44

Glass M, Dragunow M, Faull RL (1997) Cannabinoid receptors in the human brain: a detailed anatomical and quantitative autoradiographic study in the fetal, neonatal and adult human brain. Neuroscience 77(2):299–318

Goldstein RZ, Volkow ND (2002) Drug addiction and its underlying neurobiological basis: neuroimaging evidence for the involvement of the frontal cortex. Am J Psychiatry 159(10): 1642–1652

Gong JP, Onaivi ES, Ishiguro H, Liu QR, Tagliaferro PA, Brusco A, Uhl GR (2006) Cannabinoid CB2 receptors: immunohistochemical localization in rat brain. Brain Res 1071(1):10–23

Gray KM, LaRowe SD, Upadhyaya HP (2008) Cue reactivity in young marijuana smokers: a preliminary investigation. Psychol Addict Behav 22(4):582–586

Haj-Dahmane S, Shen RY, (2010) Regulation of plasticity of glutamate synapses by endocannabinoids and the cyclic-AMP/protein kinase A pathway in midbrain dopamine neurons. Journal Physiol 588(14):2589–2604

Haughey HM, Marshall E, Schacht JP, Louis A, Hutchison KE (2008) Marijuana withdrawal and craving: influence of the cannabinoid receptor 1 (CNR1) and fatty acid amide hydrolase (FAAH) genes. Addiction 103(10):1678–1686

Heifets BD, Castillo PE, (2009) Endocannabinoid signaling and long-term synaptic plasticity. Annu Rev Physiol 71:283–306

Heishman SJ, Singleton EG, Liguori A (2001) Marijuana Craving Questionnaire: development and initial validation of a self-report instrument. Addiction, England 96(7):1023–1034

Herkenham M, Lynn AB, Little MD, Johnson MR, Melvin LS, Costa BR de, Rice KC (1990) Cannabinoid receptor localization in brain. Proc Natl Acad Sci U S A 87(5):1932–1936

Herkenham M, Groen BG, Lynn AB, De Costa BR, Richfield EK. (1991). Neuronal localization of cannabinoid receptors and second messengers in mutant mouse cerebellum. Brain Res 552(2):301–310
Hoffman AF, Lupica CR (2001) Direct actions of cannabinoids on synaptic transmission in the nucleus accumbens: a comparison with opioids. J Neurophysiol 85(1):72–83
Hommer DW, Bjork JM, Gilman JM. (2011). Imaging brain response to reward in addictive disorders. Ann N Y Acad Sci 1216 50–61
Howlett AC (2002) The cannabinoid receptors. Prostaglandins & other lipid mediators 68–69:619–631
Howlett AC, Barth F, Bonner TI, Cabral G, Casellas P, Devane WA, Felder CC, Herkenham M, Mackie K, Martin BR, Mechoulam R, Pertwee RG (2002) International Union of Pharmacology. XXVII. Classification of cannabinoid receptors. Pharmacol Rev 54(2):161–202
Hyman SE, Malenka RC, Nestler EJ (2006) Neural mechanisms of addiction: the role of reward-related learning and memory. Annu Rev Neurosci 29:565–598
Johnson SW, North RA (1992) Two types of neurone in the rat ventral tegmental area and their synaptic inputs. J Physiol 450:455–468
Kano M, Ohno-Shosaku T, Hashimotodani Y, Uchigashima M, Watanabe M (2009) Endocannabinoid-mediated control of synaptic transmission. Physiological reviews 89:309–380
Koob GF, Volkow ND (2010) Neurocircuitry of addiction. Neuropsychopharmacology 35(1): 217–238
Kreitzer AC, Regehr WG. (2001). Cerebellar depolarization-induced suppression of inhibition is mediated by endogenous cannabinoids. J Neurosci 21(20) RC174
Laberg JC, Ellertsen B (1987) Psychophysiological indicators of craving in alcoholics: effects of cue exposure. Br J Addict 82(12):1341–1348
Lanciego JL, Barroso-Chinea P, Rico AJ, Conte-Perales L, Callen L, Roda E, Franco R (2011) Expression of the mRNA coding the cannabinoid receptor 2 in the pallidal complex of Macaca fascicularis. J Psychopharmacol, England 25(1):97–104
Lang PJ, Davis M (2006). Emotion, motivation, and the brain: reflex foundations in animal and human research. Prog Brain Res 156 3–29
Lubman DI, Allen NB, Peters LA, Deakin JF (2007) Electrophysiological evidence of the motivational salience of drug cues in opiate addiction. Psychol Med 37(8):1203–1209
Lupica CR, Riegel AC, Hoffman AF (2004) Marijuana and cannabinoid regulation of brain reward circuits. Br J Pharmacol 143(2):227–234
Mackie K, Hille B, (1992) Cannabinoids inhibit N-type calcium channels in neuroblastoma-glioma cells. Proc Natl Acad Sci U S A 89(9):3825–3829
Mackie K, Lai Y, Westenbroek R, Mitchell R, (1995) Cannabinoids activate an inwardly rectifying potassium conductance and inhibit Q-type calcium currents in AtT20 cells transfected with rat brain cannabinoid receptor. J Neurosci 15(10):6552–6561
Manzoni OJ, Bockaert J (2001). Cannabinoids inhibit GABAergic synaptic transmission in mice nucleus accumbens. Eur J Pharmacol 412(2) 3–5
Marsicano G, Wotjak CT, Azad SC, Bisogno T, Rammes G, Cascio MG, Hermann H, Tang J, Hofmann C, Zieglgansberger W, Di Marzo V, Lutz B (2002) The endogenous cannabinoid system controls extinction of aversive memories. Nature 418(6897):530–534
Matsuda LA, Bonner TI, Lolait SJ (1993) Localization of cannabinoid receptor mRNA in rat brain. J Comp Neurol 327(4):535–550
Matsuda LA, Lolait SJ, Brownstein MJ, Young AC, Bonner TI. (1990). Structure of a cannabinoid receptor and functional expression of the cloned cDNA. Nature 346(6284):561–564
McClernon FJ, Hiott FB, Huettel SA, Rose JE (2005) Abstinence-induced changes in self-report craving correlate with event-related FMRI responses to smoking cues. Neuropsychopharmacology 30(10):1940–1947
McRae-Clark AL, Carter RE, Price KL, Baker NL, Thomas S, Saladin ME, Brady KT. (2011). Stress- and cue-elicited craving and reactivity in marijuana-dependent individuals. Psychopharmacology

Melis M, Pistis M, Perra S, Muntoni AL, Pillolla G, Gessa GL (2004) Endocannabinoids mediate presynaptic inhibition of glutamatergic transmission in rat ventral tegmental area dopamine neurons through activation of CB1 receptors. J Neurosci 24(1):53–62

Morgan NH, Stanford IM, Woodhall GL (2009) Functional CB2 type cannabinoid receptors at CNS synapses. Neuropharmacology 57(4):356–368

Munro S, Thomas KL, Abu-Shaar M (1993) Molecular characterization of a peripheral receptor for cannabinoids. Nature 365(6441):61–65

Nickerson LD, Ravichandran C, Lundahl LH, Rodolico J, Dunlap S, Trksak GH, Lukas SE (2011) Cue reactivity in cannabis-dependent adolescents. Psychol Addict Behav 25(1):168–173

Ohno-Shosaku T, Maejima T, Kano M (2001) Endogenous cannabinoids mediate retrograde signals from depolarized postsynaptic neurons to presynaptic terminals. Neuron 29(3):729–738

Onaivi ES, Ishiguro H, Gong JP, Patel S, Meozzi PA, Myers L, Perchuk A, Mora Z, Tagliaferro PA, Gardner E, Brusco A, Akinshola BE, Liu QR, Chirwa SS, Hope B, Lujilde J, Inada T, Iwasaki S, Macharia D, Teasenfitz L, Arinami T, Uhl GR (2008) Functional expression of brain neuronal CB2 cannabinoid receptors are involved in the effects of drugs of abuse and in depression. Ann N Y Acad Sci 1139 434–449

Polich J, Criado JR (2006) Neuropsychology and neuropharmacology of P3a and P3b. Int J Psychophysiol 60(2):172–185

Robbe D, Alonso G, Duchamp F, Bockaert J, Manzoni OJ (2001) Localization and mechanisms of action of cannabinoid receptors at the glutamatergic synapses of the mouse nucleus accumbens. J Neurosci 21(1):109–116

Schacht JP, Selling RE, Hutchison KE (2009) Intermediate cannabis dependence phenotypes and the FAAH C385A variant: an exploratory analysis. Psychopharmacology (Berl) 203(3): 511–517. doi: 10.1007/s00213–008-1397-z

Schmidt W, Schafer F, Striggow V, Frohlich K, Striggow F (2012). Cannabinoid receptor subtypes 1 and 2 mediate long-lasting neuroprotection and improve motor behavior deficits after transient focal cerebral ischemia. Neuroscience 227 313–326. doi: 10.1016/j.neuroscience.2012.09.080

Schupp HT, Cuthbert BN, Bradley MM, Cacioppo JT, Ito T, Lang PJ, (2000) Affective picture processing: the late positive potential is modulated by motivational relevance. Psychophysiology 37(2):257–261

Schupp HT, Flaisch T, Stockburger J, Junghofer M (2006) Emotion and attention: event-related brain potential studies. Prog Brain Res 156:31–51

Sidhpura N, Parsons LH (2011) Endocannabinoid-mediated synaptic plasticity and addiction-related behavior. Neuropharmacology 61(7):1070–1087

Singleton EG, Trotman AJ, Zavahir M, Taylor RC, Heishman SJ (2002) Determination of the reliability and validity of the Marijuana Craving Questionnaire using imagery scripts. Exp Clin Psychopharmacol 10(1):47–53

Szabo B, Siemes S, Wallmichrath I (2002) Inhibition of GABAergic neurotransmission in the ventral tegmental area by cannabinoids. Eur J Neurosci 15(12):2057–2061

Tanda G, Pontieri FE, Di Chiara G (1997) Cannabinoid and heroin activation of mesolimbic dopamine transmission by a common mu1 opioid receptor mechanism. Science, New York 276(5321): 2048–2050

Theunissen EL, Kauert GF, Toennes SW, Moeller MR, Sambeth A, Blanchard MM, Ramaekers JG (2012) Neurophysiological functioning of occasional and heavy cannabis users during THC intoxication. Psychopharmacology 220(2):341–50

Tsou K, Mackie K, Sanudo-Pena MC, Walker JM (1999) Cannabinoid CB1 receptors are localized primarily on cholecystokinin-containing GABAergic interneurons in the rat hippocampal formation. Neuroscience 93(3):969–975

Van Sickle MD, Duncan M, Kingsley PJ, Mouihate A, Urbani P, Mackie K, Sharkey KA (2005) Identification and functional characterization of brainstem cannabinoid CB2 receptors. Science, New York 310(5746):329–332

Volkow ND, Fowler JS, Wang GJ, Goldstein RZ (2002) Role of dopamine, the frontal cortex and memory circuits in drug addiction: insight from imaging studies. Neurobiol Learn Mem 78(3):610–624

Volkow ND, Fowler JS, Wang GJ (2004) The addicted human brain viewed in the light of imaging studies: brain circuits and treatment strategies. Neuropharmacology 47 Suppl 1:3–13

Weiss F (2005) Neurobiology of craving, conditioned reward and relapse. Curr Opin Pharmacol 5(1):9–19

Wenger T, Moldrich G, Furst S, (2003) Neuromorphological background of cannabis addiction. Brain Res Bull 61(2):125–128

Wilson RI, Nicoll RA (2001) Endogenous cannabinoids mediate retrograde signalling at hippocampal synapses. Nature 410(6828):588–592

Wolfling K, Flor H, Grusser SM (2008) Psychophysiological responses to drug- associated stimuli in chronic heavy cannabis use. Eur J Neurosci 27(4):976–983

Chapter 5
Cannabinoid Modulation of Dopaminergic Circuits in Neurodegenerative and Neuropsychiatric Disorders

Julien Matricon and Andrea Giuffrida

Abstract The endocannabinoid system timely orchestrates a variety of cerebral physiological processes by modulating brain neurotransmitters, and in particular the dopamine system. Both endocannabinoid and dopamine receptors are highly abundant and often coexpressed in the basal ganglia and mesolimbic pathways, where they regulate motor functions and motivational aspects of behavior. Understanding the interrelationship between these two systems is crucial to gain new insight on the pathophysiology of brain disorders characterized by a dysregulation of dopamine, such as Parkinson's disease and schizophrenia. This review aims at: (1) presenting the complex functional interactions between these two neurotransmitter systems at the anatomical, pharmacological, cellular and electrophysiological levels, and (2) addressing the contribution of disturbances of cannabinoid-dopamine interactions to neurodegenerative and psychiatric disorders.

Abbreviations

2-AG	2-arachidonoylglycerol
CNS	central nervous system
GABA	γ-aminobutyric acid
NAPE-PLD	N-acyl phosphatidylethanolamine phospholipase D
FAAH	fatty acid amide hydrolase
MAGL	monoacylglycerol lipase
DAGL	diacylglycerol lipase
DSE	depolarization-induced suppression of excitation
DSI	depolarization-induced suppression of inhibition
STD	short-term depression
PFC	prefrontal cortex
LTP	long-term potentiation
LTD	long-term depression

A. Giuffrida (✉) · J. Matricon
Department of Pharmacology, UT Health Science Center San Antonio,
7703 Floyd Curl Drive, San Antonio, TX 78229-3900, USA
e-mail: giuffrida@uthscsa.edu

J. Matricon
e-mail: matricon@uthscsa.edu

E. J. Van Bockstaele (ed.), *Endocannabinoid Regulation of Monoamines in Psychiatric and Neurological Disorders*, DOI 10.1007/978-1-4614-7940-6_5,

THC	tetrahydrocannabinol
CB1	cannabinoid receptor 1
D1R	dopamine receptor 1
PD	Parkinson's disease
D2R	dopamine receptor 1
KO	knock-out
LID	L-dopa induced dyskinesia
6-OHDA	6-hydroxydopamine
MPTP	(1-methyl-4-phenyl-1,2,3,6-tetrahydropyridine)
VTA	ventral tegmental area
ECS	endocannabinoid system

5.1 Introduction

The endocannabinoid system (ECS) plays a key role in the mammalian central nervous system (CNS) as it regulates major functions, such as learning and memory (Chevaleyre et al. 2006; Marsicano and Lafenetre 2009), motor control (Morera-Herreras et al. 2012), emotion and stress responses (Haring et al. 2012; McLaughlin and Gobbi 2012), energy balance (Matias and Di Marzo 2007), and immunity (Pandey et al. 2009).

The ECS is densely expressed in dopaminergic brain areas, in particular in the basal ganglia and mesocorticolimbic pathway, which modulate motor functions and motivational aspects of behavior. Within these areas, the ECS modulates the dopamine system by regulating neurotransmitter release at GABAergic and/or glutamatergic terminals projecting onto dopamine neurons (Szabo et al. 2000; Gerdeman and Lovinger 2001; Wallmichrath and Szabo 2002; Julian et al. 2003). Endocannabinoid-dopamine interactions are required for the coordination and fine-tuning of movement (Fernandez-Ruiz and Gonzales 2005), for attributing appropriate salience to sensory stimuli underlying rewarding properties of drugs (Parolaro and Rubino 2008) and behaviors such as food intake, social play, or sex (Fattore et al. 2010), and for memory and cognitive processes (Wotjak 2005; Heifets and Castillo 2009).

In the last decade, new experimental evidence has shed light on how these two systems may functionally interact and how diseases can develop from the dysregulation of this cross talk. In this review, we will cover: (1) the interactions between the ECS and the dopamine systems at the anatomical, cellular, functional and system level, and (2) their role in the pathophysiology of neurodegenerative and psychiatric disorders.

5.2 The Endocannabinoid System

The ECS consists of a family of endogenous lipid signaling molecules (the endocannabinoids), their metabolizing enzymes, and several metabotropic (cannabinoid), ionotropic, and nuclear receptors activated by the endocannabinoids.

To date, a large number of endocannabinoid ligands have been identified (Lopez-Moreno et al. 2008; Hudson et al. 2010a), of which the arachidonic acid derivatives arachidonoyl ethanolamine (also known as anandamide) and 2-arachidonoyl glycerol (2-AG) are the two most studied. The endocannabinoids exert their physiological actions by activating G protein-coupled cannabinoid receptors (CB1 and CB2) and/or a variety of other receptors, such as some members of the transient receptor potential (TRP) family receptors (TRPV1, TRPA1, TRPV4), and the peroxisome proliferator-activated receptors, for which they show lower affinity (Console-Bram et al. 2012). Endocannabinoids can also serve as allosteric modulators of the orphan receptor 55 (GPR55) (McPartland et al. 2007; Henstridge et al. 2011).

In humans and rodents, CB1 is highly expressed in the peripheral and central nervous system (CNS) (Glass et al. 1997; Moldrich and Wenger 2000; Mackie 2005b; Haring et al. 2012), while CB2 expression is restricted to the immune system and the associated lymphoid organs and microglia (Munro et al. 1993; Galiègue et al. 1995; Schatz et al. 1997; Nunez et al. 2004). Recent studies, however, have suggested the presence of CB2 in the CNS as well (Van Sickle et al. 2005; Onaivi et al. 2006; Suarez et al. 2009; Garcia-Gutierrez et al. 2010). The emerging role of CB2 in the CNS has been covered by several comprehensive studies (Onaivi 2006; Patel et al. 2010; Onaivi et al. 2011) and will not be addressed in this review.

Within the prefrontal cortex, hippocampus, and basal ganglia, CB1 receptors are mainly expressed on presynaptic neurons, including GABAergic (Wallmichrath and Szabo 2002), cholinergic (Degroot et al. 2006), glutamatergic (Tsou et al. 1998; Gerdeman and Lovinger 2001; Brown et al. 2003; Köfalvi et al. 2005), noradrenergic (Oropeza et al. 2007), and serotonergic (Haring et al. 2007) terminals. CB1 receptors are retrogradely activated by endocannabinoids, which are released on demand from lipid precursors in the postsynaptic membrane (Piomelli 2003; Chevaleyre et al. 2006; Kano et al. 2009; Turu and Hunyady 2010). Specifically, anandamide is synthesized from N-arachidonoyl phosphatidyl ethanolamine (NAPE) via multiple pathways (Basavarajappa 2007; Ahn et al. 2008), while 2-AG is synthesized from acyl arachidonoyl glycerols by diacyl glycerol lipases (DAGLα et β). Anandamide is hydrolyzed by the fatty acid amide hydrolase (FAAH) in the postsynaptic element (Cravatt et al. 1996; Wei et al. 2006). Although FAAH can also hydrolyze 2-AG, the biological actions of this lipid in the brain are mainly terminated by a monoacyl glycerolipase (MAGL) localized in the presynaptic element (Dinh et al. 2002).

Activation of CB1 inhibits the release of excitatory and inhibitory neurotransmitters in several brain areas involved in psychomotor function, including glutamate (Gerdeman and Lovinger 2001; Robbe et al. 2001), gamma-aminobutyric acid (GABA) (Szabo et al. 1998; Sidlo et al. 2008), acetylcholine (Gessa et al. 1997; Gessa et al. 1998a), noradrenaline (Kathmann et al. 1999), and serotonin (Nakazi et al. 2000; Balazsa et al. 2008; Ferreira et al. 2012).

Retrogradely released endocannabinoids mediate three common forms of transient or sustained changes in synaptic strength via activation of presynaptic CB1: (1) short-term depolarization-induced suppression of excitatory (DSE) or inhibitory (DSI) transmission, (2) short-term depression (STD) and (3) long-term depression (LTD) (Chevaleyre et al. 2006; Heifets and Castillo 2009; Kano et al. 2009;

Cachope 2012). CB1-dependent synaptic plasticity has been observed in the PFC, hippocampus, amygdala, VTA, cerebellum, and striatum, where it is necessary for the expression of psychomotor behaviors (Freund et al. 2003; van der Stelt and Di Marzo 2003; Chevaleyre et al. 2006; Heifets and Castillo 2009; Kano et al. 2009; Cachope 2012; Katona and Freund 2012; Lovinger and Mathur 2012).

5.3 Endocannabinoid–Dopamine Interactions

5.3.1 Neuroanatomy

CB1 receptors are present at high density in dopamine-innervated brain regions, such as the cingulate, frontal, and limbic cortices (i.e. the entorhinal cortex), hippocampus (especially in the dentate gyrus), cerebellum, and basal ganglia (especially in the globus pallidus and striatum) (Herkenham et al. 1990; Herkenham et al. 1991a, Herkenham et al. 1991b; Mailleux et al. 1992; Mailleux and Vanderhaeghen 1992b, 1992a, Moldrich and Wenger 2000; Mackie 2005b; Martin et al. 2008). Moderate CB1 densities have been found in the basal forebrain, amygdala, nucleus accumbens, periaqueductal gray, and hypothalamus, while low densities are in the midbrain, pons, medulla, primary motor cortex, thalamus (mainly in the mediodorsal nucleus) and spinal cord (Herkenham et al. 1990; Moldrich and Wenger 2000; Mackie 2005b; Pazos et al. 2005; McPartland et al. 2007; Haring et al. 2012).

In contrast to the widespread distribution of CB1 in the CNS, D1-like and D2-like dopamine receptors are mainly restricted to mesolimbic and mesocortical areas (Missale et al. 1998; Pivonello et al. 2007; Obeso et al. 2008; Cave and Baker 2009; Beaulieu and Gainetdinov 2011; Morera-Herreras et al. 2012).

Although some studies have shown that CB1 colocalizes with tyrosine hydroxylase or the dopamine transporter in midbrain neurons (Wenger et al. 2003; Köfalvi et al. 2005; Haring et al. 2007; Lau and Schloss 2008), several neuroanatomical reports indicate that CB1 are not present in dopaminergic neurons (Matsuda et al. 1993; Katona et al. 1999; Rodriguez et al. 2001; Pistis et al. 2002; Julian et al. 2003; Matyas et al. 2006). CB1 immunoreactivity has been detected in GABAergic, glutamatergic, and opioidergic terminals located in the close proximity to dopaminergic neurons (Herkenham et al. 1991a; Mailleux and Vanderhaeghen 1992b; Matsuda et al. 1993; Tsou et al. 1998), suggesting that CB1-mediated effects on dopamine transmission are mostly indirect and exerted via modulation of inhibitory or excitatory inputs (Hermann et al. 2002; Degroot et al. 2006; Pickel et al. 2006; Kortleven et al. 2011).

5.3.2 *Signaling Pathways*

CB1 activation leads to inhibition of adenylate cyclase activity and reduction in cyclic AMP production, followed by a subsequent decrease in protein kinase A activity (Glass and Felder 1997; Maneuf and Brotchie 1997; Rhee et al. 2000; Howlett 2002, 2004). Multiple downstream pathways have been described, including activation of inward rectifying potassium channels (Mackie et al. 1995; Vasquez et al. 2003), inhibition of voltage-gated calcium (Pan et al. 1996; Twitchell et al. 1997) and sodium channels (Nicholson et al. 2003), and phosphorylation of multiple members of the mitogen-activated protein kinase family, such as extracellular signal-regulated kinase 1 and 2 (ERK1/2) (Bouaboula et al. 1995; Galve-Roperh et al. 2002; Derkinderen et al. 2003), c-Jun N-terminal kinase (JNK) and p38 kinase (Rueda et al. 2000). In addition to Gi/o, CB1 receptors can also interact with Gs or Gq/11 after pertussis toxin treatment (Lauckner et al. 2005; Demuth and Molleman 2006; Pertwee 2006; McIntosh et al. 2007; Hudson et al. 2010a).

CB1 receptors can also exist as homodimers (Wager-Miller et al. 2002; Mackie 2005a) or heterodimers in combination with other receptors, including D2 (Kearn et al. 2005; Mackie 2005a), orexin-1 (Ellis et al. 2006), A2A (Carriba et al. 2007), β2AR (Hudson et al. 2010b), and μ, δ, and κ opioid receptors (Rios et al. 2006; Hojo et al. 2008). In primary cultures of striatal neurons, pharmacological costimulation of CB1 and D2R switched CB1 signaling from a Gi/o- to a Gs-coupled response (Glass and Felder 1997). Interestingly, the coexpression of D2R alone was later found to be sufficient to switch CB1 coupling toward Gs, even in the absence of a D2R agonist (Jarrahian et al. 2004). This phenomenon can be attributed to the fact that CB1-D2R heteromerization may modify CB1 functional selectivity or the strength of ligand-directed effects (Hudson et al. 2010a). More recent studies showed the actual physical interaction of CB1 and D2R in HEK293 cells by coimmunoprecipitation (Kearn et al. 2005) and FRET/BRET approaches (Marcellino et al. 2008; Navarro et al. 2008). Finally, stimulation of both receptors appears to increase heterodimer formation (Meschler and Howlett 2001; Kearn et al. 2005; Marcellino et al. 2008).

5.3.3 *Electrophysiological and Microdialysis Studies*

The ECS exerts a major modulatory action on dopamine neurons firing activity (Melis and Pistis 2007) and participates in several dopamine-dependent long-term forms of synaptic plasticity (i.e., LTD or LTP) in the VTA, PFC, and striatum, (Laviolette and Grace 2006; Fattore et al. 2010; Esteban and Garcia-Sevilla 2012; Mathur and Lovinger 2012; Morera-Herreras et al. 2012).

In the VTA, acute or chronic treatment with cannabinoid agonists has been shown to enhance the spontaneous firing rate of dopamine neurons (French et al. 1997; Gessa et al. 1998b; Wu and French 2000; Cheer et al. 2003) via a CB1-dependent mechanism. Reciprocally, stimulation of PFC glutamatergic afferent to

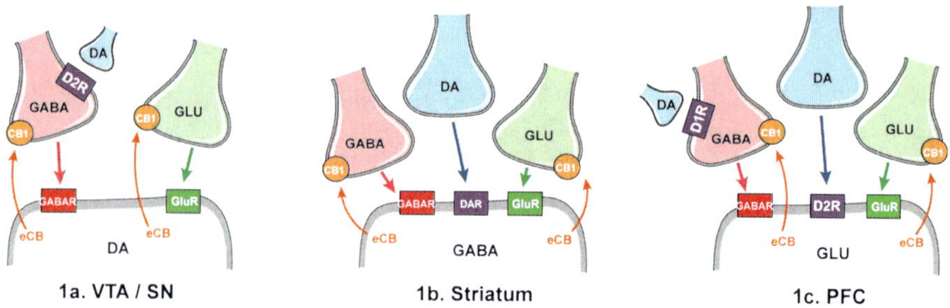

Fig. 5.1 CB1-mediated synaptic modulation in motor and corticolimbic pathways. Activation of CB1 receptor inhibits glutamate (GLU) or GABA release from terminals projecting to: dopaminergic neurons (DA) in the ventral tegmental area (VTA) and substantia nigra (SN) (**a**), GABAergic medium spiny neurons in the striatum (**b**) and glutamatergic pyramidal cells in the prefrontal cortex (PFC) (**c**). *eCB* endocannabinoids; *GluR* glutamate receptor; *DAR* dopamine receptor; *D1R/D2R* D1-like/D2-like dopamine receptors; *GABAR* GABA receptor

the VTA leads to local endocannabinoid release, resulting in CB1-dependent presynaptic inhibition of glutamatergic stimulation (DSE) of VTA dopamine neurons (Melis et al. 2004a; Melis et al. 2004b; Riegel and Lupica 2004). Endocannabinoids can also mediate long-term plasticity by inhibiting LTP of glutamatergic neurons (Kortleven et al. 2011) or by facilitating LTD of GABAergic neurons (Pan et al. 2008) projecting onto VTA dopamine neurons. Endocannabinoid-mediated LTD of inhibitory synapses (I-LTD) is presumably D2R-dependent (Pan et al. 2008). As the excitatory actions of cannabinoid agonists on VTA dopamine neurons are blocked by pharmacological inhibition of $GABA_A$ receptors (Szabo et al. 2002), it is likely that endocannabinoids activate these neurons through inhibition of their GABAergic afferents. Endocannabinoid-mediated disinhibition of PFC glutamatergic outputs to VTA could also account for increased dopamine neurons activity in this brain area (Fig. 5.1a).

Endocannabinoids, as well as acute or chronic THC treatment, have been shown to enhance neuronal firing of substantia nigra dopamine neurons (Melis et al. 2000 Wu and French 2000; Morera-Herreras et al. 2012). In particular, anandamide can facilitate glutamate release in the substantia nigra pars compacta by activating TRPV1 (Marinelli et al. 2003). Another endocannabinoid, N-arachidonoyl-dopamine, may also affect glutamatergic transmission in this area via TRPV1- and CB1-mediated mechanisms, and decrease GABAergic transmission via CB1 stimulation (Marinelli et al. 2007).

In the dorsal striatum, several studies have shown that endocannabinoids and CB agonists can decrease dopamine synthesis (Moranta et al. 2004, 2009). However, the ability of cannabinoid agonists to affect dopamine release in the caudate putamen remains controversial as some studies have reported a decrease (Cadogan et al. 1997; Kathmann et al. 1999; Sidlo et al. 2008) or an increase (Chen et al. 1990a; Chen et al. 1990b; Tanda et al. 1997; Malone and Taylor 1999; Fadda et al. 2006; Solinas et al. 2006), or no effect (Szabo et al. 1999; Köfalvi et al. 2005). These inconsistencies may be attributable to the different experimental settings (slices vs.

freely moving animals, dose of drugs, time of measurements, etc.), as well as type of agonist (THC vs. WIN or CP), model (rodent vs. human) and methodology (microdialysis, autoradiography, voltammetry) used.

Activation of D1- or D2-like receptors seems to affect striatal endocannabinoid levels in opposite ways: while D1-like agonists decrease anandamide (Patel et al. 2003), D2-like agonists can increase it (Giuffrida et al. 1999; Centonze et al. 2004). These effects may result from the ability of D1R and D2R agonists to increase or decrease excitatory postsynaptic currents (EPSC), respectively, in medium spiny neurons. As these effects were prevented by CB1 activation (Andre et al. 2010), it can be speculated that dopamine receptors control endocannabinoid mobilization, which in turn modulates glutamate or GABA release from terminals projecting to striatal neurons expressing D1R and D2R (Fig. 5.1b).

In the ventral striatum, acute administration of THC and WIN has been shown to increase dopamine release (Chen et al. 1990b; Tanda et al. 1997). These findings, however, have not been replicated by other groups (Szabo et al. 1999). The observation that amphetamine-induced dopamine release in the shell region of the rat nucleus accumbens is CB1-dependent suggests that the ECS can indirectly affect dopamine transmission in the ventral striatum (Kleijn et al. 2012), possibly via endocannabinoid-induced depression of GABAergic inhibitory inputs to dopamine neurons (Manzoni and Bockaert 2001).

In the rat medial PFC, intravenous administration of THC and WIN 55,212-2 suppresses the inhibitory effect of VTA stimulation on pyramidal cells and increases the firing rate of neurons projecting back to the VTA (Pistis et al. 2002, 2001). In slices and synaptosomes of human neocortex, the CB agonist CP55,940 inhibited electrically evoked dopamine release, whereas the CB1 antagonist AM251 had the opposite effect (Steffens et al. 2004). Although the mechanisms underlying the endocannabinoid–dopamine interplay in the PFC need further elucidation, a recent study has shown that suppression of GABA release onto layer 5 pyramidal cells occurs via a particular type of I-LTD that is heterosynaptic in nature and requires coactivation of CB1 and D2R (Chiu et al. 2010). Thus, these data suggest that ECS-dopamine cross talk at GABAergic terminals is necessary to tune up pyramidal cells (Fig. 5.1c).

5.3.4 Pharmacological and Behavioral Studies

5.3.4.1 Regulation of Motor Function

Administration of THC or synthetic cannabinoids typically has a dose-dependent bidirectional effect on locomotion, with low doses producing hyperlocomotion and high doses being sedative (Prescott et al. 1992; Anderson et al. 1996; de Lago et al. 2004; Polissidis et al. 2012). Enhancement of anandamide levels via inhibition of its transport or hydrolysis produces CB1-dependent hypokinesia in rats (Compton and Martin 1997; González et al. 1999). CB1-KO mice exhibit reduced locomotor activity (Zimmer et al. 1999), suppression of THC-induced hypomotility

(Ledent et al. 1999; Zimmer et al. 1999), and suppression of cocaine-enhanced locomotion (Li et al. 2009). By contrast, CB1 antagonism does not affect locomotion (Compton et al. 1996; Polissidis et al. 2012).

The cannabinoid-mediated motor effects likely involve modulation of the dopamine system. Indeed, CB1 agonists and antagonists have been shown to affect motor behaviors induced by D1R-D2R ligands (Anderson et al. 1996; Giuffrida et al. 1999; Marcellino et al. 2008), as well as by amphetamines (Gorriti et al. 1999; Tzavara et al. 2003; Cortright et al. 2011). In addition, in dopamine-depleted 6-hydroxydopamine (6-OHDA)-treated rodents, motor dysfunction is alleviated by inhibiting endocannabinoid metabolism via systemic injection of a FAAH inhibitor (Kreitzer and Malenka 2007) or direct CB agonism (Ferrer et al. 2003; Morgese et al. 2007). Finally, mice lacking the dopamine transporter (DAT) present striatal hyperdopaminergia associated with attenuated ECS activity, suggesting a cross talk between the two systems (Tzavara et al. 2006). This scenario is further complicated by a TRPV1-dependent modulatory component as suggested by the observation that inhibitors of anandamide reuptake or hydrolysis can reduce hyperlocomotion in DAT-KO mice via a TRPV1-dependent and CB1-independent mechanism (Tzavara et al. 2006). On the other hand, in 6-OHDA-treated rats undergoing chronic treatment with levodopa, TRPV1 blockade is necessary to unmask the antidyskinetic effects of FAAH inhibitors (Morgese et al. 2007), suggesting that CB1 and TRPV1 receptors have opposite actions on aberrant motor behaviors.

5.3.4.2 Regulation of Emotion and Cognitive Function

Little is known on possible endocannabinoid/dopamine interactions in the modulation of neural circuitries related to emotion and anxiety responses. High doses of cannabinoid agonists can produce powerful anxiogenic effects which are also observed after administration of psychostimulant drugs that activate dopamine transmission (Hayase et al. 2005; Ruehle et al. 2012; Tambaro and Bortolato 2012), Mice with genetic deletion of CB1 on DR-1 positive neurons show anhedonia-like behavior in the sucrose preference test (Terzian et al. 2011). However, the same mutant mice do not differ from controls in a variety of other anxiety-related paradigms (Terzian et al. 2011), suggesting that receptors other than D1R might be involved in the expression of anxious behavior. In agreement with these observations, D1 and D2 receptors located in the amygdala have been shown to mediate the anxiolytic-like effect induced by the CB1 agonist arachydonil cyclopropylamide (Zarrindast et al. 2011a). Other studies have shown that pharmacological blockade of CB1 in mice increases anxiety in the elevated-plus maze and open field tests (Thiemann et al. 2009). This anxiogenic effect was associated with changes of dopamine turnover in different brain areas, such as striatum, hippocampus, and frontal cortex (Thiemann et al. 2009).

Cannabinoids have a deleterious effect on short-term memory, working memory, and decision-making (Kalant 2004; Iversen 2005; Ranganathan and D'Souza 2006; Crean et al. 2011). In line with the modulatory role of dopamine on higher executive functions (Cools 2011), an endocannabinoid–dopamine cross talk has been described in the hippocampus (Thiemann et al. 2008; Nasehi et al. 2009; Thiemann et al. 2009), where dopamine-dependent and endocannabinoid-mediated plasticity (DSE, DSI, LTD) is likely to occur as previously described for the VTA, PFC, or striatum (see the previous text). In addition, human studies suggest that long-term cannabis exposure may enhance dopaminergic drive in the VTA and mesolimbic areas (Ameri 1999; Egerton et al. 2006; Ranganathan and D'Souza 2006).

Increased dopamine tone in rodent PFC (Jentsch et al. 1997), amygdala (Hernandez-Tristan et al. 2000) or dorsal hippocampus (Nava et al. 2000; Zarrindast et al. 2010; Zarrindast et al. 2011b) can underlie cannabinoid-mediated memory impairments. For example, two studies assessing memory consolidation following an inhibitory avoidance task indicated that D1R and D2R agonists prevent anandamide-induced memory deficits (Castellano et al. 1997; Costanzi et al. 2004). Further support to the hypothesis of a role played by endocannabinoid–dopamine interactions in memory modulation comes from the well-established link between cannabinoids and dopamine in addiction/reward (van der Stelt and Di Marzo 2003; Fernandez-Ruiz and Gonzales 2005; Laviolette and Grace 2006; Lopez-Moreno et al. 2008; Fattore et al. 2010), which may be considered a form of learning and memory.

Details on cannabinoid–dopamine contributions to cognition are also provided in the section on schizophrenia (see Sect. 4.2).

5.4 Neurodegenerative and Psychiatric Disorders

The interactions between the endocannabinoid and dopamine systems have spurred interest in studying the role played by endocannabinoid transmission in neurodegenerative and psychiatric disorders associated with dopamine dysfunction, such as Parkinson's disease and schizophrenia.

5.4.1 Parkinson's disease

Parkinson's disease (PD) is a neurodegenerative disorder characterized by progressive loss of nigrostriatal neurons, maladaptive striatal plasticity and disabling motor disturbances. The presence of CB1 receptors in the basal ganglia and the ECS ability to modulate neurotransmission and synaptic plasticity in these circuitries strongly suggest a role for endocannabinoids in PD pathophysiology and the possibility to develop endocannabinoid-based therapeutic strategies.

While numerous studies have found abnormal endocannabinoid levels in animal models and PD patients, there is no consensus on the direction of the endocannabinoid fluctuations. Increased endocannabinoid tone has been associated with the progression of the disease in PD patients (Pisani et al. 2005; Pisani et al. 2011). Also, CB1 levels were found increased in the striatum of parkinsonian monkeys and PD patients (Lastres-Becker et al. 2001; Van Laere et al. 2012), but significantly decreased in the substantia nigra (Van Laere et al. 2012).

In reserpine-treated rats, striatal endocannabinoid levels were increased (Di Marzo et al. 2000; Gubellini et al. 2002), whereas studies in animals treated with the neurotoxin 6-OHDA showed decreased anandamide concentrations in this area (Ferrer et al. 2003; Morgese et al. 2007). In addition, PD-like symptoms in rodents have been associated with increased (Zeng et al. 1999; Romero et al. 2000; Gubellini et al. 2002; Maccarrone et al. 2003; Gonzalez et al. 2006), or decreased (Silverdale et al. 2001; Ferrer et al. 2003; Hurley et al. 2003; Walsh et al. 2010) CB1 levels. Differences in the type of PD model, extension of the lesion and/or time points used for endocannabinoid measurements may explain these discrepancies.

Upregulation of CB1-signaling in PD might represent a compensatory mechanism balancing the reduced dopamine tone associated with the disease (Mailleux and Vanderhaeghen 1993; Di Marzo et al. 2000; Lastres-Becker et al. 2001; Gubellini et al. 2002; Maccarrone et al. 2003; Pisani et al. 2005; Garcia-Arencibia et al. 2009). The loss of striatal dopamine might also impair corticostriatal endocannabinoid-dependent LTD, and possibly contribute to PD motor symptoms. In this context, many studies have addressed the question whether pharmacological modulation of endocannabinoid tone or CB1 function could alleviate PD-like motor symptoms or correct these anomalies. Unfortunately, the use of CB ligands in preclinical studies has yielded unclear results. CP55,940 and THC have been shown to reduce tremor and stereotypies after local injections in the basal ganglia of 6-OHDA-treated rats or 1-methyl-4-phenyl-1,2,3,6-tetrahydropyridine (MPTP)-lesioned marmosets, respectively (Sanudo-Pena et al. 1998; van Vliet et al. 2008). However, systemic injection of WIN55,212-2 has been shown to prevent the beneficial effects of D1R and D2R agonists on akinesia in the reserpine rat model of PD (Maneuf et al. 1997). Positive effects against disease progression have been also observed following CB2 receptor activation in MPTP- (Price et al. 2009) and 6-OHDA-treated rodents (Garcia et al. 2011). This neuroprotective effect on nigrostriatal neurodegeneration seems to involve the inhibition of microglial infiltration (Price et al. 2009; Garcia et al. 2011), which is consistent with the enhanced expression of CB2R on activated microglia (Cabral and Marciano-Cabral 2005; Stella 2010). Other reports have shown CB1-independent neuroprotective effects of cannabinoids in 6-OHDA-treated rats (Lastres-Becker et al. 2005; Garcia-Arencibia et al. 2007), further suggesting a possible involvement of CB2R. Nonetheless, neuroprotective contribution by CB1 cannot be ruled out since CB1-KO mice have been shown to be more vulnerable to the 6-OHDA lesion than wild-type animals (Perez-Rial et al. 2011), and cannabinoids can exert CB1-mediated anti-oxidant effects (Marsicano et al. 2002; van der Stelt et al. 2002).

CB1 antagonists were also shown to improve motor symptoms in rodent and non-human primate models of PD (Fernandez-Espejo et al. 2005; van der Stelt et al. 2005; Gonzalez et al. 2006; Kelsey et al. 2009). However, the antiparkinsonian effects of CB1 antagonists rarely ameliorated the complete range of motor impairments observed in PD (Cao et al. 2007; Kelsey et al. 2009). These limitations, along with conflicting results from other studies (Di Marzo et al. 2000; Meschler et al. 2001), raise questions on the therapeutic potential of CB1 antagonists in PD. Possibly, as proposed by some authors, CB1 blockade has beneficial effects only under specific conditions, such as in advanced phases of PD (Fernandez-Espejo et al. 2005), or when using low doses (Gonzalez et al. 2006; Kelsey et al. 2009).

Cannabinoids have been also investigated as a treatment for dyskinesias, which are chorea-like movements induced by prolonged exposure to levodopa (levodopa-induced dyskinesia or LID), the gold standard for PD therapy. Studies in 6-OH-DA-treated rats (Ferrer et al. 2003; Morgese et al. 2007; Martinez et al. 2012) and MPTP-treated marmosets (Fox et al. 2002) showed LID attenuation following systemic administration of cannabinoid agonists. Interestingly, indirect stimulation of CB1 via administration of the FAAH inhibitor URB597 did not produce a similar effect, presumably because of anandamide ability to activate both CB1 and TRPV1 receptors (Morgese et al. 2007). Indeed, coadministration of the TRPV1 antagonist capsazepine with URB597 significantly reduced dyskinesias, suggesting that CB1 and TRPV1 receptors play opposing roles in LID (Morgese et al. 2007). Nevertheless, the potential antidyskinetic effects of cannabinoids require further elucidation, as they have not been confirmed by other studies. For example, genetic deletion of CB1 prevented the development of severe dyskinetic movements in mice (Perez-Rial et al. 2011), suggesting a putative prodyskinetic role of CB1. In line with these observations, CB1 antagonists have been shown to reduce and/or delay LID (Silverdale et al. 2001; Cao et al. 2007; Fabbrini et al. 2007; Huot et al. 2013). Furthermore, no correlation between CB1 expression and severity of dyskinetic symptoms has been observed in PD patients (Fernandez-Ruiz 2009; Garcia-Arencibia et al. 2009). Finally, the few clinical studies carried out so far have provided equivocal conclusions on the putative antidyskinetic properties of endocannabinoid- or CB1-based therapies (Frankel et al. 1990; Sieradzan et al. 2001; Carroll et al. 2004; Mesnage et al. 2004). Possible methodological confounders (e.g., limitations in mimicking the biological and pathological aspects of PD in different animal models, compensatory neurodevelopment adjustments in knockout mice, use of different scales or self-reported questionnaires for LID assessment in clinical studies, etc.) may account for these conflicting results, underlying the need for new and larger-scale clinical studies.

5.4.2 *Schizophrenia*

Schizophrenia is a severe mental illness characterized by 3 major categories of symptoms: positive (e.g., hallucinations, delusions), negative (e.g., social withdrawal, anhedonia) and cognitive deficits (e.g., impaired working memory and attention) (van Os and Kapur 2009). Dopamine alterations have been observed in schizophrenic patients (Meador-Woodruff et al. 1997; Stefanis et al. 1998) leading to the hypothesis that mesocortical hypodopaminergia and mesolimbic hyperdopaminergia constitute the neurochemical substrates of schizophrenia pathophysiology (Laruelle et al. 2003a; Toda and Abi-Dargham 2007). Given the previously mentioned endocannabinoid–dopamine cross talk, it has been suggested that the ECS could also play a role in schizophrenia by interfering with dopamine transmission (Muller-Vahl and Emrich 2008).

In humans, cannabis consumption can trigger schizophrenia-like states in normal individuals, exacerbate psychotic symptoms in schizophrenic patients, and increase the risk of developing schizophrenia in predisposed individuals (Ujike and Morita 2004; Koethe et al. 2009b; Sewell et al. 2009). Epidemiological investigations provided the first evidence for an association between cannabis intake and schizophrenia. Consumption of large amounts of cannabis can result in a "cannabinoid psychosis" characterized by schizophrenia-like symptoms such as blunted affect, distorted sensory perceptions, and cognitive deficits (D'Souza et al. 2004; Ujike and Morita 2004 D'Souza 2007; Sewell et al. 2009). Metaanalysis studies estimate that cannabis use in adolescents may account for 8–14 % of schizophrenia cases (Henquet et al. 2005). Moreover, cannabis use has been also associated with more frequent and earlier relapses (Costain 2008; D'Souza et al. 2009; San et al. 2012).

Cannabinoids may cause or exacerbate psychotic symptoms by inhibiting GABA and/or glutamate transmission leading to: (1) disruption of synaptic plasticity and disturbance of neuronal synchrony maintained by GABAergic interneurons, which may contribute to cognitive symptoms; (2) facilitation of dopamine transmission in the mesolimbic pathway, which may favor the expression of positive symptoms; (3) alterations in the activity of monoaminergic and cholinergic subcortical pathways (van der Stelt and Di Marzo 2003; D'Souza 2007; Muller-Vahl and Emrich 2008; Curley and Lewis 2012).

The occurrence of cannabis-induced psychoses led to the formulation of the "cannabinoid hypothesis of schizophrenia" (Muller-Vahl and Emrich 2008), which postulates that overactivity of the brain ECS might contribute to the etiology of this mental disorder. In agreement with this hypothesis, initial studies found increased levels of anandamide in the cerebrospinal fluid (CSF) of drug-naive paranoid schizophrenia patients (Leweke et al. 1999; Giuffrida et al. 2004; Leweke et al. 2007; Koethe et al. 2009a), as well as in the prodromal phase of the disease (Koethe et al. 2009a). Similarly, autoradiographic postmortem studies examining CB1 density in schizophrenic patients showed increased CB1 binding in the dorsolateral PFC (Dean et al. 2001; Dalton et al. 2011) and cingulate cortex (Zavitsanou et al. 2004; Newell et al. 2006). A similar trend has been also reported in living patients using the PET tracer [(11)C]OMAR (JHU75528) (Wong et al. 2010). However, new experimental evidence is challenging the idea that an overactive ECS might contribute to schizophrenia. First, the CB1 antagonist Rimonabant has failed as antipsychotic in clinical trials (Meltzer et al. 2004) and has been withdrawn from the market because of its association with increased rates of depression and suicides (Nissen et al. 2008). Second, direct measures of CB1 mRNA or protein have not confirmed CB1 upregulation in the anterior cingulate cortex (Koethe et al. 2007), but instead found decreased CB1 density in the dorsolateral PFC (Eggan et al. 2008; Uriguen et al. 2009; Eggan et al. 2010). These discrepancies might be due to the fact that the changes in CB1 expression are susceptible to: (1) the subclinical type of schizophrenia, as indicated by the specific CB1 increase in paranoid (Dalton et al. 2011) and hebephrenic patients (Wong et al. 2010) only; (2) medication regimen, as suggested by reduced CB1 levels found in antipsychotic-treated patients but not in drug-free subjects (Eggan et al. 2008; Uriguen et al. 2009; Zuardi et al. 2011); (3) the time course and severity of symptoms (Wong et al. 2010); and (4) the neuronal population studied (Eggan et al. 2008).

Third, schizophrenic patients with elevated CSF anandamide suffer from fewer psychotic symptoms (Giuffrida et al. 2004) and the negative correlation between anandamide concentrations and symptom severity has been also confirmed in prodromal states (Koethe et al. 2009a), suggesting that endocannabinoids do play a protective role in schizophrenia.

The idea that anandamide, unlike THC, might ameliorate schizophrenic symptoms is also supported by several studies showing that increased dopaminergic transmission—which has been reported in schizophrenia (Laruelle et al. 2003a; Laviolette and Grace 2006—elevates anandamide levels in rat striatum (Giuffrida et al. 1999), and that this elevation may reflect a homeostatic adaptation to counteract dopamine-mediated psychomotor activation (Giuffrida et al. 1999; Beltramo et al. 2000).

Solid evidence for the involvement (and the beneficial effects) of endocannabinoids in psychosis comes from preclinical studies carried out in the phencyclidine (PCP) rat model of schizophrenia (Jentsch et al. 1997; Jentsch et al. 1998). Subchronic treatment with PCP has been shown to increase CB1 density in rat VTA, globus pallidus and cerebellum, and to reduce CB1 activity as assessed by GTPγS binding (Vigano et al. 2009; Guidali et al. 2011). These findings, however, were not confirmed by other studies using a different PCP dose and regimen (Seillier et al. 2010), suggesting that differences in PCP exposure may significantly impact endocannabinoid transmission.

Although the CB1 antagonist AM251 reversed the PCP-induced memory deficit and avolition (Seillier et al. 2010; Guidali et al. 2011), AM251 was ineffective in rescuing PCP-induced social withdrawal. On the other hand, elevation of anandamide tone via administration of the FAAH inhibitor URB597 reversed social deficits in this animal model (Seillier et al. 2010). Like URB597, direct CB1 agonism attenuated PCP-induced deficits in sociability (Seillier et al. 2010; Spano et al. 2010). These findings indicate that the effects of cannabinoids in PCP-treated rats vary depending on the type of schizophrenia-like behavior considered, and suggest that positive and negative symptoms are associated to enhanced or deficient CB1 activation, respectively.

5.4.3 Social Behavior

Endocannabinoids seem to play a significant role in social behavior, which involves modulation of salient sensory and emotional information, two processes closely tied to dopamine neurotransmission (Robinson et al. 2006; Furmark 2009; Fattore et al. 2010). Low D2R-binding has been found in people with social anxiety (Schneier et al. 2000; Schneier et al. 2008; Schneier et al. 2009), and the negative symptoms of schizophrenia (such as social withdrawal) are thought to be related to subcortical hypodopaminergia (Laruelle et al. 2003b; van der Stelt and Di Marzo 2003; Toda and Abi-Dargham 2007; Muller-Vahl and Emrich 2008).

Although cannabis is widely used as a recreational drug because of its anxiolytic properties, the most commonly cited reason for its discontinuation is increased anxiety and panic reactions (Szuster et al. 1988; Reilly et al. 1998; Green et al. 2003). A possible explanation for this paradoxical effect is that anxiety modulation by cannabis greatly depends on dose (low doses being anxiolytic and high doses being anxiogenic) and environmental context (anxiogenic effects are driven by stress and aversive situations). For example, CB1-KO mice present more aggressive behavior in their home-cage, but reduced social behavior in an unfamiliar cage, especially under bright light conditions (Haller et al. 2004). They also show diminished response to social stress (Dubreucq et al. 2012) and resistance to the anxiolytic effect of buspirone (Uriguen et al. 2004). These results suggest that CB1 stimulation is required to prevent the expression of aggressive, as well as normal, social behavior. The dual effects of cannabinoids on anxiety/sociability might be attributed to activation of different CB1-mediated signaling pathways and/or distinct interactions with specific dopamine receptor subtypes. For instance, conditional CB1-KO in D1R expressing neurons produced anxiety-like phenotypes during social behavior, suggesting that CB1-D1R cross talk is implicated in negative social affect (Terzian et al. 2011). This hypothesis is supported by pharmacological studies showing that normal social behavior requires the coactivation of D1R and CB1 in the piriform cortex of rats (Zenko et al. 2011).

Further evidence for the ECS involvement in sociability has been provided by studies showing that administration of anandamide enhances social behavior (Umathe et al. 2009). In adolescent rats, administration of the anandamide transporter inhibitor VDM11 in the hippocampus (Trezza and Vanderschuren 2009) or of URB597 in the nucleus accumbens or basolateral amygdala enhanced social play (Trezza et al. 2012). In addition, stimulation of CB1 with methanandamide or low doses of URB597 injected into the PFC had anxiolytic-like effects in the elevated-plus maze (Rubino et al. 2008). Interestingly, administration of high doses of direct and indirect cannabinoid agonists, or the decrease of local anandamide levels via lentivirus-mediated FAAH overexpression, produced an anxiogenic response (Rubino et al. 2008).

5.5 Concluding Remarks

Given the brain-wide modulatory actions of endocannabinoids and the primary role played by dopamine in regulating motor, cognitive and emotional functions, deciphering the details of the ECS-dopamine cross talk is of great importance to understand the pathophysiology of disorders characterized by dopamine dysfunction, such as PD and schizophrenia.

Direct and indirect cannabinoid agonists have shown promising antiparkinsonian, antidyskinetic and antipsychotic-like properties in animal models; however, large-scale clinical studies that may translate this knowledge into new therapies are still lacking.

Future research efforts should also consider new approaches to modulate the ECS system, including the: (1) enhancement of CB1 functional selectivity; (2) assessment of the therapeutic potentials of endocannabinoid-enhancing drugs and pharmacological manipulation of non-CB1 targets, such as TRP channels and PPAR receptors; (3) study of therapeutic potentials of polyunsaturated fatty acid (PUFA), and in particular of n-3 long chain PUFA-derived conjugates with ethanolamine, dopamine, serotonin, or other amines, which are capable of modulating the ECS (Meijerink et al. 2012).

Funding Sources NIMH RO1MH91130-01A1 and NS 050401-07 (to A.G.)

Conflict of Interest We declare no conflict of interest.

References

Ahn K, McKinney MK, Cravatt BF (2008) Enzymatic pathways that regulate endocannabinoid signaling in the nervous system. Chem Rev 108:1687–1707

Ameri A (1999) The effects of cannabinoids on the brain. Prog Neurobiol 58:315–348

Anderson JJ, Kask AM, Chase TN (1996) Effects of cannabinoid receptor stimulation and blockade on catalepsy produced by dopamine receptor antagonists. Eur J Pharmacol 295:163–168

Andre VM, Cepeda C, Cummings DM, Jocoy EL, Fisher YE, William Yang X, Levine MS (2010) Dopamine modulation of excitatory currents in the striatum is dictated by the expression of D1 or D2 receptors and modified by endocannabinoids. Eur J Neurosci 31:14–28

Balazsa T, Biro J, Gullai N, Ledent C, Sperlagh B (2008) CB1-cannabinoid receptors are involved in the modulation of non-synaptic [3H]serotonin release from the rat hippocampus. Neurochem Int 52:95–102

Basavarajappa BS (2007) Critical enzymes involved in endocannabinoid metabolism. Protein Pept Lett 14:237–246

Beaulieu JM, Gainetdinov RR (2011) The physiology, signaling, and pharmacology of dopamine receptors. Pharmacol Rev 63:182–217

Beltramo M, Fonseca FR de, Navarro M, Calignano A, Gorriti MA, Grammatikopoulos G, Sadile AG, Giuffrida A, Piomelli D (2000) Reversal of dopamine D(2) receptor responses by an anandamide transport inhibitor. J Neurosci 20:3401–3407

Bouaboula M, Poinot-Chazel C, Bourrie B, Canat X, Calandra B, Rinaldi-Carmona M, Le Fur G, Casellas P (1995) Activation of mitogen-activated protein kinase by stimulation of the central cannabinoid receptor CB1. Biochem J 312:637–641

Brown SP, Brenowitz SD, Regehr WG (2003) Brief presynaptic bursts evoke synapse-specific retrograde inhibition mediated by endogenous cannabinoids. Nat Neurosci 6:1048–1057

Cabral GA, Marciano-Cabral F (2005) Cannabinoid receptors in microglia of the central nervous system: immune functional relevance. J Leukoc Biol 78:1192–1197

Cachope R (2012) Functional diversity on synaptic plasticity mediated by endocannabinoids. Philos Trans R Soc Lond B Biol Sci 367:3242–3253

Cadogan AK, Alexander SP, Boyd EA, Kendall DA (1997) Influence of cannabinoids on electrically evoked dopamine release and cyclic AMP generation in the rat striatum. J Neurochem 69:1131–1137

Cao X, Liang L, Hadcock JR, Iredale PA, Griffith DA, Menniti FS, Factor S, Greenamyre JT, Papa SM (2007) Blockade of cannabinoid type 1 receptors augments the antiparkinsonian action of levodopa without affecting dyskinesias in 1-methyl-4-phenyl-1,2,3,6-tetrahydropyridine-treated rhesus monkeys. J Pharmacol Exp Ther 323:318–326

Carriba P, Ortiz O, Patkar K, Justinova Z, Stroik J, Themann A, Muller C, Woods AS, Hope BT, Ciruela F, Casado V, Canela EI, Lluis C, Goldberg SR, Moratalla R, Franco R, Ferre S (2007) Striatal adenosine A2A and cannabinoid CB1 receptors form functional heteromeric complexes that mediate the motor effects of cannabinoids. Neuropsychopharmacology 32:2249–2259

Carroll CB, Bain PG, Teare L, Liu X, Joint C, Wroath BA, Parkin SG, Fox P, Wright D, Hobart J, Zajicek JP (2004) Cannabis for dyskinesia in Parkinson disease. Neurology 63:1245–1250

Castellano C, Cabib S, Palmisano A, Di Marzo V, Puglisi-Allegra S (1997) The effects of anandamide on memory consolidation in mice involve both D1 and D2 dopamine receptors. Behav Pharmacol 8:707–712

Cave JW, Baker H (2009) Dopamine systems in the forebrain. Adv Exp Med Biol 651:15–35

Centonze D, Battista N, Rossi S, Mercuri NB, Finazzi-Agro A, Bernardi G, Calabresi P, Maccarrone M (2004) A critical interaction between dopamine D2 receptors and endocannabinoids mediates the effects of cocaine on striatal gabaergic Transmission. Neuropsychopharmacology 29:1488–1497

Cheer JF, Kendall DA, Mason R, Marsden CA (2003) Differential cannabinoid-induced electrophysiological effects in rat ventral tegmentum. Neuropharmacology 44:633–641

Chen J, Paredes W, Lowinson JH, Gardner EL (1990a) Delta 9-tetrahydrocannabinol enhances presynaptic dopamine efflux in medial prefrontal cortex. Eur J Pharmacol 190:259–262

Chen JP, Paredes W, Li J, Smith D, Lowinson J, Gardner EL (1990b) Delta 9-tetrahydrocannabinol produces naloxone-blockable enhancement of presynaptic basal dopamine efflux in nucleus accumbens of conscious, freely-moving rats as measured by intracerebral microdialysis. Psychopharmacology (Berl) 102:156–162

Chevaleyre V, Takahashi KA, Castillo PE (2006) Endocannabinoid-mediated synaptic plasticity in the CNS. Annu Rev Neurosci 29:37–76

Chiu CQ, Puente N, Grandes P, Castillo PE (2010) Dopaminergic modulation of endocannabinoid-mediated plasticity at GABAergic synapses in the prefrontal cortex. J Neurosci 30:7236–7248

Compton DR, Aceto MD, Lowe J, Martin BR (1996) In vivo characterization of a specific cannabinoid receptor antagonist (SR141716A): inhibition of D^9-tetrahydrocannabinol-induced responses and apparent agonist activity. J Pharmacol Exp Ther 277:586–594

Compton DR, Martin BR (1997) The effect of the enzyme inhibitor phenylmethylsulfonyl fluoride on the pharmacological effect of anandamide in the mouse model of cannabimimetic activity. J Pharmacol Exp Ther 283:1138–1143

Console-Bram L, Marcu J, Abood ME (2012) Cannabinoid receptors: nomenclature and pharmacological principles. Prog Neuropsychopharmacol Biol Psychiatry 38:4–15

Cools R (2011) Dopaminergic control of the striatum for high-level cognition. Curr Opin Neurobiol 21:402–407

Cortright JJ, Lorrain DS, Beeler JA, Tang WJ, Vezina P (2011) Previous exposure to delta9-tetrahydrocannibinol enhances locomotor responding to but not self-administration of amphetamine. J Pharmacol Exp Ther 337:724–733

Costain WF (2008) The effects of cannabis abuse on the symptoms of schizophrenia: patient perspectives. Int J Ment Health Nurs 17:227–235

Costanzi M, Battaglia M, Rossi-Arnaud C, Cestari V, Castellano C (2004) Effects of anandamide and morphine combinations on memory consolidation in cd1 mice: involvement of dopaminergic mechanisms. Neurobiol Learn Mem 81:144–149

Cravatt BF, Giang DK, Mayfield SP, Boger DL, Lerner RA, Gilula NB (1996) Molecular characterization of an enzyme that degrades neuromodulatory fatty-acid amides. Nature 384:83–87

Crean RD, Crane NA, Mason BJ (2011) An evidence based review of acute and long-term effects of cannabis use on executive cognitive functions. J Addict Med 5:1–8

Curley AA, Lewis DA (2012) Cortical basket cell dysfunction in schizophrenia. J Physiol 590:715–724

D'Souza DC (2007) Cannabinoids and psychosis. Int Rev Neurobiol 78:289–326

D'Souza DC, Perry E, MacDougall L, Ammerman Y, Cooper T, Wu YT, Braley G, Gueorguieva R, Krystal JH (2004) The psychotomimetic effects of intravenous delta-9-tetrahydrocannabinol in healthy individuals: implications for psychosis. Neuropsychopharmacology 29:1558–1572

D'Souza DC, Sewell RA, Ranganathan M (2009) Cannabis and psychosis/schizophrenia: human studies. Eur Arch Psychiatry Clin Neurosci 259:413–431

Dalton VS, Long LE, Weickert CS, Zavitsanou K (2011) Paranoid schizophrenia is characterized by increased CB1 receptor binding in the dorsolateral prefrontal cortex. Neuropsychopharmacology 36:1620–1630

Lago E de, Miguel R de, Lastres-Becker I, Ramos JA, Fernandez-Ruiz J (2004) Involvement of vanilloid-like receptors in the effects of anandamide on motor behavior and nigrostriatal dopaminergic activity: in vivo and in vitro evidence. Brain Res 1007:152–159

Dean B, Sundram S, Bradbury R, Scarr E, Copolov D (2001) Studies on [3H]CP-55940 binding in the human central nervous system: regional specific changes in density of cannabinoid-1 receptors associated with schizophrenia and cannabis use. Neuroscience 103:9–15

Degroot A, Kofalvi A, Wade MR, Davis RJ, Rodrigues RJ, Rebola N, Cunha RA, Nomikos GG (2006) CB1 receptor antagonism increases hippocampal acetylcholine release: site and mechanism of action. Mol Pharmacol 70:1236–1245

Demuth DG, Molleman A (2006) Cannabinoid signalling. Life Sci 78:549–563

Derkinderen P, Valjent E, Toutant M, Corvol JC, Enslen H, Ledent C, Trzaskos J, Caboche J, Girault JA (2003) Regulation of extracellular signal-regulated kinase by cannabinoids in hippocampus. J Neurosci 23:2371–2382

Di Marzo V, Hill MP, Bisogno T, Crossman AR, Brotchie JM (2000) Enhanced levels of endogenous cannabinoids in the globus pallidus are associated with a reduction in movement in an animal model of Parkinson's disease. FASEB J 14:1432–1438

Dinh TP, Carpenter D, Leslie FM, Freund TF, Katona I, Sensi SL, Kathuria S, Piomelli D (2002) Brain monoglyceride lipase participating in endocannabinoid inactivation. Proc Natl Acad Sci U S A 99:10819–10824

Dubreucq S, Matias I, Cardinal P, Haring M, Lutz B, Marsicano G, Chaouloff F (2012) Genetic dissection of the role of cannabinoid type-1 receptors in the emotional consequences of repeated social stress in mice. Neuropsychopharmacology 37:1885–1900

Egerton A, Allison C, Brett RR, Pratt JA (2006) Cannabinoids and prefrontal cortical function: insights from preclinical studies. Neurosci Biobehav Rev 30:680–695

Eggan SM, Hashimoto T, Lewis DA (2008) Reduced cortical cannabinoid 1 receptor messenger RNA and protein expression in schizophrenia. Arch Gen Psychiatry 65:772–784

Eggan SM, Stoyak SR, Verrico CD, Lewis DA (2010) Cannabinoid CB1 receptor immunoreactivity in the prefrontal cortex: comparison of schizophrenia and major depressive disorder. Neuropsychopharmacology 35:2060–2071

Ellis J, Pediani JD, Canals M, Milasta S, Milligan G (2006) Orexin-1 receptor-cannabinoid CB1 receptor heterodimerization results in both ligand-dependent and -independent coordinated alterations of receptor localization and function. J Biol Chem 281:38812–38824

Esteban S, Garcia-Sevilla JA (2012) Effects induced by cannabinoids on monoaminergic systems in the brain and their implications for psychiatric disorders. Prog Neuropsychopharmacol Biol Psychiatry 38:78–87

Fabbrini G, Brotchie JM, Grandas F, Nomoto M, Goetz CG (2007) Levodopa-induced dyskinesias. Mov Disord 22:1379–1389; quiz 1523

Fadda P, Scherma M, Spano MS, Salis P, Melis V, Fattore L, Fratta W (2006) Cannabinoid self-administration increases dopamine release in the nucleus accumbens. Neuroreport 17:1629–1632

Fattore L, Melis M, Fadda P, Pistis M, Fratta W (2010) The endocannabinoid system and nondrug rewarding behaviours. Exp Neurol 224:23–36

Fernandez-Espejo E, Caraballo I, Rodriguez de Fonseca F, El Banoua F, Ferrer B, Flores JA, Galan-Rodriguez B (2005) Cannabinoid CB1 antagonists possess antiparkinsonian efficacy only in rats with severe nigral lesion in experimental parkinsonism. Neurobiol Dis 18:591–601

Fernandez-Ruiz J (2009) The endocannabinoid system as a target for the treatment of motor dysfunction. Br J Pharmacol 156:1029–1040

Fernandez-Ruiz J, Gonzales S (2005) Cannabinoid control of motor function at the basal ganglia. Handb Exp Pharmacol 168:479–507

Ferreira SG, Teixeira FM, Garcao P, Agostinho P, Ledent C, Cortes L, Mackie K, Kofalvi A (2012) Presynaptic CB(1) cannabinoid receptors control frontocortical serotonin and glutamate release–species differences. Neurochem Int 61:219–226

Ferrer B, Asbrock N, Kathuria S, Piomelli D, Giuffrida A (2003) Effects of levodopa on endocannabinoid levels in rat basal ganglia: implications for the treatment of levodopa-induced dyskinesias. Eur J Neurosci 18:1607–1614

Fox SH, Henry B, Hill M, Crossman A, Brotchie J (2002) Stimulation of cannabinoid receptors reduces levodopa-induced dyskinesia in the MPTP-lesioned nonhuman primate model of Parkinson's disease. Mov Disord 17:1180–1187

Frankel JP, Hughes A, Lees AJ, Stern GM (1990) Marijuana for Parkinsonian tremor. J Neurol Neurosurg Psychiatry 53:436

French ED, Dillon K, Wu X (1997) Cannabinoids excite dopamine neurons in the ventral tegmentum and substantia nigra. Neuroreport 8:649–652

Freund TF, Katona I, Piomelli D (2003) Role of endogenous cannabinoids in synaptic signaling. Physiol Rev 83:1017–1066

Furmark T (2009) Neurobiological aspects of social anxiety disorder. Isr J Psychiatry Relat Sci 46:5–12

Galiègue S, Mary S, Marchand J, Dussossoy D, Carrière D, Carayon P, Bouaboula M, Shire D, Le Fur G, Casellas P (1995) Expression of central and peripheral cannabinoid receptors in human immune tissues and leukocyte subpopulations. Eur J Biochem 232:54–61

Galve-Roperh I, Rueda D, Gomez del Pulgar T, Velasco G, Guzman M (2002) Mechanism of extracellular signal-regulated kinase activation by the CB(1) cannabinoid receptor. Mol Pharmacol 62:1385–1392

Garcia C, Palomo-Garo C, Garcia-Arencibia M, Ramos J, Pertwee R, Fernandez-Ruiz J (2011) Symptom-relieving and neuroprotective effects of the phytocannabinoid Delta(9)-THCV in animal models of Parkinson's disease. Br J Pharmacol 163:1495–1506

Garcia-Arencibia M, Garcia C, Fernandez-Ruiz J (2009) Cannabinoids and Parkinson's disease. CNS Neurol Disord Drug Targets 8:432–439

Garcia-Arencibia M, Gonzalez S, Lago E de, Ramos JA, Mechoulam R, Fernandez-Ruiz J (2007) Evaluation of the neuroprotective effect of cannabinoids in a rat model of Parkinson's disease: importance of antioxidant and cannabinoid receptor-independent properties. Brain Res 1134:162–170

Garcia-Gutierrez MS, Perez-Ortiz JM, Gutierrez-Adan A, Manzanares J (2010) Depression-resistant endophenotype in mice overexpressing cannabinoid CB(2) receptors. Br J Pharmacol 160:1773–1784

Gerdeman G, Lovinger DM (2001) CB1 cannabinoid receptor inhibits synaptic release of glutamate in rat dorsolateral striatum. J Neurophysiol 85:468–471

Gessa GL, Casu MA, Carta G, Mascia MS (1998a) Cannabinoids decrease acetylcholine release in the medial-prefrontal cortex and hippochampus, reversal by SR 141716A. Eur J Pharmacol 355:119–124

Gessa GL, Mascia MS, Casu MA, Carta G (1997) Inhibition of hippocampal acetylcholine release by cannabinoids: reversal by SR 141716A. Eur J Pharmacol 327:R1–R2

Gessa GL, Melis M, Muntoni AL, Diana M (1998b) Cannabinoids activate mesolimbic dopamine neurons by an action on cannabinoid CB1 receptors. Eur J Pharmacol 341:39–44

Giuffrida A, Leweke FM, Gerth CW, Schreiber D, Koethe D, Faulhaber J, Klosterkotter J, Piomelli D (2004) Cerebrospinal anandamide levels are elevated in acute schizophrenia and are inversely correlated with psychotic symptoms. Neuropsychopharmacology 29:2108–2114

Giuffrida A, Parsons LH, Kerr TM, Rodriguez de Fonseca F, Navarro M, Piomelli D (1999) Dopamine activation of endogenous cannabinoid signaling in dorsal striatum. Nat Neurosci 2:358–363
Glass M, Dragunow M, Faull RL (1997) Cannabinoid receptors in the human brain: a detailed anatomical and quantitative autoradiographic study in the fetal, neonatal and adult human brain. Neuroscience 77:299–318
Glass M, Felder CC (1997) Concurrent stimulation of cannabinoid CB1 and dopamine D2 receptors augments cAMP accumulation in striatal neurons: evidence for a Gs linkage to the CB1 receptor. J Neurosci 17:5327–5333
González S, Romero J, Miguel R de, Lastres-Becker I, Villanua MA, Makriyannis A, Ramos JA, Fernández-Ruiz JJ (1999) Extrapyramidal and neuroendocrine effects of AM404, an inhibitor of the carrier-mediated transport of anandamide. Life Sci 65:327–336
Gonzalez S, Scorticati C, Garcia-Arencibia M, Miguel R de, Ramos JA, Fernandez-Ruiz J (2006) Effects of rimonabant, a selective cannabinoid CB1 receptor antagonist, in a rat model of Parkinson's disease. Brain Res 1073–1074:209–219
Gorriti MA, Rodriguez de Fonseca F, Navarro M, Palomo T (1999) Chronic (-)-delta9-tetrahydrocannabinol treatment induces sensitization to the psychomotor effects of amphetamine in rats. Eur J Pharmacol 365:133–142
Green B, Kavanagh D, Young R (2003) Being stoned: a review of self-reported cannabis effects. Drug Alcohol Rev 22:453–460
Gubellini P, Picconi B, Bari M, Battista N, Calabresi P, Centonze D, Bernardi G, Finazzi-Agro A, Maccarrone M (2002) Experimental parkinsonism alters endocannabinoid degradation: implications for striatal glutamatergic transmission. J Neurosci 22:6900–6907
Guidali C, Vigano D, Petrosino S, Zamberletti E, Realini N, Binelli G, Rubino T, Di Marzo V, Parolaro D (2011) Cannabinoid CB1 receptor antagonism prevents neurochemical and behavioural deficits induced by chronic phencyclidine. Int J Neuropsychopharmacol 14:17–28
Haller J, Varga B, Ledent C, Barna I, Freund TF (2004) Context-dependent effects of CB1 cannabinoid gene disruption on anxiety-like and social behaviour in mice. Eur J Neurosci 19:1906–1912
Haring M, Guggenhuber S, Lutz B (2012) Neuronal populations mediating the effects of endocannabinoids on stress and emotionality. Neuroscience 204:145–158
Haring M, Marsicano G, Lutz B, Monory K (2007) Identification of the cannabinoid receptor type 1 in serotonergic cells of raphe nuclei in mice. Neuroscience 146:1212–1219
Hayase T, Yamamoto Y, Yamamoto K (2005) Persistent anxiogenic effects of a single or repeated doses of cocaine and methamphetamine: interactions with endogenous cannabinoid receptor ligands. Behav Pharmacol 16:395–404
Heifets BD, Castillo PE (2009) Endocannabinoid signaling and long-term synaptic plasticity. Annu Rev Physiol 71:283–306
Henquet C, Murray R, Linszen D, Os J van (2005) The environment and schizophrenia: the role of cannabis use. Schizophr Bull 31:608–612
Henstridge CM, Balenga NA, Kargl J, Andradas C, Brown AJ, Irving A, Sanchez C, Waldhoer M (2011) Minireview: recent developments in the physiology and pathology of the lysophosphatidylinositol-sensitive receptor GPR55. Mol Endocrinol 25:1835–1848
Herkenham M, Lynn AB, Costa BR de, Richfield EK (1991a) Neuronal localization of cannabinoid receptors in the basal ganglia of the rat. Brain Res 547:267–274
Herkenham M, Lynn AB, Johnson MR, Melvin LS, Costa BR de, Rice KC (1991b) Characterization and localization of cannabinoid receptors in rat brain: a quantitative in vitro autoradiographic study. J Neurosci 11:563–583
Herkenham M, Lynn AB, Little MD, Johnson MR, Melvin LS, Costa BR de, Rice KC (1990) Cannabinoid receptor localization in brain. Proc Natl Acad Sci U S A 87:1932–1936
Hermann H, Marsicano G, Lutz B (2002) Coexpression of the cannabinoid receptor type 1 with dopamine and serotonin receptors in distinct neuronal subpopulations of the adult mouse forebrain. Neuroscience 109:451–460

Hernandez-Tristan R, Arevalo C, Canals S, Leret ML (2000) The effects of acute treatment with delta9-THC on exploratory behaviour and memory in the rat. J Physiol Biochem 56:17–24
Hojo M, Sudo Y, Ando Y, Minami K, Takada M, Matsubara T, Kanaide M, Taniyama K, Sumikawa K, Uezono Y (2008) mu-Opioid receptor forms a functional heterodimer with cannabinoid CB1 receptor: electrophysiological and FRET assay analysis. J Pharmacol Sci 108:308–319
Howlett AC (2002) The cannabinoid receptors. Prostaglandins Other Lipid Mediat 68–69:619–631
Howlett AC (2004) Efficacy in CB1 receptor-mediated signal transduction. Br J Pharmacol 142:1209–1218
Hudson BD, Hebert TE, Kelly ME (2010a) Ligand- and heterodimer-directed signaling of the CB(1) cannabinoid receptor. Mol Pharmacol 77:1–9
Hudson BD, Hebert TE, Kelly ME (2010b) Physical and functional interaction between CB1 cannabinoid receptors and beta2-adrenoceptors. Br J Pharmacol 160:627–642
Huot P, Johnston TH, Koprich JB, Fox SH, Brotchie JM (2013) The pharmacology of L-dopa-induced dyskinesia in Parkinson's disease. Pharmacol Rev 65:171–222
Hurley MJ, Mash DC, Jenner P (2003) Expression of cannabinoid CB1 receptor mRNA in basal ganglia of normal and parkinsonian human brain. J Neural Transm 110:1279–1288
Iversen L (2005) Long-term effects of exposure to cannabis. Curr Opin Pharmacol 5:69–72
Jarrahian A, Watts VJ, Barker EL (2004) D2 dopamine receptors modulate Galpha-subunit coupling of the CB1 cannabinoid receptor. J Pharmacol Exp Ther 308:880–886
Jentsch JD, Elsworth JD, Taylor JR, Redmond DE Jr, Roth RH (1998) Dysregulation of mesoprefrontal dopamine neurons induced by acute and repeated phencyclidine administration in the nonhuman primate: implications for schizophrenia. Adv Pharmacol 42:810–814
Jentsch JD, Tran A, Le D, Youngren KD, Roth RH (1997) Subchronic phencyclidine administration reduces mesoprefrontal dopamine utilization and impairs prefrontal cortical-dependent cognition in the rat. Neuropsychopharmacology 17:92–99
Julian MD, Martin AB, Cuellar B, Rodriguez De Fonseca F, Navarro M, Moratalla R, Garcia-Segura LM (2003) Neuroanatomical relationship between type 1 cannabinoid receptors and dopaminergic systems in the rat basal ganglia. Neuroscience 119:309–318
Kalant H (2004) Adverse effects of cannabis on health: an update of the literature since 1996. Prog Neuropsychopharmacol Biol Psychiatry 28:849–863
Kano M, Ohno-Shosaku T, Hashimotodani Y, Uchigashima M, Watanabe M (2009) Endocannabinoid-mediated control of synaptic transmission. Physiol Rev 89:309–380
Kathmann M, Bauer U, Schlicker E, Gothert M (1999) Cannabinoid CB1 receptor-mediated inhibition of NMDA- and kainate-stimulated noradrenaline and dopamine release in the brain. Naunyn Schmiedebergs Arch Pharmacol 359:466–470
Katona I, Freund TF (2012) Multiple functions of endocannabinoid signaling in the brain. Ann Rev Neurosci 35:529-58
Katona I, Sperlagh B, Sik A, Kafalvi A, Vizi ES, Mackie K, Freund TF (1999) Presynaptically located CB1 cannabinoid receptors regulate GABA release from axon terminals of specific hippocampal interneurons. J Neurosci 19:4544–4558
Kearn CS, Blake-Palmer K, Daniel E, Mackie K, Glass M (2005) Concurrent stimulation of cannabinoid CB1 and dopamine D2 receptors enhances heterodimer formation: a mechanism for receptor cross-talk? Mol Pharmacol 67:1697–1704
Kelsey JE, Harris O, Cassin J (2009) The CB(1) antagonist rimonabant is adjunctively therapeutic as well as monotherapeutic in an animal model of Parkinson's disease. Behav Brain Res 203:304–307
Kleijn J, Wiskerke J, Cremers TI, Schoffelmeer AN, Westerink BH, Pattij T (2012) Effects of amphetamine on dopamine release in the rat nucleus accumbens shell region depend on cannabinoid CB1 receptor activation. Neurochem Int 60:791–798
Koethe D, Giuffrida A, Schreiber D, Hellmich M, Schultze-Lutter F, Ruhrmann S, Klosterkotter J, Piomelli D, Leweke FM (2009a) Anandamide elevation in cerebrospinal fluid in initial prodromal states of psychosis. Br J Psychiatry 194:371–372
Koethe D, Hoyer C, Leweke FM (2009b) The endocannabinoid system as a target for modelling psychosis. Psychopharmacology (Berl) 206:551–561

Koethe D, Llenos IC, Dulay JR, Hoyer C, Torrey EF, Leweke FM, Weis S (2007) Expression of CB1 cannabinoid receptor in the anterior cingulate cortex in schizophrenia, bipolar disorder, and major depression. J Neural Transm 114:1055–1063

Köfalvi A, Rodrigues RJ, Ledent C, Mackie K, Vizi ES, Cuhna RA, Sperlagh B (2005) Involvement of cannabinoid receptors in the regulation of neurotransmitter release in the rodent striatum: a combined immunochemical and pharmacological analysis. J Neurosci 25:2874–2884

Kortleven C, Fasano C, Thibault D, Lacaille JC, Trudeau LE (2011) The endocannabinoid 2-arachidonoylglycerol inhibits long-term potentiation of glutamatergic synapses onto ventral tegmental area dopamine neurons in mice. Eur J Neurosci 33:1751–1760

Kreitzer AC, Malenka RC (2007) Endocannabinoid-mediated rescue of striatal LTD and motor deficits in Parkinson's disease models. Nature 445:643–647

Laruelle M, Kegeles LS, Abi-Dargham A (2003a) Glutamate, dopamine, and schizophrenia: from pathophysiology to treatment. Ann N Y Acad Sci 1003:138–158

Laruelle M, Kegeles LS, Abi-Dargham A (2003b) Glutamate, dopamine, and schizophrenia: from pathophysiology to treatment. Ann N Y Acad Sci 1003:138–158

Lastres-Becker I, Cebeira M, Ceballos ML de, Zeng BY, Jenner P, Ramos JA, Fernandez-Ruiz JJ (2001) Increased cannabinoid CB1 receptor binding and activation of GTP-binding proteins in the basal ganglia of patients with Parkinson's syndrome and of MPTP-treated marmosets. Eur J Neurosci 14:1827–1832

Lastres-Becker I, Molina-Holgado E, Ramos JA, Mechoulam R, Fernandez-Ruiz J (2005) Cannabinoids provide neuroprotection against 6-hydroxydopamine toxicity in vivo and in vitro: relevance to Parkinson's disease. Neurobiol Dis 19:96–107

Lau T, Schloss P (2008) The cannabinoid CB1 receptor is expressed on serotonergic and dopaminergic neurons. Eur J Pharmacol 578:137–141

Lauckner JE, Hille B, Mackie K (2005) The cannabinoid agonist WIN55,212–2 increases intracellular calcium via CB1 receptor coupling to Gq/11 G proteins. Proc Natl Acad Sci U S A 102:19144–19149

Laviolette SR, Grace AA (2006) The roles of cannabinoid and dopamine receptor systems in neural emotional learning circuits: implications for schizophrenia and addiction. Cell Mol Life Sci 63:1597–1613

Ledent C, Valverde O, Cossu G, Petitet F, Aubert JF, Beslot F, Bohme GA, Imperato A, Pedrazzini T, Roques BP, Vassart G, Fratta W, Parmentier M (1999) Unresponsiveness to cannabinoids and reduced addictive effects of opiates in CB1 receptor knockout mice. Science 283:401–404

Leweke FM, Giuffrida A, Koethe D, Schreiber D, Nolden BM, Kranaster L, Neatby MA, Schneider M, Gerth CW, Hellmich M, Klosterkotter J, Piomelli D (2007) Anandamide levels in cerebrospinal fluid of first-episode schizophrenic patients: impact of cannabis use. Schizophr Res 94:29–36

Leweke FM, Giuffrida A, Wurster U, Emrich HM, Piomelli D (1999) Elevated endogenous cannabinoids in schizophrenia. Neuroreport 10:1665–1669

Li X, Hoffman AF, Peng XQ, Lupica CR, Gardner EL, Xi ZX (2009) Attenuation of basal and cocaine-enhanced locomotion and nucleus accumbens dopamine in cannabinoid CB1-receptor-knockout mice. Psychopharmacology (Berl) 204:1–11

Lopez-Moreno JA, Gonzalez-Cuevas G, Moreno G, Navarro M (2008) The pharmacology of the endocannabinoid system: functional and structural interactions with other neurotransmitter systems and their repercussions in behavioral addiction. Addict Biol 13:160–187

Lovinger DM, Mathur BN (2012) Endocannabinoids in striatal plasticity. Parkinsonism Relat Disord 18(Suppl 1):S132–134

Maccarrone M, Gubellini P, Bari M, Picconi B, Battista N, Centonze D, Bernardi G, Finazzi-Agro A, Calabresi P (2003) Levodopa treatment reverses endocannabinoid system abnormalities in experimental parkinsonism. J Neurochem 85:1018–1025

Mackie K (2005a) Cannabinoid receptor homo- and heterodimerization. Life Sci 77:1667–1673

Mackie K (2005b) Distribution of cannabinoid receptors in the central and peripheral nervous system. Handb Exp Pharmacol 299–325

Mackie K, Lai Y, Westenbroek R, Mitchell R (1995) Cannabinoids activate an inwardly rectifying potassium conductance and inhibit Q-type calcium currents in AtT20 cells transfected with rat brain cannabinoid receptor. J Neurosci 15:6552–6561

Mailleux P, Parmentier M, Vanderhaeghen JJ (1992) Distribution of cannabinoid receptor messenger RNA in the human brain: an in situ hybridization histochemistry with oligonucleotides. Neurosci Lett 143:200–204

Mailleux P, Vanderhaeghen JJ (1992a) Distribution of neuronal cannabinoid receptor in the adult rat brain: a comparative receptor binding radioautography and in situ hybridization histochemistry. Neuroscience 48:655–668

Mailleux P, Vanderhaeghen JJ (1992b) Localization of cannabinoid receptor in the human developing and adult basal ganglia. Higher levels in the striatonigral neurons. Neurosci Lett 148:173–176

Mailleux P, Vanderhaeghen JJ (1993) Dopaminergic regulation of cannabinoid receptor mRNA levels in the rat caudate-putamen: an in situ hybridization study. J Neurochem 61:1705–1712

Malone DT, Taylor DA (1999) Modulation by fluoxetine of striatal dopamine release following Delta9-tetrahydrocannabinol: a microdialysis study in conscious rats. Br J Pharmacol 128:21–26

Maneuf YP, Brotchie JM (1997) Paradoxical action of the cannabinoid WIN 55,212–2 in stimulated and basal cyclic AMP accumulation in rat globus pallidus slices. Br J Pharmacol 120:1397–1398

Maneuf YP, Crossman AR, Brotchie JM (1997) The cannabinoid receptor agonist WIN 55,212–2 reduces D2, but not D1, dopamine receptor-mediated alleviation of akinesia in the reserpine-treated rat model of Parkinson's disease. Exp Neurol 148:265–270

Manzoni OJ, Bockaert J (2001) Cannabinoids inhibit GABAergic synaptic transmission in mice nucleus accumbens. Eur J Pharmacol 412:R3–R5

Marcellino D, Carriba P, Filip M, Borgkvist A, Frankowska M, Bellido I, Tanganelli S, Muller CE, Fisone G, Lluis C, Agnati LF, Franco R, Fuxe K (2008) Antagonistic cannabinoid CB1/dopamine D2 receptor interactions in striatal CB1/D2 heteromers. A combined neurochemical and behavioral analysis. Neuropharmacology 54:815–823

Marinelli S, Di Marzo V, Berretta N, Matias I, Maccarrone M, Bernardi G, Mercuri NB (2003) Presynaptic facilitation of glutamatergic synapses to dopaminergic neurons of the rat substantia nigra by endogenous stimulation of vanilloid receptors. J Neurosci 23:3136–3144

Marinelli S, Di Marzo V, Florenzano F, Fezza F, Viscomi MT, Stelt M van der, Bernardi G, Molinari M, Maccarrone M, Mercuri NB (2007) N-arachidonoyl-dopamine tunes synaptic transmission onto dopaminergic neurons by activating both cannabinoid and vanilloid receptors. Neuropsychopharmacology 32:298–308

Marsicano G, Lafenetre P (2009) Roles of the endocannabinoid system in learning and memory. Curr Top Behav Neurosci 1:201–230

Marsicano G, Moosmann B, Hermann H, Lutz B, Behl C (2002) Neuroprotective properties of cannabinoids against oxidative stress: role of the cannabinoid receptor CB1. J Neurochem 80:448–456

Martin AB, Fernandez-Espejo E, Ferrer B, Gorriti MA, Bilbao A, Navarro M, Rodriguez de Fonseca F, Moratalla R (2008) Expression and function of CB1 receptor in the rat striatum: localization and effects on D1 and D2 dopamine receptor-mediated motor behaviors. Neuropsychopharmacology 33:1667–1679

Martinez A, Macheda T, Morgese MG, Trabace L, Giuffrida A (2012) The cannabinoid agonist WIN55212–2 decreases L-DOPA-induced PKA activation and dyskinetic behavior in 6-OHDA-treated rats. Neurosci Res 72:236–242

Mathur BN, Lovinger DM (2012) Endocannabinoid-dopamine interactions in striatal synaptic plasticity. Front Pharmacol 3:66

Matias I, Di Marzo V (2007) Endocannabinoids and the control of energy balance. Trends Endocrinol Metab 18:27–37

Matsuda LA, Bonner TI, Lolait SJ (1993) Localization of cannabinoid receptor mRNA in rat brain. J Comp Neurol 327:535–550

Matyas F, Yanovsky Y, Mackie K, Kelsch W, Misgeld U, Freund TF (2006) Subcellular localization of type 1 cannabinoid receptors in the rat basal ganglia. Neuroscience 137:337–361

McIntosh BT, Hudson B, Yegorova S, Jollimore CA, Kelly ME (2007) Agonist-dependent cannabinoid receptor signalling in human trabecular meshwork cells. Br J Pharmacol 152:1111–1120

McLaughlin RJ, Gobbi G (2012) Cannabinoids and emotionality: a neuroanatomical perspective. Neuroscience 204:134–144

McPartland JM, Glass M, Pertwee RG (2007) Meta-analysis of cannabinoid ligand binding affinity and receptor distribution: interspecies differences. Br J Pharmacol 152:583–593

Meador-Woodruff JH, Haroutunian V, Powchik P, Davidson M, Davis KL, Watson SJ (1997) Dopamine receptor transcript expression in striatum and prefrontal and occipital cortex. Focal abnormalities in orbitofrontal cortex in schizophrenia. Arch Gen Psychiatry 54:1089–1095

Meijerink J, Balvers M, Witkamp R (2012) N-acyl amines of docosahexaenoic acid and other n-3 polyunsatured fatty acids—From fishy endocannabinoids to potential leads. Br J Pharmacol 169(4):772–83 (2013 Jun)

Melis M, Gessa GL, Diana M (2000) Different mechanisms for dopaminergic excitation induced by opiates and cannabinoids in the rat midbrain. Prog Neuro-Psychopharmacol Biol Psychiatry 24:993–1006

Melis M, Perra S, Muntoni AL, Pillolla G, Lutz B, Marsicano G, Di Marzo V, Gessa GL, Pistis M (2004a) Prefrontal cortex stimulation induces 2-arachidonoyl-glycerol-mediated suppression of excitation in dopamine neurons. J Neurosci 24:10707–10715

Melis M, Pistis M, Perra S, Muntoni AL, Pillola G, Gessa GL (2004b) Endocannabinoids mediate presynaptic inhibition of glutamatergic transmission in rat ventral tegmental area dopamine neurons through activation of CB1 receptors. J Neurosci 24:53–62

Melis M, Pistis P (2007) Endocannabinoid signaling in midbrain dopamine neurons: more than physiology? Curr Neuropharmacol 5:268–277

Meltzer HY, Arvanitis L, Bauer D, Rein W (2004) Placebo-controlled evaluation of four novel compounds for the treatment of schizophrenia and schizoaffective disorder. Am J Psychiatry 161:975–984

Meschler JP, Howlett AC (2001) Signal transduction interactions between CB1 cannabinoid and dopamine receptors in the rat and monkey striatum. Neuropharmacology 40:918–926

Meschler JP, Howlett AC, Madras BK (2001) Cannabinoid receptor agonist and antagonist effects on motor function in normal and 1-methyl-4-phenyl-1,2,5,6-tetrahydropyridine (MPTP)-treated non-human primates. Psychopharmacology (Berl) 156:79–85

Mesnage V, Houeto JL, Bonnet AM, Clavier I, Arnulf I, Cattelin F, Le Fur G, Damier P, Welter ML, Agid Y (2004) Neurokinin B, neurotensin, and cannabinoid receptor antagonists and Parkinson disease. Clin Neuropharmacol 27:108–110

Missale C, Nash SR, Robinson SW, Jaber M, Caron MG (1998) Dopamine receptors: from structure to function. Physiol Rev 78:189–225

Moldrich G, Wenger T (2000) Localization of the CB1 cannabinoid receptor in the rat brain. An immunohistochemical study. Peptides 21:1735–1742

Moranta D, Esteban S, Garcia-Sevilla JA (2004) Differential effects of acute cannabinoid drug treatment, mediated by CB1 receptors, on the in vivo activity of tyrosine and tryptophan hydroxylase in the rat brain. Naunyn Schmiedebergs Arch Pharmacol 369:516–524

Moranta D, Esteban S, Garcia-Sevilla JA (2009) Chronic treatment and withdrawal of the cannabinoid agonist WIN 55,212–2 modulate the sensitivity of presynaptic receptors involved in the regulation of monoamine syntheses in rat brain. Naunyn Schmiedebergs Arch Pharmacol 379:61–72

Morera-Herreras T, Miguelez C, Aristieta A, Ruiz-Ortega JA, Ugedo L (2012) Endocannabinoid modulation of dopaminergic motor circuits. Front Pharmacol 3:110

Morgese MG, Cassano T, Cuomo V, Giuffrida A (2007) Anti-dyskinetic effects of cannabinoids in a rat model of Parkinson's disease: role of CB(1) and TRPV1 receptors. Exp Neurol 208:110–119

Muller-Vahl KR, Emrich HM (2008) Cannabis and schizophrenia: towards a cannabinoid hypothesis of schizophrenia. Expert Rev Neurother 8:1037–1048

Munro S, Thomas KL, Abu-Shaar M (1993) Molecular characterization of a peripheral receptor for cannabinoids. Nature 365:61–65

Nakazi M, Bauer U, Nickel T, Kathmann M, Schlicker E (2000) Inhibition of serotonin release in the mouse brain via presynaptic cannabinoid CB1 receptors. Naunyn Schmiedebergs Arch Pharmacol 361:19–24

Nasehi M, Sahebgharani M, Haeri-Rohani A, Zarrindast MR (2009) Effects of cannabinoids infused into the dorsal hippocampus upon memory formation in 3-days apomorphine-treated rats. Neurobiol Learn Mem 92:391–399

Nava F, Carta G, Battasi AM, Gessa GL (2000) D(2) dopamine receptors enable delta(9)-tetrahydrocannabinol induced memory impairment and reduction of hippocampal extracellular acetylcholine concentration. Br J Pharmacol 130:1201–1210

Navarro G, Carriba P, Gandia J, Ciruela F, Casado V, Cortes A, Mallol J, Canela EI, Lluis C, Franco R (2008) Detection of heteromers formed by cannabinoid CB1, dopamine D2, and adenosine A2A G-protein-coupled receptors by combining bimolecular fluorescence complementation and bioluminescence energy transfer. ScientificWorldJournal 8:1088–1097

Newell KA, Deng C, Huang XF (2006) Increased cannabinoid receptor density in the posterior cingulate cortex in schizophrenia. Exp Brain Res 172:556–560

Nicholson RA, Liao C, Zheng J, David LS, Coyne L, Errington AC, Singh G, Lees G (2003) Sodium channel inhibition by anandamide and synthetic cannabimimetics in brain. Brain Res 978:194–204

Nissen SE, Nicholls SJ, Wolski K, Rodes-Cabau J, Cannon CP, Deanfield JE, Despres JP, Kastelein JJ, Steinhubl SR, Kapadia S, Yasin M, Ruzyllo W, Gaudin C, Job B, Hu B, Bhatt DL, Lincoff AM, Tuzcu EM (2008) Effect of rimonabant on progression of atherosclerosis in patients with abdominal obesity and coronary artery disease: the STRADIVARIUS randomized controlled trial. JAMA 299:1547–1560

Nunez E, Benito C, Pazos MR, Barbachano A, Fajardo O, Gonzalez S, Tolon RM, Romero J (2004) Cannabinoid CB2 receptors are expressed by perivascular microglial cells in the human brain: an immunohistochemical study. Synapse 53:208–213

Obeso JA, Rodriguez-Oroz MC, Benitez-Temino B, Blesa FJ, Guridi J, Marin C, Rodriguez M (2008) Functional organization of the basal ganglia: therapeutic implications for Parkinson's disease. Mov Disord 23(Suppl 3):S548–S559

Onaivi ES (2006) Neuropsychobiological evidence for the functional presence and expression of cannabinoid CB2 receptors in the brain. NeuropsychoBiology 54:231–246

Onaivi ES, Ishiguro H, Gong JP, Patel S, Perchuk A, Meozzi PA, Myers L, Mora Z, Tagliaferro P, Gardner E, Brusco A, Akinshola BE, Liu QR, Hope B, Iwasaki S, Arinami T, Teasenfitz L, Uhl GR (2006) Discovery of the presence and functional expression of cannabinoid CB2 receptors in brain. Ann N Y Acad Sci 1074:514–536

Onaivi ES, Ishiguro H, Gu S, Liu QR (2011) CNS effects of CB2 cannabinoid receptors: beyond neuro-immuno-cannabinoid activity. J Psychopharmacol 26(1):92-103

Oropeza VC, Mackie K, Van Bockstaele EJ (2007) Cannabinoid receptors are localized to noradrenergic axon terminals in the rat frontal cortex. Brain Res 1127:36–44

Pan B, Hillard CJ, Liu QS (2008) D2 dopamine receptor activation facilitates endocannabinoid-mediated long-term synaptic depression of GABAergic synaptic transmission in midbrain dopamine neurons via cAMP-protein kinase A signaling. J Neurosci 28:14018–14030

Pan X, Ikeda SR, Lewis DL (1996) Rat brain cannabinoid receptor modulates N-type Ca2+ channels in a neuronal expression system. Mol Pharmacol 49:707–714

Pandey R, Mousawy K, Nagarkatti M, Nagarkatti P (2009) Endocannabinoids and immune regulation. Pharmacol Res 60:85–92

Parolaro D, Rubino T (2008) The role of the endogenous cannabinoid system in drug addiction. Drug News Perspect 21:149–157

Patel KD, Davison JS, Pittman QJ, Sharkey KA (2010) Cannabinoid CB(2) receptors in health and disease. Curr Med Chem 17:1393–1410

Patel S, Rademacher DJ, Hillard CJ (2003) Differential regulation of the endocannabinoids anandamide and 2-arachidonylglycerol within the limbic forebrain by dopamine receptor activity. J Pharmacol Exp Ther 306:880–888

Pazos MR, Nunez E, Benito C, Tolon RM, Romero J (2005) Functional neuroanatomy of the endocannabinoid system. Pharmacol Biochem Behav 81:239–247

Perez-Rial S, Garcia-Gutierrez MS, Molina JA, Perez-Nievas BG, Ledent C, Leiva C, Leza JC, Manzanares J (2011) Increased vulnerability to 6-hydroxydopamine lesion and reduced development of dyskinesias in mice lacking CB1 cannabinoid receptors. Neurobiol Aging 32:631–645

Pertwee RG (2006) The pharmacology of cannabinoid receptors and their ligands: an overview. Int J Obes (Lond) 30(Suppl 1):S13–18

Pickel VM, Chan J, Kearn CS, Mackie K (2006) Targeting dopamine D2 and cannabinoid-1 (CB1) receptors in rat nucleus accumbens. J Comp Neurol 495:299–313

Piomelli D (2003) The molecular logic of endocannabinoid signalling. Nat Rev Neurosci 4:873–884

Pisani A, Fezza F, Galati S, Battista N, Napolitano S, Finazzi-Agro A, Bernardi G, Brusa L, Pierantozzi M, Stanzione P, Maccarrone M (2005) High endogenous cannabinoid levels in the cerebrospinal fluid of untreated Parkinson's disease patients. Ann Neurol 57:777–779

Pisani V, Madeo G, Tassone A, Sciamanna G, Maccarrone M, Stanzione P, Pisani A (2011) Homeostatic changes of the endocannabinoid system in Parkinson's disease. Mov Disord 26:216–222

Pistis M, Ferraro L, Pira L, Flore G, Tanganelli S, Gessa GL, Devoto P (2002) Delta(9)-tetrahydrocannabinol decreases extracellular GABA and increases extracellular glutamate and dopamine levels in the rat prefrontal cortex: an in vivo microdialysis study. Brain Res 948:155–158

Pistis M, Porcu G, Melis M, Diana M, Gessa GL (2001) Effects of cannabinoids on prefrontal neuronal responses to ventral tegmental area stimulation. Eur J Neurosci 14:96–102

Pivonello R, Ferone D, Lombardi G, Colao A, Lamberts SW, Hofland LJ (2007) Novel insights in dopamine receptor physiology. Eur J Endocrinol 156(Suppl 1):S13–S21

Polissidis A, Galanopoulos A, Naxakis G, Papahatjis D, Papadopoulou-Daifoti Z, Antoniou K (2012) The cannabinoid CB1 receptor biphasically modulates motor activity and regulates dopamine and glutamate release region dependently. Int J Neuropsychopharmacol 1–11

Prescott WR, Gold LH, Martin BR (1992) Evidence for separate neuronal mechanisms for the discriminative stimulus and catalepsy induced by delta 9-THC in the rat. Psychopharmacology (Berl) 107:117–124

Price DA, Martinez AA, Seillier A, Koek W, Acosta Y, Fernandez E, Strong R, Lutz B, Marsicano G, Roberts JL, Giuffrida A (2009) WIN55,212–2, a cannabinoid receptor agonist, protects against nigrostriatal cell loss in the 1-methyl-4-phenyl-1,2,3,6-tetrahydropyridine mouse model of Parkinson's disease. Eur J Neurosci 29:2177–2186

Ranganathan M, D'Souza DC (2006) The acute effects of cannabinoids on memory in humans: a review. Psychopharmacology (Berl) 188:425–444

Reilly D, Didcott P, Swift W, Hall W (1998) Long-term cannabis use: characteristics of users in an Australian rural area. Addiction 93:837–846

Rhee MH, Nevo I, Avidor-Reiss T, Levy R, Vogel Z (2000) Differential superactivation of adenylyl cyclase isozymes after chronic activation of the CB(1) cannabinoid receptor. Mol Pharmacol 57:746–752

Riegel AC, Lupica CR (2004) Independent presynaptic and postsynaptic mechanisms regulate endocannabinoid signaling at multiple synapses in the ventral tegmental area. J Neurosci 24:11070–11078

Rios C, Gomes I, Devi LA (2006) mu opioid and CB1 cannabinoid receptor interactions: reciprocal inhibition of receptor signaling and neuritogenesis. Br J Pharmacol 148:387–395

Robbe D, Alonso G, Duchamp F, Bockaert J, Manzoni OJ (2001) Localization and mechanisms of action of cannabinoid receptors at the glutamatergic synapses of the mouse nucleus accumbens. J Neurosci 21:109–116

Robinson HM, Hood SD, Bell CJ, Nutt DJ (2006) Dopamine and social anxiety disorder. Rev Bras Psiquiatr 28:263–264

Rodriguez JJ, Mackie K, Pickel VM (2001) Ultrastructural localization of the CB1 cannabinoid receptor in mu-opioid receptor patches of the rat Caudate putamen nucleus. J Neurosci 21:823–833

Romero J, Berrendero F, Perez-Rosado A, Manzanares J, Rojo A, Fernandez-Ruiz JJ, Yebenes JG de, Ramos JA (2000) Unilateral 6-hydroxydopamine lesions of nigrostriatal dopaminergic neurons increased CB1 receptor mRNA levels in the caudate-putamen. Life Sci 66:485–494

Rubino T, Realini N, Castiglioni C, Guidali C, Vigano D, Marras E, Petrosino S, Perletti G, Maccarrone M, Di Marzo V, Parolaro D (2008) Role in anxiety behavior of the endocannabinoid system in the prefrontal cortex. Cereb Cortex 18:1292–1301

Rueda D, Galve-Roperh I, Haro A, Guzman M (2000) The CB(1) cannabinoid receptor is coupled to the activation of c-Jun N-terminal kinase. Mol Pharmacol 58:814–820

Ruehle S, Rey AA, Remmers F, Lutz B (2012) The endocannabinoid system in anxiety, fear memory and habituation. J Psychopharmacol 26:23–39

San L, Bernardo M, Gomez A, Pena M (2012) Factors associated with relapse in patients with schizophrenia. Int J Psychiatry Clin Pract 17(1):2-9

Sanudo-Pena MC, Patrick SL, Khen S, Patrick RL, Tsou K, Walker JM (1998) Cannabinoid effects in basal ganglia in a rat model of Parkinson's disease. Neurosci Lett 248:171–174

Schatz AR, Lee M, Condie RB, Pulaski JT, Kaminski NE (1997) Cannabinoid receptors CB1 and CB2: a characterization of expression and adenylate cyclase modulation within the immune system. Toxicol Appl Pharmacol 142:278–287

Schneier FR, Abi-Dargham A, Martinez D, Slifstein M, Hwang DR, Liebowitz MR, Laruelle M (2009) Dopamine transporters, D2 receptors, and dopamine release in generalized social anxiety disorder. Depress Anxiety 26:411–418

Schneier FR, Liebowitz MR, Abi-Dargham A, Zea-Ponce Y, Lin SH, Laruelle M (2000) Low dopamine D(2) receptor binding potential in social phobia. Am J Psychiatry 157:457–459

Schneier FR, Martinez D, Abi-Dargham A, Zea-Ponce Y, Simpson HB, Liebowitz MR, Laruelle M (2008) Striatal dopamine D(2) receptor availability in OCD with and without comorbid social anxiety disorder: preliminary findings. Depress Anxiety 25:1–7

Seillier A, Advani T, Cassano T, Hensler JG, Giuffrida A (2010) Inhibition of fatty-acid amide hydrolase and CB1 receptor antagonism differentially affect behavioural responses in normal and PCP-treated rats. Int J Neuropsychopharmacol 13:373–386

Sewell RA, Ranganathan M, D'Souza DC (2009) Cannabinoids and psychosis. Int Rev Psychiatry 21:152–162

Sidlo Z, Reggio PH, Rice ME (2008) Inhibition of striatal dopamine release by CB1 receptor activation requires nonsynaptic communication involving GABA, H2O2, and KATP channels. Neurochem Int 52:80–88

Sieradzan KA, Fox SH, Hill M, Dick JP, Crossman AR, Brotchie JM (2001) Cannabinoids reduce levodopa-induced dyskinesia in Parkinson's disease: a pilot study. Neurology 57:2108–2111

Silverdale MA, McGuire S, McInnes A, Crossman AR, Brotchie JM (2001) Striatal cannabinoid CB1 receptor mRNA expression is decreased in the reserpine-treated rat model of Parkinson's disease. Exp Neurol 169:400–406

Solinas M, Justinova Z, Goldberg SR, Tanda G (2006) Anandamide administration alone and after inhibition of fatty acid amide hydrolase (FAAH) increases dopamine levels in the nucleus accumbens shell in rats. J Neurochem 98:408–419

Spano MS, Fadda P, Frau R, Fattore L, Fratta W (2010) Cannabinoid self-administration attenuates PCP-induced schizophrenia-like symptoms in adult rats. Eur Neuropsychopharmacol 20:25–36

Stefanis NC, Bresnick JN, Kerwin RW, Schofield WN, McAllister G (1998) Elevation of D4 dopamine receptor mRNA in postmortem schizophrenic brain. Brain Res Mol Brain Res 53:112–119

Steffens M, Engler C, Zentner J, Feuerstein TJ (2004) Cannabinoid CB1 receptor-mediated modulation of evoked dopamine release and of adenylyl cyclase activity in the human neocortex. Br J Pharmacol 141:1193–1203

Stella N (2010) Cannabinoid and cannabinoid-like receptors in microglia, astrocytes, and astrocytomas. Glia 58:1017–1030

Suarez J, Llorente R, Romero-Zerbo SY, Mateos B, Bermudez-Silva FJ, Fonseca FR de, Viveros MP (2009) Early maternal deprivation induces gender-dependent changes on the expression of hippocampal CB(1) and CB(2) cannabinoid receptors of neonatal rats. Hippocampus 19:623–632

Szabo B, Dorner L, Pfreundtner C, Norenberg W, Starke K (1998) Inhibition of GABAergic inhibitory postsynaptic currents by cannabinoids in rat corpus striatum. Neuroscience 85:395–403

Szabo B, Muller T, Koch H (1999) Effects of cannabinoids on dopamine release in the corpus striatum and the nucleus accumbens in vitro. J Neurochem 73:1084–1089

Szabo B, Siemes S, Wallmichrath I (2002) Inhibition of GABAergic neurotransmission in the ventral tegmental area by cannabinoids. Eur J Neurosci 15:2057–2061

Szabo B, Wallmichrath I, Mathonia P, Pfreundtner C (2000) Cannabinoids inhibit excitatory neurotransmission in the substantia nigra pars reticulata. Neuroscience 97:89–97

Szuster RR, Pontius EB, Campos PE (1988) Marijuana sensitivity and panic anxiety. J Clin Psychiatry 49:427–429

Tambaro S, Bortolato M (2012) Cannabinoid-related agents in the treatment of anxiety disorders: current knowledge and future perspectives. Recent Pat CNS Drug Discov 7:25–40

Tanda G, Pontieri FE, Di Chiara G (1997) Cannabinoid and heroin activation of mesolimbic dopamine transmission by a common mu1 opioid receptor mechanism. Science 276:2048–2050

Terzian AL, Drago F, Wotjak CT, Micale V (2011) The dopamine and cannabinoid interaction in the modulation of emotions and cognition: assessing the role of cannabinoid CB1 receptor in neurons expressing dopamine D1 receptors. Front Behav Neurosci 5:49

Thiemann G, Di Marzo V, Molleman A, Hasenohrl RU (2008) The CB(1) cannabinoid receptor antagonist AM251 attenuates amphetamine-induced behavioural sensitization while causing monoamine changes in nucleus accumbens and hippocampus. Pharmacol Biochem Behav 89:384–391

Thiemann G, Watt CA, Ledent C, Molleman A, Hasenohrl RU (2009) Modulation of anxiety by acute blockade and genetic deletion of the CB(1) cannabinoid receptor in mice together with biogenic amine changes in the forebrain. Behav Brain Res 200:60–67

Toda M, Abi-Dargham A (2007) Dopamine hypothesis of schizophrenia: making sense of it all. Curr Psychiatry Rep 9:329–336

Trezza V, Damsteegt R, Manduca A, Petrosino S, Van Kerkhof LW, Pasterkamp RJ, Zhou Y, Campolongo P, Cuomo V, Di Marzo V, Vanderschuren LJ (2012) Endocannabinoids in amygdala and nucleus accumbens mediate social play reward in adolescent rats. J Neurosci 32:14899–14908

Trezza V, Vanderschuren LJ (2009) Divergent effects of anandamide transporter inhibitors with different target selectivity on social play behavior in adolescent rats. J Pharmacol Exp Ther 328:343–350

Tsou K, Nogueron MI, Muthian S, Sanudo-Pena MC, Hillard CJ, Deutsch DG, Walker JM (1998) Fatty acid amide hydrolase is located preferentially in large neurons in the rat central nervous system as revealed by immunohistochemistry. Neurosci Lett 254:137–140

Turu G, Hunyady L (2010) Signal transduction of the CB1 cannabinoid receptor. J Mol Endocrinol 44:75–85

Twitchell W, Brown S, Mackie K (1997) Cannabinoids inhibit N- and P/Q-type calcium channels in cultured rat hippocampal neurons. J Neurophysiol 78:43–50

Tzavara ET, Davis RJ, Perry KW, Li X, Salhoff C, Bymaster FP, Witkin JM, Nomikos GG (2003) The CB1 receptor antagonist SR141716A selectively increases monoaminergic neurotransmission in the medial prefrontal cortex: implications for therapeutic actions. Br J Pharmacol 138:544–553

Tzavara ET, Li DL, Moutsimilli L, Bisogno T, Di Marzo V, Phebus LA, Nomikos GG, Giros B (2006) Endocannabinoids activate transient receptor potential vanilloid 1 receptors to reduce hyperdopaminergia-related hyperactivity: therapeutic implications. Biol Psychiatry 59:508–515

Ujike H, Morita Y (2004) New perspectives in the studies on endocannabinoid and cannabis: cannabinoid receptors and schizophrenia. J Pharmacol Sci 96:376–381

Umathe SN, Manna SS, Utturwar KS, Jain NS (2009) Endocannabinoids mediate anxiolytic-like effect of acetaminophen via CB1 receptors. Prog Neuropsychopharmacol Biol Psychiatry 33:1191–1199

Uriguen L, Garcia-Fuster MJ, Callado LF, Morentin B, La Harpe R, Casado V, Lluis C, Franco R, Garcia-Sevilla JA, Meana JJ (2009) Immunodensity and mRNA expression of A2A adenosine, D2 dopamine, and CB1 cannabinoid receptors in postmortem frontal cortex of subjects with schizophrenia: effect of antipsychotic treatment. Psychopharmacology (Berl) 206:313–324

Uriguen L, Perez-Rial S, Ledent C, Palomo T, Manzanares J (2004) Impaired action of anxiolytic drugs in mice deficient in cannabinoid CB1 receptors. Neuropharmacology 46:966–973

Stelt M van der, Di Marzo V (2003) The endocannabinoid system in the basal ganglia and in the mesolimbic reward system: implications for neurological and psychiatric disorders. Eur J Pharmacol 480:133–150

Stelt M van der, Fox SH, Hill M, Crossman AR, Petrosino S, Di Marzo V, Brotchie JM (2005) A role for endocannabinoids in the generation of parkinsonism and levodopa-induced dyskinesia in MPTP-lesioned non-human primate models of Parkinson's disease. FASEB J 19:1140–1142

Stelt M van der, Veldhuis WB, Maccarrone M, Bar PR, Nicolay K, Veldink GA, Di Marzo V, Vliegenthart JF (2002) Acute neuronal injury, excitotoxicity, and the endocannabinoid system. Mol Neurobiol 26:317–346

Van Laere K, Casteels C, Lunskens S, Goffin K, Grachev ID, Bormans G, Vandenberghe W (2012) Regional changes in type 1 cannabinoid receptor availability in Parkinson's disease in vivo. Neurobiol Aging 33:620 e 621–628

Os J van, Kapur S (2009) Schizophrenia. Lancet 374:635–645

Van Sickle MD, Duncan M, Kingsley PJ, Mouihate A, Urbani P, Mackie K, Stella N, Makriyannis A, Piomelli D, Davison JS, Marnett LJ, Di Marzo V, Pittman QJ, Patel KD, Sharkey KA (2005) Identification and functional characterization of brainstem cannabinoid CB2 receptors. Science 310:329–332

Vliet SA van, Vanwersch RA, Jongsma MJ, Olivier B, Philippens IH (2008) Therapeutic effects of Delta9-THC and modafinil in a marmoset Parkinson model. Eur Neuropsychopharmacol 18:383–389

Vasquez C, Navarro-Polanco RA, Huerta M, Trujillo X, Andrade F, Trujillo-Hernandez B, Hernandez L (2003) Effects of cannabinoids on endogenous K + and Ca2 + currents in HEK293 cells. Can J Physiol Pharmacol 81:436–442

Vigano D, Guidali C, Petrosino S, Realini N, Rubino T, Di Marzo V, Parolaro D (2009) Involvement of the endocannabinoid system in phencyclidine-induced cognitive deficits modelling schizophrenia. Int J Neuropsychopharmacol 12:599–614

Wager-Miller J, Westenbroek R, Mackie K (2002) Dimerization of G protein-coupled receptors: CB1 cannabinoid receptors as an example. Chem Phys Lipids 121:83–89

Wallmichrath I, Szabo B (2002) Cannabinoids inhibit striatonigral GABAergic neurotransmission in the mouse. Neuroscience 113:671–682

Walsh S, Gorman AM, Finn DP, Dowd E (2010) The effects of cannabinoid drugs on abnormal involuntary movements in dyskinetic and non-dyskinetic 6-hydroxydopamine lesioned rats. Brain Res 1363:40–48

Wei BQ, Mikkelsen TS, McKinney MK, Lander ES, Cravatt BF (2006) A second fatty acid amide hydrolase with variable distribution among placental mammals. J Biol Chem 281:36569–36578

Wenger T, Moldrich G, Furst S (2003) Neuromorphological background of cannabis addiction. Brain Res Bull 61:125–128

Wong DF, Kuwabara H, Horti AG, Raymont V, Brasic J, Guevara M, Ye W, Dannals RF, Ravert HT, Nandi A, Rahmim A, Ming JE, Grachev I, Roy C, Cascella N (2010) Quantification of cerebral cannabinoid receptors subtype 1 (CB1) in healthy subjects and schizophrenia by the novel PET radioligand [11C]OMAR. Neuroimage 52:1505–1513

Wotjak CT (2005) Role of endogenous cannabinoids in cognition and emotionality. Mini Rev Med Chem 5:659–670

Wu X, French ED (2000) Effects of chronic delta9-tetrahydrocannabinol on rat midbrain dopamine neurons: an electrophysiological assessment. Neuropharmacology 39:391–398

Zarrindast MR, Dorrani M, Lachinani R, Rezayof A (2010) Blockade of dorsal hippocampal dopamine receptors inhibits state-dependent learning induced by cannabinoid receptor agonist in mice. Neurosci Res 67:25–32

Zarrindast MR, Mahboobi S, Sadat-Shirazi MS, Ahmadi S (2011a) Anxiolytic-like effect induced by the cannabinoid CB1 receptor agonist, arachydonilcyclopropylamide (ACPA), in the rat amygdala is mediated through the D1 and D2 dopaminergic systems. J Psychopharmacol 25:131–140

Zarrindast MR, Navaeian M, Nasehi M (2011b) Influence of three-day morphine-treatment upon impairment of memory consolidation induced by cannabinoid infused into the dorsal hippocampus in rats. Neurosci Res 69:51–59

Zavitsanou K, Garrick T, Huang XF (2004) Selective antagonist [3H]SR141716A binding to cannabinoid CB1 receptors is increased in the anterior cingulate cortex in schizophrenia. Prog Neuropsychopharmacol Biol Psychiatry 28:355–360

Zeng BY, Dass B, Owen A, Rose S, Cannizzaro C, Tel BC, Jenner P (1999) Chronic L-DOPA treatment increases striatal cannabinoid CB1 receptor mRNA expression in 6-hydroxydopamine-lesioned rats. Neurosci Lett 276:71–74

Zenko M, Zhu Y, Dremencov E, Ren W, Xu L, Zhang X (2011) Requirement for the endocannabinoid system in social interaction impairment induced by coactivation of dopamine D1 and D2 receptors in the piriform cortex. J Neurosci Res 89:1245–1258

Zimmer A, Zimmer AM, Hohmann AG, Herkenham M, Bonner TI (1999) Increased mortality, hypoactivity, and hypoalgesia in cannabinoid CB1 receptor knockout mice. Proc Natl Acad Sci 96:5780–5785

Zuardi AW, Guimaraes FS, Hallak JE, Crippa JA (2011) Is the highest density of CB1 receptors in paranoid schizophrenia a correlate of endocannabinoid system functioning? Expert Rev Neurother 11:1111–1114

Chapter 6
Pathophysiology of Mood Disorders and Mechanisms of Action of Antidepressants and Mood Stabilizers

Zdeněk Fišar

Abstract The present chapter summarizes information on the pathophysiology of mood disorders and mechanisms of action of antidepressants and mood stabilizers with focus on an endocannabinoid regulation of monoamines in depression and bipolar disorder. Leading role in neurochemistry and pathophysiology of mood disorders could be awarded to disturbed monoamine neurotransmission, dysfunction in energy metabolism of neurons, modulation of inflammatory and neuroendocrine pathways, and changes in activities of transcription factors, neurotrophic factors and other components involved in neuroplasticity. A role of endocannabinoid system in pathophysiology of mood disorders is supposed, but little known. In the light of new findings, there is potential for pharmacological regulation of endocannabinoid system in treatment of depressive and bipolar disorder.

6.1 Introduction

Attention in the research of biological basis of mood disorders has been devoted to an overlapping set of molecular and cellular mechanisms of mood disorders, antidepressant response, neuroplasticity, and chronic stress (Duman et al. 1997; Duman 2002; Zarate et al. 2006a; Einat and Manji 2006; Pittenger and Duman 2008), including changes in neuroprogression, inflammatory and cell-mediated immune response, antioxidant capacity, oxidative and nitrosative stress (Maes et al. 2012), and mitochondrial functions (Stork and Renshaw 2005; Kato and Kato 2000; Kato 2007, 2008; Quiroz et al. 2008; Hroudová and Fišar 2011). Therefore, changes in the activities of compounds of these intracellular signaling pathways are studied with the aim of discovering new biological markers of mood disorders or predictors of response to antidepressant treatment (Fišar and Raboch 2008; Fišar and Hroudová 2010a, 2010b; Schmidt et al. 2011).

There is now evidence that depression is an inflammatory disorder that is accompanied by cell-mediated immune activation. There are also evidences that damage

Z. Fišar (✉)
Department of Psychiatry, First Faculty of Medicine, Charles University in Prague and General University Hospital in Prague, Prague, Czech Republic
e-mail: zfisar@lf1.cuni.cz

E. J. Van Bockstaele (ed.), *Endocannabinoid Regulation of Monoamines in Psychiatric and Neurological Disorders*, DOI 10.1007/978-1-4614-7940-6_6,

by oxidative and nitrosative stress and decreased antioxidant defenses are involved in the pathophysiology of depression (Maes et al. 2011; Leonard and Maes 2012). Finally, there is evidence that depression is accompanied by mitochondrial dysfunctions and that some patients display characteristics of a neuroprogressive disorder with recurrent episodes and a more chronic course (Maes et al. 2012). Corresponding signaling pathways are studied as drug targets in treatment of depression.

Monoamine neurotransmitters such as dopamine, norepinephrine or serotonin are the most important neurotransmitters both in pathophysiology of major mental disorders and in mechanisms of action of many psychotropic drugs, including drug abuse. Monoamines, energy metabolism, and inflammatory pathways are interrelated in many complex manners. Brain monoamines are involved in a range of the same processes affected by neuropsychiatric disorders and by cannabinoids. Recent evidences include association of disturbances in neuroplasticity with mitochondrial dysfunctions and inflammation in pathogenesis of mental disorders, and influencing of these processes by cannabinoids.

Interconnections of cannabinoid and monoamine pathways (Fišar 2012) include effects of cannabinoids on monoaminergic neurotransmission through (1) regulation of synthesis of monoamine neurotransmitters, (2) regulation of catabolism of monoamine neurotransmitters, (3) inhibition of release of neurotransmitters, such as gamma-aminobutyric acid (GABA), glutamate or acetylcholine, which are coupled to monoaminergic systems, (4) interconnections of cannabinoid and monoamine signaling pathways.

By retrograde signaling, endocannabinoids play important role in modulation of synaptic plasticity in the central nervous system (CNS) (endocannabinoid-mediated short-term synaptic plasticity includes both depolarization-induced suppression of inhibition, which is due to reduction of GABA release, and depolarization-induced suppression of excitation, which results from inhibition of glutamate release). Modulation of the endocannabinoid system may produce neuroprotective effects (Fowler et al. 2005; Fišar 2009).

6.2 Mood Disorders

Mood disorder is the term designating two groups of diagnoses; there are depressive disorders, which are most researched in major depressive disorder (MDD), and bipolar disorder (BD), which is characterized by intermittent episodes of mania (or hypomania) and depression. The World Health Organization (WHO) defines depression as a common mental disorder characterized by sadness, loss of interest or pleasure, feelings of guilt or low self-worth, disturbed sleep or appetite, feelings of tiredness, and poor concentration (World Health Organization 2012). Mania is the opposite of depression; it is a state of abnormally elevated or irritable mood, arousal, and/or energy levels. Individuals have a single episode of depression/mania within their life or repeated episodes occur or may become chronic. MDD is the leading cause of years lost due to disability for people of ages 15–44 years old worldwide (World Health Organization 2004). MDD is one of the most common psychiatric disorders, with a worldwide lifetime prevalence rate of 10–20 % in women and a slightly lower rate in men.

Treatments that can be effective for treatment of major depression include pharmacotherapy, psychotherapy, electroconvulsive therapy, transcranial magnetic stimulation, and deep brain stimulation. There are problems in pharmacotherapy of depression with side effects, late onset of therapeutic effects, treatment-resistant depression, and prevention of recurrence and relapse of depressive symptoms. Biochemical effects of antidepressants and mood stabilizers are studied with the aim to discover both molecular mechanism of their therapeutic efficiency and neurochemical nature of mood disorders. Both complex clinical pattern of mood disorders and adaptive changes in activity or availability of a large number of components of signaling pathways after long-term treatment with antidepressants is responsible for the fact, that definite molecular mechanisms responsible for therapeutic action of drugs are not known.

Depression and mania are thought to be heterogeneous illnesses that can result from dysfunction of several neurotransmitters or metabolic systems. Despite extensive biological research, the pathophysiology of mood disorders is still little known, and treatments that target the causal factors of these disorders are not yet available. The pathophysiological mechanisms giving rise to mental disorders are complex, i.e., interplay between the environment, genetics, anatomy, neurobiology, neurophysiology, and psychological mechanisms give rise to these disorders (Fava and Kendler 2000). Our understanding of normal brain mechanisms mediating complex behaviors underlies advances in understanding of pathophysiological mechanisms underlying mood disorders. Biological markers of mood disorders are searched that might improve both diagnosis of the illness and prediction of therapeutic response or nonresponse to pharmacotherapy. Most of the data are related to pathophysiology of major depression and BD. Potential mechanisms associated with the pathogenesis of this disorder include monoamine deficits, hypothalamic-pituitary-adrenal (HPA) axis dysfunctions, and inflammatory and neurodegenerative alterations.

6.3 Mechanisms of Action of Antidepressants and Mood Stabilizers

6.3.1 Antidepressants

Antidepressants are psychiatric medication used for the treatment of mood disorders, such as major depression, dysthymia, and anxiety disorders. More than 40 drugs are currently used as antidepressants worldwide and many other drugs are administered as supportive therapy or holding course.

Clinical effects of antidepressants are obviously caused by their ability to induce adaptive changes in neurotransmission, mainly serotonergic and noradrenergic. Changes in the availability of neurotransmitters and also in the density and sensitivity of their receptors and transporters are not sufficient to explain either origin and course of the mood disorders, or the mechanisms of action of antidepressants and mood stabilizers. It was supposed that intracellular processes included in apoptotic, neurodegenerative, and inflammatory pathways are responsible for final therapeutic effects of antidepressants (Porcelli et al. 2011).

Treatment with antidepressants is generally associated with delay in onset of therapeutic response. Receptor hypotheses of mood disorders supposing causal association between clinical response and drug-induced adaptive changes in density or sensitivity of neurotransmitter receptors were not confirmed. More complex response to antidepressant treatment, comprehensive of changes and adaptation both on cellular level (intracellular signaling) and in interconnection of brain areas (brain circuits), seems to be responsible for their effects; e.g., chronic treatment with fluoxetine (selective serotonin reuptake inhibitor) induced both changes in serotonin neurotransmission and changes in brain glucose metabolism (clinical improvement was associated with limbic and striatal decreases and brain stem and dorsal cortical increases) (Mayberg et al. 2000).

Brain effects of antidepressants have been studied using several imaging methods including both resting positron emission tomography (PET) and functional magnetic resonance imaging (fMRI). Increased neuronal activity following antidepressant treatment was found in several neocortical areas (dorsolateral, dorsomedial, and ventrolateral prefrontal cortices), and decreased activation was found in several limbic regions, including hippocampus, parahippocampal region, amygdala, insula, ventral anterior cingulate cortex, and orbitofrontal cortex (Delaveau et al. 2011). Resting PET studies have shown that antidepressants tend to restore a normal brain function (Mayberg et al. 2000; Mayberg 2003) while improving depressive symptoms. Normalization of hypometabolism in neocortical regions (prefrontal and parietal cortex) and hypermetabolism in limbic and paralimbic areas has been reported (Fitzgerald et al. 2008). Thus, antidepressants could modulate the activity of brain regions involved in both resting state and emotional processing; it can be hypothesized that the antidepressants could decrease the hypersensitivity to negative emotional stimuli and could facilitate the visual positive emotional processing in patients with MDD (Delaveau et al. 2011). It seems that there are common pathways in both pharmacologic and nonpharmacologic treatments (cognitive behavioral therapy, interpersonal therapy, and deep brain stimulation) based on modulation of cortico-limbic balance (Mayberg 2003).

Antidepressants affect learning and memory in animal models and enhance structural plasticity and hippocampal neurogenesis (Drzyzga et al. 2009; Kasper and McEwen 2008; Warner-Schmidt and Duman 2006). Antidepressants can directly modulate glutamatergic neurotransmission through ionotropic glutamate *N*-methyl-D-aspartate (NMDA) or 2-amino-3-(3-hydroxy-5-methyl-isoxazol-4-yl) propanoic acid (AMPA) receptors; it is likely that an intimate relationship exists between regulation of monoaminergic and glutamatergic neurotransmission and

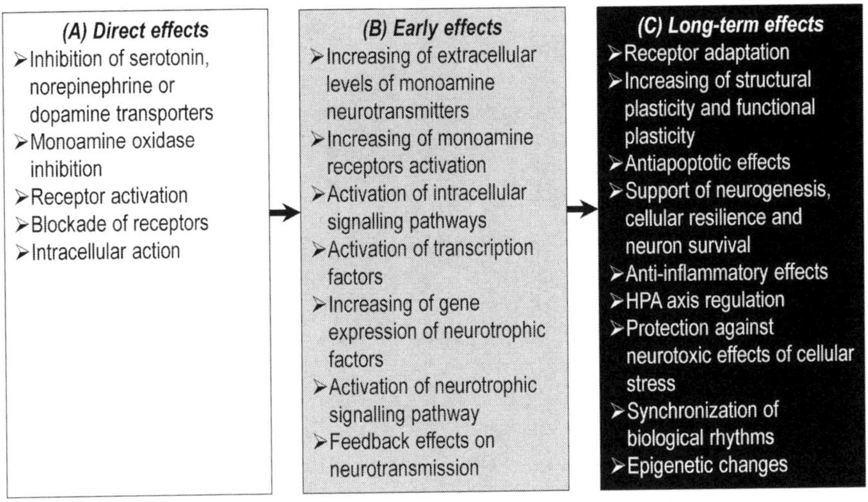

Fig. 6.1 Sequence of events induced by antidepressants (Fišar et al. 2012)

antidepressant effects (Paul and Skolnick 2003). An inhibition of an excessive release of glutamate appears to be important for mechanisms of neuroprotective or mood affecting action of some drugs (Zarate et al. 2006a). Mechanisms underlying the rapid antidepressant action of ketamine, noncompetitive glutamate NMDA receptor antagonist (Zarate et al. 2006b; Diazgranados et al. 2010) correlate with rapid synaptogenesis and spine formation in the prefrontal cortex (Duman et al. 2012). These effects suggest that depressive symptoms can be improved by altering the action of glutamate (Krishnan and Nestler 2008). However, formulating of a glutamatergic hypothesis of depression may be grossly simplistic (Krishnan and Nestler 2010).

6.3.2 *Sequence of Cellular Events Induced by Antidepressants*

Sequence of biochemical events induced by antidepressants is crucial for discovery of molecular mechanisms associated with their therapeutic effects. Neurochemical events for antidepressant action classified into direct effects, early effects, and long-term effects are summarized in the Fig. 6.1:

a. **Direct (immediate) biochemical effects** of antidepressants related to their therapeutic action include:
    1. Inhibition of reuptake of monoamine neurotransmitters.
    2. Inhibition of metabolism of monoamine neurotransmitters, e.g., monoamine oxidase inhibition.

3. Neurotransmitter receptor activation (e.g., postsynaptic serotonin receptors type 1A or sigma receptors) or inhibition (e.g., postsynaptic serotonin receptors type 2A and 2C or presynaptic α_2-adrenoceptors).
4. Direct inhibition or activation of several intracellular components of signaling pathways participant on neurotransmission.

b. **Early (intermediate) events** following immediate biochemical effects include:

1. Increasing of availability and extracellular concentrations of monoamine neurotransmitters.
2. Increasing of monoamine receptors activation.
3. Activation of intracellular signaling pathways (adenylate cyclase, phosphoinositide, or calcium pathway).
4. Activation of transcription factors such as cyclic adenosine monophosphate (cAMP) response element-binding protein (CREB) and increasing of gene expression of neurotrophic factors, such as brain-derived neurotrophic factor (BDNF) etc.
5. Feedback effects on neurotransmission.

c. **Long-term (delayed) adaptive changes** in central neurotransmission are responsible for therapeutic effects of antidepressants:

1. Neurochemical events include receptor adaptation (desensitization or down-regulation, sensitization or up-regulation), increasing of structural plasticity (synaptogenesis, formation or changes of axons, synapses, dendrites and dendritic spines) and functional plasticity (long-term potentiation, long-term depression, strength of synapse), antiapoptotic effects, support of neurogenesis, cellular resilience and neuron survival, and protection against neurotoxic effects of cellular stress.
2. Neuroimmune approach is based on observations that MDD is an inflammatory disorder with an overproduction of proinflammatory cytokines (Maes et al. 2009) and anti-inflammatory effects of antidepressants participate on their pharmacological effects (Janssen et al. 2010).
3. Neuroendocrine hypotheses suppose that therapeutic effects of antidepressants consist in regulation of HPA axis, which can be overactivated during depression (Nikisch 2009).
4. Chronobiological hypotheses connect therapeutic effects of antidepressants with synchronization of biological rhythms disturbed in depression (Bunney and Potkin 2008; Mendlewicz 2009; Schulz and Steimer 2009).
5. Genetic factors contribute for about 50 % of the antidepressant response (Crisafulli et al. 2011). Epigenetic changes are studied both in relation to gene-environment interactions (G × E) (Caspi et al. 2003; Uher 2008) and in animal models of stress, depression and antidepressant treatment (Schroeder et al. 2007; Tsankova et al. 2007; Yasuda et al. 2009).

6.3.3 Mood Stabilizers

Mood stabilizers are psychiatric medication used in treatment of mood disorders, which are characterized by intense and sustained mood shifts (e.g., BD). Most of mood stabilizers are anticonvulsants (valproate, carbamazepine, and lamotrigine), with an important exception of lithium, which is the oldest and the best known mood stabilizing drug. Some atypical antipsychotics (olanzapine, quetiapine, aripiprazole, risperidone, and ziprasidone) have mood stabilizing effects, as well. It is also suggested that ω-3 fatty acids, which act as endogenous inhibitors of second messenger-regulated protein kinases, may have a mood stabilizing effect (Mirnikjoo et al. 2001; McNamara et al. 2006).

The mood stabilizers have no direct biochemical effects on monoamine neurotransmitter systems, i.e., on synthesis or metabolism of neurotransmitters, and density and sensitivity of monoamine receptors and transporters. Virtually they act intracellulary, affecting activities of several intracellular enzymes (Hroudová and Fišar 2011). Main targets of mood stabilizers are neurotrophin BDNF, mitogen-activated protein kinase/extracellular signal-regulated kinase (MAPK/ERK) pathway, and pathways modulated by glycogen synthase kinase 3 (GSK-3) or antiapoptotic factor Bcl-2 (Gould and Manji 2005; Einat and Manji 2006; Shaltiel et al. 2007).

6.3.4 Mitochondrial Effects of Antidepressants and Mood Stabilizers

Neuroprotective and antiapoptotic effect of several antidepressants and mood stabilizers could be related to their effects on mitochondrial functions, e.g., inhibition of the mitochondrial permeability transition (MPT) (Stavrovskaya et al. 2004; Nahon et al. 2005; Zhang et al. 2008), stimulation of the state 3 and 4 respiration rates after chronic treatment (Katyare and Rajan 1995) and increasing of mitochondrial energy generation (Bachmann et al. 2009; Maurer et al. 2009; Valvassori et al. 2010; Abdel-Razaq et al. 2011). Electroconvulsive therapy and transcranial magnetic stimulation also enhance mitochondrial function (Dragicevic et al. 2011). Enhancing mitochondrial function may represent a critical component for the optimal treatment of mood disorders (Quiroz et al. 2008). However, data are inconsistent; several studies showed that antidepressants of different structures have common antimitochondrial effects, i.e., inhibition of complexes of electron transport chain (Hroudová and Fišar 2010, 2012). These antimitochondrial effects could underlie unwanted side effects and/or they could initiate some of the adaptive responses underlying their clinical effects (Abdel-Razaq et al. 2011). In any case, more information about action of antidepressants on mitochondrial function is needed.

6.4 Pathophysiology of Mood Disorders

Approaches of biological psychiatry to the study of mood disorders can be divided into:

1. Neurobiological approach involving neuroanatomical changes, genetic effects, effects of stress, and effects of disturbed chronobiology.
2. Neurochemical approach involving properties of neurotransmitter systems, i.e., neurotransmitters, receptors, and coupled signaling pathways. The role of neurochemical hypotheses of mood disorders is to suggest the relationship between symptoms of the disease, changes in signaling pathways, and mechanisms of action of psychotropic drugs.
3. Neuro-immuno-endocrinological approach, which is based both on observation of increased activity HPA axis during depression and on changes in immune and inflammatory system.

6.4.1 Neuroanatomy of Depression

It is unlikely that depression is caused by dysfunction in a single neurotransmitter system of brain region. Many studies used functional brain imaging techniques to define the role of cortical and subcortical brain areas involved in the development, maintenance, response to treatment, and treatment-refractoriness of depressive illness (Mayberg 2009).

Neuroanatomical evidences for cellular alterations in depression are focused on certain cortical and hippocampal regions, which show smaller neuronal size, fewer glial cells, and shorter dendrites. Data support the hypothesis of altered cell plasticity in depression and suicide occurring mainly in fronto-limbic areas (Hercher et al. 2009). Morphometric changes in depression and suicide confirm that several neuromodulatory systems are affected, notably GABAergic, serotonergic, noradrenergic, and glutamatergic pathways.

Brain circuit models of depression derived primarily from PET scan measures of regional glucose metabolism and blood flow; fMRI scans have also proven to be sensitive to antidepressant-induced changes in brain function. PET and fMRI have shown how depressive behavior can be correlated with hypermetabolism of the subgenual cingulate cortex and amygdala as well as hypometabolism of the dorsal prefrontal cortex and striatal regions.

Formulation of neural "depression circuits" offers framework for further research into pathophysiology of depression (Krishnan and Nestler 2010):

1. **Amygdala-centric model** of potential pathophysiological changes in bipolar disorder and MDD (Savitz and Drevets 2009) proposes that the emotional symptoms of depression can be brought by functional impairment of the striatum or the prefrontal and orbital cortex, resulting in disinhibition of amygdala and

downstream structures. Alternatively, they can arise from functional hypersensitivity of deeper limbic structures and/or amygdala which gives rise to dysregulation of prefrontal cortical structures.

2. **Limbic-cortical circuits model** (Mayberg 1997, 2003, 2009) supposes that failure of regional network, composed of cortical, limbic, and subcortical compartments, can explain clinical symptoms of depression. The model proposes that decrease in dorsal neocortical regions and relative increase in ventral limbic and paralimbic areas are associated with depressive disorder. In turn, remission occurs when there is appropriate modulation of dysfunctional limbic-cortical interactions by treatment. Limbic-cortical model is directed to four clusters of brain regions with strong anatomical connections to each other, i.e., it compartmentalizes depressive endophenotypes into exteroceptive (cognitive such as attention, appraisal, action), interoceptive (visceral-motor, incl. drive states, autonomic function, circadian rhythms), mood-regulating (self-relevance, priorization, contingencies, reinforcement), and mood-monitoring functions (novelty, salience, habit).

6.4.2 *Genetics of Depression*

Depressive symptoms occur in several disorders, firstly in major depressive, bipolar, dysthymic, and schizo-affective disorder. These disorders are considered to have a genetic based predisposition under certain environmental influences. Twin studies have provided evidence that genetics explains 50–70 % of the etiology of mood disorders. As many as 32 potential candidate genes have been investigated to identify genetic predisposition to depression, including genetic polymorphisms of key components of the major pathways of neurotransmitters and neurotrophins involved, such as serotonin transporter, dopamine transporter, neurotrophin BDNF, tryptophan hydroxylase, and catechol-*O*-methyltranspherase (Kato 2007; Elder and Mosack 2011). Moreover, gene-expression studies are also useful to understand mechanisms of action of antidepressants and mood stabilizers, and pharmacogenetics helps us to explain different responses to pharmacotherapy of mood disorders.

Gene-environment interactions are studied with the aim to explain how relatively weak genetic vulnerability may trigger a psychiatric disorder. However, there is significant discordance of depression between monozygotic twins who share the same environment, slow progress in identifying genetic risk factors, and twofold prevalence of depression in females. It suggests the presence of nongenetic and nonevironmental factors in pathophysiology of depression. Epigenetic modifications (including inherited and acquired modifications of deoxyribonucleic acid (DNA) and histones that regulate various genomic functions without a change in nuclear DNA sequence, and micro ribonucleic acid (miRNA) that regulate gene expression) have been implicated as such factors (Mill and Petronis 2007; Tsankova et al. 2007; Im and Kenny 2012).

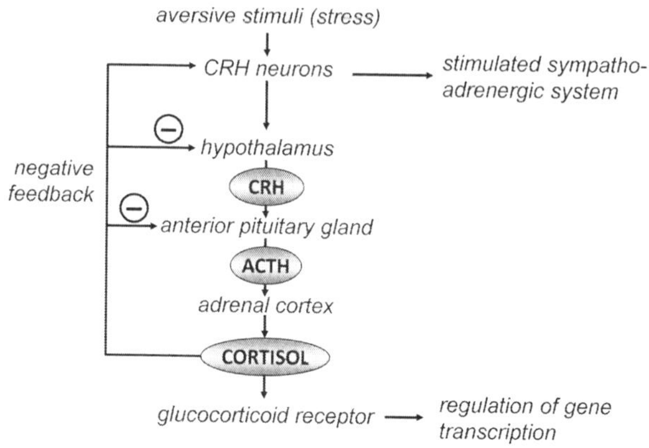

Fig. 6.2 Hypothalamic-pituitary-adrenal (HPA) axis and feedback controls by cortisol. Alarm reactions are primarily linked to a stimulated sympatho-adrenergic system and involve an activation of brainstem nuclei, the vagal nerve, and the medulla of the adrenal gland. Such events lead to the release of noradrenaline and adrenaline into the blood. In contrast, when aversively perceived encounters cannot be controlled by fight or flight, animals show passive coping strategies that are primarily associated with an activation of the HPA axis. Chronic stress frequently results in a hypersecretion of adrenal glucocorticoids and desensitization of the central glucocorticoid receptors and a resistance to feedback inhibition. *CRH* corticotropin-releasing hormone, *ACTH* adrenocorticotropic hormone

6.4.3 Neuroendocrinology and Neuroimmunology of Depression

MDD is associated with subtle cellular and molecular alterations in a complex neuronal network. The affected brain regions display dynamic neuroplastic adaptations to neuroendocrine and neuroimmune stimuli arising both from within and outside the brain (Krishnan and Nestler 2010). Initially, depression is associated frequently with stressful lifetime events. Stress-induced hypercortisolemia leads to the downregulation of central glucocorticoid receptors, impairing negative feedbacks mediated by cortisol, and enhancing concentrations of corticotropin-releasing hormone (CRH) and adrenocorticotropic hormone (ACTH). The chronic mild stress evokes an array of neurobiological changes that mirror those seen in depressive disorders and may be a suitable tool to investigate novel systems that could be disturbed in depression (Engelmann et al. 2004; Leonard 2005; Hill et al. 2012).

Research of endocrinology and immunology of depression focuses on a central role of dysregulation of the HPA axis (Krishnan and Nestler 2008). CRH-containing parvocellular neurons of the paraventricular nucleus of the hypothalamus integrate information relevant to stress. CRH is released into hypophyseal portal system and acts on the corticotropes in the anterior pituitary that produces ACTH. ACTH reaches the adrenal cortex via the bloodstream and stimulates the release of glucocorticoids (Fig. 6.2).

Glucocorticoids, such as cortisol, regulate metabolism and immunity but also neuronal survival and neurogenesis. Glucocorticoids, among other functions, repress CRH and ACTH synthesis and release; thus, they inhibit their own production. At higher concentrations, glucocorticoids also damage the hippocampus and reduce rate of neurogenesis, which could be related to hypercortisolemia in some cases of depressive disorder. Several systems seem to be involved interacting with each other; thus, the simple model cannot be unequivocally accepted that glucocorticoids induce neurodegeneration, but rather that elevated cytokines, in the context of glucocorticoid resistance, are probably the offenders (Zunszain et al. 2011).

The HPA axis and the inflammatory response system have been suggested as pathophysiological mechanisms implicated in the etiology of MDD. Excessive activation of the HPA axis is observed in approximately half of individuals with depression that results in increased release of CRH and in some cases sustained elevation of cortisol (Bao et al. 2008). Dexamethasone suppression test (Holsboer et al. 1982) and combined dexamethasone suppression/CRH stimulation test (Ising et al. 2005) were proposed to detect HPA axis dysfunctions. Evidence indicate that increased activity of the HPA axis characterize not only individuals during the acute episodes of depression and psychosis but also those at risk to develop these disorders in the future (Pariante 2009; Vreeburg et al. 2009). It was confirmed that chronic forms of the melancholic depression and atypical depression are associated with different biological correlates with inflammatory and metabolic dysregulation in atypical depression and HPA axis hyperactivity in melancholic depression (Lamers et al. 2013). Hypo- and hypercortisolemic depression may represent different subtypes of depression (Bremmer et al. 2007). Thus, chronic forms of depressive subtypes differ not only in their symptom presentation, but also in their biological correlates.

Depression is highly prevalent in infectious, autoimmune, and neurodegenerative diseases and at the same time, depressed patients show higher levels of pro-inflammatory cytokines. Since communication occurs between the endocrine system, immune system, and CNS (Haddad et al. 2002), an activation of the inflammatory responses can affect neuroendocrine processes, and vice versa (Zunszain et al. 2011).

Now there is evidence that the activation of the immune system is associated with the symptoms of depression (Leonard and Myint 2009; Catena-Dell'Osso et al. 2011) and major depression is accompanied by immune dysregulation and activation of the inflammatory response system, i.e., that depression belongs to the spectrum of inflammatory and degenerative disorders (Dowlati et al. 2010; Maes et al. 2011). The **inflammatory and neurodegenerative hypothesis** of depression (Maes et al. 2009) supposes that depression is associated with both inflammatory processes, as well as with neurodegeneration and reduced neurogenesis. According to this hypothesis, cell-mediated-immune activation and inflammation contribute to depressive symptoms by increased levels of pro-inflammatory cytokines (e.g., interleukin-1, interleukin-6, and tumor necrosis factor α), and Th-1-derived cytokines (e.g., interleukin-2 and interferon γ). Peripheral inflammation and cell-mediated immune activation are translated to the brain to cause neuroinflammation and microglial activation (Maes et al. 2011). Several signaling pathways are affected by cell-mediated-immune activation and inflammation in depression, leading to depletion of plasma tryptophan, increased production of tryptophan catabolites along the indoleamine-2,3-dioxygenase pathway, up-regulation of the serotonin transporter, cellular damages due to increased oxidative and nitrosative stress and/or decreased antioxidative protection, decrease of ω -3 polyunsaturated fatty acids, changes in the expression or functions of brain serotonin receptors and glutamate ionotropic receptors (Leonard and Maes 2012; Maes et al. 2012). It is supposed that these factors cause neuroprogression (a combination of neurodegeneration, neuronal apoptosis, and lowered neurogenesis and neuroplasticity), which play a role in the pathophysiology of depression.

6.4.4 Neurochemistry of Mood Disorders

For decades, the monoamine hypothesis of depression has guided research into the etiology of depression. The serotonin hypothesis of depression relies primarily on the fact that brain serotonin extracellular levels are increased by most antidepressants. Accordingly, the development of new antidepressants has been limited by a focus on modulation of serotonin receptors and serotonin reuptake. The current understanding of the pathophysiology of depression is based on the role of growth factors and neurogenesis. Neurotrophic, neuroplasticity, and network hypotheses were formulated, which reflect key role of neurotrophin BDNF; but also serotonin transporter, catechol-*O*-methyl transferase and monoamine oxidase type A, play an important role in recent biochemical hypotheses.

Neuroplasticity is a fundamental mechanism of neuronal adaptation to environmental inputs. The term neuroplasticity (also known as brain plasticity, cortical plasticity, and cortical re-mapping) is used for description of either functional or structural changes of neurons and glial cells that occur in developing brain as well as in the adult brain in order to adjust to external or internal stimuli (Mesulam 1999; Nestler et al. 2002; Citri and Malenka 2008). The most widely recognized forms of plasticity are learning, memory, and recovery from nervous system injury, which may happen through the change in the strength of connections among brain cells, by adding or removing connections, or by adding new cells.

6.4.4.1 Homocysteine Hypothesis

Homocysteine is biosynthesized from methionine via *S*-adenosyl methionine, which transfers the methyl group to an acceptor molecule during numerous reactions in the body, including DNA methylation or synthesis of monoamine neurotransmitters and essential phospholipids (Fig. 6.3). Homocysteine is converted back to methionine (Bottiglieri 2005). The metabolism of homocysteine depends on several B vitamins, including B_9 (folate, folic acid), B_{12} (cobalamin), B_6 (pyridoxine), and B_2 (riboflavin). Deficiencies of these vitamins can lead to high homocysteine levels (Savage et al. 1994; Green 2011). Homocysteine is both a marker of folate or B_{12} deficiency (Bottiglieri et al. 2000) and a cause of many adverse effects on neurons resulting in disturbed biosynthesis of neurotransmitters and neurodegenerative damage

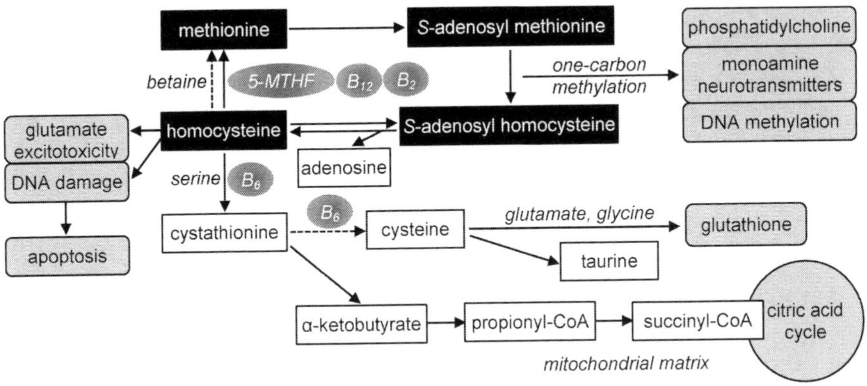

Fig. 6.3 Homocysteine's main biochemical roles. Homocysteine can be synthesized from methionine and then converted back to methionine via the *S*-adenosyl methionine cycle or used to create cysteine and α-ketobutyrate. *5-MTHF* 5-methyltetrahydrofolate

(Mattson and Shea 2003). Changes in homocysteine metabolism, inadequate intake of vitamin B, impaired renal function, increasing age, male sex, smoking, excessive alcohol intake, lack of physical activity, and high coffee consumption lead to increased total homocysteine blood levels (Bottiglieri 2005; Refsum et al. 2006).

It has been shown in several large studies that elevated plasma or serum homocysteine concentrations and folate and vitamin B_{12} deficiency are all significantly associated with depressive disorders (Tiemeier et al. 2002; Gu et al. 2012) or with increased risk of depression (Tolmunen et al. 2004; Refsum et al. 2006; Gilbody et al. 2007; Almeida et al. 2008). Approximately one-third of depressive patients showed low concentrations of folate and elevated concentrations of homocysteine in serum or erythrocytes (Carney et al. 1990; Bottiglieri et al. 2000). Moreover, several studies have supported the use of folate and vitamin B_{12} supplementation in the treatment of depression (Bottiglieri 2005; Coppen and Bolander-Gouaille 2005; Lazarou and Kapsou 2010).

The vascular depression hypothesis proposes that cerebrovascular disease can predispose, precipitate, or perpetuate some geriatric depressive syndromes (Alexopoulos et al. 1997). The **homocysteine hypothesis** of depression presumes that genetic and environmental factors elevate homocysteine levels, which cause vascular disease of the brain and/or neurotransmitter alterations, which in turn cause depression (Folstein et al. 2007). However, another important pathophysiological mechanism should be taken into account when discussing the homocysteine hypothesis of depression, namely the role of homocysteine and its metabolites in DNA methylation (Miller 2008) or in gene-environment interactions. These processes play a role in the pathogenesis of different psychiatric disorders, not only depression but also schizophrenia, eating disorders, and addiction. The hypothesis that blood homocysteine is a predictor for incident depression was moderately supported in elderly women only (Forti et al. 2010).

6.4.4.2 Monoamine Hypothesis

Inadvertently induced depression by reserpine (irreversible blocker of the vesicular monoamine transporter) and effectiveness of first antidepressants (inhibitors of reuptake of norepinephrine and serotonin by presynaptic neuron; inhibitors of monoamine oxidase) led to the **monoamine hypothesis** supposing that affective disorders are due to catecholamine (Schildkraut 1965) and/or indolamine (Coppen 1967) deficiency at functionally important receptor sites in the CNS and that therapeutic effects of antidepressants result from increased stimulation of norepinephrine and/or serotonin receptors due to elevation of these monoamine neurotransmitters in the extracellular space. The monoamine-deficiency hypothesis was broadened by presumption of disturbances in other neurotransmitter systems and their mutual interactions.

In order to test monoamine hypothesis, a series of studies was conducted to evaluate effects of diet-induced **monoamine depletion** on depressive symptoms in depressed patients and in healthy controls. Relapse after serotonin depletion or catecholamine depletion was found to be specific to the type of antidepressant treatment and type of depletion. Serotonin or norepinephrine/dopamine depletion did not decrease mood in healthy controls and slightly lowered mood in healthy controls with a family history of MDD. In drug-free patients with MDD in remission, a moderate mood decrease was found for acute tryptophan depletion only. However, acute tryptophan depletion induced relapse in patients in remission who used serotonergic antidepressants (Delgado et al. 1999). Depletion studies failed to demonstrate a causal relation between serotonin and norepinephrine with depressive disorder (Ruhé et al. 2007; Cowen 2008; Mendelsohn et al. 2009). These findings forced formulation of the **revised monoamine theory of depression** (Heninger et al. 1996; aan het Rot et al. 2009), which supposes that monoamine systems are only modulating other brain neurobiological systems that have more primary role in depression.

The mitochondrial enzyme monoamine oxidase (MAO), which regulates metabolic degradation of catecholamines and serotonin in neural and other target tissues, participate both on regulation of monoamine neurotransmission and on reactive oxygen species (ROS) production. Density and activity of membrane transporters for monoamine neurotransmitters determine extracellular neurotransmitter concentrations and strongly regulate synaptic signal transduction. Serotonin transporter (SERT, 5HTT), as target of many antidepressants, is the center of interest; persons with reduced expression of SERT are more sensitive to stress-induced depression (Caspi et al. 2003). According to **advanced monoamine theory** (Meyer et al. 2006), serotonin or norepinephrine levels in the brain are mainly regulated by MAO type A (MAO-A) activity , and severity of symptoms of depression is related to changes in the activity of monoamine transporters in specific brain regions. Thus, both increased MAO-A activity and modified density of transporters are included in the pathophysiology of affective disorders. The advanced monoamine hypothesis was supported by observation that during a major depressive episode, both MAO-A density is elevated, resulting in greater metabolism of monoamines in the brain (Meyer et al. 2006, 2009), and brain SERT binding is diminished (Selvaraj et al. 2011).

PET and other imaging methods enabled in vivo study of neurotransmitter receptors and transporters in human brain. These imaging studies focused on serotonin (5-HT_{1A}, 5-HT_{2A}, 5-HT_{1B}), dopamine and norepinephrine receptors, serotonin or dopamine transporters, monoamine oxidase A, and muscarinic acetylcholine receptor 2 (Meyer 2012). The role of these receptors and transporters in pathophysiology of depression are discussed with respect to results of animal studies, genetic studies and alterations in inflammation, endocrine function, and neurocircuitry (Savitz and Drevets 2013), e.g., acetylcholine neurotransmission has been linked to the regulation of mood, sleep, and neuroendocrine functions. The central acetylcholine system has been implicated in pathophysiology of mood disorders by findings that muscarinic acetylcholine receptor agonists or acetylcholinesterase inhibitors exacerbate depressive symptoms in bipolar or unipolar depressive disorder and reduces manic symptoms in BD (Cannon et al. 2006). The increased muscarinic sensitivity evidenced in individuals with mood disorders conceivably may contribute to the altered perceptions of emotionally-valenced events reported in these conditions (Phillips et al. 2003). Recently, reduced cholinergic-muscarinic 2 receptor distribution volume in depression has been associated with genetic variation within receptor gene (Cannon et al. 2011).

Increased serotonergic neurotransmission in response to stress or depression-related inflammation may result in a chronic increase in serotonin transporter function, contributing to decrease of serotonergic signaling in specific brain areas. The hypothesis is discussed that 5-HT_{1A} receptor dysfunction represents one potential mechanism underpinning MDD and other stress-related disorders. Stress-induced secretion of corticosteroids may lead to down-regulation or desensitization of both somatodendritic and postsynaptic 5-HT_{1A} receptors. In contrast, 5-HT_{2A} receptors may be upregulated by glucocorticoid secretion, facilitating the secretion of CRH during stress and contributing to the chronic elevation of cortisol (Savitz and Drevets 2013). Antidepressants are hypothesized to increase serotonergic signaling at the postsynaptic 5-HT_{1A} receptor through either direct or indirect effects; e.g., by desensitizing the somatodendritic 5-HT_{1A} autoreceptor in the raphe nuclei or by facilitating the activation of G-proteins by the postsynaptic 5-HT_{1A} receptor. Normalized signaling at the postsynaptic 5-HT_{1A} receptor reduces cortisol and CRH release, restores endocrine function, and improves mood (Savitz et al. 2009). Antidepressant-induced antagonism, down-regulation or desensitization of corticolimbic 5-HT_{2A} receptors may participate on overcoming of the serotonin signaling abnormality in depression (Savitz and Drevets 2013).

In summary, monoamine depletion studies, PET studies, and genetic association studies (polymorphisms of monoaminergic genes) supported a role of monoaminergic neurotransmission in the pathophysiology of depression but not evidenced the primary role of monoaminergic system in development of the disorder. Moreover, the prerequisite that the mechanism of action of antidepressants is opposite of disease pathophysiology may not be valid generally (Krishnan and Nestler 2010). Because direct measurements of monoamine neurotransmission did not yield definitive evidences in relation to depression, the downstream effects of monoamine neurotransmission were explored (Belmaker and Agam 2008). The role of neuronal atrophy owing to stress or reduced growth factor support has been discussed.

6.4.4.3 Neurotrophic, Neuroplasticity and Neurogenesis Hypotheses

Difficulties in interpretation of molecular mechanisms of action of mood stabilizers and the introduction of new antidepressants without direct monoaminergic action led to more complex biological hypotheses of depression. Disturbances are searched in five major signaling pathways supposing to be involved in pathophysiology of mood disorders or in mechanisms of action of antidepressants and mood stabilizers: (1) adenylate cyclase pathway, (2) phosphoinositide pathway, (3) calcium signaling pathway, (4) Wnt pathway, and (5) tyrosine kinase (neurotrophic) pathway (Fišar and Hroudová 2010b).

The **neurotrophic hypothesis** of depression (Duman et al. 1997; Duman 2002; Einat and Manji 2006; Zarate et al. 2006a) states that a deficiency in neurotrophic support may contribute to hippocampal pathology during the development of depression and that reversal of this deficiency by antidepressants or mood stabilizers may contribute to the resolution of depressive symptoms. The hypothesis supposed that vulnerability to depression can arise as a result of neuronal damage, e.g., after chronic stress, long-term increased levels of glucocorticoids, hypoglycemia, ischemia, certain viral infections, effects of neurotoxins, etc. The therapeutic effects of antidepressants may consist in increased function of the monoaminergic system leading to higher expression of neurotrophin BDNF and its receptor TrkB, and consequently to increased neuronal plasticity and resumption of cellular functions. However, the scheme may be too simplified (Carlezon et al. 2005). A model has been suggested whereby the effects of antidepressant treatments could be explained by a reactivation of activity-dependent and BDNF-mediated cortical plasticity, which in turn leads to the adjustment of neuronal networks to better adapt to environmental challenges (Castrén and Rantamäki 2010).

The **neurogenesis hypothesis** of depression proposes that depression can arise from impaired hippocampal neurogenesis and that an array of antidepressants ultimately works by stimulating such neurogenesis. The first component of this hypothesis is not tenable, and the evidence for second component is conflicting (Sapolsky 2004). Coupling of hippocampal neurogenesis to pathophysiology of depression requires further research to be confirmed (Santarelli et al. 2003; Gass and Riva 2007).

In summary, neurodegeneration or impaired neurogenesis may be important mechanism of depression, and neurotrophins are key regulators of neurogenesis and neuroplasticity (Pittenger and Duman 2008). Recovery of brain networks through increase of neuroplasticity induces antidepressant effect. It is supposed that structural and functional brain abnormalities in patients with depressive disorder may be associated with low levels of BDNF, abnormal function of HPA axis and glutamatergic toxicity (aan het Rot et al. 2009; Krishnan and Nestler 2008; Mathew et al. 2008).

6.4.4.4 Mitochondrial Dysfunction Hypothesis

A growing body of evidence suggests that mitochondrial dysfunction is important in patients with mood disorders, and other psychiatric disorders, such as schizophrenia, anxiety disorders, and borderline personality disorders (Fattal et al. 2006; Shao et al. 2008; Jou et al. 2009). Evidences supporting the role of mitochondrial dysfunctions in pathophysiology of mood disorders include (Clay et al. 2011): (1) decrease of mitochondrial respiration; (2) changes in mitochondrial morphology; (3) increase in mtDNA polymorphisms or mutations; (4) down-regulation of nuclear mRNA molecules and proteins involved in mitochondrial respiration; (5) decrease of high-energy phosphates and decrease of pH in the brain; and (6) affective symptoms and cognitive decline in mitochondrial disorders. Effects of psychotropic drugs on mitochondrial functions (Quiroz et al. 2008; Hroudová and Fišar 2012) also contribute to the role of mitochondria in psychiatric disorders.

A hypothesis of **mitochondrial dysfunction** in BD (Stork and Renshaw 2005) was proposed and involved impaired oxidative phosphorylation, a resultant shift toward glycolytic energy production, a decrease in total energy production (decreased adenosine triphosphate (ATP) production) and/or substrate availability, and changed concentrations of phosphomonoesters and altered phospholipid metabolism. Changes in cerebral concentrations of *N*-acetyl aspartate, glutamate/glutamine, choline-containing compounds, *myo*-inositol, lactate, phosphocreatine, phosphomonoesters, and intracellular pH in bipolar subjects were described (Yildiz-Yesiloglu and Ankerst 2006).

Neuronal calcium homeostasis and calcium signaling regulate multiple neuronal functions, including synaptic transmission, neuronal plasticity and survival. The idea, that altered intracellular calcium signaling may be crucial for the molecular mechanisms leading to both schizophrenia and mood disorders was suggested (Jimerson et al. 1979). **Calcium and mitochondrial dysfunction hypothesis** of BD offers that mtDNA polymorphisms/mutations or mtRNA deletions caused by nuclear gene mutations can cause mitochondrial dysregulation of calcium leading to symptoms of BD (Kato and Kato 2000; Kato 2007, 2008).

Hypothesis of mitochondrial dysfunctions allow for the role of neurotoxicity and/or oxidative stress in pathophysiology of depression. The brain is extremely vulnerable to oxidative stress damage and mitochondria are the major sources of ROS. Thus, damages due to increased oxidative and nitrosative stress and/or lowered levels of antioxidant protections are probably involved in the pathophysiology of depression (Maes et al. 2009, 2011, 2012; Berk et al. 2011).

In summary, current view is that depression is accompanied by neurodegeneration, neuronal apoptosis, reduced neurogenesis, disturbed neuroplasticity, mitochondrial dysfunctions, cell-mediated immune activation, and inflammatory processes. These pathways are new drug targets in depression.

6.5 Cannabinoid System and Mood Disorders

Since the discovery of the endocannabinoid system numerous studies tried to address its role in anxiety and depression. Several evidences support the assumption that CB_1 receptors are responsible for CNS effects of cannabinoids: (1) CB_1 knockout mice displayed increased anxiety-like behavior (Martin et al. 2002); (2) subjects with a life-time diagnosis of major depression had increased CB_1 receptor mRNA in the dorsolateral prefrontal cortex (Hungund et al. 2004); (3) rimonabant, a CB_1 inverse agonist, induced anxiety and depression in some patients (Buggy et al. 2011).

The CB_1 receptor was originally considered to be mainly localized in the brain. It is now known that they are present in many tissues and organs. The CB_2 receptors are most abundantly expressed in cells of the immune system and in cells of immune origin. In these cells CB_2 receptors mostly mediate immunosuppressive effects (Pacher and Mechoulam 2011). The CB_2 receptors were initially presumed to be missing in the brain; however, recently they were found in microglia. CB_2 receptors are coupled to $G_{i/o}$ protein; thus CB_2 activation leads to inhibition of adenylyl cyclase. CB_1 receptors are not only coupled to $G_{i/o}$ protein, but also to G_s and $G_{q/11}$ proteins.

The best known effect of CB_1 activation is psychoactivity. High density of CB_1 receptors in the whole brain, including amygdala, cortex and hippocampus, i.e., in brain structures involved in regulation of cognition and mood, implies that the endocannabinoid system is probably involved in pathophysiology of mood disorders and other stress-related disorders. The role of endocannabinoid signaling system in controlling of affective state was documented by behavioral (mood elevating) effects of plant cannabinoids. Moreover, rimonabant has induced anxiety and depression-like symptoms.

There is not sufficient data about the role of CB_2 receptors in depression. It is accepted that depression can be associated with inflammatory processes and many antidepressants show anti-inflammatory effects. Endocannabinoid signaling through CB_2 receptors may participate in biological protective systems against nonprotein attacks (Pacher and Mechoulam 2011). Inflammation triggers elevation in local endocannabinoid levels, which regulate signaling response in immune and other cells. Thus, it is possible that increased expression and activation of CB_2 receptors could provide antidepressant response through their anti-inflammatory action. However, specific effect of CB_2 receptor activation on emotional behavior may be related to CB_2 localized on microglia (Gorzalka and Hill 2011).

There are divergent opinions on the role of cannabis use in development of mood disorders. Some studies indicate that frequent cannabis use in adolescence predisposes to higher rates of depression and anxiety in young adulthood (Patton et al. 2002). Other findings suggest that exposure to cannabis by itself does not lead to depression but that it may be associated with later suicidal thoughts and attempts (Pedersen 2008). The association between early-onset cannabis use and later risk of a depressive episode was found only modest in large WHO study (de Graaf et al. 2010). Recent studies confirmed that daily adolescent cannabis use is associated consistently with anxiety, but not depressive disorder, in adolescence and late young adulthood (Degenhardt et al. 2013). Opinion that moderate consumption of cannabis in adult age increases positive and reduces negative mood was supported by epidemiological study, which has found that frequent adult users of cannabis exhibit less depressed mood and more positive affect than nonconsumers of cannabis (Denson and Earleywine 2006). These results suggest that adults apparently do not increase their risk for depression by using cannabis and that some users of cannabis may be self-medicating a depression with the drug.

Evidences for the hypothesis that **deficient endocannabinoid signaling** (disruptions in the signaling capacity of CB_1 receptor) may produce a vulnerability to, or directly contribute to, the development of a depressive episode (Gorzalka and Hill 2011) include:

1. Endocannabinoid signaling is present throughout cortico-limbic structures, which are involved in the regulation of mood, emotional behavior and reward sensitivity.
2. Suppression of endocannabinoid signaling is sufficient to induce a depression-like state.
3. Impairments in CB_1 receptor signaling increase HPA axis activity.
4. Administration of CB_1 receptor inverse agonist rimonabant to humans resulted in symptoms of anxiety and depression.
5. Single nucleotide polymorphisms in the CB_1 receptor can increase the presence of neurotism, vulnerability to develop stress-induced depressive episode, and risk of antidepressant resistance.
6. Blunted endocannabinoid signaling has been documented in patients with depressive episode.
7. Both direct and indirect activation of CB_1 receptors have been found to produce antidepressant-like responses on the biochemical as well as behavioral level.
8. Many forms of antidepressant treatment significantly alter endocannabinoid signaling, which is involved in the neuroadaptive effects of these treatments.

6.5.1 Endocannabinoid and Monoaminergic Systems

The functional interactions between endocannabinoid and monoamine systems in the brain indicate role of cannabinoid signaling in pathophysiology of mood disorders. Acute stimulation of CB_1 receptors increases the activity of noradrenergic, serotonergic, and dopaminergic neurons and release of monoamine neurotransmitters in specific brain regions. Monoaminergic systems are regulated by CB_1 receptors by direct or indirect effects depending on their localization on monoaminergic, GABAergic, or glutamatergic neurons (Esteban and García-Sevilla 2012). Inhibition of MAO by cannabinoids could contribute to their effect on monoaminergic systems (Fišar 2010, 2012).

The interest in interaction of cannabinoids with serotonin system was renewed in context of the role of serotonergic neurons in mediating cannabinoid effects such as antiemesis, hypothermia, analgesia, sleep, and appetite stimulation (Sagredo et al. 2006). The role of cannabinoids in the control of pain and emesis may be explained by direct inhibition by cannabinoids of 5-HT_3 receptors (Barann et al. 2002; Przegaliński et al. 2005). New studies examined the effects of long-term cannabinoid administration on the responsivity of 5-HT_{1A} and 5-HT_{2A} receptors (Russo et al. 2005; Hill et al. 2006). Both in vitro and in vivo effects of cannabinoids on the function of serotonin transporter were confirmed (Velenovská and Fišar 2007). It was concluded that activity of serotonin transporter is acutely affected by cannabinoids at relatively high drug concentrations; this effect is inhibitory, noncompetitive and indirect, and can be partially accounted for the changes in the membrane microviscosity.

Noradrenergic system plays a significant role in the modulation of emotional state (related to stress, arousal, and anxiety) and in pathophysiology of mood disorders. It was confirmed that endocannabinoid system mediates stress response and emotional homeostasis partially by targeting noradrenergic circuits; cannabinoids can modulate noradrenergic transmission in both noradrenergic nuclei and target regions (Carvalho and Van Bockstaele 2012).

Motivational and motor effects of endocannabinoids are ascribed, in part, to modulation of dopamine neurotransmission. Cannabinoids modulate dopamine transmission in both the nigrostriatal and mesocortical systems by indirect means (Fitzgerald et al. 2012). This modulation is of direct relevance for complex behaviors.

6.5.2 Endocannabinoid System in Depression

The evidences and literature on the role of endocannabinoid system in the neurobiology of depression has been accumulated from 2005 (Hill and Gorzalka 2005). A large amount of data come from animal models of depression and studies using CB_1 knockout mice (Parolaro et al. 2010). Mice lacking CB_1 receptor became more vulnerable to chronic stress (Martin et al. 2002), showed an increase in passive coping behavior in the forced swim test (Steiner et al. 2008), and had longer immobility time in the tail suspension test (Aso et al. 2008). They also showed hyperactivity of the HPA axis (Urigüen et al. 2004).

CB_1 receptor density was found increased in rat prefrontal cortex both in chronic stress induced depression (Hill et al. 2008) and after bilateral olfactory bulbectomy (Rodríguez-Gaztelumendi et al. 2009). Decreased density of CB_1 receptors in model depression was observed in midbrain (Bortolato et al. 2007), hippocampus, hypothalamus, and ventral striatum (Hill et al. 2008). Adolescent rat exposure to tetrahydrocannabinol (THC) was associated with decreased CB_1 density in nucleus accumbens, amygdala, and ventral tegmental area (Rubino et al. 2008). In human, increased CB_1 density was found in dorsolateral prefrontal cortex of depressed suicide victims (Hungund et al. 2004). Basal serum concentrations of anandamide and

2-arachidonoyglycerol were significantly reduced in women with major depression relative to matched controls, indicating a deficit in peripheral endocannabinoid activity (Hill et al. 2009). In response to chronic mild stress, anandamide concentrations were decreased (Hill et al. 2008) or unchanged (Bortolato et al. 2007) in several brain structures. 2-arachidonoylglycerol was found decreased in the hippocampus (Hill et al. 2005) but increased or unchanged in several other structures.

In summary, endocannabinoid system seems downregulated in most brain areas of chronically stressed rats. Endocannabinoid signaling is an important stress buffer and modulates emotional and cognitive functions. Genetic polymorphisms in the human gene for CB_1 receptor and fatty acid amide hydrolase (FAAH) have been found to contribute to occurrence of several mental disorders (Hillard et al. 2012).

6.5.3 Antidepressant-Like Effect of Cannabinoids

Long lasting impairment of CB_1 receptor function led to the development of depression-like symptoms, as well as to decrease in neuroplasticity in the prefrontal cortex (Rubino et al. 2008, 2009). Assumption that a dysfunction of endocannabinoid system plays a role in the ethiopathology of depression led to study of antidepressant effects due to activation of this system. Antidepressant-like effects of chronic treatment with CB_1 receptor agonists have been evident in the forced swim test or tail suspension test (Jiang et al. 2005; Morrish et al. 2009). Chronic treatment with an inhibitor of FAAH, the enzyme responsible for metabolism of anandamide, also exerted antidepressant-like effects (Gobbi et al. 2005). Thus, facilitating endocannabinoid neurotransmission has antidepressant-like effects in animal models of depression. Moreover, clinical use of the rimonabant was ended in 2009 due to induction of psychiatric adverse effects, depression mainly, e.g., in patients without previous history of psychiatric illness, there were more depressive episodes after starting treatment with rimonabant (Buggy et al. 2011). However, acute blocking of the cannabinoid receptors also induced an antidepressant-like response (Shearman et al. 2003; Griebel et al. 2005) and antidepressants like properties of both repeated THC and rimonabant were observed in olfactory bulbectomised rats used as a model of depression (El Batsh et al. 2012). Additional studies are required to clarify the effect of cannabinoids in human depression.

There are many similarities in biochemical mechanisms of action of antidepressants and cannabinoids:

1. Most antidepressants increase extracellular serotonin and/or norepinephrine. Both CB_1 receptor agonists and inhibitors of anandamide hydrolysis increase firing activity of serotoninergic neurons in dorsal raphe (Gobbi et al. 2005; Bambico et al. 2007) or noradrenergic neurons in locus coeruleus and the release of norepinephrine in the prefrontal cortex (Gobbi et al. 2005; Oropeza et al. 2005).
2. Cannabinoids are capable of inhibiting monoamine oxidase activity (Fišar 2010, 2012) and can affect monoamine neurotransmitter concentrations in this way.

Moreover, long-term cannabis use could lead to adaptive changes in function of the serotonin transporter (Velenovská and Fišar 2007).

3. Inhibitors of endocannabinoid uptake or metabolism attenuates the neuroendocrine response to psychological stressors (Patel et al. 2004; Gorzalka et al. 2008), similarly to antidepressants.
4. Cannabinoid agonists, inhibitors of FAAH or inhibitors of reuptake of endocannabinoids have been shown to increase hippocampal neurogenesis (Jiang et al. 2005; Hill et al. 2006; Marchalant et al. 2009), process that can affect onset and treatment of depression.
5. THC treatment increases activation of extracellular signal-regulated kinase (Derkinderen et al. 2003) and expression of neurotrophin BDNF in the hippocampus, nucleus accumbens and other specific brain regions (Butovsky et al. 2005; Rubino et al. 2008, 2009), which is effect supposed to be associated with antidepressant action (Duman and Monteggia 2006).

These data indicate that increasing cannabinoid signaling exerts antidepressant properties through the same or similar mechanisms as conventional antidepressants (Bambico et al. 2009; Parolaro et al. 2010).

At the biochemical level, adaptive changes induced by long-term treatment with antidepressants or mood stabilizers include an increase in monoaminergic neurotransmission, a reduction in HPA axis activity, and an increase in neuroplasticity, neurogenesis, and cellular resilience. The data demonstrate that the enhancing of CB_1 receptor signaling produces all these biochemical effects, i.e., increasing CB_1 receptor activity is sufficient to produce antidepressant-like effects (Gorzalka and Hill 2011).

6.6 Conclusion

Regulation of critical intracellular signaling pathways plays a critical role in higher-order brain functions, which are altered in mood disorders, suggesting the involvement of dysfunctions of signaling pathways in the pathophysiology and the treatment of mood disorders (Gould et al. 2007). It can be concluded that both some structural deviations in neural networks and disturbances of signal transduction in certain neurons participate in development of mood disorders.

It is suggested that environmental stress and genetic risk variants interact with each other in a complex manner to alter neural circuitry and evoke illness. Developmental and structural changes in neural networks could be considered as a necessary condition for vulnerability to the development of pathological states of mood, whereas disturbances in signal transduction pathways in chemical synapses could be related to the onset of specific symptoms of mood disorder. Identification of compounds of signal transduction, which are primarily responsible for shifts of mood, remains incomplete. Mild dysfunction of some mitochondrial functions might be basis for homeostatic imbalance in synapses during episodes of depression, hypomania, mania, or the appearance of mixed states.

Both complex clinical pattern of mood disorders and adaptive changes in activity or availability of a large number of components of signaling pathways after long-term treatment with antidepressants is responsible for the fact, that definite molecular mechanisms responsible for therapeutic action of drugs are not known. The long-term administration of antidepressants leads to the effects similar to neurotrophic, as seen through the activation of transcription factors and increased gene expression of neurotrophins. Mood stabilizers, such as lithium and valproate, strongly activate the neurotrophic signal cascades and affects other signaling pathways.

Preclinical and clinical data indicate that leading role in neurochemistry of mood disorders could be awarded to disturbed monoamine neurotransmission, dysfunction in energy metabolism of neurons (Kato and Kato 2000; Kato 2007, 2008; Stork and Renshaw 2005), modulation of inflammatory pathway (Maes et al. 2009), and changes in activities of transcription factors, neurotrophic factors and other components involved in neuroplasticity and apoptosis (Duman et al. 1997; Duman 2002, 2009; Duman and Monteggia 2006). Different neurotransmitters may be responsible for different symptoms of depression, depending on which brain region is affected: (1) changes in norepinephrine, dopamine, glutamate, and GABA in cortical regions may contribute to depression, cognitive dysfunction, anhedonia, and apathy; (2) dysfunctional transmission of dopamine and norepinephrine in the prefrontal cortex can impair concentration and decisiveness; (3) anxiety, guilt, and negative emotions are influenced by serotonergic activity in the limbic system as well as by lack of glutamate reuptake or metabolism in amygdala, etc.

Neural circuits in the brain that may contribute to depressive symptoms involve hippocampus, frontal and anterior temporal cortex, and several subcortical structures implicated in reward, fear and motivation (these include the nucleus accumbens, amygdala, and hypothalamus) (Mayberg 1997; Nestler et al. 2002). Limbic-cortical dysregulation model (Mayberg 2003) may be advanced by assumption that changes in brain CB_1 receptor pathways participate in development of symptoms of depressive disorder due to dysregulation of cortical, limbic and subcortical compartments. Moreover, CB_2 receptors localized on cell of immune system may influence neuroimmune processes accompanying depressive disorder.

Neuromodulation of synapses by cannabinoids is proving to have a wide range of functional effects, making them potential targets as medical preparations in a variety of illnesses, including some neurodegenerative and mental disorders. A new class of presynaptic plasticity that requires signaling by endocannabinoids has been identified in several brain structures (Chevaleyre et al. 2006). Moreover, cannabinoids appear to be the drug whose capacity to produce increased hippocampal neurogenesis is positively correlated with its antidepressant effects (Jiang et al. 2005). Pharmacological modulation of the endocannabinoid system has been proposed as a novel therapeutical strategy for the treatment of stress-related mood disorders such as anxiety and depression.

Acknowledgment This research was supported by project MSM0021620849 given by Ministry of Education, Youth and Sports, and by project PRVOUK-P26/LF1/4 given by Charles University in Prague, Czech Republic.

References

aan het Rot M, Mathew SJ, Charney DS (2009) Neurobiological mechanisms in major depressive disorder. Can Med Assoc J 180(3):305–313

Abdel-Razaq W, Kendall DA, Bates TE (2011) The effects of antidepressants on mitochondrial function in a model cell system and isolated mitochondria. Neurochem Res 36(2):327–338

Alexopoulos GS, Meyers BS, Young RC, Campbell S, Silbersweig D, Charlson M (1997) 'Vascular depression' hypothesis. Arch Gen Psychiatry 54:915–922

Almeida OP, McCaul K, Hankey GJ, Norman P, Jamrozik K, Flicker L (2008) Homocysteine and depression in later life. Arch Gen Psychiatry 65:1286–1294

Aso E, Ozaita A, Valdizán EM, Ledent C, Pazos A, Maldonado R, Valverde O (2008) BDNF impairment in the hippocampus is related to enhanced despair behavior in CB1 knockout mice. J Neurochem 105(2):565–572

Bachmann RF, Wang Y, Yuan P, Zhou R, Li X, Alesci S, Du J, Manji HK (2009) Common effects of lithium and valproate on mitochondrial functions: protection against methamphetamine-induced mitochondrial damage. Int J Neuropsychopharmacol 12(6):805–822

Bambico FR, Katz N, Debonnel G, Gobbi G (2007) Cannabinoids elicit antidepressant-like behavior and activate serotonergic neurons through the medial prefrontal cortex. J Neurosci 27(43):11700–11711

Bambico FR, Duranti A, Tontini A, Tarzia G, Gobbi G (2009) Endocannabinoids in the treatment of mood disorders: evidence from animal models. Curr Pharm Des 15(14):1623–1646

Bao AM, Meynen G, Swaab DF (2008) The stress system in depression and neurodegeneration: focus on the human hypothalamus. Brain Res Rev 57:531–553

Barann M, Molderings G, Brüss M, Bönisch H, Urban BW, Göthert M (2002) Direct inhibition by cannabinoids of human 5-HT_{3A} receptors: probable involvement of an allosteric modulatory site. Br J Pharmacol 137(5):589–596

Belmaker RH, Agam G (2008) Major depressive disorder. N Engl J Med 358(1):55–68

Berk M, Kapczinski F, Andreazza AC, Dean OM, Giorlando F, Maes M, Yücel M, Gama CS, Dodd S, Dean B, Magalhães PV, Amminger P, McGorry P, Malhi GS (2011) Pathways underlying neuroprogression in bipolar disorder: focus on inflammation, oxidative stress and neurotrophic factors. Neurosci Biobehav Rev 35(3):804–817

Bortolato M, Mangieri RA, Fu J, Kim JH, Arguello O, Duranti A, Tontini A, Mor M, Tarzia G, Piomelli D (2007) Antidepressant-like activity of the fatty acid amide hydrolase inhibitor URB597 in a rat model of chronic mild stress. Biol Psychiatry 62(10):1103–1110

Bottiglieri T (2005) Homocysteine and folate metabolism in depression. Prog Neuropsychopharmacol Biol Psychiatry 29:1103–1112

Bottiglieri T, Laundy M, Crellin R, Toone BK, Carney MW, Reynolds EH (2000) Homocysteine, folate, methylation, and monoamine metabolism in depression. J Neurol Neurosurg Psychiatry 69:228–232

Bremmer MA, Deeg DJ, Beekman AT, Penninx BW, Lips P, Hoogendijk WJ (2007) Major depression in late life is associated with both hypo- and hypercortisolemia. Biol Psychiatry 62:479–486

Buggy Y, Cornelius V, Wilton L, Shakir SA (2011) Risk of depressive episodes with rimonabant: a before and after modified prescription event monitoring study conducted in England. Drug Saf 34(6):501–509

Bunney JN, Potkin SG (2008) Circadian abnormalities, molecular clock genes and chronobiological treatments in depression. Br Med Bull 86:23–32

Butovsky E, Juknat A, Goncharov I, Elbaz J, Eilam R, Zangen A, Vogel Z (2005) In vivo up-regulation of brain-derived neurotrophic factor in specific brain areas by chronic exposure to Δ^9-tetrahydrocannabinol. J Neurochem 93:802–811

Cannon DM, Carson RE, Nugent AC, Eckelman WC, Kiesewetter DO, Williams J, Rollis D, Drevets M, Gandhi S, Solorio G, Drevets WC (2006) Reduced muscarinic type 2 receptor binding in subjects with bipolar disorder. Arch Gen Psychiatry 63(7):741–747

Cannon DM, Klaver JK, Gandhi SK, Solorio G, Peck SA, Erickson K, Akula N, Savitz J, Eckelman WC, Furey ML, Sahakian BJ, McMahon FJ, Drevets WC (2011) Genetic variation in cholinergic muscarinic-2 receptor gene modulates muscarinic$_2$-receptor binding in vivo and accounts for reduced binding in bipolar disorder. Mol Psychiatry 16(4):407–418

Carlezon WA Jr, Duman RS, Nestler EJ (2005) The many faces of CREB. Trends Neurosci 28(8):436–445

Carney MW, Chary TK, Laundy M, Bottiglieri T, Chanarin I, Reynolds EH, Toone B (1990) Red cell folate concentrations in psychiatric patients. J Affect Disord 19:207–213

Carvalho AF, Van Bockstaele EJ (2012) Cannabinoid modulation of noradrenergic circuits: implications for psychiatric disorders. Prog Neuropsychopharmacol Biol Psychiatry 38(1):59–67

Caspi A, Sugden K, Moffitt TE, Taylor A, Craig IW, Harrington H, McClay J, Mill J, Martin J, Braithwaite A, Poulton R (2003) Influence of life stress on depression: moderation by a polymorphism in the 5-HTT gene. Science 301(5631):386–389

Castrén E, Rantamäki T (2010) Role of brain-derived neurotrophic factor in the aetiology of depression: implications for pharmacological treatment. CNS Drugs 24(1):1–7

Catena-Dell'Osso M, Bellantuono C, Consoli G, Baroni S, Rotella F, Marazziti D (2011) Inflammatory and neurodegenerative pathways in depression: a new avenue for antidepressant development? Curr Med Chem 18(2):245–255

Chevaleyre V, Takahashi KA, Castillo PE (2006) Endocannabinoid-mediated synaptic plasticity in the CNS. Annu Rev Neurosci 29:37–76

Citri A, Malenka RC (2008) Synaptic plasticity: multiple forms, functions, and mechanisms. Neuropsychopharmacology 33(1):18–41

Clay HB, Sillivan S, Konradi C (2011) Mitochondrial dysfunction and pathology in bipolar disorder and schizophrenia. Int J Dev Neurosci 29(3):311–324

Coppen A (1967) The biochemistry of affective disorders. Br J Psychiatry 113(504):1237–1264

Coppen A, Bolander-Gouaille C (2005) Treatment of depression: time to consider folic acid and vitamin B_{12}. J Psychopharmacol 19:59–65

Cowen PJ (2008) Serotonin and depression: pathophysiological mechanism or marketing myth? Trends Pharmacol Sci 29(9):433–436

Crisafulli C, Fabbri C, Porcelli S, Drago A, Spina E, De Ronchi D, Serretti A (2011) Pharmacogenetics of antidepressants. Front Pharmacol 2(Art 6):1–21

Graaf R de, Radovanovic M, Laar M van, Fairman B, Degenhardt L, Aguilar-Gaxiola S, Bruffaerts R, Girolamo G de, Fayyad J, Gureje O, Haro JM, Huang Y, Kostychenko S, Lépine JP, Matschinger H, Mora ME, Neumark Y, Ormel J, Posada-Villa J, Stein DJ, Tachimori H, Wells JE, Anthony JC (2010) Early cannabis use and estimated risk of later onset of depression spells: epidemiologic evidence from the population-based World Health Organization World Mental Health Survey Initiative. Am J Epidemiol 172(2):149–159

Degenhardt L, Coffey C, Romaniuk H, Swift W, Carlin JB, Hall WD, Patton GC (2013) The persistence of the association between adolescent cannabis use and common mental disorders into young adulthood. Addiction. 108(1):124–133

Delaveau P, Jabourian M, Lemogne C, Guionnet S, Bergouignan L, Fossati P (2011) Brain effects of antidepressants in major depression: a meta-analysis of emotional processing studies. J Affect Disord 130(1–2):66–74

Delgado PL, Miller HL, Salomon RM, Licinio J, Krystal JH, Moreno FA, Heninger GR, Charney DS (1999) Tryptophan-depletion challenge in depressed patients treated with desipramine or fluoxetine: implications for the role of serotonin in the mechanism of antidepressant action. Biol Psychiatry 46(2):212–220

Denson TF, Earleywine M (2006) Decreased depression in marijuana users. Addict Behav 31(4):738–742
Derkinderen P, Valjent E, Toutant M, Corvol JC, Enslen H, Ledent C, Trzaskos J, Caboche J, Girault JA (2003) Regulation of extracellular signal-regulated kinase by cannabinoids in hippocampus. J Neurosci 23(6):2371–2382
Diazgranados N, Ibrahim L, Brutsche NE, Newberg A, Kronstein P, Khalife S, Kammerer WA, Quezado Z, Luckenbaugh DA, Salvadore G, Machado-Vieira R, Manji HK, Zarate CA Jr (2010) A randomized add-on trial of an N-methyl-D-aspartate antagonist in treatment-resistant bipolar depression. Arch Gen Psychiatry 67(8):793–802
Dowlati Y, Herrmann N, Swardfager W, Liu H, Sham L, Reim EK, Lanctôt KL (2010) A meta-analysis of cytokines in major depression. Biol Psychiatry 67(5):446–457
Dragicevic N, Bradshaw PC, Mamcarz M, Lin X, Wang L, Cao C, Arendash GW (2011) Long-term electromagnetic field treatment enhances brain mitochondrial function of both Alzheimer's transgenic mice and normal mice: a mechanism for electromagnetic field-induced cognitive benefit? Neuroscience 185:135–149
Drzyzga ŠR, Marcinowska A, Obuchowicz E (2009) Antiapoptotic and neurotrophic effects of antidepressants: a review of clinical and experimental studies. Brain Res Bull 79(5):248–257
Duman RS (2002) Synaptic plasticity and mood disorders. Mol Psychiatry 7(Suppl 1):S29–S34
Duman RS (2009) Neuronal damage and protection in the pathophysiology and treatment of psychiatric illness: stress and depression. Dialogues Clin Neurosci 11(3):239–255
Duman RS, Monteggia LM (2006) A neurotrophic model for stress-related mood disorders. Biol Psychiatry 59(12):1116–1127
Duman RS, Heninger GR, Nestler EJ (1997) A molecular and cellular theory of depression. Arch Gen Psychiatry 54:597–606
Duman RS, Li N, Liu RJ, Duric V, Aghajanian G (2012) Signaling pathways underlying the rapid antidepressant actions of ketamine. Neuropharmacology 62(1):35–41
Einat H, Manji HK (2006) Cellular plasticity cascades: genes-to-behavior pathways in animal models of bipolar disorder. Biol Psychiatry 59(12):1160–1171
ElBatsh MM, Moklas MA, Marsden CA, Kendall DA (2012) Antidepressant-like effects of Δ^9-tetrahydrocannabinol and rimonabant in the olfactory bulbectomised rat model of depression. Pharmacol Biochem Behav 102(2):357–365
Elder BL, Mosack V (2011) Genetics of depression: an overview of the current science. Issues Ment Health Nurs 32(4):192–202
Engelmann M, Landgraf R, Wotjak CT (2004) The hypothalamic-neurohypophysial system regulates the hypothalamic-pituitary-adrenal axis under stress: an old concept revisited. Front Neuroendocrinol 25(3–4):132–149
Esteban S, Garcàa-Sevilla JA (2012) Effects induced by cannabinoids on monoaminergic systems in the brain and their implications for psychiatric disorders. Prog Neuropsychopharmacol Biol Psychiatry 38(1):78–87
Fattal O, Budur K, Vaughan AJ, Franco K (2006) Review of the literature on major mental disorders in adult patients with mitochondrial diseases. Psychosomatics 47(1):1–7
Fava M, Kendler KS (2000) Major depressive disorder. Neuron 28(2):335–341
Fišar Z (2009) Phytocannabinoids and endocannabinoids. Curr. Drug Abuse Rev 2(1):51–75
Fišar Z (2010) Inhibition of monoamine oxidase activity by cannabinoids. Naunyn Schmiedebergs Arch Pharmacol 381(6):563–572
Fišar Z (2012) Cannabinoids and monoamine neurotransmission with focus on monoamine oxidase. Prog Neuropsychopharmacol Biol Psychiatry 38(1):68–77
Fišar Z, Hroudová J (2010a) Common aspects of neuroplasticity, stress, mood disorders and mitochondrial functions. Act Nerv Super Rediviva 52(1):3–20
Fišar Z, Hroudová J (2010b) Intracellular signalling pathways and mood disorders. Folia Biol 56(4):135–148
Fišar Z, Raboch J (2008) Depression, antidepressants, and peripheral blood components. Neuro Endocrinol Lett 29(1):17–28

Fišar Z, Hroudová J, Raboch J (2012) Neurotransmission in mood disorders. In: Mario Francisco Juruena (ed) Clinical, research and treatment approaches to affective disorders. InTech, Rijeka, pp 191–234. Available from: http://www.intechopen.com/articles/show/title/neurotransmission-in-mood-disorders

Fitzgerald ML, Shobin E, Pickel VM (2012) Cannabinoid modulation of the dopaminergic circuitry: implications for limbic and striatal output. Prog Neuropsychopharmacol Biol Psychiatry 38(1):21–29

Fitzgerald PB, Laird AR, Maller J, Daskalakis ZJ (2008) A meta-analytic study of changes in brain activation in depression. Hum Brain Mapp 29(6):683–695

Folstein M, Liu T, Peter I, Buell J, Arsenault L, Scott T, Qiu WW (2007) The homocysteine hypothesis of depression. Am J Psychiatry 164:861–867

Forti P, Rietti E, Pisacane N, Olivelli V, Dalmonte E, Mecocci P, Ravaglia G (2010) Blood homocysteine and risk of depression in the elderly. Arch Gerontol Geriatr 51:21–25

Fowler CJ, Holt S, Nilsson O, Jonsson KO, Tiger G, Jacobsson SO (2005) The endocannabinoid signaling system: pharmacological and therapeutic aspects. Pharmacol Biochem Behav 81(2):248–262

Gass P, Riva MA (2007) CREB, neurogenesis and depression. Bioessays 29:957–961

Gilbody S, Lightfoot T, Sheldon T (2007) Is low folate a risk factor for depression? A meta-analysis and exploration of heterogeneity. J Epidemiol Community Health 61:631–637

Gobbi G, Bambico FR, Mangieri R, Bortolato M, Campolongo P, Solinas M, Cassano T, Morgese MG, Debonnel G, Duranti A, Tontini A, Tarzia G, Mor M, Trezza V, Goldberg SR, Cuomo V, Piomelli D (2005) Antidepressant-like activity and modulation of brain monoaminergic transmission by blockade of anandamide hydrolysis. Proc Natl Acad Sci U S A 102(51):18620–18625

Gorzalka BB, Hill MN (2011) Putative role of endocannabinoid signaling in the etiology of depression and actions of antidepressants. Prog Neuropsychopharmacol Biol Psychiatry 35(7):1575–1585

Gorzalka BB, Hill MN, Hillard CJ (2008) Regulation of endocannabinoid signaling by stress: implications for stress-related affective disorders. Neurosci Biobehav Rev 32(6):1152–1160

Gould TD, Manji HK (2005) Glycogen synthase kinase-3: a putative molecular target for lithium mimetic drugs. Neuropsychopharmacology 30:1223–1237

Gould TD, Dow ER, O'Donnell KC, Chen G, Manji HK (2007) Targeting signal transduction pathways in the treatment of mood disorders: recent insights into the relevance of the Wnt pathway. CNS Neurol Disord Drug Targets 6(3):193–204

Green R (2011) Indicators for assessing folate and vitamin B-12 status and for monitoring the efficacy of intervention strategies. Am J Clin Nutr 94:666S–672S

Griebel G, Stemmelin J, Scatton B (2005) Effects of the cannabinoid CB1 receptor antagonist rimonabant in models of emotional reactivity in rodents. Biol Psychiatry 57(3):261–267

Gu P, Defina LF, Leonard D, John S, Weiner MF, Brown ES (2012) Relationship between serum homocysteine levels and depressive symptoms: the Cooper Center Longitudinal Study. J Clin Psychiatry 73:691–695

Haddad JJ, Saadé NE, Safieh-Garabedian B (2002) Cytokines and neuro-immune-endocrine interactions: a role for the hypothalamic-pituitary-adrenal revolving axis. J Neuroimmunol 133(1–2):1–19

Heninger GR, Delgado PL, Charney DS (1996) The revised monoamine theory of depression: a modulatory role for monoamines, based on new findings from monoamine depletion experiments in humans. Pharmacopsychiatry 29(1):2–11

Hercher C, Turecki G, Mechawar N (2009) Through the looking glass: examining neuroanatomical evidence for cellular alterations in major depression. J Psychiatr Res 43(11):947–961

Hill MN, Gorzalka BB (2005) Is there a role for the endocannabinoid system in the etiology and treatment of melancholic depression? Behav Pharmacol 16(5–6):333–352

Hill MN, Patel S, Carrier EJ, Rademacher DJ, Ormerod BK, Hillard CJ, Gorzalka BB (2005) Down-regulation of endocannabinoid signaling in the hippocampus following chronic unpredictable stress. Neuropsychopharmacology 30(3):508–515

Hill MN, Sun JC, Tse MT, Gorzalka BB (2006) Altered responsiveness of serotonin receptor subtypes following long-term cannabinoid treatment. Int J Neuropsychopharmacol 9(3):277–286

Hill MN, Kambo JS, Sun JC, Gorzalka BB, Galea LA (2006) Endocannabinoids modulate stress-induced suppression of hippocampal cell proliferation and activation of defensive behaviours. Eur J Neurosci 24(7):1845–1849

Hill MN, Carrier EJ, McLaughlin RJ, Morrish AC, Meier SE, Hillard CJ, Gorzalka BB (2008) Regional alterations in the endocannabinoid system in an animal model of depression: effects of concurrent antidepressant treatment. J Neurochem 106(6):2322–2336

Hill MN, Miller GE, Carrier EJ, Gorzalka BB, Hillard CJ (2009) Circulating endocannabinoids and *N*-acyl ethanolamines are differentially regulated in major depression and following exposure to social stress. Psychoneuroendocrinology 34(8):1257–1262

Hill MN, Hellemans KG, Verma P, Gorzalka BB, Weinberg J (2012) Neurobiology of chronic mild stress: parallels to major depression. Neurosci Biobehav Rev 36(9):2085–2117

Hillard CJ, Weinlander KM, Stuhr KL (2012) Contributions of endocannabinoid signaling to psychiatric disorders in humans: genetic and biochemical evidence. Neuroscience 204:207–229

Holsboer F, Liebl R, Hofschuster E (1982) Repeated dexamethasone suppression test during depressive illness. Normalisation of test result compared with clinical improvement. J Affect Disord 4:93–101

Hroudova J, Fisar Z (2010) Activities of respiratory chain complexes and citrate synthase influenced by pharmacologically different antidepressants and mood stabilizers. Neuro Endocrinol Lett 31(3):336–342

Hroudová J, Fišar Z (2011) Connectivity between mitochondrial functions and psychiatric disorders. Psychiatry Clin Neurosci 65(2):130–141

Hroudová J, Fišar Z (2012) In vitro inhibition of mitochondrial respiratory rate by antidepressants. Toxicol Lett 213(3):345–352

Hungund BL, Vinod KY, Kassir SA, Basavarajappa BS, Yalamanchili R, Cooper TB, Mann JJ, Arango V (2004) Up-regulation of CB_1 receptors and agonist-stimulated [^{35}S]GTPγS binding in the prefrontal cortex of depressed suicide victims. Mol Psychiatry 9(2):184–190

Im HI, Kenny PJ (2012) MicroRNAs in neuronal function and dysfunction. Trends Neurosci 35(5):325–334

Ising M, Künzel HE, Binder EB, Nickel T, Modell S, Holsboer F (2005) The combined dexamethasone/CRH test as a potential surrogate marker in depression. Prog Neuropsychopharmacol Biol Psychiatry 29:1085–1093

Janssen DG, Caniato RN, Verster JC, Baune BT (2010) A psychoneuroimmunological review on cytokines involved in antidepressant treatment response. Hum Psychopharmacol 25(3):201–215

Jiang W, Zhang Y, Xiao L, Van Cleemput J, Ji SP, Bai G, Zhang X (2005) Cannabinoids promote embryonic and adult hippocampus neurogenesis and produce anxiolytic- and antidepressant-like effects. J Clin Invest 115(11):3104–3116

Jimerson DC, Post RM, Carman JS, Kammen DP van, Wood JH, Goodwin FK, Bunney WE Jr (1979) CSF calcium: clinical correlates in affective illness and schizophrenia. Biol Psychiatry 14(1):37–51

Jou S-H, Chiu N-Y, Liu C-S (2009) Mitochondrial dysfunction and psychiatric disorders. Chang Gung Med J 32(4):370–379

Kasper S, McEwen BS (2008) Neurobiological and clinical effects of the antidepressant tianeptine. CNS Drugs 22(1):15–26

Kato T (2007) Mitochondrial dysfunction as the molecular basis of bipolar disorder: therapeutic implications. CNS Drugs 21(1):1–11

Kato T (2008) Role of mitochondrial DNA in calcium signaling abnormality in bipolar disorder. Cell Calcium 44(1):92–102

Kato T, Kato N (2000) Mitochondrial dysfunction in bipolar disorder. Bipolar Disord 2(3 Pt 1):180–190

Katyare SS, Rajan RR (1995) Effect of long-term in vivo treatment with imipramine on the oxidative energy metabolism in rat brain mitochondria. Comp Biochem Physiol C Pharmacol Toxicol Endocrinol 112(3):353–357

Krishnan V, Nestler EJ (2008) The molecular neurobiology of depression. Nature 455(7215): 894–902

Krishnan V, Nestler EJ (2010) Linking molecules to mood: new insight into the biology of depression. Am J Psychiatry 167(11):1305–1320

Lamers F, Vogelzangs N, Merikangas KR, Jonge P de, Beekman AT, Penninx BW (2013) Evidence for a differential role of HPA-axis function, inflammation and metabolic syndrome in melancholic versus atypical depression. Mol Psychiatry. 18(6):692–699

Lazarou C, Kapsou M (2010) The role of folic acid in prevention and treatment of depression: an overview of existing evidence and implications for practice. Complement Ther Clin Pract 16:161–166

Leonard BE (2005) The HPA and immune axes in stress: the involvement of the serotonergic system. Eur Psychiatry 20(Suppl 3):S302–S306

Leonard BE, Myint A (2009) The psychoneuroimmunology of depression. Hum Psychopharmacol 24(3):165–175

Leonard B, Maes M (2012) Mechanistic explanations how cell-mediated immune activation, inflammation and oxidative and nitrosative stress pathways and their sequels and concomitants play a role in the pathophysiology of unipolar depression. Neurosci Biobehav Rev 36(2):764–785

Maes M, Yirmyia R, Noraberg J, Brene S, Hibbeln J, Perini G, Kubera M, Bob P, Lerer B, Maj M (2009) The inflammatory & neurodegenerative (I & ND) hypothesis of depression: leads for future research and new drug developments in depression. Metab Brain Dis 24(1):27–53

Maes M, Kubera M, Obuchowiczwa E, Goehler L, Brzeszcz J (2011) Depression's multiple comorbidities explained by (neuro)inflammatory and oxidative & nitrosative stress pathways. Neuro Endocrinol Lett 32(1):7–24

Maes M, Fišar Z, Medina M, Scapagnini G, Nowak G, Berk M (2012) New drug targets in depression: inflammatory, cell-mediated immune, oxidative and nitrosative stress, mitochondrial, antioxidant, and neuroprogressive pathways. And new drug candidates—Nrf2 activators and GSK-3 inhibitors. Inflammopharmacology 20(3):127–150

Marchalant Y, Brothers HM, Norman GJ, Karelina K, DeVries AC, Wenk GL (2009) Cannabinoids attenuate the effects of aging upon neuroinflammation and neurogenesis. Neurobiol Dis 34(2):300–307

Martin M, Ledent C, Parmentier M, Maldonado R, Valverde O (2002) Involvement of CB_1 cannabinoid receptors in emotional behaviour. Psychopharmacology (Berl) 159(4):379–387

Mathew SJ, Manji HK, Charney DS (2008) Novel drugs and therapeutic targets for severe mood disorders. Neuropsychopharmacology 33(9):2080–2092

Mattson MP, Shea TB (2003) Folate and homocysteine metabolism in neural plasticity and neurodegenerative disorders. Trends Neurosci 26:137–146

Maurer IC, Schippel P, Volz H-P (2009) Lithium-induced enhancement of mitochondrial oxidative phosphorylation in human brain tissue. Bipolar Disord 11(5):515–522

Mayberg HS (1997) Limbic-cortical dysregulation: a proposed model of depression. J Neuropsychiatry Clin Neurosci 9(3):471–481

Mayberg HS (2003) Modulating dysfunctional limbic-cortical circuits in depression: towards development of brain-based algorithms for diagnosis and optimised treatment. Br Med Bull 65:193–207

Mayberg HS (2009) Targeted electrode-based modulation of neural circuits for depression. J Clin Investigation 119:717–725

Mayberg HS, Brannan SK, Tekell JL, Silva JA, Mahurin RK, McGinnis S, Jerabek PA (2000) Regional metabolic effects of fluoxetine in major depression: serial changes and relationship to clinical response. Biol Psychiatry 48(8):830–843

McNamara RK, Ostrander M, Abplanalp W, Richtand NM, Benoit SC, Clegg DJ (2006) Modulation of phosphoinositide-protein kinase C signal transduction by omega-3 fatty acids: implications for the pathophysiology and treatment of recurrent neuropsychiatric illness. Prostaglandins Leukot Essent Fatty Acids 75(4–5):237–257

Mendelsohn D, Riedel WJ, Sambeth A (2009) Effects of acute tryptophan depletion on memory, attention and executive functions: a systematic review. Neurosci Biobehav Rev 33(6):926–952

Mendlewicz J (2009) Disruption of the circadian timing systems: molecular mechanisms in mood disorders. CNS Drugs 23(Suppl 2):15–26

Mesulam MM (1999) Neuroplasticity failure in Alzheimer's disease: bridging the gap between plaques and tangles. Neuron 24(3):521–529

Meyer JH (2012) Neuroimaging markers of cellular function in major depressive disorder: implications for therapeutics, personalized medicine, and prevention. Clin Pharmacol Ther 91(2):201–214

Meyer JH, Ginovart N, Boovariwala A, Sagrati S, Hussey D, Garcia A, Young T, Praschak-Rieder N, Wilson AA, Houle S (2006) Elevated monoamine oxidase a levels in the brain: an explanation for the monoamine imbalance of major depression. Arch Gen Psychiatry 63(11):1209–1216

Meyer JH, Wilson AA, Sagrati S, Miler L, Rusjan P, Bloomfield PM, Clark M, Sacher J, Voineskos AN, Houle S (2009) Brain monoamine oxidase A binding in major depressive disorder: relationship to selective serotonin reuptake inhibitor treatment, recovery, and recurrence. Arch Gen Psychiatry 66(12):1304–1312

Mill J, Petronis A (2007) Molecular studies of major depressive disorder: the epigenetic perspective. Mol Psychiatry 12(9):799–814

Miller AL (2008) The methylation, neurotransmitter, and antioxidant connections between folate and depression. Altern Med Rev 13:216–226

Mirnikjoo B, Brown SE, Kim HF, Marangell LB, Sweatt JD, Weeber EJ (2001) Protein kinase inhibition by ω-3 fatty acids. J Biol Chem 276(14):10888–10896

Morrish AC, Hill MN, Riebe CJ, Gorzalka BB (2009) Protracted cannabinoid administration elicits antidepressant behavioral responses in rats: role of gender and noradrenergic transmission. Physiol Behav 98(1–2):118–124

Nahon E, Israelson A, Abu-Hamad S, Varda SB (2005) Fluoxetine (Prozac) interaction with the mitochondrial voltage-dependent anion channel and protection against apoptotic cell death. FEBS Lett 579(22):5105–5110

Nestler EJ, Barrot M, DiLeone RJ, Eisch AJ, Gold SJ, Monteggia LM (2002) Neurobiology of depression. Neuron 34(1):13–25

Nikisch G (2009) Involvement and role of antidepressant drugs of the hypothalamic-pituitary-adrenal axis and glucocorticoid receptor function. Neuro Endocrinol Lett 30(1):11–16

Oropeza VC, Page ME, Van Bockstaele EJ (2005) Systemic administration of WIN 55,212–2 increases norepinephrine release in the rat frontal cortex. Brain Res 1046(1–2):45–54

Pacher P, Mechoulam R (2011) Is lipid signaling through cannabinoid 2 receptors part of a protective system? Prog Lipid Res 50(2):193–211

Pariante CM (2009) Risk factors for development of depression and psychosis. Glucocorticoid receptors and pituitary implications for treatment with antidepressant and glucocorticoids. Ann N Y Acad Sci 1179:144–152

Parolaro D, Realini N, Vigano D, Guidali C, Rubino T (2010) The endocannabinoid system and psychiatric disorders. Exp Neurol 224(1):3–14

Patel S, Roelke CT, Rademacher DJ, Cullinan WE, Hillard CJ (2004) Endocannabinoid signaling negatively modulates stress-induced activation of the hypothalamic-pituitary-adrenal axis. Endocrinology 145(12):5431–5438

Patton GC, Coffey C, Carlin JB, Degenhardt L, Lynskey M, Hall W (2002) Cannabis use and mental health in young people: cohort study. BMJ 325(7374):1195–1198

Paul IA, Skolnick P (2003) Glutamate and depression: clinical and preclinical studies. Ann N Y Acad Sci 1003:250–272

Pedersen W (2008) Does cannabis use lead to depression and suicidal behaviours? A population-based longitudinal study. Acta Psychiatr Scand 118(5):395–403

Phillips ML, Drevets WC, Rauch SL, Lane R (2003) Neurobiology of emotion perception II: implications for major psychiatric disorders. Biol Psychiatry 54(5):515–528

Pittenger C, Duman RS (2008) Stress, depression, and neuroplasticity: a convergence of mechanisms. Neuropsychopharmacology 33(1):88–109

Porcelli S, Drago A, Fabbri C, Serretti A (2011) Mechanisms of antidepressant action: an integrated dopaminergic perspective. Prog Neuropsychopharmacol Biol Psychiatry 35(7):1532–1543

Przegaliński E, Göthert M, Frankowska M, Filip M (2005) WIN 55,212-2-induced reduction of cocaine hyperlocomotion: possible inhibition of 5-HT_3 receptor function. Eur J Pharmacol 517(1–2):68–73

Quiroz JA, Gray NA, Kato T, Manji HK (2008) Mitochondrially mediated plasticity in the pathophysiology and treatment of bipolar disorder. Neuropsychopharmacology 33(11):2551–2565

Refsum H, Nurk E, Smith AD, Ueland PM, Gjesdal CG, Bjelland I, Tverdal A, Tell GS, Nygård O, Vollset SE (2006) The Hordaland Homocysteine Study: a community-based study of homocysteine, its determinants, and associations with disease. J Nutr 136:1731S–1740S

Rodràguez-Gaztelumendi A, Rojo ML, Pazos A, Dàaz A (2009) Altered CB_1 receptor-signaling in prefrontal cortex from an animal model of depression is reversed by chronic fluoxetine. J Neurochem 108(6):1423–1433

Rubino T, Vigano' D, Realini N, Guidali C, Braida D, Capurro V, Castiglioni C, Cherubino F, Romualdi P, Candeletti S, Sala M, Parolaro D (2008) Chronic Δ^9-tetrahydrocannabinol during adolescence provokes sex-dependent changes in the emotional profile in adult rats: behavioral and biochemical correlates. Neuropsychopharmacology 33(11):2760–2771

Rubino T, Realini N, Braida D, Alberio T, Capurro V, Viganò D, Guidali C, Sala M, Fasano M, Parolaro D (2009) The depressive phenotype induced in adult female rats by adolescent exposure to THC is associated with cognitive impairment and altered neuroplasticity in the prefrontal cortex. Neurotox Res 15(4):291–302

Ruhé HG, Mason NS, Schene AH (2007) Mood is indirectly related to serotonin, norepinephrine and dopamine levels in humans: a meta-analysis of monoamine depletion studies. Mol Psychiatry 12(4):331–359

Russo EB, Burnett A, Hall B, Parker KK (2005) Agonistic properties of cannabidiol at 5-HT1a receptors. Neurochem Res 30(8):1037–1043

Sagredo O, Ramos JA, Fernández-Ruiz J, Rodràguez ML, Miguel R de (2006) Chronic Δ^9-tetrahydrocannabinol administration affects serotonin levels in the rat frontal cortex. Naunyn Schmiedebergs Arch Pharmacol 372:313–317

Santarelli L, Saxe M, Gross C, Surget A, Battaglia F, Dulawa S, Weisstaub N, Lee J, Duman R, Arancio O, Belzung C, Hen R (2003) Requirement of hippocampal neurogenesis for the behavioral effects of antidepressants. Science 301:805–809

Sapolsky RM (2004) Is impaired neurogenesis relevant to the affective symptoms of depression? Biol Psychiatry 56(3):137–139

Savage DG, Lindenbaum J, Stabler SP, Allen RH (1994) Sensitivity of serum methylmalonic acid and total homocysteine determinations for diagnosing cobalamin and folate deficiencies. Am J Med 96:239–246

Savitz J, Drevets WC (2009) Bipolar and major depressive disorder: neuroimaging the developmental-degenerative divide. Neurosci Biobehav Rev 33(5):699–771

Savitz J, Lucki I, Drevets WC (2009) 5-HT1A receptor function in major depressive disorder. Prog Neurobiol 88(1):17–31

Savitz JB, Drevets WC (2013) Neuroreceptor imaging in depression. Neurobiol Dis 52:49–65

Schildkraut JJ (1965) The catecholamine hypothesis of affective disorders: a review of supporting evidence. Am J Psychiatry 122(5):509–522

Schmidt HD, Shelton RC, Duman RS (2011) Functional biomarkers of depression: diagnosis, treatment, and pathophysiology. Neuropsychopharmacology 36(12):2375–2394

Schroeder FA, Lin CL, Crusio WE, Akbarian S (2007) Antidepressant-like effects of the histone deacetylase inhibitor, sodium butyrate, in the mouse. Biol Psychiatry 62(1):55–64

Schulz P, Steimer T (2009) Neurobiology of circadian systems. CNS Drugs 23(Suppl 2):3–13

Selvaraj S, Murthy NV, Bhagwagar Z, Bose SK, Hinz R, Grasby PM, Cowen PJ (2011) Diminished brain 5-HT transporter binding in major depression: a positron emission tomography study with [^{11}C]DASB. Psychopharmacology (Berl) 213(2–3):555–562

Shaltiel G, Chen G, Manji HK (2007) Neurotrophic signaling cascades in the pathophysiology and treatment of bipolar disorder. Curr Opin Pharmacol 7:22–26

Shao L, Martin MV, Watson SJ, Schatzberg A, Akil H, Myers RM, Jones EG, Bunney WE, Vawter MP (2008) Mitochondrial involvement in psychiatric disorders. Ann Med 40(4):281–295

Shearman LP, Rosko KM, Fleischer R, Wang J, Xu S, Tong XS, Rocha BA (2003) Antidepressant-like and anorectic effects of the cannabinoid CB_1 receptor inverse agonist AM251 in mice. Behav Pharmacol 14(8):573–582

Stavrovskaya IG, Narayanan MV, Zhang W, Krasnikov BF, Heemskerk J, Young SS, Blass JP, Brown AM, Beal MF, Friedlander RM, Kristal BS (2004) Clinically approved heterocyclics act on a mitochondrial target and reduce stroke-induced pathology. J Exp Med 200(2):211–222

Steiner MA, Wanisch K, Monory K, Marsicano G, Borroni E, Bächli H, Holsboer F, Lutz B, Wotjak CT (2008) Impaired cannabinoid receptor type 1 signaling interferes with stress-coping behavior in mice. Pharmacogenomics J 8(3):196–208

Stork C, Renshaw PF (2005) Mitochondrial dysfunction in bipolar disorder: evidence from magnetic resonance spectroscopy research. Mol Psychiatry 10(10):900–919

Tiemeier H, Tuijl HR van, Hofman A, Meijer J, Kiliaan AJ, Breteler MM (2002) Vitamin B_{12}, folate, and homocysteine in depression: the Rotterdam Study. Am J Psychiatry 159:2099–2101

Tolmunen T, Hintikka J, Voutilainen S, Ruusunen A, Alfthan G, Nyyssönen K, Viinamäki H, Kaplan GA, Salonen JT (2004) Association between depressive symptoms and serum concentrations of homocysteine in men: a population study. Am J Clin Nutr 80:1574–1578

Tsankova N, Renthal W, Kumar A, Nestler EJ (2007) Epigenetic regulation in psychiatric disorders. Nat Rev Neurosci 8(5):355–367

Uher R (2008) The implications of gene-environment interactions in depression: will cause inform cure? Mol Psychiatry 13(12):1070–1078

Urigüen L, Pérez-Rial S, Ledent C, Palomo T, Manzanares J (2004) Impaired action of anxiolytic drugs in mice deficient in cannabinoid CB_1 receptors. Neuropharmacology 46(7):966–973

Valvassori SS, Rezin GT, Ferreira CL, Moretti M, Gonçalves CL, Cardoso MR, Streck EL, Kapczinski F, Quevedo J (2010) Effects of mood stabilizers on mitochondrial respiratory chain activity in brain of rats treated with d-amphetamine. J Psychiatr Res 44(14):903–909

Velenovská M, Fišar Z (2007) Effect of cannabinoids on platelet serotonin uptake. Addict Biol 12:158–166

Vreeburg SA, Hoogendijk WJ, Pelt J van, Derijk RH, Verhagen JC, Dyck R van, Smit JH, Zitman FG, Penninx BW (2009) Major depressive disorder and hypothalamic-pituitary-adrenal axis activity: results from a large cohort study. Arch Gen Psychiatry 66:617–626

Warner-Schmidt JL, Duman RS (2006) Hippocampal neurogenesis: opposing effects of stress and antidepressant treatment. Hippocampus 16(3):239–249

World Health Organization (WHO) (2004) The global burden of disease, 2004 update, part 3: Disease incidence, prevalence and disability. http://www.who.int/healthinfo/global_burden_disease/GBD_report_2004update_part3.pdf

World Health Organization (2012) Depression. http://www.who.int/mental_health/management/depression/en/

Yasuda S, Liang MH, Marinova Z, Yahyavi A, Chuang DM (2009) The mood stabilizers lithium and valproate selectively activate the promoter IV of brain-derived neurotrophic factor in neurons. Mol Psychiatry 14(1):51–59

Yildiz-Yesiloglu A, Ankerst DP (2006) Neurochemical alterations of the brain in bipolar disorder and their implications for pathophysiology: a systematic review of the in vivo proton magnetic resonance spectroscopy findings. Prog Neuropsychopharmacol Biol Psychiatry 30(6):969–995

Zarate CA Jr, Singh J, Manji HK (2006a) Cellular plasticity cascades: targets for the development of novel therapeutics for bipolar disorder. Biol Psychiatry 59(11):1006–1020

Zarate CA Jr , Singh JB, Carlson PJ, Brutsche NE, Ameli R, Luckenbaugh DA, Charney DS, Manji HK (2006b) A randomized trial of an N-methyl-D-aspartate antagonist in treatment-resistant major depression. Arch Gen Psychiatry 63(8):856–864

Zhang WH, Wang H, Wang X, Narayanan MV, Stavrovskaya IG, Kristal BS, Friedlander RM (2008) Nortriptyline protects mitochondria and reduces cerebral ischemia/hypoxia injury. Stroke 39(2):455–462

Zunszain PA, Anacker C, Cattaneo A, Carvalho LA, Pariante CM (2011) Glucocorticoids, cytokines and brain abnormalities in depression. Prog Neuropsychopharmacol Biol Psychiatry 35(3):722–729

Chapter 7
Anatomical, Biochemical, and Behavioral Evidence for Cannabinoid Modulation of Noradrenergic Circuits: Role of Norepinephrine in Cannabinoid-Induced Aversion

Ana Franky Carvalho and Elisabeth J. Van Bockstaele

Abstract The noradrenergic system plays a critical role in the modulation of emotional state, primarily related to anxiety, arousal, and stress. Recent evidence suggests that the endocannabinoid system mediates stress responses and emotional homeostasis, in part, by targeting noradrenergic circuits. This chapter summarizes our current knowledge regarding the anatomical substrates underlying regulation of noradrenergic circuitry by the endocannabinoid system. It then presents biochemical and functional evidence showing an important effect of cannabinoid modulation on adrenergic receptor signaling. Finally, the impact of this interaction with respect to specific behaviors is explored, demonstrating that norepinephrine is a critical determinant of cannabinoid-induced aversion, which adds another dimension to how central noradrenergic circuitry is regulated by the cannabinoid system.

7.1 Introduction

The widespread effects of cannabis use on human physiology and behavior have led to a significant interest in understanding substrates of the endocannabinoid system. The identification of major components of the endocannabinoid system, such as characterization of its main receptors (cannabinoid-1 receptor (CB1r) and cannabinoid-2 receptor (CB2r)) (Matsuda et al. 1990; Munro et al. 1993) and endogenous ligands (such as anandamide (Devane et al. 1992) and 2-arachidonoylglycerol

A. F. Carvalho (✉)
Life and Health Science Research Institute (ICVS), School of Health Sciences, University of Minho, Braga, Portugal

ICVS/3B's – PT Government Associate Laboratory, Braga/Guimarães, Portugal
e-mail: arfranky@gmail.com

E. J. Van Bockstaele
Department of Neuroscience, Farber Institute for Neurosciences, Thomas Jefferson University, 900 Walnut Street, Suite 417, Philadelphia, PA 19107, USA
e-mail: ejv101@jefferson.edu

E. J. Van Bockstaele (ed.), *Endocannabinoid Regulation of Monoamines in Psychiatric and Neurological Disorders*, DOI 10.1007/978-1-4614-7940-6_7,

(2-AG) (Mechoulam et al. 1995; Sugiura et al. 1995)), has provided important insight into this novel modulatory system. One of the most striking effects of cannabis involves its psychotropic effects, such as, but not limited to, euphoria, sedation, altered perception, and impairments of memory and motor control. These behavioral observations were accompanied by the anatomical identification of the widespread distribution of cannabinoid receptors in the central nervous system (Herkenham et al. 1991). The recognition of a diverse expression profile of cannabinoid receptors in the brain has refocused and provided novel insights into several neuropsychiatric disorders. Moreover, the description of the endocannabinoid system as a "retrograde signaling system," first described by Llano et al. (1991), underlies the potential role of this ligand as a modulator of other neurotransmitter systems including acetylcholine, glutamate, dopamine, norepinephrine (NE), and serotonin. NE is considered an important brain neuromodulatory system (Sara 2009) implicated in brain arousal and critical for attention, cognition, and memory consolidation (Aston-Jones et al. 1991). Noradrenergic dysfunction is often seen in neuropsychiatric disorders that arise following chronic stress (Southwick et al. 1993). The noradrenergic system continues to be an important target in the development of new therapies because of its critical role in the modulation of emotional state and regulation of arousal and stress responses (Heninger and Charney 1988; Charney et al. 1989; Carrasco and Van de Kar 2003). On the other hand, the use of synthetic cannabinoid receptor agonists/antagonists or compounds targeting endocannabinoid synthesis/metabolism in the brain has received widespread attention as these approaches may hold some therapeutic potential for psychiatric disorders (Witkin et al. 2005a, b;). Cannabinoid ligands have been shown to alleviate depressive- and anxiety-like behaviors in preclinical studies (Gobbi et al. 2005; Hill and Gorzalka 2005a). However, the cannabinoid receptor antagonist, rimonabant, was withdrawn from clinical trials because of an unacceptably high incidence of psychiatric side effects (Nissen et al. 2008; Després et al. 2009; Janero and Makriyannis 2009). This has generated significant interest in understanding the regulation of endogenous cannabinoid signaling in psychiatric disorders and has stimulated investigations into manipulating endogenous cannabinoids for potential clinical benefit (Lee et al. 2009; Moreira et al. 2009).

This chapter reviews results of anatomical, biochemical, electrophysiological, and behavioral studies that demonstrate significant cannabinoid–adrenergic interactions in the coeruleocortical and limbic circuits that may provide the basis for the development of novel treatment strategies for psychiatric disorders.

7.2 The Interaction Between the Endocannabinoid and Noradrenergic Systems

A growing body of evidence suggests that the endocannabinoid and noradrenergic systems interact at a behavioral level because of common effects on mood and cognition and at a cellular level on the basis of putative common signaling pathways.

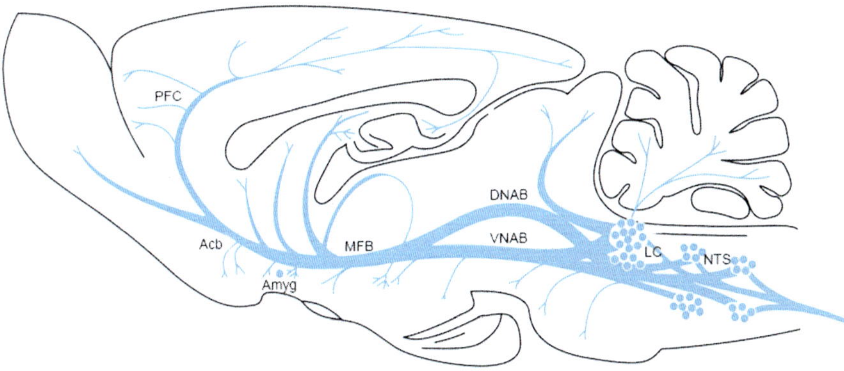

Fig. 7.1 Distribution of central noradrenergic nuclei and projections in rat brain. The locus coeruleus (*LC*) is situated within the dorsal pons and sends its efferent projections via the dorsal noradrenergic ascending bundle (*DNAB*). The nucleus of the solitary tract (*NTS*), localized within the caudal brainstem, sends its projections via the ventral noradrenergic ascending bundle (*VNAB*). *Acb* nucleus accumbens, *Amyg* amygdala, *MFB* medial forebrain bundle, *PFC* prefrontal cortex. (Modified with permission from Koob 2008)

Briefly, increasing endocannabinoid tone has been shown to improve mood just as increasing noradrenergic tone with antidepressants shows improvements in mood (Bond et al. 2008; Morrish et al. 2009). In this study, the antidepressant effects of chronic CB1r agonist administration, as measured by a reduction in immobility and an increase in climbing behavior in the forced swim test (FST), was dependent on the activity of both α1-adrenergic receptor (AR) and β-AR (Morrish et al. 2009). Moreover, overactivation of the endocannabinoid system can cause mania (Henquet et al. 2006), a side effect that has been reported in patients using antidepressants (Peet 1994; Bond et al. 2008; Tondo et al. 2010). Next, the anatomical basis for putative cannabinoid–adrenergic interactions is discussed followed by evidence supporting functional interactions between the two systems.

7.2.1 *Anatomical Localization of the Endocannabinoid System in Noradrenergic Circuits*

7.2.1.1 CB1r

The noradrenergic system has its cell bodies grouped in nuclei in the brainstem, namely the locus coeruleus (LC) and the nucleus of the solitary tract (NTS) (Foote et al. 1983; Weinshenker and Schroeder 2007; Itoi and Sugimoto 2010) (Fig. 7.1). While the LC is a homogeneous nucleus in which most cells are noradrenergic (Foote et al. 1983), the NTS contains several other neurotransmitters (Barraco et al. 1992). The noradrenergic neurons of the NTS are distributed throughout the caudal NTS (subpostremal and commissural NTS) (Barraco et al. 1992). The LC, located within the dorsal wall of the rostral pons, in the lateral floor of the fourth ventricle, is the largest noradrenergic nucleus in the brain (Foote et al. 1983) and, through the dorsal noradrenergic bundle, is the sole source of NE in the prefrontal cortex (PFC) (Sara 2009). The LC is seen as the "arousal" center, important for regulation of sleep and vigilance, and activation of the LC is important for selective attention (Southwick et al. 1999; Sara 2009). On the other hand, the NTS works as relay station for sensory signals arising from the viscera, integrating visceral information with other regulatory information coming from the brainstem, diencephalon, and forebrain (Barraco et al. 1992; Itoi and Sugimoto 2010). The NTS is known to send not only efferents to autonomic centers in the brainstem but also ascending efferents to higher levels of the neuroaxis (Barraco et al. 1992), through the ventral noradrenergic bundle.

In this respect, autoradiographic binding studies have shown the existence of a moderate density of CB1r protein and mRNA in the LC and NTS (Herkenham et al. 1991; Mailleux and Vanderhaeghen 1992; Matsuda et al. 1993; Derbenev et al. 2004; Jelsing et al. 2008). Some studies have shown, by dual immunohistochemistry with dopamine beta hydroxylase (DBH) or tyrosine hydroxylase (TH), that some of the CB1r-positive neurons in the LC (Scavone et al. 2006, 2010) and NTS (Carvalho et al. 2010a) are noradrenergic. Moreover, electron microscopy analysis revealed that most of CB1r found in the LC are postsynaptic (Scavone et al. 2010). The role of postsynaptic CB1r is not yet fully understood, although it has been described that postsynaptic CB1r can inhibit cortical interneurons in an autocrine manner (Bacci et al. 2004). Interestingly, most of postsynaptic CB1r were found in the cytoplasm, which may include trafficking of CB1r to dendritic processes or LC terminals in target regions, such as the PFC ((Scavone et al. 2010), see following text). In fact, it has been described that CB1r show an endocytosis and recycling cycle at the somatodendritic compartment of hippocampal neurons and that it is required for the proper axonal targeting of CB1r (Leterrier et al. 2006). In the study by Scavone et al. (2010), it is also shown that CB1r localized to postsynaptic profiles receive mostly asymmetric (excitatory) synapses. One can speculate that upon activation by excitatory (glutamatergic) terminals, cannabinoids are produced by and act on postsynaptic CB1r, thus directly inhibiting transmission without altering glutamate transmission. CB1r was also detected within presynaptic profiles in the LC, where the synaptic specializations were more commonly of the symmetric (inhibitory) type. Symmetric (inhibitory) synapses are thought to be γ-aminobutyric acid–ergic (GABAergic), thus suggesting that cannabinoids in the LC can have a greater impact on GABAergic transmission than on glutamatergic transmission. It seems that cannabinoids in the LC may mediate different signal transduction pathways depending on CB1r localization, pre- vs. postsynaptic localization. Regarding the NTS, there are many studies supporting the importance of cannabinoids in this nucleus (see Sect. 7.2.2.2). Many studies have shown the presence of CB1r mRNA and protein in the NTS (Herkenham et al. 1991; Mailleux and Vanderhaeghen 1992; Matsuda et al. 1993; Derbenev et al. 2004; Jelsing et al.

2008; Carvalho et al. 2010a). However, not many studies have characterized the exact neuronal population positive for CB1r (whether they are catecholaminergic, serotoninergic, dopaminergic, GABAergic, or cholinergic neurons).

Interestingly, the PFC and the nucleus accumbens (Acb), two brain regions that receive highly processed information involved in some of the symptoms of psychiatric disorders and receive noradrenergic afferents from the LC and NTS, respectively, show a very different pattern of CB1r distribution with respect to noradrenergic terminals. In the PFC, CB1r can be found in noradrenergic terminals (approximately 30% of CB1r-positive fibers were noradrenergic) (Oropeza et al. 2007), whereas in the Acb, the percentage of colocalization of CB1r and DBH is very low (Carvalho et al. 2010a). This may reflect differential modulation of NE by endocannabinoids in these two regions. In support of this, systemic WIN 55,212-2 administration differentially impacts AR expression in the PFC and Acb (Carvalho et al. 2010a) (see following text).

In the Acb, CB1r shows an interesting topographical distribution, with higher CB1r expression being found in the shell of the Acb at midrostral levels and higher CB1r expression in the core of the Acb at caudal levels (Carvalho et al. 2010a). The heterogeneous distribution of CB1r throughout the Acb may reflect different abilities of the endocannabinoid system to modulate behavior in the Acb. It is proposed that the Acb subregions (shell and core) can be further subdivided with respect to function (Zahm 1999). For instance, anatomical and behavioral studies support a rostrocaudal gradient for appetitive vs. aversive behaviors (Reynolds and Berridge 2001, 2002, 2003). It is tempting to speculate that on the basis of the heterogeneous distribution of CB1r, certain behaviors are more impacted than others.

7.2.1.2 Other Targets of the Endocannabinoid System

Transient receptor potential vanilloid type 1 (TRPV1) receptors have been shown to be activated by anandamide (Zygmunt et al. 1999), albeit with less efficacy than CB1r activation. High levels of TRPV1 receptors are found in the dorsal root ganglia and lower levels in other tissues such as the brain (Sanchez et al. 2001). In the brain, TRPV1 receptors have been described in the hippocampus, cortex, cerebellum, olfactory bulb, thalamus, LC, and NTS (Mezey et al. 2000; Sanchez et al. 2001; Tóth et al. 2005). In the NTS, TRPV1 are associated with unmyelinated c-fiber afferents (Doyle et al. 2002) and differentially impact glutamate release within the NTS (Peters et al. 2011). Thus, if cannabinoids have the ability to activate TRPV1 receptors (Zygmunt et al. 1999), future studies should take into consideration a participation of these receptors in cannabinoid-induced effects.

Regarding the endocannabinoid-degrading enzymes, such as fatty acid amide hydrolase (FAAH, the main degrading enzyme of anandamide) and monoacylglycerol lipase (MAGL, the main degrading enzyme of 2-AG), only in the NTS, FAAH was found at protein level (Van Sickle et al. 2001).

7.2.2 *Effects of Cannabinoids on Noradrenergic Transmission*

Impairments of the noradrenergic system have been implicated in some of the symptoms of psychiatric disorders, including schizophrenia, anxiety, depression, and post-traumatic stress disorder (PTSD) (Friedman et al. 1999; Southwick et al. 1999; Nutt 2002, 2006; Itoi and Sugimoto 2010). Together with the serotoninergic, cholinergic, and dopaminergic systems, the noradrenergic system is typically viewed as a neuromodulatory system (Sara 2009). NE can interact with three families of ARs: α1, α2, and β(1–3) receptors. α1-ARs are coupled to Gq proteins, hence activating phospholipase C and phosphatidylinositol intracellular pathway, resulting in activation of protein kinase C and release of intracellular calcium (Duman and Nestler 1995). α2-ARs, found pre- and postsynaptically (MacDonald et al. 1997), are coupled to Gi proteins, which can lead to a decrease in intracellular cAMP (Duman and Nestler 1995). Presynaptic localized α2-ARs work as autoreceptors, as activation of these receptors will decrease intracellular cAMP and Ca^{2+}, inhibiting the release of neurotransmitters. β-ARs are coupled to Gs proteins, activating adenylyl cyclase and increasing intracellular cAMP (Duman and Nestler 1995). Several studies have revealed alterations in the levels of AR expression in depressed suicide victims. α2-AR density is increased in brains of depressed suicide victims (Meana et al. 1992; De Paermentier et al. 1997; Callado et al. 1998), whereas β1-AR density is decreased (De Paermentier et al. 1990). These changes were not found throughout the brain, suggesting that specific areas of the brain may be involved in the pathophysiology of mood disorders. Moreover, pharmacological depletion of monoamines (e.g., reserpine) produces depressive-like behaviors in animal models, suggesting a role for monoamines (including NE) in the pathophysiology of depression (Nutt 2006). Additionally, most antidepressant drugs act by increasing the levels of synaptic monoamines. Hence, low levels of NE seem to account for the expression of depressive symptoms. In fact, higher levels of plasma NE were correlated with longer periods of remission to a new depression episode in patients that had suffered their first major depression episode, suggesting a protective effect of NE (Johnston et al. 1999). However, it has also been described that patients with melancholic depression show dysregulation of the hypothalamic–pituitary–adrenal (HPA) axis, with high levels of plasma cortisol and cerebrospinal fluid (CSF) NE being found (Wong et al. 2000). Thus, although the molecular mechanisms underlying depression are still largely unclear, abnormalities of noradrenergic transmission take part on the pathophysiology of depression.

7.2.2.1 Cannabinoid Effects on LC Activity

Several studies have reported an effect of cannabinoids on LC activity. Namely, cannabinoids have been shown to increase LC spontaneous firing (Mendiguren and Pineda 2004, 2006; Muntoni et al. 2006). Patel and Hillard (2003)) showed increased

Fos labeling in noradrenergic neurons in the LC following systemic injection of CP55940 and WIN 55,212-2. In this study, it is also shown that both CB1r agonists increase Fos expression in A10 dopaminergic neurons. However, the activation of dopaminergic neurons by cannabinoid agonists is blocked by an α1-AR antagonist and by an α2-AR agonist, suggesting that CP55940 and WIN 55,212-2 may be activating dopaminergic neurons by acting on LC-NE neurons. In another study, Oropeza et al. (2005)) showed that systemic WIN 55,212-2 induces Fos expression in noradrenergic neurons of the LC. This effect was blocked in the presence of the CB1r antagonist SR141716A, suggesting an effect mediated by CB1r. Recordings from LC-NE neurons in anesthetized rats have shown that systemic and central administration of cannabinoids, dose-dependently, increased the firing rate of the LC (Mendiguren and Pineda 2006; Muntoni et al. 2006). This effect was blocked by administration of the CB1r antagonist SR141716A. Interestingly, administration of SR141716A alone caused a significant reduction of LC spontaneous firing, suggesting that LC is under the control of an endogenous cannabinoid tone. This hypothesis is further supported by evidence showing that URB597, a selective inhibitor of FAAH (the enzyme responsible for degradation of endocannabinoid anandamide) is able to enhance the spontaneous firing rate of LC-NE neurons (Gobbi et al. 2005). These excitatory effects by cannabinoids may be due to inactivation of GABAergic inputs, as CB1r have been identified in inhibitory terminals in the LC (Scavone et al. 2010), see previous text).

Cannabinoids have also been shown to inhibit KCl-evoked excitation of the LC (Mendiguren and Pineda 2007), indicating that cannabinoids may have a protective role in the LC by preventing overactivation of neuronal activity. Hyperactivity of the LC has been proposed to alter behavioral flexibility and disable focused or selective attention (Aston-Jones et al. 1999a, b; Usher et al. 1999). Hyperactivation of the LC may occur during stress situations, which have been shown to impact behavioral flexibility (Cerqueira et al. 2007), and cannabinoids may play a role in refraining LC activation under stress conditions. The inhibition of KCl-evoked excitation of the LC by cannabinoids seems to occur through inhibition of glutamatergic transmission (Mendiguren and Pineda 2007). As most presynaptic CB1r in the LC are thought to be in GABAergic terminals (Scavone et al. 2010) (see previous text), it is still not clear whether the effects of cannabinoids in glutamatergic transmission are through direct actions on CB1r located on glutamatergic terminals, through activation of postsynaptic receptors or indirectly acting through CB1r located in GABAergic terminals. On the other hand, the phasic firing of the LC is important for optimal performance on tasks that require focused attention. Hence, an excess in inhibition by cannabinoids may lead to a decrease in the phasic activation of the LC, which could result in an overall disruption of attention in both animals and humans (Jentsch et al. 1997; Solowij et al. 2002; Arguello and Jentsch 2004). Thus, the homeostasis of the cannabinoid system within the LC seems critical for optimal learning capacity as well as ability to cope with stress.

7.2.2.2 Effects on NTS Activity

There is compelling evidence for complex actions of cannabinoids in the NTS. In the NTS, not all neurons are sensitive to cannabinoids (Himmi et al. 1996, 1998). About 50 % of the neurons of the NTS are responsive to cannabinoid analogs, a response apparently mediated by CB1r. Interestingly, some NTS neurons have their activity increased following cannabinoid treatment, whereas others exhibit decreased neuronal activity. Moreover, both WIN 55,212-2 and the antagonist rimonabant were able to increase Fos expression in the NTS, albeit apparently in different sets of neurons (Jelsing et al. 2009). In a different study, analyzing cardiovascular regulation by the NTS, a subset of NTS neurons with baroreceptive properties was found to increase discharge after application of endocannabinoid anandamide and the endocannabinoid uptake inhibitor AM404 (Seagard et al. 2005), similar to conditions in which there is an increase in blood pressure, showing that cannabinoids functionally impact NTS. The different responses to cannabinoid analogs observed in the NTS may be due to the fact that the NTS is a heterogeneous nucleus containing a large variety of neurotransmitters and neuropeptides. Catecholaminergic, serotoninergic, dopaminergic, GABAergic, and cholinergic neurons can be found within similar subregions of the NTS (Barraco et al. 1992). As most studies fail to identify the neurochemical properties of the neuronal population analyzed, it is hard to speculate regarding the functional implications of these findings. In any case, the different studies reveal that cannabinoids can strongly influence activity of NTS neurons. With respect to NTS-NE neurons, it has been shown that noradrenergic neurons in the NTS are positive for CB1r (Carvalho et al. 2010a), providing anatomical evidence for a potential action of cannabinoids on noradrenergic neurons. In addition, some Δ9-tetrahydrocannabinol (THC)-sensitive neurons were depressed when clonidine, a α2-AR agonist, was coadministered, suggesting that these neurons are likely noradrenergic (Himmi et al. 1996).

7.2.2.3 The Effects of Cannabinoids on NE Release in Target Regions—Focus on LC-PFC and NTS-Acb Projections

In the previous sections, we have provided evidence for the effects of cannabinoids within noradrenergic nuclei. However, impacting cell bodies will also affect noradrenergic transmission in target regions. In fact, several studies have reported that systemic and local administration of cannabinoid analogs alters the release of NE in specific areas of the brain. Systemic administration of WIN 55,212-2 or Δ9-THC has been shown to increase the release of NE in the PFC and in the Acb (Jentsch et al. 1997; Oropeza et al. 2005; Page et al. 2007). Jentsch et al. (1997) showed an increase in NE turnover in the PFC and Acb of rats after systemic injection of Δ9-THC. They also showed that Δ9-THC increased dopamine turnover but only in the PFC; no effects were observed in serotonin turnover. Oropeza et al. (2005) reported an increase of NE release in the PFC with concomitant Fos activation in noradrenergic neurons of the LC; importantly, these effects were blocked by the CB1r antagonist, SR141716A. In another study, repeated administration of WIN 55,212-2

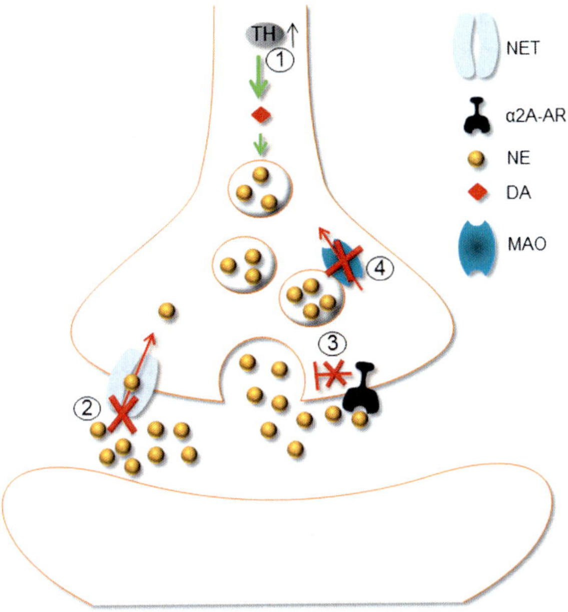

Fig. 7.2 The effect of cannabinoid receptor ligands on noradrenergic transmission in the prefrontal cortex (PFC). Acute cannabinoid treatment has been shown to increase norepinephrine (*NE*) content within the PFC. Putative mechanisms underlying cannabinoid-induced increases in cortical NE may involve increases in the level of the rate-limiting enzyme, tyrosine hydroxylase (*TH*), in LC neurons projecting to the PFC, resulting in increase production of NE (*1*) or decreases in the expression of the cortical NE transporter (*NET*) (*2*) resulting in higher content of synaptic NE in the PFC. Desensitization of the adrenergic receptor (*α2A-AR*) may also lead to an increase in NE release (*3*). Finally, evidence exists that cannabinoids are able to inhibit the activity of monoamine oxidase (*MAO*), the enzyme responsible for the metabolism of NE and dopamine (*DA*), resulting in an increase in intraneuronal NE. See text for details

increased the release of NE in PFC with increased TH expression in the LC (Page et al. 2007). Consistent with this, an increased activity rate of TH in rats given Δ9-THC and WIN 55,212-2 has been reported, resulting in increased levels of NE in the LC, hippocampus, cortex, hypothalamus, and cerebellum (Moranta et al. 2004). In addition, decreased synthesis of serotonin and dopamine were observed upon Δ9-THC and WIN 55,212-2 administration. Interestingly, an in vitro study has showed that cannabinoids have the ability to inhibit the activity of monoamine oxidase (MAO), the enzyme responsible for the metabolism of monoamine neurotransmitters, such as NE and dopamine (Fisar 2010), which could be another mechanism by which cannabinoids can increase NE levels. In line with the increased release of NE in the PFC and Acb, another study has reported alterations in the expression of ARs, as well as in the NE transporter (NET) (Reyes et al. 2009). Reyes et al. have showed that acute administration of WIN 55,212-2 decreases NET expression in the PFC, which in addition to LC activation (Oropeza et al. 2005), increased TH activity in the LC (Moranta et al. 2004; Page et al. 2007); inhibition of MAO (Fisar 2010) may account for the increased release of NE (Fig. 7.2). Furthermore, repeated systemic administration of WIN 55,212-2 was shown to decrease the levels of β1-AR in the PFC (Reyes et al. 2009). On the contrary, abstinence from WIN 55,212-2 induced an upregulation of β1-AR, which can be interpreted as a rebound effect attributed to a return to basal levels following a period of abstinence. Although no changes were observed in α2A-AR levels, a recent study has shown that acute WIN 55,212-2 is able to desensitize α2A-AR in the PFC (Reyes et al. 2012), effect that correlated with increased immobility time in the FST, suggesting that impairment of α2A-AR prevents proper coping with stress. The exact mechanism by which this desensitization occurs is not certain; nonetheless, because CB1r are coupled to Gi/o as are α2A-AR, it is possible that cross-regulation takes place, as it has been proposed that activation of CB1r can sequester G proteins, making them unavailable for other G protein-coupled receptors (GPCRs) such as α2A-AR and somatostatin receptors (Vasquez and Lewis 1999). Interestingly, chronic WIN 55,212-2 administration desensitized α2A-AR but did not disrupt α2A-AR signaling when animals were exposed to an acute swim stress (Reyes et al. 2012), suggesting that chronic cannabinoid administration may have a protective role in preventing overactivation of the PFC without changing the ability to respond to an acute stressor.

Interestingly, the effects of WIN 55,212-2 administration in AR expression in the Acb differ from the ones in the PFC. In the Acb, it has been shown that β1-AR expression was decreased with acute or repeated administration of WIN 55,212-2 (Carvalho et al. 2010a). Additionally, α2A-AR was decreased but only after repeated administration; this effect persisted with abstinence from WIN 55,212-2 (Carvalho et al. 2010a). The lower levels of β1-AR may represent an adaptive mechanism following increases in extracellular NE in the Acb after WIN 55,212-2 treatment. The decreased in α2A-AR expression only after repeated exposure to WIN 55,212-2 may reflect a secondary mechanism to increase NE release as activation of α2A-AR is known to decrease cAMP production in the axon terminal, decreasing the release of vesicular NE (Wozniak et al. 2000).

Intriguingly, some reports have also shown that the CB1r antagonist, SR141716A is capable of increasing NE release in the PFC (Tzavara et al. 2003) and in the hypothalamus (Tzavara et al. 2001), and the administration of SR141716A is accompanied by antidepressant effects in the FST. However, in another study, SR141716A alone did not trigger an effect in the levels of NE compared with vehicle-treated animals. By contrast, in this study, it was observed that SR141716A blocked the effects of WIN 55,212-2-induced NE release (Oropeza et al. 2005). These contradictory effects can be explained in part by the different doses used in these studies. In the latter study, SR141716A was used at 0.2 mg/kg, whereas in the former study, the doses applied ranged from 1 mg/kg to 10 mg/kg. The findings from studies involving CB1r antagonism can also reflect the existence of a basal tone of endocannabinoids in these regions. On the basis of the reported effects of cannabinoids on NE transmission, it is of great interest to understand the functional consequences of NE on cannabinoid-induced behavior, namely aversion and anxiety.

7.3 Cannabinoids, Norepinephrine, and Mood Regulation

Contradictory reports regarding the effects of cannabinoids on mood have been published. For instance, studies have shown that activation of the endocannabinoid system by cannabinoid agonists (Gobbi et al. 2005; Hill and Gorzalka 2005b; Morrish et al. 2009) exerts an antidepressant effect in animal models. Paradoxically, the same effects have been achieved by inactivation of the endocannabinoid system by cannabinoid antagonists (Shearman et al. 2003; Tzavara et al. 2003; Griebel et al. 2005). This is also true for anxiety-like behaviors, in which cannabinoid agonists/antagonists have been shown to exert anxiolytic effects in some studies and anxiogenic effects in others (Haller et al. 2004b; Degroot 2008; Moreira and Lutz 2008; Carvalho et al. 2010b). In human studies, combined effects have also been reported. Occasional users often report that cannabis increases well-being, euphoria, and contentment (Velez et al. 1989). However, increased anxiety, dysphoria, and depressive mood have also been reported following moderate cannabis use (Reilly et al. 1998), and the use of cannabis seems to exacerbate psychotic symptoms, such as delusions and hallucinations (Negrete et al. 1986; Cleghorn et al. 1991; Baigent et al. 1995), and increase anxiety (Morrison et al. 2009).

Regarding the noradrenergic system, emerging studies have revealed an important role for NE in cognitive and limbic function. While, for many decades, the LC-NE system was seen as the main source of forebrain NE and was intensely investigated for its role in attention, memory, and behavior, increased interest in the NTS has contributed to increasing the complexity of how this neuromodulator regulates forebrain targets. Several studies have reported the existence of direct ascending projections from the NTS to limbic areas such as the bed nucleus of the stria terminalis (BNST) and central nucleus of the amygdala (Ricardo and Koh 1978; Reyes and Van Bockstaele 2006) or Acb (Delfs et al. 1998), and these ascending projections have been shown to significantly impact motivated behaviors (Aston-Jones et al. 1999a; Delfs et al. 2000). Activation of the NTS with increased NE release in the Acb leads to memory enhancement, whereas blockade of α-AR within the Acb prevented this enhancement (Kerfoot and Williams 2011). In addition to α-AR inhibition, blockade of β-ARs is known to impair memory, decrease anxiety, and increase depressive symptoms (Gottschalk et al. 1974; Sternberg et al. 1986; Patten 1990) by targeting structures such as the hippocampus, PFC, amygdala, or BNST (Delfs et al. 2000; Tully and Bolshakov 2010). Thus, the effects of NE rely on highly intricate neurocircuitries within cortical and limbic systems.

The ability of cannabinoids to modulate LC and NTS activity can impact noradrenergic transmission in critical regions for regulation of mood and cognition. The anatomical and functional studies reviewed earlier reveal a potential mechanism by which cannabinoids exert their effects on mood and cognition. The next section details the behavioral impact of cannabinoids on selected NE circuits.

7.3.1 Cannabinoid-Induced Aversion

Cannabinoid agents have been shown to produce both preference and aversion in the place conditioning paradigm. Murray and Bevins (2010) recently considered the variability in behaviors associated with cannabinoid exposure and found that the most consistent factor to affect test outcome was the dose of the cannabinoid agent used. Low doses have a tendency to induce place preference, whereas high doses have a tendency to induce place aversion. Place conditioning is a classical conditioning paradigm in which animals learn to associate the effects of a drug (or other discrete treatment) with particular environmental (contextual) cues. Place conditioning can identify both conditioned place preference (CPP) and conditioned place aversion (CPA), and thus it can be used to study both rewarding and aversive drug effects (Bardo and Bevins 2000; Carlezon 2003). Place conditioning is useful in probing neural circuits involved in reward and aversion. For example, microinjection of amphetamine into the Acb produces CPP, whereas microinjection of amphetamine into the area postrema produces a conditioned taste aversion (CTA) (Carr and White 1983, 1986). Other studies have shown that microinjection of μ-opioid receptor ligands into the VTA produces CPP, whereas microinjection of κ-opioid receptor ligands into the VTA, Acb, medial PFC, or lateral hypothalamus produces CPA (Shippenberg and Elmer 1998). Hence, place conditioning studies allow parsing out of neural circuits involved in drug reward and aversion while shedding light on receptor subtypes being targeted. Accordingly, monoaminergic transmission in several limbic structures (e.g., amygdala, PFC, BNST, and Acb) has been reported to be important for the expression of aversive behaviors (Aston-Jones et al. 1999a; Delfs et al. 2000; Ventura et al. 2007; Kerfoot et al. 2008).

Putative neural circuitry involved in mediating cannabinoid-induced aversion was recently elucidated (Fig. 7.3) (Carvalho et al. 2010b; Carvalho and Van Bockstaele 2011). Both the Acb and BNST receive direct noradrenergic projections from the NTS (Delfs et al. 1998; Forray et al. 2000; Forray and Gysling 2004). Activation of the NTS has been shown to occur when CTA acquisition and expression occur (Sakai and Yamamoto 1997; Swank 2000). Although these studies did not provide a detailed neurochemical characterization of the activated neurons, the possibility exists that some of the activated neurons are noradrenergic considering that the highest neuronal activation was seen in the caudal and intermediate NTS, a region enriched with noradrenergic neurons. The localization of CB1r to noradrenergic neurons in the NTS (Carvalho et al. 2010a) and the ability of WIN 55,212-2 to induce neuronal activation in the NTS (Jelsing et al. 2009) support the hypothesis that WIN 55,212-2 induces aversion by increasing NE release in target regions. It has been shown that NE in the Acb, but not in the BNST, is critical for WIN 55,212-2-induced aversion, as decreasing NE signaling in the Acb, either by immunotoxin depletion of noradrenergic fibers (Carvalho et al. 2010b) or by blockade of β1-ARs (Carvalho and Van Bockstaele 2011), impaired its expression. In addition, it is known that blockade of β1-AR reduces the excitability of accumbal neurons that may trigger aversion (Kombian et al. 2006; Carlezon Jr and Thomas 2009). Interestingly, blockade of β1-AR did not impair lithium chloride-induced aversion (Carvalho and Van

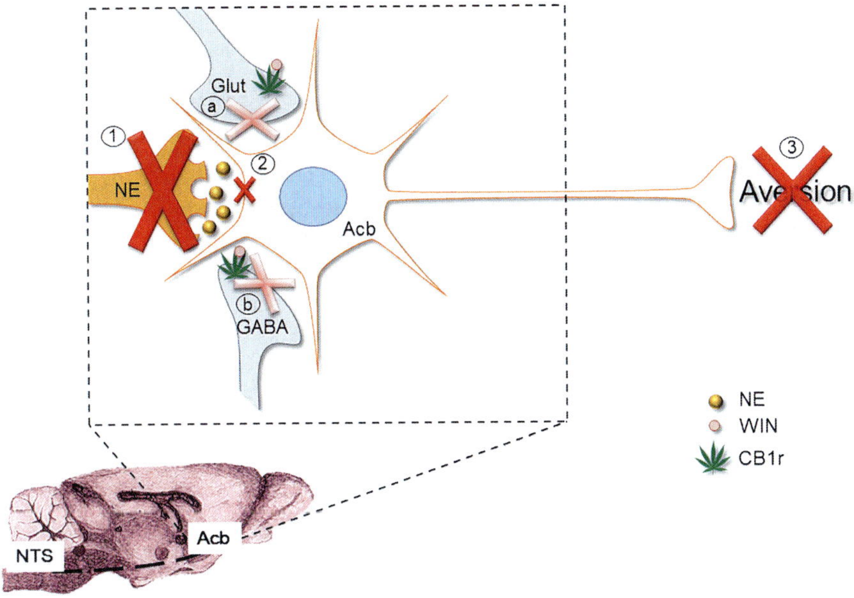

Fig. 7.3 Role of NE in cannabinoid-induced aversion. Schematic diagram depicting glutamatergic (*Glut*), GABAergic (*GABA*), and noradrenergic (*NE*) innervation of Acb neurons. CB1r is primarily associated with GABA and Glut axon terminals in this region, and few NE terminals express CB1r. In the presence of a cannabinoid receptor agonist (e.g., WIN 55,212-2 (*WIN*)), glutamate release is reduced (*a* Robbe et al. 2001) together with a reduction in GABA (*b* Manzoni and Bockaert 2001). WIN 55,212-2 causes a concomitant increase in NE (Jentsch et al. 1997). Blocking NE transmission either by depleting NE (*1*) or by blocking β1-adrenergic receptors (*2*), prevents the expression of WIN 55,212-2-induced aversion (*3*) (Carvalho et al. 2010a; Carvalho and Van Bockstaele 2011). (Modified with permission from Carvalho and Van Bockstaele 2012)

Bockstaele 2011), suggesting that noradrenergic transmission may be specific to aversion to cannabinoid-based agents. Moreover, the lack of effect of betaxolol in lithium chloride-induced aversion suggests that the β1-AR blocker did not impact learning. Noradrenergic transmission in the BNST has been implicated in the signaling of aversion in opiate withdrawal (Delfs et al. 2000; Cecchi et al. 2007) and visceral pain (Deyama et al. 2009; Minami 2009). However, NE in the BNST does not seem to be critical for WIN 55,212-2-induced aversion (Carvalho et al. 2010b).

7.3.2 Cannabinoid-Induced Anxiety

Cannabinoids have been shown to induce both anxiolytic and anxiogenic effects using the elevated plus-maze (EPM) or the elevated zero-maze (EZM). The EZM is a modification of the well-established EPM. Both EPM and EZM are based on the natural conflict of rodents to explore a novel environment and their innate aversion to open, elevated, and brightly lit spaces. As a consequence of the aversive properties of the open arms, subjects spend a greater amount of time on the closed arms and the proportion of total exploration in the open arms provides a measure of anxiety, such that increases in percent time spent on the open arms is considered to be indicative of anxiolytic drug action (Handley and Mithani 1984; Pellow and File 1986). Conversely, decreases in percent time spent on open arms reflect an anxiogenic effect of the drug.

The differential results on anxiety following exposure to cannabinoid agents may be due to some of the following variables: prior drug use, dose used, basal anxiety levels, and regional endocannabinoid basal tone (Degroot 2008). Generally, the anxiogenic properties of cannabinoid agents occur more frequently in drug-naïve subjects and in novel/stressful environments (Haller et al. 2004a; Viveros et al. 2005; Degroot 2008). This suggests that basal endocannabinoid tone is important in the response to exogenous cannabinoids. It has been shown that increases in endocannabinoid levels in specific brain areas are important for coping with anxiety-provoking stimuli (Marsicano et al. 2002). In this scenario, endocannabinoids are thought to work to restore homeostasis. While under certain physiological situations, increases in endocannabinoids may be restricted to specific brain regions (e.g., amygdala (Marsicano et al. 2002), widespread activation of cannabinoid receptors by exogenous/systemic cannabinoid ligands may trigger an anxiogenic effect. Although decreased NE tone in the Acb was able to reverse WIN 55,212-2-induced aversion, it was not sufficient to block WIN 55,212-2-induced anxiety (Carvalho et al. 2010b). Decreasing NE tone in the BNST also failed to prevent WIN 55,212-2-induced aversion. These results suggest that WIN 55,212-2-induced anxiety is not mediated by NE input to the Acb or the BNST. These findings are not surprising, as the Acb has not been implicated in the development of anxiety-like behaviors. On the other hand, the results obtained from NE depletion from the BNST are quite fascinating. The BNST is seen as an important nucleus for the expression of anxiety (Davis 1998, 2006; Walker et al. 2003) and is one of the richest areas in NE in the CNS (Forray and Gysling 2004). Although NE in the BNST has been shown to mediate anxiety to certain stressors, it does not mediate anxiety in response to all types of stressors (Cecchi et al. 2002). Considering this, it has been proposed that NE effects on anxiety are stimuli-specific. Moreover, other neurotransmitters have also been implicated in signaling anxiety in the BNTS, such as corticotropin-releasing factor (CRF) (Smith and Aston-Jones 2008). It has been suggested that anxiogenic effects of endocannabinoids can be mediated by TRPV1 receptor activation (Campos and Guimarães 2009; Micale et al. 2009), as anandamide, but not 2-AG, is a TRPV1 receptor agonist (Zygmunt et al. 1999). In addition, TRPV1 knockout (KO) mice show reduced anxiety-like behavior in the EPM (Marsch et al. 2007). It is not clear whether WIN 55,212-2 has the ability to directly modulate TRPV1. Interestingly, WIN 55,212-2 has been shown to inhibit TRPV1 in trigeminal ganglion neurons (Patwardhan et al. 2006; Wang et al. 2012), but the role of TRPV1 in WIN 55,212-2-induced anxiety has not been investigated. Taken together, the results indicate that WIN 55,212-2-induced anxiety is most likely independent of noradrenergic transmission in the Acb and BNST.

7.4 Conclusion

Growing evidence suggests significant interactions between the cannabinoid and noradrenergic systems with significant functional and behavioral implications. It is clear that the noradrenergic system plays a role in many psychiatric disorders. Thus, it is crucial to understand how the two systems adapt to pathological conditions and how this interaction is affected. As this interaction seems to be circuit-specific and may depend on the basal status of the cannabinoid and NE levels, it is possible that, under certain conditions, one circuit is more affected than others, giving rise to a specific change in behavior. If this holds true, it is important to recognize that manipulation of these two systems has widespread effects within the brain. In light of the reported effects of cannabinoids on noradrenergic transmission, it is tempting to speculate that the development of drugs that target the endocannabinoid system may provide an effective tool to modulate and reverse impairments in noradrenergic transmission. However, numerous safety issues persist with cannabinoid-based agents that may preclude their widespread utility. The question also arises as to whether prevention of side effects induced by cannabinoid-based agents may involve a combination of cannabinoid-based agents and modulators of the noradrenergic system. Continued investigations into the understanding of interactions between the two systems will no doubt lead to novel approaches for psychiatric disorders.

References

Arguello PA, Jentsch JD (2004) Cannabinoid CB1 receptor-mediated impairment of visuospatial attention in the rat. Psychopharmacology 177:141–150

Aston-Jones G, Chiang C, Alexinsky T (1991) Discharge of noradrenergic locus coeruleus neurons in behaving rats and monkeys suggests a role in vigilance. Prog Brain Res 88:501–520

Aston-Jones G, Delfs JM, Druhan J, Zhu Y (1999a) The bed nucleus of the stria terminalis. A target site for noradrenergic actions in opiate withdrawal. Ann N Y Acad Sci 877:486–498

Aston-Jones G, Rajkowski J, Cohen J (1999b) Role of locus coeruleus in attention and behavioral flexibility. Biol Psychiatry 46:1309–1320

Bacci A, Huguenard JR, Prince DA (2004) Long-lasting self-inhibition of neocortical interneurons mediated by endocannabinoids. Nature 431:312–316

Baigent M, Holme G, Hafner RJ (1995) Self reports of the interaction between substance abuse and schizophrenia. Aust N Z J Psychiatry 29:69–74

Bardo MT, Bevins RA (2000) Conditioned place preference: what does it add to our preclinical understanding of drug reward? Psychopharmacology 153:31–43

Barraco R, Ergene MEE, Parizon M, Bradley D (1992) All atlas of the rat subpostremal nucleus-tractus-solitarius. Brain Res Bull 29:703–765

Bond DJ, Noronha MM, Kauer-Sant'Anna M, Lam RW, Yatham LN (2008) Antidepressant-associated mood elevations in bipolar II disorder compared with bipolar I disorder and major depressive disorder: a systematic review and meta-analysis. J Clin Psychiatry 69:1589–1601

Callado LF, Meana JJ, Grijalba B, Pazos A, Sastre M, Garcia-Sevilla JA (1998) Selective increase of alpha2A-adrenoceptor agonist binding sites in brains of depressed suicide victims. J Neurochem 70:1114–1123

Campos AC, Guimarães FS (2009) Evidence for a potential role for TRPV1 receptors in the dorsolateral periaqueductal gray in the attenuation of the anxiolytic effects of cannabinoids. Prog Neuropsychopharmacol Biol Psychiatry 33:1517–1521

Carlezon WA Jr, Thomas MJ (2009) Biological substrates of reward and aversion: a nucleus accumbens activity hypothesis. Neuropharmacology 56(Suppl 1):122–132

Carlezon WA (2003) Place conditioning to study drug reward and aversion. Methods Mol Med 84:243–249

Carr GD, White NM (1983) Conditioned place preference from intra-accumbens but not intra-caudate amphetamine injections. Life Sci 33:2551–2557

Carr GD, White NM (1986) Anatomical disassociation of amphetamine's rewarding and aversive effects: an intracranial microinjection study. Psychopharmacology 89:340–346

Carrasco GA, Van de Kar LD (2003) Neuroendocrine pharmacology of stress. Eur J Pharmacol 463:235–272

Carvalho AF, Van Bockstaele EJ (2011) Direct intra-accumbal infusion of a beta-adrenergic receptor antagonist abolishes WIN 55,212-2-induced aversion. Neurosci Lett 500:82–85

Carvalho AF, Van Bockstaele EJ (2012) Cannabinoid modulation of noradrenergic circuits: implications for psychiatric disorders. Prog Neuropsychopharmacol Biol Psychiatry 38:59–67

Carvalho AF, Mackie K, Van Bockstaele EJ (2010a) Cannabinoid modulation of limbic forebrain noradrenergic circuitry. Eur J Neurosci 31:286–301

Carvalho AF, Reyes A-RS, Sterling RC, Unterwald E, Van Bockstaele EJ (2010b) Contribution of limbic norepinephrine to cannabinoid-induced aversion. Psychopharmacology 211:479–491

Cecchi M, Capriles N, Watson SJ, Akil H (2007) Beta1 adrenergic receptors in the bed nucleus of stria terminalis mediate differential responses to opiate withdrawal. Neuropsychopharmacology 32:589–599

Cecchi M, Khoshbouei H, Javors M, Morilak DA (2002) Modulatory effects of norepinephrine in the lateral bed nucleus of the stria terminalis on behavioral and neuroendocrine responses to acute stress. Neuroscience 112:13–21

Cerqueira JJ, Mailliet F, Almeida OFX, Jay TM, Sousa N (2007) The prefrontal cortex as a key target of the maladaptive response to stress. J Neurosci 27:2781–2787

Charney DS, Woods SW, Heninger GR (1989) Noradrenergic function in generalized anxiety disorder: effects of yohimbine in healthy subjects and patients with generalized anxiety disorder. Psychiatry Res 27:173–182

Cleghorn JM, Kaplan RD, Szechtman B, Szechtman H, Brown GM, Franco S (1991) Substance abuse and schizophrenia: effect on symptoms but not on neurocognitive function. J Clin Psychiatry 52:26–30

Davis M (1998) Are different parts of the extended amygdala involved in fear versus anxiety? Biol Psychiatry 44:1239–1247

Davis M (2006) Neural systems involved in fear and anxiety measured with fear-potentiated startle. Am Psychol 61:741–756

Degroot A (2008) Role of cannabinoid receptors in anxiety disorders. In: Kofalvi A (ed) Cannabinoids and the brain. Springer, New York, pp 559–572

Delfs JM, Zhu Y, Druhan JP, Aston-Jones G (2000) Noradrenaline in the ventral forebrain is critical for opiate withdrawal-induced aversion. Nature 403:430–434

Delfs JM, Zhu Y, Druhan JP, Aston-Jones GS (1998) Origin of noradrenergic afferents to the shell subregion of the nucleus accumbens: anterograde and retrograde tract-tracing studies in the rat. Brain Res 806:127–140

Derbenev AV, Stuart TC, Smith BN (2004) Cannabinoids suppress synaptic input to neurones of the rat dorsal motor nucleus of the vagus nerve. J Physiol 559:923–938

Després J-P, Ross R, Boka G, Alméras N, Lemieux I (2009) Effect of rimonabant on the high-triglyceride/low-HDL-cholesterol dyslipidemia, intraabdominal adiposity, and liver fat: the ADAGIO-Lipids trial. Arterioscler Thromb Vasc Biol 29:416–423

Devane WA, Hanus L, Breuer A, Pertwee RG, Stevenson LA, Griffin G, Gibson D, Mandelbaum A, Etinger A, Mechoulam R (1992) Isolation and structure of a brain constituent that binds to the cannabinoid receptor. Science 258:1946–1949

Deyama S, Katayama T, Kondoh N, Nakagawa T, Kaneko S, Yamaguchi T, Yoshioka M, Minami M (2009) Role of enhanced noradrenergic transmission within the ventral bed nucleus of the stria terminalis in visceral pain-induced aversion in rats. Behav Brain Res 197:279–283

Doyle MW, Bailey TW, Jin Y-H, Andresen MC (2002) Vanilloid receptors presynaptically modulate cranial visceral afferent synaptic transmission in nucleus tractus solitarius. J Neurosci 22:8222–8229

Duman RS, Nestler EJ (1995) Signal transduction pathways for catecholamine receptors In: Bloom FE, Kupfer DJ (eds) Psychopharmacology: the fourth generation of progress. Raven Press, New York, p 303

Fisar Z (2010) Inhibition of monoamine oxidase activity by cannabinoids. Naunyn-Schmiedeberg's Arch Pharmacol 381:563–572

Foote SL, Bloom FE, Aston-Jones G (1983) Nucleus locus ceruleus: new evidence of anatomical and physiological specificity. Physiol Rev 63:844–914

Forray MI, Gysling K (2004) Role of noradrenergic projections to the bed nucleus of the stria terminalis in the regulation of the hypothalamic-pituitary-adrenal axis. Brain Res 47:145–160

Forray MI, Gysling K, Andres ME, Bustos G, Araneda S (2000) Medullary noradrenergic neurons projecting to the bed nucleus of the stria terminalis express mRNA for the NMDA-NR1 receptor. Brain Res Bull 52:163–169

Friedman JI, Adler DN, Davis KL (1999) The role of norepinephrine in the pathophysiology of cognitive disorders: potential applications to the treatment of cognitive dysfunction in schizo phrenia and Alzheimer's disease. Biol Psychiatry 46:1243–1252

Gobbi G, Bambico FR, Mangieri R, Bortolato M, Campolongo P, Solinas M, Cassano T, Morgese MG, Debonnel G, Duranti A, Tontini A, Tarzia G, Mor M, Trezza D, Goldberg SR, Cuomo V, Piomelli D (2005) Antidepressant-like activity and modulation of brain monoaminergic transmission by blockade of anandamide hydrolysis. Proc Natl Acad Sci U S A 102:18620–18625

Gottschalk LA, Stone WN, Gleser GC (1974) Peripheral versus central mechanisms accounting for antianxiety effects of propranolol. Psychosom Med 36:47–56

Griebel G, Stemmelin J, Scatton B (2005) Effects of the cannabinoid CB1 receptor antagonist rimonabant in models of emotional reactivity in rodents. Biol Psychiatry 57:261–267

Haller J, Varga B, Ledent C, Barna I, Freund TF (2004a) Context-dependent effects of CB1 cannabinoid gene disruption on anxiety-like and social behaviour in mice. Eur J Neurosci 19:1906–1912

Haller J, Varga B, Ledent C, Freund TF (2004b) CB1 cannabinoid receptors mediate anxiolytic effects: convergent genetic and pharmacological evidence with CB1-specific agents. Behav Pharmacol 15:299–304

Handley SL, Mithani S (1984) Effects of alpha-adrenoceptor agonists and antagonists in a maze-exploration model of "fear"-motivated behaviour. Naunyn-Schmiedeberg's Arch Pharmacol 327:1–5

Heninger GR, Charney DS (1988) Monoamine receptor systems and anxiety disorders. Psychiatr Clin North Am 11:309–326

Henquet C, Krabbendam L, De Graaf R, Ten Have M, Van Os J (2006) Cannabis use and expression of mania in the general population. J Affect Disord 95:103–110

Herkenham M, Lynn AB, Johnson MR, Melvin LS, De Costa BR, Rice KC (1991) Characterization and localization of cannabinoid receptors in rat brain: a quantitative in vitro autoradiographic study. J Neurosci 11:563–583

Hill MN, Gorzalka BB (2005a) Pharmacological enhancement of cannabinoid CB1 receptor activity elicits an antidepressant-like response in the rat forced swim test. Eur Neuropsychopharmacol 15:593–599

Hill MN, Gorzalka BB (2005b) Is there a role for the endocannabinoid system in the etiology and treatment of melancholic depression? Behav Pharmacol 16:333–352

Himmi T, Dallaporta M, Perrin J, Orsini J (1996) Neuronal Responses to delta 9-tetrahydrocannabinol in the solitary tract nucleus. Eur J Pharmacol 312:273–279

Himmi T, Perrin J, Ouazzani TE, Orsini JC (1998) Neuronal responses to cannabinoid receptor ligands in the solitary tract nucleus. Eur J Pharmacol 359:49–54

Itoi K, Sugimoto N (2010) The brainstem noradrenergic systems in stress, anxiety and depression. J Neuroendocrinol 22:355–361
Janero DR, Makriyannis A (2009) Cannabinoid receptor antagonists: pharmacological opportunities, clinical experience, and translational prognosis. Expert Opin Emerg Drugs 14:43–65
Jelsing J, Galzin AM, Guillot E, Pruniaux MP, Larsen PJ, Vrang N (2009) Localization and phenotypic characterization of brainstem neurons activated by rimonabant and WIN55,212-2. Brain Res Bull 78:202–210
Jelsing J, Larsen PJ, Vrang N (2008) Identification of cannabinoid type 1 receptor expressing cocaine amphetamine-regulated transcript neurons in the rat hypothalamus and brainstem using in situ hybridization and immunohistochemistry. Neuroscience 154:641–652
Jentsch JD, Andrusiak E, Tran A, Bowers MBJ, Roth RH (1997) Delta 9-tetrahydrocannabinol increases prefrontal cortical catecholaminergic utilization and impairs spatial working memory in the rat: blockade of dopaminergic effects with HA966. Neuropsychopharmacology 16:426–432
Johnston TG, Kelly CB, Stevenson MR, Cooper SJ (1999) Plasma norepinephrine and prediction of outcome in major depressive disorder. Biol Psychiatry 46:1253–1258
Kerfoot EC, Chattillion EA, Williams CL (2008) Functional interactions between the nucleus tractus solitarius (NTS) and nucleus accumbens shell in modulating memory for arousing experiences. Neurobiol Learn Mem 89:47–60
Kerfoot EC, Williams CL (2011) Interactions between brainstem noradrenergic neurons and the nucleus accumbens shell in modulating memory for emotionally arousing events. Learn Mem 18:405–413
Kombian SB, Ananthalakshmi KV, Edafiogho IO (2006) Enaminones and norepinephrine employ convergent mechanisms to depress excitatory synaptic transmission in the rat nucleus accumbens in vitro. Eur J Neurosci 24:2781–2788
Koob GF (2008) A role for brain stress systems in addiction. Neuron 59:11–34
Lee S, Kim DH, Yoon S-H, Ryu JH (2009) Sub-chronic administration of rimonabant causes loss of antidepressive activity and decreases doublecortin immunoreactivity in the mouse hippocampus. Neurosci Lett 467:111–116
Leterrier C, Lainé J, Darmon M, Boudin H, Rossier J, Lenkei Z (2006) Constitutive activation drives compartment-selective endocytosis and axonal targeting of type 1 cannabinoid receptors. J Neurosci 26:3141–3153
Llano I, Leresche N, Marty A (1991) Calcium entry increases the sensitivity of cerebellar Purkinje cells to applied GABA and decreases inhibitory synaptic currents. Neuron 6:565–574
MacDonald E, Kobilka BK, Scheinin M (1997) Gene targeting—homing in on alpha 2-adrenoceptor-subtype function. Trends Pharmacol Sci 18:211–219
Mailleux P, Vanderhaeghen JJ (1992) Distribution of neuronal cannabinoid receptor in the adult rat brain: a comparative receptor binding radioautography and in situ hybridization histochemistry. Neuroscience 48:655–668
Manzoni OJ, Bockaert J (2001) Cannabinoids inhibit GABAergic synaptic transmission in mice nucleus accumbens. Eur J Pharmacol 412(2):R3–5
Marsch R, Foeller E, Rammes G, Bunck M, Kössl M, Holsboer F, Zieglgänsberger W, Landgraf R, Lutz B, Wotjak CT (2007) Reduced anxiety, conditioned fear, and hippocampal long-term potentiation in transient receptor potential vanilloid type 1 receptor-deficient mice. J Neurosci 27:832–839
Marsicano G, Wotjak CT, Azad SC, Bisogno T, Rammes G, Cascio MG, Hermann H, Tang J, Hofmann C, Zieglgansberger W, Di Marzo V, Lutz B (2002) The endogenous cannabinoid system controls extinction of aversive memories. Nature 418:530–534
Matsuda LA, Bonner TI, Lolait SJ (1993) Localization of cannabinoid receptor mRNA in rat brain. J Comp Neurol 327:535–550
Matsuda LA, Lolait SJ, Brownstein MJ, Young AC, Bonner TI (1990) Structure of a cannabinoid receptor and functional expression of the cloned cDNA. Nature 346:561–564
Meana JJ, Barturen F, Garcia-Sevilla JA (1992) Alpha 2-adrenoceptors in the brain of suicide victims: increased receptor density associated with major depression. Biol Psychiatry 31:471–490

Mechoulam R, Benshabat S, Hanus L, Ligumsky M, Kaminski NE, Schatz AR, Gopher A, Almog S, Martin BR, Compton DR, Pertwee RG, Griffin G, Bayewitch M, Barg J, Vogel Z (1995) Identification of an endogenous 2-monoglyceride, present in canine gut, that binds to cannabinoid receptors. Biochem Pharmacol 50:83–90

Mendiguren A, Pineda J (2004) Cannabinoids enhance N-methyl-D-aspartate-induced excitation of locus coeruleus neurons by CB1 receptors in rat brain slices. Neurosci Lett 363:1–5

Mendiguren A, Pineda J (2006) Systemic effect of cannabinoids on the spontaneous firing rate of locus coeruleus neurons in rats. Eur J Pharmacol 534:83–88

Mendiguren A, Pineda J (2007) CB(1) cannabinoid receptors inhibit the glutamatergic component of KCl-evoked excitation of locus coeruleus neurons in rat brain slices. Neuropharmacology 52:617–625

Mezey E, Tóth ZE, Cortright DN, Arzubi MK, Krause JE, Elde R, Guo A, Blumberg PM, Szallasi A (2000) Distribution of mRNA for vanilloid receptor subtype 1 (VR1), and VR1-like immunoreactivity, in the central nervous system of the rat and human. Proc Natl Acad Sci U S A 97:3655–3660

Micale V, Cristino L, Tamburella A, Petrosino S, Leggio GM, Drago F, Di Marzo V (2009) Anxiolytic effects in mice of a dual blocker of fatty acid amide hydrolase and transient receptor potential vanilloid type-1 channels. Neuropsychopharmacology 34:593–606

Minami M (2009) Neuronal mechanisms for pain-induced aversion behavioral studies using a conditioned place aversion test. Int Rev Neurobiol 85:135–144

Moranta D, Esteban S, Garcia-sevilla JA (2004) Differential effects of acute cannabinoid drug treatment, mediated by CB 1 receptors, on the in vivo activity of tyrosine and tryptophan hydroxylase in the rat brain. Naunyn Schmiedebergs Arch Pharmacol 369:516–524

Moreira FA, Grieb M, Lutz B (2009) Central side-effects of therapies based on CB1 cannabinoid receptor agonists and antagonists: focus on anxiety and depression. Best Pract Res Clin Endocrinol Metab 23:133–144

Moreira FA, Lutz B (2008) The endocannabinoid system: emotion, learning and addiction. Addict Biol 13:196–212

Morrish AC, Hill MN, Riebe CJ, Gorzalka BB (2009) Protracted cannabinoid administration elicits antidepressant behavioral responses in rats: role of gender and noradrenergic transmission. Physiology Behavior 98:118–124

Morrison PD, Zois V, McKeown DA, Lee TD, Holt DW, Powell JF, Kapur S, Murray RM (2009) The acute effects of synthetic intravenous Delta9-tetrahydrocannabinol on psychosis, mood and cognitive functioning. Psychol Med 39:1607–1616

Munro S, Thomas KL, Abu-Shaar M (1993) Molecular characterization of a peripheral receptor for cannabinoids. Nature 365:61–65

Muntoni AL, Pillolla G, Melis M, Perra S, Gessa GL, Pistis M (2006) Cannabinoids modulate spontaneous neuronal activity and evoked inhibition of locus coeruleus noradrenergic neurons. Eur J Neurosci 23:2385–2394

Murray JE, Bevins RA (2010) Cannabinoid conditioned reward and aversion: behavioral and neural processes. ACS Chem Neurosci 1:265–278

Negrete JC, Knapp WP, Douglas DE, Smith WB (1986) Cannabis affects the severity of schizophrenic symptoms: results of a clinical survey. Psychol Med 16:515–520

Nissen SE, Nicholls SJ, Wolski K, Rodes-Cabau J, Cannon CP, Deanfield JE, Despres JP, Kastelein JJ, Steinhubl SR, Kapadia S, Yasin M, Ruzyllo W, Gaudin C, Job B, Hu B, Bhatt DL, Lincoff AM, Tuzcu EM, Investigators S (2008) Effect of rimonabant on progression of atherosclerosis in patients with abdominal obesity and coronary artery disease: the STRADIVARIUS randomized controlled trial. JAMA 299:1547–1560

Nutt DJ (2002) The neuropharmacology of serotonin and noradrenaline in depression. Int Clin Psychopharmacol 17(Suppl 1):S1–12

Nutt DJ (2006) The role of dopamine and norepinephrine in depression and antidepressant treatment. J Clin Psychiatry 67(Suppl 6):3–8

Oropeza VC, Mackie K, Van Bockstaele EJ (2007) Cannabinoid receptors are localized to noradrenergic axon terminals in the rat frontal cortex. Brain Res 1127:36–44

Oropeza VC, Page ME, Van Bockstaele EJ (2005) Systemic administration of WIN 55,212-2 increases norepinephrine release in the rat frontal cortex. Brain Res 1046:45–54
De Paermentier F, Cheetham SC, Crompton MR, Katona CL, Horton RW (1990) Brain beta-adrenoceptor binding sites in antidepressant-free depressed suicide victims. Brain Res 525:71–77
De Paermentier F, Mauger JM, Lowther S, Crompton MR, Katona CL, Horton RW (1997) Brain alpha-adrenoceptors in depressed suicides. Brain Res 757:60–68
Page ME, Oropeza VC, Sparks SE, Qian Y, Menko AS, Van Bockstaele EJ (2007) Repeated cannabinoid administration increases indices of noradrenergic activity in rats. Pharmacol Biochem Behav 86:162–168
Patel S, Hillard CJ (2003) Cannabinoid-induced Fos expression within A10 dopaminergic neurons. Brain Res 963:15–25
Patten SB (1990) Propranolol and depression: evidence from the antihypertensive trials. Can J Psychiatry 35:257–259
Patwardhan AM, Jeske NA, Price TJ, Gamper N, Akopian AN, Hargreaves KM (2006) The cannabinoid WIN 55,212-2 inhibits transient receptor potential vanilloid 1 (TRPV1) and evokes peripheral antihyperalgesia via calcineurin. Proc Natl Acad Sci U S A 103:11393–11398
Peet M (1994) Induction of mania with selective serotonin re-uptake inhibitors and tricyclic antidepressants. Br J Psychiatry 164:549–550
Pellow S, File SE (1986) Anxiolytic and anxiogenic drug effects on exploratory activity in an elevated plus-maze: a novel test of anxiety in the rat. Pharmacol Biochem Behav 24:525–529
Peters JH, McDougall SJ, Fawley JA, Andresen MC (2011) TRPV1 marks synaptic segregation of multiple convergent afferents at the rat medial solitary tract nucleus. PloS One 6:e25015
Reilly D, Didcott P, Swift W, Hall W (1998) Long-term cannabis use: characteristics of users in an Australian rural area. Addiction 93:837–846
Reyes BAS, Van Bockstaele EJ (2006) Divergent projections of catecholaminergic neurons in the nucleus of the solitary tract to limbic forebrain and medullary autonomic brain regions. Brain Res 1117:69–79
Reyes BAS, Rosario JC, Piana EJ, Van Bockstaele EJ (2009) Cannabinoid modulation of cortical adrenergic receptors and transporters. J Neurosci Res 87:3671–3678
Reyes BAS, Szot P, Sikkema C, Cathel AM, Kirby LG, Van Bockstaele EJ (2012) Stress-induced sensitization of cortical adrenergic receptors following a history of cannabinoid exposure. Exp Neurol 236:327–335
Reynolds SM, Berridge KC (2001) Fear and feeding in the nucleus accumbens shell: rostrocaudal segregation of GABA-elicited defensive behavior versus eating behavior. J Neurosci 21:3261–3270
Reynolds SM, Berridge KC (2002) Positive and negative motivation in nucleus accumbens shell: bivalent rostrocaudal gradients for GABA-elicited eating, taste "liking"/"disliking" reactions, place preference/avoidance, and fear. J Neurosci 22:7308–7320
Reynolds SM, Berridge KC (2003) Glutamate motivational ensembles in nucleus accumbens: rostrocaudal shell gradients of fear and feeding. Eur J Neurosci 17:2187–2200
Ricardo JA, Koh ET (1978) Anatomical evidence of direct projections from the nucleus of the solitary tract to the hypothalamus, amygdala, and other forebrain structures in the rat. Brain Res 153:1–26
Robbe D, Alonso G, Duchamp F, Bockaert J, Manzoni OJ (2001) Localization and mechanisms of action of cannabinoid receptors at the glutamatergic synapses of the mouse nucleus accumbens. J Neurosci 21(1):109–116
Sakai N, Yamamoto T (1997) Conditioned taste aversion and c-fos expression in the rat brainstem after administration of various USs. Neuroreport 8:2215–2220
Sanchez JF, Krause JE, Cortright DN (2001) The distribution and regulation of vanilloid receptor VR1 and VR1 5′ splice variant RNA expression in rat. Neuroscience 107:373–381
Sara SJ (2009) The locus coeruleus and noradrenergic modulation of cognition. Nat Rev Neuroscience 10:211–223
Scavone JL, Mackie K, Van Bockstaele EJ (2006) Distribution and trafficking of the cannabinoid receptor CB1 in the rat noradrenergic locus coeruleus. Soc Neurosci Abs

Scavone JL, Mackie K, Van Bockstaele EJ (2010) Characterization of cannabinoid-1 receptors in the locus coeruleus: relationship with mu-opioid receptors. Brain Res 1312:18–31

Seagard JL, Hopp FA, Hillard CJ, Dean C (2005) Effects of endocannabinoids on discharge of baroreceptive NTS neurons. Neurosci Lett 381:334–339

Shearman LP, Rosko KM, Fleischer R, Wang J, Xu S, Tong XS, Rocha BA (2003) Antidepressant-like and anorectic effects of the cannabinoid CB1 receptor inverse agonist AM251 in mice. Behav Pharmacol 14:573–582

Shippenberg TS, Elmer GI (1998) The neurobiology of opiate reinforcement. Crit Rev Neurobiol 12:267–303

Van Sickle MD, Oland LD, Ho W, Hillard CJ, Mackie K, Davison JS, Sharkey KA (2001) Cannabinoids inhibit emesis through CB1 receptors in the brainstem of the ferret. Gastroenterology 121:767–774

Smith RJ, Aston-Jones G (2008) Noradrenergic transmission in the extended amygdala: role in increased drug-seeking and relapse during protracted drug abstinence. Brain Struct Funct 213:43–61

Solowij N, Stephens RS, Roffman RA, Babor T, Kadden R, Miller M, Christiansen K, McRee B, Vendetti J, Group MTPR (2002) Cognitive functioning of long-term heavy cannabis users seeking treatment. JAMA 287:1123–1131

Southwick SM, Bremner JD, Rasmusson A, Morgan CA 3rd, Arnsten A, Charney DS (1999) Role of norepinephrine in the pathophysiology and treatment of posttraumatic stress disorder. Biol Psychiatry 46:1192–1204

Southwick SM, Krystal JH, Morgan CA, Johnson D, Nagy LM, Nicolaou A, Heninger GR, Charney DS (1993) Abnormal noradrenergic function in posttraumatic stress disorder. Arch Gen Psychiatry 50:266–274

Sternberg DB, Korol D, Novack GD, McGaugh JL (1986) Epinephrine-induced memory facilitation: attenuation by adrenoceptor antagonists. Eur J Pharmacol 129:189–193

Sugiura T, Kondo S, Sukagawa A, Nakane S, Shinoda A, Itoh K, Yamashita A, Waku K (1995) 2-Arachidonoylglycerol: a possible endogenous cannabinoid receptor ligand in brain. Biochem Biophys Res Commun 215:89–97

Swank MW (2000) Conditioned c-Fos in mouse NTS during expression of a learned taste aversion depends on contextual cues. Brain Res 862:138–144

Tondo L, Vázquez G, Baldessarini RJ (2010) Mania associated with antidepressant treatment: comprehensive meta-analytic review. Acta Psychiatr Scand 121:404–414

Tully K, Bolshakov VY (2010) Emotional enhancement of memory: how norepinephrine enables synaptic plasticity. Mol Brain 3:15

Tzavara ET, Davis RJ, Perry KW, Li X, Salhoff C, Bymaster FP, Witkin JM, Nomikos GG (2003) The CB1 receptor antagonist SR141716A selectively increases monoaminergic neurotransmission in the medial prefrontal cortex: implications for therapeutic actions. Br J Pharmacology 138:544–553

Tzavara ET, Perry KW, Rodriguez DE, Bymaster FP, Nomikos GG (2001) The cannabinoid CB(1) receptor antagonist SR141716A increases norepinephrine outflow in the rat anterior hypothalamus. Eur J Pharmacol 426:R3–R4

Tóth A, Boczán J, Kedei N, Lizanecz E, Bagi Z, Papp Z, Edes I, Csiba L, Blumberg PM (2005) Expression and distribution of vanilloid receptor 1 (TRPV1) in the adult rat brain. Brain Res Mol Brain Res 135:162–168

Usher M, Cohen JD, Servan-Schreiber D, Rajkowski J, Aston-Jones G (1999) The role of locus coeruleus in the regulation of cognitive performance. Science 283:549–554

Vasquez C, Lewis DL (1999) The CB1 cannabinoid receptor can sequester G-proteins, making them unavailable to couple to other receptors. J Neurosci 19:9271–9280

Velez CN, Johnson J, Cohen P (1989) A longitudinal analysis of selected risk factors for childhood psychopathology. J Am Acad Child Adolesc Psychiatry 28:861–864

Ventura R, Morrone C, Puglisi-Allegra S (2007) Prefrontal/accumbal catecholamine system determines motivational salience attribution to both reward- and aversion-related stimuli. Proc Natl Acad Sci U S A 104:5181–5186

Viveros MP, Marco EM, File SE (2005) Endocannabinoid system and stress and anxiety responses. Pharmacol Biochem Behav 81:331–342

Walker DL, Toufexis DJ, Davis M (2003) Role of the bed nucleus of the stria terminalis versus the amygdala in fear, stress, and anxiety. Eur J Pharmacol 463:199–216

Wang W, Cao X, Liu C, Liu L (2012) Cannabinoid WIN 55,212-2 inhibits TRPV1 in trigeminal ganglion neurons via PKA and PKC pathways. Neurol Sci 33:79–85

Weinshenker D, Schroeder JP (2007) There and back again: a tale of norepinephrine and drug addiction. Neuropsychopharmacology 32:1433–1451

Witkin JM, Tzavara ET, Davis RJ, Li X, Nomikos GG (2005a) A therapeutic role for cannabinoid CB1 receptor antagonists in major depressive disorders. Trends Pharmacol Sci 26:609–617

Witkin JM, Tzavara ET, Nomikos GG (2005b) A role for cannabinoid CB 1 receptors in mood and anxiety disorders. Behav Pharmacol 16:315–331

Wong ML, Kling MA, Munson PJ, Listwak S, Licinio J, Prolo P, Karp B, McCutcheon IE, Geracioti TD, DeBellis MD, Rice KC, Goldstein DS, Veldhuis JD, Chrousos GP, Oldfield EH, McCann SM, Gold PW (2000) Pronounced and sustained central hypernoradrenergic function in major depression with melancholic features: relation to hypercortisolism and corticotropin-releasing hormone. Proc Natl Acad Sci U S A 97:325–330

Wozniak M, Schramm N, Limbird L (2000) The noradrenergic receptor subtypestle. In: Bloom F, Kupfer D (eds) Psychopharmacology: the fourth generation of progress. Raven Press, New York

Zahm DS (1999) Functional-anatomical implications of the nucleus accumbens core and shell subterritories. Ann N Y Acad Sci 877:113–128

Zygmunt PM, Petersson J, Andersson DA, Chuang H, Sørgård M, Di Marzo V, Julius D, Högestätt ED (1999) Vanilloid receptors on sensory nerves mediate the vasodilator action of anandamide. Nature 400:452–457

Chapter 8
Gender Disparity of Depression: The Role of Endocannabinoids and Noradrenergic Function

Boris B. Gorzalka and Silvain S. Dang

Abstract Depression is a common and potentially debilitating psychiatric disorder, and is twice as prevalent in women as in men. The traditional monoamine hypothesis of depression provides one perspective into the biological basis of depression, but it is unable to explain all facets of this disease. The reason for the sex difference is currently unclear. The endocannabinoid system, a major neuromodulatory system in the brain, interacts with multiple neurotransmitter and hormone systems, including the monoamine neurotransmitter norepinephrine. Increased endocannabinoid signaling appears to cause greater levels of noradrenergic activation in the locus coeruleus and in axons projecting into other parts of the brain. Dysfunctions in both the endocannabinoid system and the noradrenergic system have been linked to the physiology of depression, with the hypothalamic–pituitary–adrenal (HPA) axis stress response being a major area of interaction. Norepinephrine acts as a "gatekeeper" to the body's stress response, mobilizing the HPA axis to react to stressors. The endocannabinoid system is also a "gatekeeper" to this response, preventing maladaptive HPA hyperactivation and potentially protecting the noradrenergic system from entering into a "burn-out" state in the face of chronic stress. Sexual dimorphism in both systems, as well as in how cells of the locus coeruleus respond to stress, may contribute to some of the sex differences seen in depression. Disruptions to these systems may underlie some cases of depression, and provide potential targets for novel antidepressant treatments.

Clinical depression is a disorder characterized by either depressed mood, including feelings of guilt, low self-esteem, and worthlessness; anhedonia, a loss of enjoyment in daily activities or previously pleasurable stimuli; or both (American Psychiatric Association 2013). Other behavioral and cognitive symptoms of depression include reduced sexual motivation and functioning; changes in metabolism, energy level, and appetite; disruption of sleep patterns; and deficits in concentration and memory (American Psychiatric Association 2013). Major depression has a lifetime prevalence of 8–12 % (Kessler et al. 1994). There is also a strong gender disparity, with

B. B. Gorzalka (✉) · S. S. Dang
Department of Psychology, The University of British Columbia, 3511-2136 West Mall, Vancouver, BC V6T 1Z4, Canada
e-mail: bgorzalka@psych.ubc.ca

E. J. Van Bockstaele (ed.), *Endocannabinoid Regulation of Monoamines in Psychiatric and Neurological Disorders*, DOI 10.1007/978-1-4614-7940-6_8,

the disorder being approximately twice as prevalent in women as in men (Kuehner 2003; Piccinelli and Wilkinson 2000). Because of its potential to lower the quality of life, as well as its prevalence, it is the psychological disorder that likely causes the greatest social and economic burden in North America (McKenna et al. 2005). Currently the underlying cause of this gender difference is not well understood. Determining the causes of this gender difference may contribute to the development of more effective treatment and prevention options.

Our current understanding of the neurobiological bases of depression places an emphasis on the importance of disruptions to monoamine neurotransmitter activity. The monoamine hypothesis of depression states that the symptoms of the depression are caused by a monoamine deficiency in the brain (Schildkraut 1965). This hypothesis was developed from the observation that drugs which increased synaptic monoamine levels could alleviate symptoms of depression and improve mood. In turn, it has guided the development of more effective pharmaceuticals targeting monoamine systems for the treatment of this disorder. However, current generation antidepressants are often unable to produce long-term improvements, require weeks to take effect, and have undesirable side effects (Gorzalka and Hill 2011).

One type of monoaminergic activity that has received substantial focus in depression research is the noradrenergic system. Norepinephrine acts as a "gatekeeper" to arousal and attention and therefore is strongly implicated in the pathological pattern of responding to negative stimuli and stress that often underlie clinical depression (Goddard et al. 2010). Another system that has received recent attention as a contributor to the etiology of depression is the endocannabinoid system. The endocannabinoid system also acts as a type of "gatekeeper" to the hypothalamic–pituitary–adrenal (HPA) axis stress response (Gorzalka and Hill 2011). The noradrenergic and endocannabinoid system interact extensively, both via direct central neurotransmission and less directly through associations with the HPA axis. Both systems also display sexual dimorphism and therefore may contribute to the gender differences in the prevalence of depression. The current chapter will review some current findings regarding how the endocannabinoid system influences noradrenergic signaling and how this interaction may contribute to the biological causes of depression. It will also focus on how sex differences in these two systems may contribute to the greater rate of depression among women.

8.1 Endocannabinoid Regulation of Noradrenergic Signaling

The endocannabinoid and noradrenergic systems interact in several key regions of the brain. Cell bodies of noradrenergic neurons in the brain are localized in the locus coeruleus (LC) (Foote et al. 1983) and the nucleus of the solitary tract (NTS) (Itoi and Sugimoto 2010). The NTS integrates peripheral inputs with central regulatory signaling (Itoi and Sugimoto 2010), whereas the LC plays a crucial role in arousal, attention, and sleep regulation (Sara 2009). Cannabinoid-1 (CB_1) receptor protein and mRNA have been shown to be present in noradrenergic neurons within the LC and NTS (Carvalho et al. 2010; Carvalho and Van Bockstaele 2012; Herkenham et al. 1991; Jelsing et al. 2008; Mailleux and Vanderhaeghen 1992; Matsuda et al. 1993; Scavone et al. 2010). A large proportion of these receptors appear to be localized postsynaptically (Bacci et al. 2004); although CB_1 receptors are generally presynaptic, the presence of postsynaptic CB_1 receptor activity suggests that they play a role in autocrine inhibition for these cells. Presynaptic CB_1 receptors are also present on axon terminals that synapse with the cell bodies of noradrenergic neurons in the LC and NTS (Scavone et al. 2010). Noradrenergic axon terminals projecting from the LC and NTS to the prefrontal cortex (PFC) (Oropeza et al. 2007) and the nucleus accumbens (NAc) (Carvalho et al. 2010) have also been found to interact with CB_1 receptors. In the PFC, CB_1 receptors are expressed directly by noradrenergic terminals as well as their associated excitatory glutamatergic and inhibitory γ-aminobutyric acid–ergic (GABAergic) terminals (Oropeza et al. 2007). In the NAc, CB_1 receptors are not present on noradrenergic terminals but are expressed by associated glutamatergic and GABAergic terminals (Carvalho et al. 2010).

Increased endocannabinoid activity has been shown to increase spontaneous firing in the LC. CB_1 receptor agonists WIN 55,212-2, CP55,940, and Δ9-tetrahydrocannabinol (THC) increase spontaneous firing (Mendiguren and Pineda 2006; Muntoni et al. 2006) and Fos expression in LC noradrenergic neurons (Patel and Hillard 2003). These effects are blocked by CB_1 receptor antagonists SR 141617A and rimonabant, showing that they are dependent on CB_1 receptor activity. Fatty acid amide hydrolase (FAAH) inhibitor URB 597 causes a similar CB_1 antagonist-attenuated increase in spontaneous firing of noradrenergic LC neurons, demonstrating that endogenous anandamide has a role in mediating norepinephrine activity in the LC (Gobbi et al. 2005). SR 141617A alone is sufficient to inhibit basal firing of noradrenergic LC neurons (Muntoni et al. 2006), whereas cannabinoid agonists appear to also decrease KCl-evoked excitation of LC neurons (Mendiguren and Pineda 2007). These results may be indicative of tonal regulation of noradrenergic activity in the LC by the endocannabinoid system, with this system potentially protecting the LC against overexcitation.

Endocannabinoid activity also influences the functioning of the NTS. Approximately half of the neurons in NTS appear to be responsive to CB_1 receptor-mediated effects of cannabinoids (Himmi et al. 1996, 1998). Some of these cells show increased activity following exposure to cannabinoids, whereas others show decreased activity. WIN 55,212-2 and rimonabant both cause increased Fos activation in different groups of neurons within the NTS (Jelsing et al. 2009). This cell-specific variation in response to cannabinoids is consistent with the NTS being a heterogeneous cluster of various neuron types.

Cannabinoids appear to increase norepinephrine release by neurons projecting from the LC in various areas of the brain. Global administration of WIN 55,212-2 and THC increase norepinephrine levels and the rate of norepinephrine turnover in the PFC and NAc (Jentsch et al. 1997; Oropeza et al. 2005; Page et al. 2007). The increase in norepinephrine levels correlates with Fos activation in noradrenergic neurons in the LC; this effect can be inhibited by SR 141716A. These findings suggest that the release of norepinephrine in the PFC and NAc in response to cannabinoids may be due to the activation of noradrenergic neurons in the LC and is mediated by CB_1 receptors. Cannabinoids appear to inhibit the activity of monoamine oxidase, an enzyme that metabolizes norepinephrine (Fisar 2010). However, the CB_1 antagonist/inverse agonist SR 141716A has also been found to increase norepinephrine levels in the PFC (Tzavara et al. 2003), as well as the hypothalamus (Tzavara et al. 2001). While these differential results may be partially because of the use of different drug doses in different studies, they also suggest that norepinephrine signaling pathways are mobilized in response to alterations in the endocannabinoid tone.

Cannabinoids have been shown to be capable of inducing long-term changes in noradrenergic activity. Chronic administration of WIN 55,212-2 increases norepinephrine release in the PFC and increases expression of tyrosine hydroxylase, an enzyme responsible for converting L-tyrosine into L-3,4-dihydroxyphenylalanine (L-DOPA) (a step in catecholamine synthesis) in the LC (Page et al. 2007). Acute exposure to WIN 55,212-2 causes increased expression of the norepinephrine transporter (Reyes et al. 2009). Conversely, chronic exposure to WIN 55,212-2 elicits a downregulation in β_1 receptors in both the PFC (Reyes et al. 2009) and the NAc (Carvalho et al. 2010). This may be a compensatory effect in response to the increase in synaptic norepinephrine induced by CB_1 receptor agonists. Chronic WIN 55,212-2 administration decreases the expression of α_{2A} receptors in the NAc (Carvalho et al. 2010). α_{2A} receptors often act as autoreceptors whose activation leads to decreased subsequent release of norepinephrine. This result may be in response to increased levels of extracellular norepinephrine, or it may be a secondary response to compensate for decreased norepinephrine signaling following cannabinoid-induced downregulation of β_1 receptors.

8.2 The Noradrenergic System and Depression

Consistent with the monoamine hypothesis of depression, deficiencies in norepinephrine levels have long been implicated in the biology of mood disorders. Noradrenergic signaling, especially via neurons with cell bodies in and axons projecting from the LC, is involved in the regulation of attention, arousal, and responses to stressors (Goddard et al. 2010). Basal noradrenergic neuronal firing in the LC regulates arousal and attention, with increased firing rates being associated with increased levels of arousal and attention (Aston-Jones and Cohen 2005; Berridge 2008). Noradrenergic signaling from the LC allows organisms to balance between the needs for vigilance and scanning with those for focused attention. The role of "attention gatekeeper" implies that norepinephrine circuits from the LC play an important role in perceiving and responding to acute and chronic stress (Goddard et al. 2010).

There is a strong body of evidence suggesting that exposure to chronic stress and associated increases in HPA axis activity play a causal role in depression. Noradrenergic signaling may contribute to depression when responses to stress are maladaptive. In animal models of acute stress, such as restraint and conditioned fear paradigms, noradrenergic LC neurons fire in phasic bursts (Abercrombie and Jacobs 1987). Many of these neurons innervate the paraventricular nucleus (PVN) of the hypothalamus and stimulate the release of corticotropin-releasing hormone (CRH) and subsequent activation of the HPA axis (Flugge et al. 2004). Acute stress appears to decrease expression of α_2 receptor activity, perhaps partially disengaging the autocrine negative feedback loop to allow for an increase in norepinephrine activity in response to stress. Following removal of acute stressors, it is expected that α_2 receptor expression returns to normal levels and basal noradrenergic activity is restored.

However, exposure to chronic stress can lead to long-term alterations in this system. Chronic exposure to social stress in rodent models increases plasma levels of norepinephrine, but reduces CNS norepinephrine concentrations and increases α_2 receptor expression (Flugge et al. 2003; Reber et al. 2007). Chronic stress also causes an upregulation of β receptors, perhaps as a compensatory mechanism for reduced central norepinephrine availability. Atrophy of noradrenergic axon projections has been seen in association with chronic stress exposure (Liu et al. 2003; Kitayama et al. 1997). Patients with depression, compared with healthy control subjects, have been shown to have lower levels of circulating norepinephrine in the internal jugular vein, which receives superior sagittal sinus blood that is somewhat more representative of conditions in the brain (Lambert et al. 2000). Postmortem studies reveal that nonmedicated depressed individuals exhibit a loss of noradrenergic neurons in the LC and noradrenergic axon terminals in the limbic system, as well as increased α_2 receptor expression in the PFC and other sites innervated by noradrenergic axons (Ordway and Klimek 2001). In association with chronic stress exposure and long-term HPA activation, noradrenergic signaling becomes inhibited at both ligand and autocrine receptor levels and perhaps enters into a "burn-out" state, thereby contributing to depression (Goddard et al. 2010).

8.3 The Endocannabinoid System and Depression

Studies in both human patients and animal models have demonstrated that the endocannabinoid system plays a key role in depression. In humans, alterations to endocannabinoid functionality have been linked to mood disorders. Women suffering from major depression exhibit significantly reduced levels of endocannabinoids in the blood (Hill et al. 2008, 2009). Presence of single nucleotide polymorphisms (SNPs) in the CB_1 receptor gene is associated with increased neuroticism, vulnerability to depressive episodes in response to life stress, decreased receptivity to the effects of antidepressant medications, and decreased neuronal responses to rewards (Domschke et al. 2008; Juhasz et al. 2009). Patients suffering from depression also exhibited a greater frequency of CB_1 receptor SNPs (Monteleone et al. 2010).

In rodents, exposure to chronic stress has been shown to produce depression-like behaviors, as well as reductions in CB_1 receptor expression and the ability of CB_1 receptor activation to inhibit neurotransmitter release in limbic structures

(Hill and Gorzalka 2005; Reich et al. 2009). Additionally, anandamide levels are reduced following exposure to chronic stress in both the limbic system and PFC (Patel et al. 2005; Rademacher et al. 2008). Conversely, in humans, postmortem studies of depressed suicide victims not taking antidepressants show increased expression of CB_1 receptors and increased CB_1 receptor-mediated G-protein activation in the PFC (Hungund et al. 2004; Valdizán et al. 2011). In depressed suicide victims who had been taking antidepressants, this effect was not seen (Valdizán et al. 2011). Overall, it appears that depression is associated with suppression of endocannabinoid activity at both the receptor and ligand level in humans and rodents, although some compensatory upregulation of CB_1 receptor levels may occur in the PFC.

Artificial disruptions to the endocannabinoid system can also precipitate depression-like symptoms. In rodent models, suppression of CB_1 receptor signaling appears to produce a behavioral and cognitive profile similar to typical, melancholic depression (Gorzalka and Hill 2011). CB_1 receptor antagonists or genetic knockout of CB_1 receptors have been shown to produce anhedonia (Sanchis-Segura et al. 2004), increased trait anxiety (Mikics et al. 2009), impaired extinction of aversive memories (Marsicano et al. 2002), reductions in feeding (Ravinet et al. 2004) and sexual motivation (Gorzalka et al. 2010), and increased passive coping toward stress (Steiner et al. 2008a). Disruptions to the endocannabinoid system in rodents appear to produce physiological changes that are associated with depression in humans. Suppression of CB_1 receptor signaling can increase basal levels of HPA axis activity (Patel et al. 2004), increased HPA activation in response to stress (Steiner et al. 2008b), and decrease habituation in the HPA axis to chronic stress (Patel et al. 2005). Corticosterone hypersecretion in response to chronic stress is attenuated by enhancing anandamide activity via inhibition of FAAH, the primary anandamide breakdown enzyme (Hill et al. 2010).

These findings are further corroborated by clinical trials for the CB_1 receptor antagonist rimonabant in humans as a treatment of obesity. Clinical use of rimonabant was halted after a significant number of participants, including those with no history of psychological illness, experienced symptoms of anxiety and depression (Christensen et al. 2007; Hill and Gorzalka 2009). Subsequent studies also found that rimonabant disrupted positive affective memories and neurophysiological responses to rewards (Horder et al 2009, 2010). These results provide double-blind, placebo-controlled data suggesting that perturbations to CB_1 receptor signaling may underlie at least some cases of depression.

CB_1 receptor agonists may therefore have therapeutic potential as novel antidepressants. Users of marijuana often report improvements in mood and a sense of euphoria. In commonly used models of depression in rodents, such as the forced swim test (FST) and the tail suspension test, administration of CB_1 receptor agonists or of endocannabinoid reuptake and metabolic protein inhibitors leads to behavioral changes similar to those of traditional antidepressants (Gorzalka and Hill 2011). For example, administration of the CB_1 receptor agonist HU-210 to rats in the FST reduced the time interval spent engaging in immobility, a depression-like passive coping behavior (Hill and Gorzalka 2005). Moreover, this was comparable to the effects of tricyclic antidepressant desipramine. Increasing synaptic levels of anandamide through inhibition of FAAH has similar antidepressant effects, and reduces anhedonia induced by exposure to chronic stress (Gobbi et al. 2005; Rademacher and Hillard 2007). FAAH inhibitors may present a particularly promising avenue of investigation, as they do not appear to share the same reinforcing properties as THC (Gobbi et al. 2005; Hill and Gorzalka 2009; Justinova et al. 2008).

8.4 Endocannabinoid and Noradrenergic Interactions in Depression

These collective findings begin to paint a picture of how depression, at least in some cases, may develop with regards to the endocannabinoid and noradrenergic systems. Response to exposure to chronic stress appears to be a major area of interaction between endocannabinoids and norepinephrine. It appears that the noradrenergic system is responsible for mobilizing cognitive resources in response to stressors (Goddard et al. 2010), whereas the endocannabinoid system is responsible for preventing hyperactivation of the HPA axis. Paradoxically, increased endocannabinoid activity appears to increase noradrenergic signaling in the brain. However, the endocannabinoid system may be very broadly viewed as acting to maintain and restore homeostasis in response to changes in an organism's external and internal environment. In this way, the endocannabinoid system guides the HPA axis toward activating the noradrenergic system to allow for optimal responding to stressors.

As previously noted, chronic stress and subsequent depression-like behavior in animal models have been associated with reduced CB_1 receptor expression and functionality and reduced central anandamide content. Furthermore, the endocannabinoid system has also been described as a "gatekeeper" to the HPA axis, with basal levels of anandamide signaling engaging in a tonic suppression of excitatory inputs into the PVN and restraining CRH secretion (Gorzalka and Hill 2009). Following exposure to chronic stress, increases in endocannabinoid 2-arachidonoylglycerol (2-AG) signaling in the limbic system allow for the development of HPA axis habituation to stressful stimuli. However, data regarding the effects of chronic stress on the endocannabinoid system suggest that chronic stress itself may reduce the effectiveness of this system in regulating HPA axis activation. It is possible that pre-existing dysfunctions in the endocannabinoid system may combine with chronic stress and lead to maladaptive HPA axis hyperactivity, therefore contributing to the development of depression.

Conversely, in response to chronic stress exposure, the facilitatory effects of CB_1 receptor activity on the noradrenergic system may compensate for a "burn-out" state and allow for a more adaptive stress response than would otherwise be possible. This model is supported by findings that reductions in depression-like behavior caused by HU-210 are attenuated by both α and β receptor antagonists, showing that cannabinoids recruit norepinephrine in their antidepressant profile (Morrish et al. 2009). Similarly, both α_2 and β receptor antagonists reduce the secretion of corticosterone caused by HU-210, suggesting that the endocannabinoid system recruits noradrenergic signaling in its regulation of adaptive HPA axis stress responses (McLaughlin et al. 2009). Under this model, depression may result when the noradrenergic system no longer mobilizes in response to HPA axis activation in an adaptive fashion and the endocannabinoid system is no longer able to bring both systems back to homeostasis. Inhibition of noradrenergic signaling following chronic stress may account in part for the monoamine hypothesis of depression and provides a possible explanation for the functionality of serotonin and noradrenaline reuptake inhibitors (SNRIs). However, as previously discussed, there is some evidence that acute administrations of CB_1 receptor antagonists can cause activation of the noradrenergic system, whereas chronic exposure to CB_1 receptor agonists can downregulate adrenergic receptor expression. These results suggest that disruptions to endocannabinoid tone itself may be stressful to the noradrenergic system.

Depression is a heterogeneous condition that may be the manifestation of a wide range of physiological and psychological conditions. From the evidence, it is clear that the interactions between the endocannabinoid system and the noradrenergic system described here represent an important but only incomplete aspect of the biology of depression. Beyond the systems described here, dysfunctions in the activity of other monoamines including serotonin (Esteban and García-Sevilla 2011), hormonal systems such as the hypothalamus–pituitary–gonadal axis (Martel et al. 2009), and processes like neurogenesis (Eisch et al. 2008) have also been implicated in depression. Psychological and social factors only increase the number of possible “causes” of depression. The complexity of this disorder suggests it could manifest from a variety of different systems and processes. For example, anxious depression is characterized by comorbidity with lifelong generalized anxiety (Goddard et al. 2010). As SNRIs are more effective than other antidepressants at treating this subtype, anxious depression may represent underlying dysfunctions in the noradrenergic system (Nelson 2008). Meanwhile, melancholic depression is often described with anhedonia as a dominant symptom and shows similarities to CB_1 receptor knockout animal models (Gorzalka and Hill 2011). It is possible that lifelong disruptions to endocannabinoid signaling contribute to this subtype of depression. Further research is needed to assess the empirical validity of depression subtypes and whether they reflect different biological contributions to pathology.

8.5 Influence of Gender

The prevalence of depression is greater in women than men, at a ratio of approximately two to one (Kuehner 2003; Piccinelli and Wilkinson 2000; Lundberg 2005). The specific causes of this disparity are currently unknown, although evidence suggests both biological and social factors play key roles. Biologically, developmental effects of sex as well as of gonadal hormones have a strong influence on the functioning of monoamine systems, including the noradrenergic system. There are also reciprocal interactions between gonadal hormones and the endocannabinoid system. Finally, there is evidence suggesting sex differences in response to stress, with a potentially increased vulnerability to stress-related disorders in women.

Sex differences have been shown in some animal models linking stress and depression (Dalla et al. 2010). In those models, females appear more vulnerable to developing depression-like behaviors. Female rats exhibit greater levels of immobility in the FST (Dalla et al. 2008a; Drossopoulou et al. 2004; Pitychoutis et al. 2009) and show greater vulnerability to swim stress compared with males (Sun and Alkon 2006). FST also leads to an increased hippocampal and hypothalamic serotonergic signaling in male rats but decreased signaling in female rats (Drossopoulou et al. 2004), while only males display reduced levels of glucocorticoid receptor mRNA in the hippocampus following repeated exposure to swim stress (Karandrea et al. 2002). In addition, in measures of associative learning following shock or swim stress, a model of cognitive deficits that can occur in depression, females exhibited impairment, whereas males show enhanced learning (Shors et al. 2007); this is associated with decreased density of dendritic spines in the CA1 area of the hippocampus of females, but increased density in males, following stress exposure (Shors et al. 2001, 2007). Chronic mild stress has been shown to produce a greater increase in basal levels of corticosterone and to disrupt hippocampal serotonergic and prefrontal dopaminergic activity more readily in females (Dalla et al. 2005). However, chronic mild stress also appears to produce a less consistent reduction in sucrose water consumption, a measure of anhedonia, in females (Dalla et al. 2005, 2008a); this may be because females naturally exhibit less consistent patterns of sucrose consumption even in the absence of stress (Dalla et al. 2008a).

In other models however, males seem to be more vulnerable than females. In contrast to shock or swim stress and associative learning, male rats exhibited impaired spatial learning in the Y-maze following acute restraint stress (Conrad et al. 2004) and in the Morris water maze after chronic restraint stress (Bowman et al. 2003; Kitraki et al. 2004). In contrast to males, females exhibited enhanced spatial learning following restraint stress. In the learned helplessness model, which has been viewed as "hopelessness" in depression, male rats more readily develop learned helplessness and decreases in neurogenesis in the dentate gyrus following exposure to repeated inescapable foot shocks, while sexual dimorphism in the alteration of monoamine activity was also seen (Shors et al. 2007; Heinsbroek et al. 1991). Some of the above-mentioned differences, such as in associative learning (Wood and Shors 1998), are dependent on gonadal hormones, whereas others, such as those in learned helplessness (Dalla et al. 2008b), are not. Overall, this suggests that males and females are vulnerable to different types of stressors on different modalities of behavior and physiological functioning. It remains to be seen which, if any, of these processes are applicable to the gender difference in depression in humans.

There is evidence for the sexual dimorphism of the noradrenergic system. Morphologically, the LC is larger in volume in females and has a greater number of neurons (Pinos et al. 2001). In humans, depressed men have circulating 3-methoxy-4-hydroxyphenylglycol (MHPG, a norepinephrine metabolite) levels similar to healthy controls, but depressed women have either elevated or suppressed circulating

MHPG levels (Halbreich et al. 1987). Estradiol implants in menopausal women have been shown to decrease anxiety and depression and reduce circulating MHPG levels (Best et al. 1992). Women have also been shown to be more responsive to SNRI antidepressants (Khan et al. 2005). In vitro, functional estrogen β receptors have been detected in LC-derived cultures of mouse cells (Rincavage et al. 2003), while estradiol stimulates the expression of the norepinephrine synthesis enzymes tyrosine hydroxylase and dopamine β-hydroxylase in the LC of rats (Servoa et al. 2002). Behaviorally, estradiol prevents the α_{2A} receptor agonist guanfacine reversal of stress-induced working memory impairments in ovariectomized rats (Shansky et al. 2009).

The response of neurons in the LC to stress also appears to vary by sex. CRH is 10–30 times more effective in inducing activation of LC neurons in female rats than male rats (Curtis et al. 2006). This effect does not depend on the gonadal hormone status of the females. In addition, females express greater levels of CRH receptors in the LC than males. Furthermore, exposure to forced swim stress abolished this sex difference. A likely mechanism for the increased sensitivity of the female LC-noradrenergic neurons to CRH is the lack of receptor internalization. Receptors internalized into the cytoplasm of a cell are not exposed to binding by extracellular ligands. Unstressed male rats localize the majority of their LC CRH receptors on the plasma membrane, whereas males administered with CRH internalize most of their CRH receptors in the cytoplasm (Bangasser et al. 2010). Association of β-arrestin2 with the CRH receptor is a crucial step in the internalization of these receptors and only occurs in males. Similar sex differences are found using mutant CRH overexpressing mice (Bangasser et al. 2013). In addition, the coupling between the CRH receptor and associated secondary messenger G_s is stronger in unstressed female compared with male rats, but surprisingly, this sex difference is eliminated by stress exposure (Bangasser et al. 2010). These results suggest that the noradrenergic system, as well as other CRH-sensitive systems, in the female brain is more sensitive to small increases in HPA axis activation and less adaptable in the face of large increases in CRH signaling. The "unstressed" state of LC neurons in females may be similar to the "stressed" state of LC neurons in males. Similar to sex differences in the other animal models presented above, these results provide a potential explanation or partial explanation for the gender differences in human depression.

Several lines of evidence highlight the sexually dimorphic nature of the endocannabinoid system. In humans, men appear to show greater CB_1 receptor activity than women, but the CB_1 receptor activity of women increases with age (Van Laere et al. 2008). In animal models, CB_1 receptor antagonist AM-251 shows a diurnal rhythm in the strength of its blockade of CB_1 receptors; this rhythm is more pronounced in male than in female rats (Atkinson et al. 2010). Developmentally, differential patterns of alternations in CB_1 and CB_2 expression are seen in male and female rats after maternal deprivation in infancy (Suárez et al. 2009; Llorente et al. 2008; López-Gallardo et al. 2008). In addition, chronic administration of THC during adolescence causes anhedonia, as measured by decreased conditioned place preference to sucrose water, and despair, as measured by increased time spent in immobility during the forced swim test, in female rats but only anhedonia and not despair in male rats (Rubino et al. 2008). This suggests that the endocannabinoid system of females is more vulnerable to permanent change caused by perturbations during development. The endocannabinoid system has also been shown to interact extensively with androgens, estrogens, and progesterone (Gorzalka and Dang 2012). These effects have not been investigated to any appreciable extent in the context of depression, although it has been shown that estradiol recruits the endocannabinoid system in producing antidepressant and anxiolytic effects (Hill et al. 2007). However, chronic administrations of the CB_1 receptor agonist HU-210 have been shown to produce antidepressant effects in both male and female rats (Morrish et al. 2009). This suggests that despite sex differences, agents that modulate endocannabinoid activity may represent potential pharmacological treatments for depression in both men and women.

8.6 Conclusion

The balance of the current evidence shows that interactions between the endocannabinoid system and the noradrenergic system play important roles in the physiological basis of depression. The two systems interact both directly in the brain, such as in the LC and NTS, as well as indirectly via the HPA axis. Maladaptive interactions between the two systems with the HPA axis stress response may underlie many if not most of the cases of depression. In healthy individuals, noradrenergic signaling appears to regulate the appropriate cognitive response to stressors, whereas the endocannabinoid system in turn regulates HPA axis activation. Depression is often associated with a breakdown in these regulatory pathways when individuals experience chronic exposure to stress. Both these systems also show a notable sexual dimorphism and interactions with gonadal hormones. The gender disparity of depression may be caused by greater vulnerability of the female noradrenergic system to stress. Notwithstanding gender differences, development of cannabinoid pharmaceutical agents, as well as improved SNRIs, may present new opportunities for the treatment of depression.

Since the development of the relatively historic and simple monoamine hypothesis of depression, our view of the biology of this disorder has become progressively more intricate and nuanced. The interaction between the endocannabinoid and noradrenergic activity represents one facet in our current understanding of depression. However, our knowledge is still incomplete in how various systems influence each other over the course of the disease and where possible vulnerabilities to dysfunction may lie. Current antidepressant medications are useful but imperfect solutions. The noradrenergic system and the endocannabinoid system together are potential targets for current and future treatment options. Therefore, further research leading to a clear understanding of their functioning has the potential to improve health outcomes and quality of life for those suffering from depression.

References

Abercrombie ED, Jacobs BL (1987) Single-unit response of noradrenergic neurons in the locus coeruleus of freely moving cats. I. Acutely presented stressful and nonstressful stimuli. J Neurosci 7:2837–2843

American Psychiatric Association (2013) Diagnostic and statistical manual of mental disorders: DSM-V. American Psychiatric Association, Washington

Aston-Jones G, Cohen JD (2005) An integrative theory of locus coeruleus-norpeinephrine function: adaptive gain and optimal performance. Annu Rev Neurosci 28:403–450

Atkinson HC, Leggett JD, Wood SA et al (2010) Regulation of the hypothalamic-pituitary-adrenal axis circadian rhythm by endocannabinoids is sexually diergic. Endocrinology 151:3720–3727

Bacci A, Huguenard JR, Prince DA (2004) Long-lasting self-inhibition of neocortical interneurons mediated by endocannabinoids. Nature 431:312–316

Bangasser DA, Curtis A, Reyes BA et al (2010) Sex differences in corticotropin-releasing factor receptor signaling and trafficking: potential role in female vulnerability to stress-related psychopathology. Mol Psychiatry 15:877–904

Bangasser DA, Reyes BA, Piel D et al (2013) Increased vulnerability of the brain norepinephrine system of females to corticotropin-releasing factor overexpression. Mol Psychiatry 18:166–173

Berridge CW (2008) Noradrenergic modulation of arousal. Brain Res Rev 58:1–17

Best NR, Rees MP, Barlow DH, Cowen PJ (1992) Effect of estradiol implant on noradrenergic function and mood in menopausal subjects. Psychoneuroendocrinology 17:87–93

Bowman RE, Beck KD, Luine VN (2003) Chronic stress effects on memory: sex differences in performance and monoaminergic activity. Horm Behav 43:48–59

Carvalho AF, Van Bockstaele EJ (2012) Cannabinoid modulation of noradrenergic circuits: implications for psychiatric disorders. Prog Neuropsychopharmacol Biol Psychiatry 38:59–67

Carvalho AF, Mackie K, Van Brockstaele EJ (2010) Cannabinoid modulation of limbic forebrain noradrenergic circuitry. Eur J Neurosci 31:286–301

Christensen R, Kristensen PK, Bartels EM, Bliddal H, Astrup A (2007) Efficacy and safety of the weight-loss drug rimonabant: a meta-analysis of randomised trials. Lancet 370:1706–1713

Conrad CD, Jackson JL, Wieczorek L et al (2004) Acute stress impairs spatial memory in male but not female rats: influence of estrous cycle. Pharmacol Biochem Behav 78:569–579

Curtis AL, Bethea T, Valentino RJ (2006) Sexually dimorphic responses of the brain norepinephrine system to stress and corticotropin-releasing factor. Neuropsychopharmacology 31:544–554

Dalla C, Antoniou K, Drossopoulou G et al (2005) Chronic mild stress impact: are females more vulnerable? Neuroscience 135:703–714

Dalla C, Antoniou K, Kokras N et al (2008a) Sex differences in the effects of two stress paradigms on dopaminergic neurotransmission. Physiol Behav 93:595–605

Dalla C, Edgecomb C, Whetstone AS, Shors TJ (2008b) Females do not express learned helplessness like males do. Neuropsychopharmacology 33:1559–1569

Dalla C, Pitychoutis PM, Kokras N, Papadopoulou-Daifoti Z (2010) Sex differences in animal models of depression and antidepressant response. Basic Clin Pharmacol Toxicol 106:226–233

Domschke K, Dannlowski U, Ohrmann P, Lawford B, Bauer J, Kugel H, Baune BT (2008) Cannabinoid receptor 1 (CNR1) gene: impact on antidepressant treatment response and emotion processing in major depression. Eur Neuropsychopharm 18:751–759

Drossopoulou G, Antoniou K, Kitraki E etal (2004) Sex differences in behavioral, neurochemical and neuroendocrine effects induced by the forced swim test in rats. Neuroscience 126:849–857

Eisch AJ, Cameron HA, Encinas JM, Meltzer LA, Ming GL, Overstreet-Wadiche LS (2008) Adult neurogenesis, mental health, and mental illness: hope or hype? J Neurosci 28:11785–11791

Esteban S, García-Sevilla JA (2011) Effects induced by cannabinoids on monoaminergic systems in the brain and their implications for psychiatric disorders. Prog Neuropsychopharmacol Biol Psychiatry 38:78–87

Fisar Z (2010) Inhibition of monoamine oxidase activity by cannabinoids. Naunyn Schmiedebergs Arch Pharmacol 381:563–572

Flugge G, Van KM, Meyer H, Fuchs E (2003) Alpha2A and alpha2C-adrenoceptor regulation in the brain: alpha2A changes persist after chronic stress. Eur J Neurosci 17:917–928

Flugge G, Van KM, Mijnster MJ (2004) Perturbations in brain monoamine systems during stress. Cell Tissue Res 315:1–14

Foote SL, Bloom FE, Aston-Jones G (1983) Nucleus locus ceruleus: new evidence of anatomical and physiological specificity. Physiol Rev 63:844–914

Gobbi G, Bambico FR, Mangieri R, Bortolato M, Campolongo P, Solinas M, Piomelli D (2005) Antidepressant-like activity and modulation of brain monoaminergic transmission by blockade of anandamide hydrolysis. Proc Natl Acad Sci U S A 102:18620–18625

Goddard AW, Ball SG, Martinez J, Robinson MJ, Yang CR, Russell JM, Shekhar A (2010) Current perspectives of the roles of the central norepinephrine system in anxiety and depression. Depress Anxiety 27:339–350

Gorzalka BB, Dang SS (2012) Endocannabinoids and gonadal hormones: bidirectional interactions in physiology and behavior. Endocrinology 153:1016–1024

Gorzalka BB, Hill MN (2009) Integration of endocannabinoid signaling into the neural network regulating stress-induced activation of the hypothalamic-pituitary-adrenal axis. Curr Top Behav Neurosci 1:289–306

Gorzalka BB, Hill MN (2011) Putative role of endocannabinoid signaling in the etiology of depression and actions of antidepressants. Prog Neuropsychopharmacol Biol Psychiatry 35:1575–1585

Gorzalka BB, Hill MN, Chang SC (2010) Male–female differences in the effects of cannabinoids on sexual behavior and gonadal hormone function. Horm Behav 58:91–99

Halbreich U, Sharpless N, Asnis GM et al (1987) Afternoon continuous plasma levels of 3-methoxy-4-hydroxphenylglycol and age. Distinctive biologic subgroups of endogenous depression? Arch Gen Psychiatry 44:804–812

Heinsbroek RP, van Haaren F, Feenstra MG et al (1991) Controllable and uncontrollable footshock and monoaminergic activity in the frontal cortex of male and female rats. Brain Res 551:247–255

Herkenham M, Lynn AB, De Costa BR, Richfield EK (1991) Neuronal localization of cannabinoid receptors in the basal ganglia of the rat. Brain Res 547:267–274

Hill MN, Gorzalka BB (2005) Pharmacological enhancement of cannabinoid CB1 receptor activity elicits an antidepressant-like response in the rat forced swim test. Eur Neuropsychopharmacology 15:593–599

Hill MN, Gorzalka BB (2009) Impairments in endocannabinoid signaling and depressive illness. JAMA 301:1165–1166

Hill MN, Karacabeyli ES, Gorzalka BB (2007) Estrogen recruits the endocannabinoid system to modulate emotionality. Psychoneuroendocrinology 32:350–357

Hill MN, Miller GE, Ho WS, Gorzalka BB, Hillard CJ (2008) Serum endocannabinoid content is altered in females with depressive disorders: a preliminary report. Pharmacopsychiatry 41:48–53

Hill MN, Miller GE, Carrier EJ, Gorzalka BB, Hillard CJ (2009) Circulating endocannabinoids and N-acyl ethanolamines are differentially regulated in major depression and following exposure to social stress. Psychoneuroendocrinology 34:1257–1262

Hill MN, McLaughlin RJ, Bingham B et al (2010) Endogenous cannabinoid signaling is essential for stress adaptation. Proc Natl Acad Sci U S A 107:9406–9411

Himmi T, Dallaporta M, Perrin J, Orsini JC (1996) Neuronal responses to delta 9-tetrahydrocannabinol in the solitary tract nucleus. Eur J Pharmacol 312:273–279

Himmi T, Perrin J, El Ouazzani T, Orsini JC (1998) Neuronal responses to cannabinoid receptor ligands in the solitary tract nucleus. Eur J Pharmacol 359:49–54

Horder J, Cowen PJ, Di Simplicio M, Browning M, Harmer CJ (2009) Acute administration of the cannabinoid CB1 antagonist rimonabant impairs positive affective memory in healthy volunteers. Psychopharmacology 205:85–91

Horder J, Harmer CJ, Cowen PJ, McCabe C (2010) Reduced neural response to reward following 7 days treatment with the cannabinoid CB(1) antagonist rimonabant in healthy volunteers. Int J Neuropsychopharmacol 13:1103–1113

Hungund BL, Vinod KY, Kassir SA et al (2004) Upregulation of CB1 receptors and agonist-stimulated [35S]GTPγS binding in the prefrontal cortex of depressed suicide victims. Mol Psychiatry 9:184–190

Itoi K, Sugimoto N (2010) The brainstem noradrenergic systems in stress, anxiety, and depression. J Neuroendocrinol 22:355–361

Jelsing J, Larson PJ, Vrang N (2008) Identification of cannabinoid type 1 receptor expressing cocaine amphetamine-regulated transcript neurons in the rat hypothalamus and brainstem using in situ hybridization and immunohistochemistry. Neuroscience 154:641–652

Jelsing J, Galzin AM, Guillot E, Pruniaux MP, Larsen PJ, Vrang N (2009) Localization and phenotypic characterization of brainstem neurons activated by rimonabant and WIN55,212-2. Brain Res Bull 78:202–210

Jentsch JD, Andrusiak E, Tran A et al (1997) Delta 9-tetrahydrocannabinol increases prefrontal cortical catecholaminergic utilization and impairs spatial working memory in the rat: blockade of dopaminergic effects with HA966. Neuropsychopharmacology 16:426–432

Juhasz G, Chase D, Pegg E et al (2009) CNR1 gene is associated with high neuroticism and low agreeableness and interacts with recent negative life events to predict current depressive symptoms. Neuropsychopharmacology 34:2019–2027

Justinova Z, Mangieri RA, Bortolato M et al (2008) Fatty acid amide hydrolase inhibition heightens anandamide signaling without reinforcing effects in primates. Biol Psychiatry 64:930–937

Karandrea D, Kittas C, Kitraki E (2002) Forced swimming differentially affects male and female brain corticosteroid receptors. Neuroendocrinology 75:217–226

Kessler RC, McGonagle KA, Zhao S, Nelson CB, Hughes M, Eshleman S, Kendler HU (1994) Lifetime and 12-month prevalence of DSM-III-R psychiatric disorders in the United States. Results from the National Comorbidity Survey. Arch Gen Psychiatry 51:8–19

Khan A, Brodhead AE, Schwartz KA, Kolts RL, Brown WA (2005) Sex differences in antidepressant response in recent antidepressant clinical trials. J Clin Psychopharmacol 25:318–324

Kitayama I, Yaga T, Kayahara T, Nakano K, Murase S, Otani M, Nomura J (1997) Long-term stress degenerates, but imipramine regenerates, noradrenergic axons in the rat cerebral cortex. Biol Psychiat 42:687–696

Kitraki E, Kremmyda O, Youlatos D et al (2004) Gender-dependent alterations in corticosteroid receptor status and spatial performance following 21 days of restraint stress. Neuroscience 125:47–55

Kuehner C (2003) Gender differences in unipolar depression: an update of epidemiological findings and possible explanations. Acta Psychiatr Scand 108:163–174

Lambert G, Johansson M, Agren H, Friberg P (2000) Reduced brain norepinephrine and dopamine release in treatment-refractory depressive illness. Arch Gen Psychiat 57:787–793

Liu Y, Ishida Y, Shinoda K, Nakamura S (2003) Effects of repeated stress on regeneration of serotonergic and noradrenergic axons in the cerebral cortex of adult rats. Neurosci Lett 339:227–230

Llorente R, Llorente-Berzal A, Petrosino S et al (2008) Gender-dependent cellular and biochemical effects of materal deprivation on the hippocampus of neonatal rats: a possible role for the endocannabinoid system. Dev Neurobiol 68:1334–1347

López-Gallardo M, Llorente R, Llorente-Berzal A et al (2008) Neuronal and glial alterations in the cerebellar cortex of maternally deprived rats: gender differences and modulatory effects of two inhibitors of endocannabinoid inactivation. Dev Neurobiol 68:1429–1440

Lundberg U (2005) Stress hormones in health and illness: the roles of work and gender. Psychoneuroendocrinology 30:1017–1021

Mailleux P, Vanderhaeghen JJ (1992) Distribution of neuronal cannabinoid receptor in the adult rat brain: a comparative receptor binding radioautography and in situ hybridization histochemistry. Neuroscience 48:655–668

Marsicano G, Wotjak CT, Azad SC et al (2002) The endogenous cannabinoid system controls extinction of aversive memories. Nature 418:530–534

Martel MM, Klump K, Nigg JT, Breedlove SM, Sisk CL (2009) Potential hormonal mechanisms of attention-deficit/hyperactivity disorder and major depressive disorder: a new perspective. Horm Behav 55:465–479

Matsuda LA, Bonner TI, Lolait SJ (1993) Localization of cannabinoid receptor mRNA in rat brain. J Comp Neurol 327:535–550

McKenna MT, Michaud CM, Murray CJ, Marks JS (2005) Assessing the burden of disease in the United States using disability-adjusted life years. Am J Prev Med 28:415–423

McLaughlin RJ, Hill MN, Gorzalka BB (2009) Monoaminergic neurotransmission contributes to cannabinoid-induced activation of the hypothalamic-pituitary-adrenal axis. Eur J Pharmacol 624:71–76

Mendiguren A, Pineda J (2006) Systemic effect of cannabinoids on the spontaneous firing rate of locus coeruleus neurons in rats. Eur J Pharmacol 534:83–88

Mendiguren A, Pineda J (2007) CB(1) cannabinoid receptors inhibit the glutamatergic component of KCl-evoked excitation of locus coeruleus neurons in rat brain slices. Neuropharmacology 52:617–625

Mikics E, Vas J, Aliczki M, Halasz J, Haller J (2009) Interactions between the anxiogenic effects of CB1 gene disruption and 5-HT3 neurotransmission. Behav Pharmacol 20:265–272

Monteleone P, Bifulco M, Maina G et al (2010) Investigation of CNR1 and FAAH endocannabinoid gene polymorphisms in bipolar disorder and major depression. Pharmacol Res 61:400–404

Morrish AC, Hill MN, Riebe CJ, Gorzalka BB (2009) Protracted cannabinoid administration elicits antidepressant behavioral responses in rats: role of gender and noradrenergic transmission. Physiol Behav 98:118–124

Muntoni AL, Pillola G, Melis M, Perra S, Gessa GL, Pistis M (2006) Cannabinoids modulate spontaneous neuronal activity and evoked inhibition of locus coeruleus noradrenergic neurons. Eur J Neurosci 23:2385–2394

Nelson JC (2008) Anxious depression and response to treatment. Am J Psychiatry 165:297–299

Ordway GA, Klimek V (2001) Noradrenergic pathology in psychiatric disorders: postmortem studies. CNS Spectr 6:697–703

Oropeza VC, Page ME, Van Bockstaele EJ (2005) Systemic administration of WIN 55,212-2 increases norepinephrine release in the rat frontal cortex. Brain Res 1046:45–54

Oropeza VC, Mackie K, Van Brockstaele EJ (2007) Cannabinoid receptors are localized to noradrenergic axon terminals in the rat frontal cortex. Brain Res 1127:36–44

Page ME, Oropeza VC, Sparks SE et al (2007) Repeated cannabinoid administration increases indices of noradrenergic activity in rats. Pharmacol Biochem Behav 86:162–168

Patel S, Hillard CJ (2003) Cannabinoid-induced fos expression within A10 dopaminergic neurons. Brain Res 963:15–25

Patel S, Roelke CT, Rademacher DJ, Cullinan WE, Hillard CJ (2004) Endocannabinoid signaling negatively modulates stress-induced activation of the hypothalamic-pituitary-adrenal axis. Endocrinology 145:5431–5438

Patel S, Roelke CT, Rademacher DJ, Hillard CJ (2005) Inhibition of restraint stress-induced neural and behavioural activation by endogenous cannabinoid signalling. Eur J Neurosci 21:1057–1069

Piccinelli M, Wilkinson G (2000) Gender differences in depression. Critical review. Br J Psychiatry 177:486–492

Pinos H, Collado P, Rodràguez-Zafra M et al (2001) The development of sex differences in the locus coeruleus of the rat. Brain Res Bull 56:73–78

Pitychoutis PM, Nakamura K, Tsonis PA et al (2009) Neurochemical and behavioral alterations in an inflammatory model of depression: sex differences exposed. Neuroscience 159:1216–1232

Rademacher DJ, Hillard CJ (2007) Interactions between endocannabinoids and stress-induced decreased sensitivity to natural reward. Prog Neuropsychopharmacol Biol Psychiatry 31:633–641

Rademacher DJ, Meier SE, Shi L, Ho WS, Jarrahian A, Hillard CJ (2008) Effects of acute and repeated restraint stress on endocannabinoid content in the amygdala, ventral striatum, and medial prefrontal cortex in mice. Neuropharmacology 54:108–116

Ravinet Trillou C, Delgorge C, Menet C, Arnone M, Soubrie P (2004) CB1 cannabinoid receptor knockout in mice leads to leanness, resistance to diet-induced obesity and enhanced leptin sensitivity. Int J Obes Relat Metab Disord 28:640–648

Reber SO, Birkeneder L, Veenema AH, Obermeier F, Falk W, Straub RH, Neumann ID (2007) Adrenal insufficiency and colonic inflammation after a novel chronic psycho-social stress paradigm in mice: implications and mechanisms. Endocrinology 148:670–682

Reich CG, Taylor ME, McCarthy MM (2009) Differential effects of chronic unpredictable stress on hippocampal CB1 receptors in male and female rats. Behav Brain Res 203:264–269

Reyes BS, Rosario JC, Piana PT, Van Bockstaele EJ (2009) Cannabinoid modulation of cortical adrenergic receptors and transporters. J Neurosci Res 87:3671–3678

Rincavage HL, McDonnell DP, Kuhn CM (2003) Expression of functional estrogen receptor beta in locus coeruleus-derived Cath.a cells. Endocrinology 144:2829–2835

Rubino T, Vigano' D, Realini N et al (2008) Chronic delta 9-tetrahydrocannabinol during adolescence provokes sex-dependent changes in the emotional profile in adult rats: behavioral and biochemical correlates. Neuropsychopharmacology 33:2760–2771

Sanchis-Segura C, Cline BH, Marsicano G, Lutz B, Spanagel R (2004) Reduced sensitivity to reward in CB1 knockout mice. Psychopharmacology 176:223–232

Sara SJ (2009) The locus coeruleus and noradrenergic modulation of cognition. Nat Rev Neurosci 10:211–223

Scavone JL, Mackie K, Van Brockstaele EJ (2010) Characterization of cannabinoid-1 receptors in the locus coeruleus: relationship with mu-opioid receptors. Brain Res 1312:18–31

Schildkraut JJ (1965) The catecholamine hypothesis of affective disorders: a review of supporting evidence. Am J Psychiatry 122:509–522

Servoa L, Rivkin M, Nakashima A, Sabban EL (2002) Estradiol stimulates gene expression of norepinephrine biosynthetic enzymes in rat locus coeruleus. Neuroendocrinology 75:193–200

Shansky RM, Bender G, Arnsten AF (2009) Estrogen prevents norepinephrine alpha-2a receptor reversal of stress-induced work memory impairment. Stress 12:457–463

Shors TJ, Chua C, Falduto J (2001) Sex differences and opposite effects of stress on dendritic spine density in the male versus female hippocampus. J Neurosci 21:6292–6297

Shors TJ, Mathew J, Sisti HM et al (2007) Neurogenesis and helplessness are mediated by controllability in males but not females. Biol Psychiatry 62:487–495

Steiner MA, Wanisch K, Monory K et al (2008a) Impaired cannabinoid receptor type 1 signaling interferes with stress-coping behavior in mice. Pharmacogenomics J 8:196–208

Steiner MA, Marsicano G, Wotjak CT, Lutz B (2008b) Conditional cannabinoid receptor type 1 mutants reveal neuron subpopulation-specific effects on behavioral and neuroendocrine stress responses. Psychoneuroendocrinology 33:1165–1170

Sun MK, Alkon DL (2006) Differential gender-related vulnerability to depression induction and converging antidepressant responses in rats. J Pharmacol Exp Ther 316:926–932

Suárez J, Llorente R, Romero-Zerbo SY et al (2009) Early maternal deprivation induces gender-dependent changes on the expression of hippocampal CB(1) and CB(2) cannabinoid receptors of neonatal rats. Hippocampus 7:623–632

Tzavara ET, Perry KW, Rodriguez DE et al (2001) The cannabinoid CB(1) receptor antagonist SR141716A increases norepinephrine outflow in the rat anterior hypothalamus. Eur J Pharmacol 426:R3–R4

Tzavara ET, Davis RJ, Perry KW et al (2003) The CB1 receptor antagonist SR141716A selectively increases monoaminergic neurotransmission in the medial prefrontal cortex: implications for therapeutic actions. Br J Pharmacol 138:544–553

Valdizán EM, Mato S, González-Maeso J et al (2011) Functionality of cannabinoid receptors in the prefrontal cortex of major depression suicide victims: influence of antidepressant treatment at the time of death. In: XIV Congreso Nacional Sociedad Española de Neurociencia (SENC), Salamanca, abstract P-105

Van Laere K, Goffin K, Casteels C et al (2008) Gender-dependent increases with healthy aging of the human cerebral cannabinoid-type 1 receptor binding using [(18)F]MK-9470 PET. Neuroimage 39:1533–1541

Wood GE, Shors TJ (1998) Stress facilitates classical conditioning in males, but impairs classical conditioning in females through activational effects of ovarian hormones. Proc Natl Acad Sci U S A 95:4066–4071

Chapter 9
Endocannabinoids, Monoamines and Stress

Cecilia J. Hillard

Abstract There is considerable evidence that signaling through cannabinoid 1 receptors (CB1Rs) contributes to the effects of stress on the brain as well as stress-adaptation. Similarly, serotonin, norepinephrine, and dopamine are important modulators and effectors of stress. The purpose of this review is to present and discuss the results of studies that have investigated the interactions between endocannabinoid-CB1 receptor signaling and each of the biogenic amines in the context of stress.

9.1 Introduction

Chronic exposure to psychological stress is a fact of life in the twenty-first century society. Chronic stress exposure contributes to neuropsychiatric diseases and disorders; including depression, anxiety, substance abuse, and schizophrenia. In addition, stress is a risk factor for the development of obesity, cardiovascular disease, gastrointestinal disorders, and functional pain disorders. Drugs that target monoamine signaling have efficacy in the treatment of several stress-related diseases, particularly depression and schizophrenia, but not all affected individuals are responsive and adverse drug effects can interfere with efficacy. In addition, pharmacotherapies targeting the monoamines can treat disease symptoms but do not reduce the impact of stress. In fact, there are very few approaches available to reduce the consequences of stress and thereby reduce the risk of developing stress-related disorders.

It could be argued that Chinese and Indian cultures discovered a stress-reducing therapeutic, *Cannabis sativa*, thousands of years ago. *Cannabis* extracts were used as medicinals by these cultures to reduce anxiety, pain, seizures, mania, and muscle spasms; and to stimulate appetite. Modern research confirms some of these benefits; for example, recreational use of *Cannabis* can be associated with elevation of mood, and euphoria along with reduced feelings of stress and anxiety (Tournier et al. 2003). The constituent of *Cannabis* that is likely responsible for the stress-reducing effects is Δ^9-tetrahydrocannabinol (THC), though another abundant phytocannabinoid, cannabidiol, is also an effective antianxiety agent (Bergamaschi et al. 2011).

C. J. Hillard (✉)
Neuroscience Research Center, Medical College of Wisconsin, 8701 Watertown Plank Road, Milwaukee, WI, 53226 US
e-mail: chillard@mcw.edu

E. J. Van Bockstaele (ed.), *Endocannabinoid Regulation of Monoamines in Psychiatric and Neurological Disorders*, DOI 10.1007/978-1-4614-7940-6_9,

THC is a partial agonist of two G-protein coupled cannabinoid receptors (CBRs), CB1R (Matsuda et al. 1990) and CB2R (Munro et al. 1993). The effects of THC on stress, anxiety, and mood are mediated by these receptors, particularly CB1R, which will be the focus of this review (Pertwee et al. 2010).

Considerable evidence indicates that CB1R are located on presynaptic, axonal terminals (Herkenham et al. 1991a) and couple to inhibition of calcium influx (Mackie and Hille 1992), thereby inhibiting neurotransmitter release. CB1R are present on glutamatergic and GABAergic terminals throughout the central nervous system (CNS) and, via regulation of release of these primary excitatory and inhibitory transmitters, exert a profound effect on postsynaptic neuronal activity (Freund et al. 2003). Through this mechanism, CB1R activity regulates activational drive on principal, outflow neurons in many brain regions. CB1R found on axon terminals of noradrenergic (Oropeza et al. 2007; Scavone et al. 2010), and serotonergic (Hermann et al. 2002) neurons can inhibit the release of the biogenic amines under some circumstances.

The endogenous ligands for CB1R (endocannabinoids (eCBs)) are two arachidonic acid derivatives: *N*-arachidonylethanolamine (AEA) (Devane et al. 1992) and 2-arachidonoylglycerol (2-AG) (Mechoulam et al. 1995; Sugiura et al. 1995). Considerable evidence indicates that the eCBs are synthesized and released from neurons postsynaptic the CB1R-containing axon terminals (Freund et al. 2003). Multiple stimuli can induce eCB mobilization, including depolarization (Wilson and Nicoll 2001) and activation of phospholipase C (PLC) by G protein coupled receptors (GPCRs) that couple to Gq family heterotrimeric G proteins (Kim et al. 2002; Varma et al. 2001). As discussed further below, glucocorticoids also mobilize eCBs. Although there are likely other functions of endocannabinoid-CB1R signaling (ECS), it is a primary regulator of synaptic plasticity via changes in presynaptic release, specifically subserving short-term, activity-driven changes in synaptic strength as well as other forms of presynaptic plasticity (Patel and Hillard 2009a).

9.2 ECS and Stress: Endocrine Aspects

Considerable evidence has accumulated to support the hypothesis that ECS is altered by stress exposure and modulates stress responses through effects on synaptic activity. Stress activates the hypothalamic-pituitary-adrenal (HPA) axis and induces the release of glucocorticoid hormones, which exert widespread effects on the body, including the brain (McEwen 2008). Exogenous administration of glucocorticoids to rats results in a rapid (i.e. within 10 min) increase in eCB contents in several limbic structures (Hill et al. 2010a). Multiple studies indicate that ECS is an effector of glucocorticoids in the brain (Hill and McEwen 2009). In the hypothalamus, glucocorticoids act through a membrane receptor to rapidly mobilize eCBs that, through CB1R on glutamatergic axons, inhibit excitatory drive onto corticotropes in the paraventricular nucleus (PVN) (Di et al. 2003; Di et al. 2005). Glucocorticoid infusion into the PVN rapidly inhibits HPA axis activation, an effect that is blocked by the CB1R antagonist, AM-251 (Evanson et al. 2010). Collectively, these data demonstrate that glucocorticoid-mediated fast feedback inhibition of the HPA axis requires mobilization of ECS in the PVN.

Both the hippocampus (Sapolsky et al. 1985) and medial prefrontal cortex (mPFC) (Diorio et al. 1993) are responsive to glucocorticoids and participate in long-loop feedback regulation of the HPA axis. Recent evidence indicates that this function of glucocorticoids also requires ECS. In the hippocampus, exposure to acute stress, in vivo glucocorticoid treatment, and direct application of glucocorticoids to slices all act through GR to increase eCB-mediated inhibition of gamma-aminobutyric acid (GABA) release (Wang et al. 2012b). In the PFC, glucocorticoid treatment of slices produces eCB-mediated inhibition of GABA release and dysinhibition of mPFC pyramidal neurons (Hill et al. 2011a). Increased pyramidal neuronal activity contributes to termination of the HPA axis response through activation of inhibitory projections from the anterior bed nucleus of the stria terminalis to the PVN (Radley and Sawchenko 2011).

ECS signaling is also involved in the behavioral effects of glucocorticoids. In these functions, ECS is required for the positive effects of glucocorticoids and therefore can be considered as a contributor to the stress response. For example, glucocorticoid-mediated enhancement of memory consolidation (Campolongo et al. 2009) and stress-induced reinstatement of cocaine seeking behavior in mice (Vaughn et al. 2012) are both inhibited by antagonism of CB1R.

In sum, these data support a critical role of ECS as a "second messenger" for glucocorticoids in the brain. Our understanding of the mechanisms by which glucocorticoids mobilize ECS is incomplete; however, it is likely that multiple mechanisms are at play with different time courses that are well matched to the function of the glucocorticoids in the specific brain region.

In spite of the evidence discussed above that acute stress and glucocorticoid treatment *increase* eCB-mediated signaling in hypothalamus, hippocampus, and PFC, acute stress exposure in rodents results in *decreased* amygdalar and PFC AEA concentrations (Hill et al. 2009b; McLaughlin et al. 2012; Patel et al. 2005; Rademacher et al. 2008). The reduction in AEA content is accompanied by an increase in the activity of fatty acid amide hydrolase (FAAH, (Hill et al. 2009b; McLaughlin et al. 2012)), the primary catabolic enzyme for AEA in brain (Cravatt et al. 1996). If the decline in amygdalar AEA is prevented, HPA axis activation by stress is reduced (Hill et al. 2009b); evidence that tonic CB1R signaling in the amygdala opposes stress-induced HPA axis activation and must be inhibited in order for the HPA stress response to occur. In other words, AEA functions as a gatekeeper for the stress response (Patel et al. 2004). The mechanism by which stress increases the activity of FAAH is unknown, but is likely a very early event in stress response cascade. Although stress decreases AEA concentrations in the amygdala, glucocorticoid treatment increases amygdalar AEA concentrations (Hill et al. 2010a). Thus, the sensory perception of stress decreases AEA, reduces CB1R tone and allows for HPA axis activation while subsequent elevation of glucocorticoids re-establishes the CB1R "gate" by elevating AEA. This, glucocorticoid-induced elevation in amygdalar AEA could be another example of ECS involvement in feedback regulation of HPA axis activity.

These data have important implications for therapeutic treatment of disorders in which hyperactive HPA axis activity contributes to disease. Preclinical models demonstrate that FAAH inhibition inhibits stress-induced increases in circulating glucocorticoids (Patel et al. 2004), reduces anxiety in adverse environments (Patel and Hillard 2006), and decreases immobility in rats in the forced swim assay (Gobbi et al. 2005; McLaughlin et al. 2007). Several authors have suggested that therapeutic agents that increase AEA (e.g., FAAH inhibitors) should be examined in humans for treatment of anxiety and depressive disorders that are characterized by excessive or prolonged HPA axis activation (Hill and Gorzalka 2009; Hill et al. 2009a; Parolaro et al. 2010).

9.3 ECS and Stress: Neural Aspects

Acute stress exposure evokes characteristic physiological and behavioral changes that are mediated by activation of the neuronal defense pathway. Among the changes evoked are sympathoexcitation, including an increase in arterial blood pressure. ECS has been shown to regulate the sympathetic response to stress at multiple sites in the CNS. First, activation of ECS in the dorsal periaqueductal gray (PAG) enhances sympathetic nerve activity through inhibition of GABA release (Dean 2011). As ECS is increased in this region by stress (Hohmann et al. 2005), it could contribute to the effects of stress to enhance sympathetic outflow via this mechanism. Second, administration of corticotropin releasing factor (CRF) into the cerebral ventricle (i.c.v.) activates the sympathetic nervous system in anesthetized rats (Shimizu et al. 2010). This response is inhibited by CB1R direct and indirect agonists and is potentiated by CB1R antagonists administered i.c.v. (Shimizu et al. 2010). The site of this suppressive CB1R mechanism is unknown. Third, CB1R signaling in the nucleus tractus solitarius (NTS) enhances baroreceptor sympathoactivation (Seagard et al. 2004) through inhibition of GABA release (Chen et al. 2010). These brain-regional effects of ECS on regulation of sympathetic outflow illustrate an important point regarding the ECS: that it is a local modulatory system. As a result, it is not unusual to find that ECS exerts non-congruent changes at individual sites within a circuit.

CB1R are expressed at the terminals of sympathetic axons innervating blood vessels and activation of presynaptic CB1R reduces the release of NE in anesthetized (Ishac et al. 1996) and pithed (Pfitzer et al. 2005) rats. While there is evidence that these receptors are not endogenously active in healthy animals (Pfitzer et al. 2005), they could contribute to blood pressure regulation during inflammation. For example, treatment of conscious rats with lipopolysaccharide (LPS) induces vasodilation and hypotension that is reversed by both blockade of CB1R and ß-adrenergic signaling (Gardiner et al. 2005). LPS releases eCBs from circulating platelets and macrophages (Varga et al. 1998), suggesting that inflammation brings ECS "on-line" to reduce blood pressure and contribute to inflammation-induced shock. Indeed, human studies support a role for ECS in hypotension that occurs during endotoxic shock (Sakamoto et al. 2008; Wang et al. 2001).

It is hypothesized that chronic inflammation contributes to major depressive illness in humans (Haroon et al. 2012). It is interesting in this regard that women with depression exhibit a positive correlation between circulating eCBs and systolic and diastolic blood pressure while there is no correlation among eCBs and blood pressure measurements in healthy women (Ho et al. 2012). One explanation of these findings is that chronic inflammation leads to increased eCBs, depression, and hypertension in a coordinated manner.

9.4 ECS and Chronic Stress

Like most other signaling pathways that utilize GPCRs, CB1R density and coupling to downstream effectors exhibit considerable plasticity in response to increased agonist availability. In particular, exogenous agonist treatment and excessive amounts of 2-AG cause desensitization and down-regulation of CB1R signaling (Schlosburg et al. 2010; Sim et al. 1996). Chronic, non-habituating stress decreases CB1R density in the hippocampus (Hill et al. 2005) and CB1R function in the ventral striatum (Wang et al. 2010); possibly as a result of sustained increases in 2-AG. On the other hand, repeated exposure of rodents to the same stress, which is accompanied by habituation, enhances ECS at the level of the eCB ligand (Patel and Hillard 2008; Patel et al. 2009). The increase in ECS is required for the dampened responsivity to stress and fear that occurs during habituation (Hill et al. 2010b; Kamprath et al. 2011; Patel et al. 2005).

An important point illustrated by these studies is that ECS can be considered a component of the stress response that provides plasticity and a molecular memory of prior stress exposures. These features, along with the function of ECS to regulate neurotransmitter release throughout the neural axis, places ECS in a vital position to regulate the impact of stress on the body through modulation of neural circuits.

9.5 Monoamine Pathways and Stress

There is considerable evidence that serotonergic, noradrenergic, and dopaminergic networks are stress-responsive and contribute to the behavioral responses to stress. In light of the role of ECS to regulate synaptic activity, and the sensitivity of ECS to stress, it is logical to ask to what extent do ECS and monoamine signaling overlap and/or act in series to regulate behavioral and endocrine responses to stress? The purpose of this review is to examine the evidence regarding the relationships between ECS and each of the three, CNS active monoamines: serotonin (5-HT), norepinephrine (NE) and dopamine (DA) to address this question.

9.6 Interactions Between ECS and 5-HT Signaling: Mechanisms

Physiological and behavioral studies demonstrate significant interactions between cannabinoid and serotonergic signaling. In some cases, ECS and serotonergic signaling produce similar effects and act in series to affect physiological or behavioral change. For example, activation of either 5-HT_{1A} or CB1R profoundly reduces body temperature (Malone and Taylor 1998) and pharmacological data suggest that the hypothermic effects of CB1R activation require increased 5-HT release in the medial raphe nucleus (Malone and Taylor 2001). Similarly, both CB1R (Patel and Hillard 2006) and 5-HT receptors (Griebel et al. 1997) regulate anxiety in rodents. Pharmacological blockade of 5-HT_{1A} receptors inhibits both the anxiolytic (Braida et al. 2007) and anxiogenic (Marco et al. 2004) effects of the cannabinoids. On the other hand, CBR agonists can produce pharmacological effects via suppression of 5-HT signaling. For example, THC-induced impairment of spatial memory is accompanied by a reduction in 5-HT release in the ventral hippocampus and is attenuated by pharmacological treatments that maintain 5-HT signaling (Egashira et al. 2002).

Molecular evidence supports the pharmacological data that ECS and serotonergic systems interact functionally in both positive and negative ways. In the forebrain, 5-HT_{1B} and CB1R are coexpressed in principal neurons, while 5-HT_3 and CB1R are colocalized on GABA interneurons (Hermann et al. 2002). CB1R are also present on serotonergic neuron cell bodies in the dorsal raphe nucleus (DRN) and on axon terminals of serotonergic neurons in projection areas, including the hippocampus and amygdala (Haring et al. 2007; Lau and Schloss 2008). CB1R on serotonergic axon terminals inhibit 5-HT release in cortical slice preparations (Nakazi et al. 2000) and in vivo (Egashira et al. 2002; Merroun et al. 2009; Moranta et al. 2004; 2009). Blockade of CB1R increases basal extracellular concentrations of 5-HT in mPFC (Darmani et al. 2003; Tzavara et al. 2003), indicating ECS exerts tonic inhibition of 5-HT release. Sustained changes in CB1R activity alter the expression and/or functional coupling of 5-HT receptors in hippocampus and PFC (Aso et al. 2009; Mato et al. 2007; Moranta et al. 2009). Since 5-HT receptors are readily up- and down-regulated by changes in 5-HT availability, these findings are consistent with the hypothesis that ECS is a physiologically important regulator of 5-HT release, and therefore, of 5-HT signaling in the terminal regions of serotonergic projection neurons.

CB1R activity also affects serotonergic signaling through changes in the firing of serotonergic neurons. In vitro studies of synaptic activity within the DRN reveal that CB1R agonists inhibit glutamatergic activation of DRN serotonergic neurons through presynaptic inhibition of glutamate release (Haj-Dahmane and Shen 2005; 2009). Extracellular recordings provide evidence that ECS also inhibits GABA release in the DRN (Mendiguren and Pineda 2009). Therefore, the net effect of ECS on serotonergic neuronal activity will depend on synaptic strength, endogenous eCB tone, and CB1R density on terminals impinging on DRN serotonergic neurons (Haj-Dahmane and Shen 2011).

Previous work has demonstrated that DRN serotonergic neuron activity is regulated by an afferent projection from the mPFC to inhibitory interneurons of the DRN (Celada et al. 2001). Activation of this important input results in reduced activity of DRN serotonergic neurons while its inhibition is associated with decreased anxiety and depressive behaviors (Celada et al. 2001). CBIR agonist administration into the ventromedial PFC results in increased firing of serotonergic neurons in the DRN; an effect abolished by transection of the medial but not lateral PFC (Bambico et al. 2007). Since ECS also regulates the activity of serotonergic neurons projecting to the frontal cortex (Haj-Dahmane and Shen 2011), it has been suggested that ECS and 5-HT participate in a reciprocal loop; increased ECS in the PFC leads to increased 5-HT neuronal activity in the DRN, resulting in increased 5-HT release in the PFC (McLaughlin et al. 2012).

The cellular localization of CB1Rs in a brain region is one critical determinant of the effect of changes in ECS on the serotonergic circuit. Another important determinant involves the mechanisms that trigger eCB mobilization. Haj-Dahmane and Shen have demonstrated that serotonergic neurons in the DRN synthesize and release eCBs (Haj-Dahmane and Shen 2005, 2009). As has been demonstrated at other synapses, eCB mobilization can be initiated by membrane depolarization and increased intracellular calcium in serotonergic neurons (Haj-Dahmane and Shen 2009). Mobilization of eCBs in DRN neurons is also triggered by activation of GPCRs that couple to Gq proteins, including orexin receptors (Haj-Dahmane and Shen 2005). Orexin receptor activation results in recruitment of ECS, likely via increased synthesis of 2-AG since it is dependent upon PLC and diacylglycerol lipase (DGL) activities (Haj-Dahmane and Shen 2005). These studies indicate that orexin-mediated regulation of serotonergic signaling is likely indirect, through changes in ECS. The result of eCB synthesis and release from serotonergic cell bodies in the DRN can be decreased glutamate or GABA release, thus either decreased or increased excitatory drive, respectively, onto the serotonergic projections.

5-HT, acting through 5-HT$_2$ receptors which are known to couple to Gq family proteins (Barnes and Sharp 1999), activates ECS in the inferior olive, a brain region that receives extensive serotonergic input (Best and Regehr 2008). As a result, serotonin inhibits glutamate release through ECS and thereby regulates both pre- and post-synaptic function in this brain region. The authors of this study speculate that this mechanism could contribute responses to other 5-HT$_2$ receptor effects, including food intake, sexual behavior, and sleep. Studies in rat cerebellar membranes suggest that, when CB1R and 5-HT$_2$ receptors are colocalized, activation of the 5-HT$_2$ receptor increases the high affinity binding site for at least one synthetic agonist (Devlin and Christopoulos 2002), another mechanism by which 5-HT regulates ECS.

Thus, currently available evidence indicates that ECS regulates the serotonergic system via multiple mechanisms, including regulation of DRN neuronal firing and of 5-HT release from axon terminals. Current data suggest that ECS regulation of 5-HT signaling occurs in many brain regions and is the primary mechanism of interaction of these two systems in the healthy, unstressed brain. However, 5-HT can reciprocally regulate ECS through 5-HT$_2$ receptors, and, as is discussed further below, studies in rodent models of depression suggest that 5-HT regulation of ECS could contribute to behaviors in these models.

9.7 Interactions Between ECS and 5-HT Signaling: Stress Context

9.7.1 HPA Axis Regulation

It is well-documented that stress increases 5-HT turnover and serotonergic signaling plays an important and complex role in the regulation of autonomic and endocrine responses to stress (Lanfumey et al. 2008; Lowry 2002). In particular, exposure of rodents to inescapable stressors increases c fos expression in DRN serotonergic neurons (Greenwood et al. 2003) and increases 5-HT release in projection areas, such as the PFC (Bland et al. 2003). Serotonergic inputs contribute to regulation of HPA axis activity in the PVN, primarily in a positive manner (Lanfumey et al. 2008). For example, activation of either 5-HT_{1A} (Osei-Owusu et al. 2005) or 5-HT_{2C} (Klaassen et al. 2002) receptors in the PVN increase ACTH secretion. Glucocorticoids inhibit DRN neuronal activity, resulting in feedback inhibition on the serotonergic drive to increase HPA axis activity (Lanfumey et al. 2008). Recent data demonstrate that glucocorticoids regulate DRN neuronal activity via recruitment of ECS in the DRN (Wang et al. 2012a). In particular, glucocorticoids act via activation of a membrane-localized, GPCR to increase ECS which results in decreased glutamatergic drive to serotonergic neurons in the DRN (Wang et al. 2012a). ECS regulation of activity in the DRN is consistent with a general function of the ECS to inhibit or dampen HPA axis activity through effects in multiple brain regions, including hypothalamus (Di et al. 2003), PFC (Hill et al. 2011a), hippocampus (Wang et al. 2012b), and amygdala (Hill et al. 2009b).

9.7.2 Behavior in the Forced Swim (FST) and Tail Suspension Tests

A large body of evidence supports the hypothesis that 5-HT_{1A} receptors play a pivotal but complex role in the mechanism of action of anti-depressant drugs (Lanfumey et al. 2008). A commonly employed rodent assay for efficacy of anti-depressant drugs involves the examination of behavioral responses to an inescapable and stressful situation, usually either a period of forced swim or suspension by the tail. The read-out of these assays is a comparison of the time spent using active behaviors that enhance chances to escape, such as climbing or swimming, and passive behaviors, such as immobility, which conserve energy. Chronic stress and other manipulations that are considered pro-depressive increase time spent immobile while treatment with antidepressants increase active behavioral repertoires (Cryan et al. 2005).

There are multiple reports of antidepressant-like effects in the FST following treatment of rats with CB1R agonists and indirect agonists (Bambico et al. 2007; Gobbi et al. 2005; Hill and Gorzalka 2005; McLaughlin et al. 2007) and emerging evidence indicates that the cannabinoids produce these effects via increased 5-HT release. In a seminal study, Bambico et al. (2007) demonstrated that low doses of the synthetic CB1R agonist, Win 55212-2, elicit antidepressant behavior in the FST and enhance serotonergic neuronal firing in the DRN through CB1R mechanism. The antidepressant efficacy of Win 55212-2 was abolished by depletion of 5-HT, supporting the hypothesis that CB1R agonists act to increase serotonergic signaling. Enhancement of DRN neuronal firing by systemic administration of Win 55212-2 was abolished by transection of the mPFC, and both the increased firing and decreased FST immobility were mimicked by direct injection mPFC of Win 55212-2. Taken together, these results support the hypothesis that ECS in the PFC increases excitatory drive on DRN serotonergic neurons, resulting in antidepressant-like efficacy. As described above, an inhibitory, multi-synaptic projection from mPFC to the DRN has been described; an untested hypothesis is that ECS decreases activation of this projection through inhibition of glutamate release in the mPFC.

Mice with a genetic deletion of FAAH (therefore, high AEA tone) exhibit reduced responsiveness of mPFC pyramidal cells to a 5-$HT_{2A/2C}$ agonist, which is consistent with increased firing of serotonergic neurons, enhanced 5-HT release in the mPFC and subsequent down-regulation of 5-$HT_{2A/2C}$ receptors (Bambico et al. 2010). Interestingly, treatment of rats with a single dose of THC does not alter FST behavior and exerts complex effects on DRN firing while repeated THC exposure (5 days) decreases immobility in the FST and enhances 5-HT_{1A} receptor activity in the hippocampus (Bambico et al. 2012). THC is a partial agonist of CB1R (Kearn et al. 1999), which could contribute to its lower effectiveness.

A recent biochemical study supports the interaction of ECS and serotonergic circuits in regulating the behavioral response to FST in rats (McLaughlin et al. 2012). These investigators found that PFC AEA content is decreased and FAAH activity increased immediately following exposure to FST. When an inhibitor of FAAH was injected into the mPFC, active coping in the FST was increased, as was firing of DRN serotonergic neurons (McLaughlin et al. 2012). The effect of FAAH inhibition to increase active coping was blocked by depletion of 5-HT. These data suggest that PFC AEA content is a critical determinant of the firing of serotonergic DRN neurons, thus, is a critical determinant of the behavioral coping mechanism employed.

The role of the CB1R in the regulation of FST behavior is less clear in mice. For example, treatment of mice with moderate doses of THC increases immobility in the FST, an effect that requires 5-HT_{1A} receptors (Egashira et al. 2008) and CB1R antagonism can produce antidepressant-like effects (Griebel et al. 2005; Shearman et al. 2003). A recent study provides some insight into the discrepant data (Haring et al. 2013). In this study, low doses of THC (0.1 and 0.5 mg/kg) decreased immobility in the FST which was blocked by inhibition of $5HT_{1A}$ receptors and depletion of 5-HT. High doses (3 and 10 mg/kg) of the CB1R antagonist, rimonabant, also decreased immobility but this effect was unaffected by suppression of 5-HT signaling. Instead, the antidepressant-like effect of CB1R blockade was reversed by inhibition of catecholamine synthesis and did not occur in mice lacking CB1Rs on forebrain GABAergic neurons. It is possible that the prevailing ECS tone within serotonergic circuits differs among animal species.

9.7.3 Anxiety and Other Emotional Behaviors

Both serotonergic and CB1R signaling are important regulators of anxiety and there is evidence that changes in serotonergic signaling contribute to the effects of ECS on anxiety. Acute treatment of rats with low doses of THC or the AEA clearance inhibitor, AM 404 (Beltramo et al. 1997) results in decreased anxiety-like behaviors that is significantly reduced by pretreatment with a 5-HT_{1A} antagonist (Braida et al. 2007). Similarly, FAAH null mice exhibit decreased anxiety compared to wild types (Bambico et al. 2010). FAAH null mice also have a greater than 30 % increase in the firing of DRN neurons that is inhibited in a subset of neurons by CB1R antagonism. In addition, these animals have increased basal 5-HT tone in the frontal cortex, but not in the hippocampus, that is reduced by CB1R antagonist treatment (Cassano et al. 2011). Taken together, these studies indicate that AEA-mediated CB1R signaling exerts antianxiety effects through interactions with serotonergic signaling in the PFC.

9.7.4 Impact of Changes in Serotonergic Signaling on ECS

The preceding section indicates that ECS modulates serotonergic signaling. There are also data consistent with the reciprocal relationship between these two systems; that is, that serotonergic signaling alters ECS. Although activation of 5-HT_2 receptors in the inferior olive induce ECS through PLC-mediated increases in 2-AG (Best and Regehr 2008), other data indicate that acute alterations in 5-HT tone do not alter ECS. For example, the behavioral effects of acute exposure to selective serotonin reuptake inhibitors (SSRIs) are preserved in $CB1R^{-/-}$ mice and in the presence of CB1R antagonists (Gobshtis et al. 2007; Steiner et al. 2008), suggesting that ECS is not required for SSRI efficacy in the FST.

On the other hand, there is evidence that long-lasting elevation of 5-HT tone, as occurs with chronic SSRI treatment, does change ECS. Treatment of rats for 21 days with imipramine (which inhibits reuptake of both NE and 5-HT) decreases CB1R binding site density in hypothalamus, midbrain, and ventral striatum (Hill et al. 2008a). Similarly, chronic fluoxetine treatment decreases CB1R messenger ribonucleic acid (mRNA) expression in the caudate-putamen (Oliva et al. 2005) and decreases CB1R function in cerebellum (Mato et al. 2010). Chronic citalopram treatment reduces CB1R coupling in hypothalamus and hippocampus (Hesketh et al. 2008). Chronic treatment with imipramine produces a significant increase in 2-AG contents in the hypothalamus and midbrain (Hill et al. 2008a), which could result in agonist-induced desensitization and/or down-regulation of the CB1R.

In contrast to the effects of SSRIs to decrease CB1R signaling in subcortical regions outlined above, CB1R binding site density is *increased* in the amygdala by chronic imipramine (Hill et al. 2008a) and in the PFC by chronic fluoxetine

(Hill et al. 2008b). Similarly, chronic fluoxetine increases CB1R function in the frontal cortex (Mato et al. 2010). Chronic fluoxetine treatment results in decreased expression of mRNA for the 5-HT_{2C} receptor in the PFC while increasing its expression in the hippocampus (Barbon et al. 2011), raising the possibility that region-specific changes in 5-HT_{2C} expression by antidepressants underlie the region-specific changes in ECS.

9.7.5 *Models of Depression*

Several studies have demonstrated that ECS is dysregulated in animal models of depression and that the dysregulation can be reversed by SSRI treatment. Chronic, unpredictable stress (CUS) results in behavioral changes that mirror those seen in depressed humans. CUS down-regulates ECS in several brain regions, including the hippocampus, hypothalamus, and ventral striatum (Hill et al. 2008a; Hill et al. 2005). Imipramine reverses the effects of CUS on ECS in all of these brain regions (Hill et al. 2008a). Similarly, CUS exposure decreases CB1R-mediated inhibition of glutamate release in accumbens slices which is reversed by in vivo treatment with fluoxetine (Wang et al. 2010). These results suggest that stress-induced decreases in 5-HT tone contribute to the down-regulation of ECS in these brain regions. These studies also suggest that decreased ECS contributes to the symptoms of depression, perhaps through a feed-forward cycle that enhances suppression of 5-HT release.

Further evidence that reduced 5-HT tone affects ECS comes from studies of rats with bilateral olfactory bulbectomy (OBX), which induces behavioral, neurochemical, and structural abnormalities that are similar to human depression. Adaptive changes in the serotonergic system play an important role in the OBX syndrome; for example, serotonergic neurons in the DRN degenerate, resulting in permanent reductions in 5-HT secretion in the hippocampus and amygdala (van der Stelt et al. 2005). Male rats with bilateral OBX exhibit significant increases in CB1R density and coupling to GDP/GTP exchange in PFC which is reversed by chronic treatment with fluoxetine (Rodriguez-Gaztelumendi et al. 2009). CUS also produces an increase in CB1R density in the PFC; however, this change is not altered by imipramine treatment while CUS-induced decreases in CB1R density were normalized by this antidepressant (Hill et al. 2008a). The PFC of human suicides also exhibit increased CB1R density and function (Hungund et al. 2004; Vinod et al. 2005), which provides relevance to these observations.

9.7.6 *Summary*

There is clear evidence that ECS regulates the release of 5-HT in limbic brain regions through multiple mechanisms. CB1R signaling regulates a bi-directional circuit between the mPFC and DRN; this is likely the substrate for the antidepressant effects of systemic CB1R agonists in rats. Chronic exposure of rats to SSRIs decreases

CB1R expression and function in subcortical, limbic regions, but increases CB1R expression in PFC and amygdala; the function of these changes in each brain region are not well understood. Paradoxically, rodents exposed to CUS and OBX, treatments that induce depressive-like behaviors, exhibit the same pattern of changes in CB1R density. Perhaps the ECS changes caused by CUS and OBX are an attempt to recover homeostasis and oppose the depressive symptoms. In any case, it is clear that ECS and serotonergic signaling are both altered significantly by chronic stress through mechanisms that are not currently clear.

9.8 Interactions Between ECS and NE Signaling: Mechanisms

There are two major sources of noradrenergic neurons in the brain: the locus coeruleus (LC), which gives rise to the dorsal noradrenergic bundle and the lateral tegmentum, which gives rise to the ventral noradrenergic bundle (Dahlstrom and Fuxe 1964; Weinshenker and Schroeder 2007). Noradrenergic neurons from the LC project widely to all limbic and cortical regions, including the amygdala, hippocampus, neocortex, cingulate, and striatum (Berridge and Waterhouse 2003). The LC noradrenergic pathway is involved in the regulation of attention, arousal, cognitive processes, and sleep as a result of NE gating and tuning postsynaptic target neurons (Aston-Jones and Cohen 2005). LC neurons receive inputs from other neurotransmitter systems, including GABAergic, glutamatergic, and serotonergic (Berridge and Waterhouse 2003) and are also regulated by negative feedback through collateral noradrenergic terminals within the LC (Aghajanian et al. 1977). Dysregulation of the LC-NE system is thought to contribute to cognitive, emotional, and attentive dysfunctions that are associated with neuropsychiatric illnesses (Berridge and Waterhouse 2003). In light of data that *Cannabis* in humans alters attention and focus and other aspects of executive function (Pattij et al. 2008; Solowij et al. 1995; Viveros et al. 2005), a "noradrenergic hypothesis" has been put forth which posits that cannabinoids impair attention and cognition via modulation of central noradrenergic transmission (Carvalho and Van Bockstaele 2012). In support of this hypothesis, pretreatment of healthy humans with the beta-adrenergic antagonist, propranolol, prevented the acute effects of *Cannabis* to impair learning (Sulkowski et al. 1977).

CB1R protein and mRNA can be detected in the LC and NTS (Derbenev et al. 2004; Herkenham et al. 1991b; Mailleux and Vanderhaeghen 1992; Matsuda et al. 1993). Analysis of the subcellular distribution of the CB1R protein in the LC reveals the presence of the receptors on both axon terminals and cell bodies, with a majority present in somato-dendritic profiles (Scavone et al. 2010). The presynaptic CB1R were more likely to be at symmetrical synapses, suggesting that ECS can inhibit GABA-mediated suppression of noradrenergic neurons. Indeed, Win 55212-2 suppresses the inhibition of LC firing induced by activation of the major GABAergic afferent to the LC (Muntoni et al. 2006). In further support of this hypothesis, systemic administration of CB1R agonists (Muntoni et al. 2006) and FAAH inhibitors (Gobbi et al. 2005) increase the firing rate of unstimulated noradrenergic neurons in the LC in a CB1R-dependent manner. CB1R agonists also increase c fos expression in the LC (Oropeza et al. 2005; Patel and Hillard 2003); enhance NMDA-induced firing of LC neurons (Mendiguren and Pineda 2004); and increase NE synthesis (Moranta et al. 2009) and release (Oropeza et al. 2005) in terminal regions. Taken together, these results are consistent with a mechanism by which activation of CB1R on GABA terminals in the LC inhibit GABA release and increase firing of noradrenergic neurons. However, neither local administration of CB1R agonists into the LC nor CB1R treatment of LC-containing brain slice preparations alter the spontaneous firing of LC neurons (Mendiguren and Pineda 2006). These findings point to an indirect effect of CB1R agonists on LC firing, perhaps through increased peripheral afferent activity into the LC. While further studies are required to elucidate the site of action, studies consistently demonstrate that CB1R agonists increase activity of noradrenergic, LC neurons.

Interestingly, more than 80 % of the CB1R immunoreactivity in the LC is post-synaptic, and this pool of CB1R is localized to the cytosol and not plasma membrane (Scavone et al. 2010). The authors of this study speculate that intracellular CB1R could function in an autocrine fashion to regulate LC activity (Carvalho et al. 2010a). They argue that since the post-synaptic CB1R in the LC are in close proximity to asymmetrical synaptic inputs, it is possible that glutamate release could drive eCB synthesis post-synaptically resulting in activation of CB1R within the noradrenergic neuron in an autocrine manner.

Cannabinoid-mediated increases in LC noradrenergic neuron activity would be predicted to increase NE release in terminal fields; and multiple studies confirm this prediction. Systemic administration of CB1R agonists increase the release of NE in frontal cortex and nucleus accumbens (Jentsch et al. 1997; Oropeza et al. 2005). Local administration into the frontal cortex of a high concentration of the CB1R antagonist, rimonabant, blocked the effect of systemic CB1R agonist to increase NE release while having no effect alone (Page et al. 2008). Systemic administration of CB1R agonists increases the activity of the rate limiting enzyme in NE synthesis, tyrosine hydroxylase (TH), in several brain regions, including LC, hippocampus, and cortex (Moranta et al. 2004). Since NE synthesis is tightly coupled to release, these data support the hypothesis that CB1R activation increases NE release. In an excellent recent review of this topic, Kirilly et al. 2012 review other mechanisms that could also contribute to cannabinoid-mediated increases in LC neuronal discharge rate.

On the other hand, there is evidence that ECS suppresses NE release through direct effects on noradrenergic axon terminals. CB1R have been located on noradrenergic axon terminals in the frontal cortex (Oropeza et al. 2007). One-third of axon terminals positive for CB1R immunoreactivity were also positive for the NE synthesis enzyme, dopamine-ß-hydroxylase (DßH); other arrangements of CB1R and DßH were also seen that support the hypothesis that NE and eCBs can regulate each other's function in the frontal cortex. In the nucleus accumbens, CB1R are present on a small fraction (less than 8 %) of axon terminals of noradrenergic neurons (identified by immunoreactivity against DßH and the noradrenergic reuptake transporter, NET) (Carvalho et al. 2010a). The same study also found that approximately 6 % of noradrenergic axons were apposed to profiles that were immunoreactive for CB1R.

Studies have examined whether CB1R on noradrenergic terminals inhibit NE release. Incubation of synaptosomes with low concentrations of THC results in small but significant reductions in NE release (Poddar and Dewey 1980). Systemic administration of rimonabant (a CB1R antagonist) increases microdialysate NE concentrations in anterior hypothalamus and mPFC but not accumbens in freely moving, male Wistar rats (Tzavara et al. 2003; Tzavara et al. 2001). Similarly, rimonabant increases electrically-evoked release of NE from human and guinea pig hippocampal slices while agonists have the opposite effect, i.e., inhibit release (Schlicker et al. 1997). These data indicate that CB1R on presynaptic terminals can inhibit NE release and that they could be tonically active.

The identity of triggers of eCB mobilization that target CB1R of NE axon terminals is unknown. One possibility is that NE induces 2-AG synthesis in postsynaptic neurons via activation of alpha1 adrenergic receptors which couple to Gq heterotrimeric proteins (Insel and Hammond 1993) and, thereby, could trigger monoacylglycerol synthesis (Kano et al. 2009). In support of this mechanism, activation of alpha1 adrenergic receptors activates CB1R signaling in a cell-based system (Turu et al. 2009). Of course, eCB synthesis and release in these brain regions could also be triggered by the "usual suspects", i.e., glutamate, membrane depolarization, and glucocorticoids.

Further support for the hypothesis that ECS modulates noradrenergic circuits in the brain comes from studies of chronic cannabinoid agonist treatment. Repeated dosing of male rats with Win 55212-2 increases TH protein expression in the LC accompanied by potentiated NE efflux in response to an acute injection of Win 55212-2 without a change in baseline NE efflux (Page et al. 2007). Chronic treatment with CB1R agonists reduces the binding site density of ß1 adreno receptors in neocortex (Hillard and Bloom 1982; Reyes et al. 2009), and both alpha2 and ß1 adrenoceptors in the accumbens (Carvalho et al. 2010a). Chronic Win 55212-2 treatment completely abolishes the ability of clonidine to induce an increase in excitability of PFC neurons (Reyes et al. 2012). Taken together, these data indicate that sustained CB1R activation results in a sustained increase in NE release which induces down-regulation of adrenergic receptors. Tolerance, at the level of the CB1R, could ultimately develop to the effects of sustained CB1R agonists treatment to stimulate NE synthesis (Moranta et al. 2009); so this process is likely highly time and dose dependent.

In summary, convergent data indicate that CB1R activation increases activity in LC neurons, resulting in increased release NE in terminal regions. However, there is also evidence for other paradigms of interaction between ECS and noradrenergic signaling. In particular, CB1Rs in the terminal regions of LC projections can inhibit NE release, likely through presynaptic CB1R on axon terminals of noradrenergic neurons. Since systemic treatment of healthy animals with CB1R antagonists increases NE release, it is possible that ECS has greater ongoing tone at terminal, release-inhibitory CB1R. Data from studies using chronic administration of tricyclic antidepressants (TCAs) suggest that noradrenergic signaling can also regulate ECS, at least when noradrenergic signaling is sustained. There is a paucity of data exploring the effects of acute changes in NE concentrations on ECS in the brain.

9.9 Interactions Between ECS and NE Signaling: Stress Context

9.9.1 ECS Mediates Contributions of Noradrenergic Signaling to Stress Effects

Like the ECS, central noradrenergic signaling is significantly and rapidly increased in response to stress (Cassens et al. 1980). Although the ECS dampens stress responses through inhibition of HPA axis activity (described above), ECS also contributes in a positive manner to behavioral changes that result from stress and glucocorticoids. Intriguingly, there is accumulating evidence that the stress-promoting effects of ECS occur through increased noradrenergic signaling. Taken together, the studies described next support the hypothesis that stress and/or glucocorticoids recruit ECS which enhances noradrenergic signaling and contributes to stress responses.

High doses of CB1R agonists increase HPA axis responsivity to stress (Murphy et al. 1998; Patel et al. 2004) and pretreatment of rats with antagonists of either ß-adrenergic or alpha1-adrenergic receptors significantly attenuates this response (McLaughlin et al. 2009). While NE signaling is not the only process involved, these data are consistent with ECS acting up-stream of noradrenergic signaling to potentiate stress-induced HPA axis activation.

Cannabinoid agonists exert biphasic effects on anxiety, with low doses of CB1R agonists generally producing anxiolytic effects and high doses increasing anxiety (Patel and Hillard 2006; Patel and Hillard 2009b). In addition, high doses of CB1R agonists induce conditioned place aversion in rodents (Mallet and Beninger 1998; McGregor et al. 1996; Pandolfo et al. 2009; Sanudo-Pena et al. 1997). A recent study demonstrated that a relatively high dose of Win 55212-2 (3 mg/kg) produced aversion in conditioned place preference that was abolished by toxin-induced depletion of NE in nucleus accumbens but not bed nucleus of the stria terminalis (BNST) (Carvalho et al. 2010b). The role of NE was specific for aversive behaviors since toxin treatment did not alter the responses to the same dose of Win 55212-2 in assays of locomotor activity, spatial memory or elevated zero maze (Carvalho et al. 2010b). Injection of a ß-adrenergic antagonist into the accumbens abolished Win 55212-2-induced conditioned place aversion (Carvalho and Van Bockstaele 2011), further support for noradrenergic signaling acting as a down-stream effector of exogenous CB1R activation to produce aversion. Since the NTS is a source of NE innervation of the accumbens (Delfs et al. 1998) and CB1R activate NTS noradrenergic neurons (Carvalho et al. 2010a; Chen et al. 2010), increased NTS outflow could be the mechanism for this effect. Interestingly, Win 55212-2 place aversion does not occur in spontaneously hypertensive rats (SHR) (Pandolfo et al. 2009) which have reduced NTS CB1R binding site density and attenuated responsiveness to cannabinoid agonists injected into the NTS (Brozoski et al. 2009). Together, these studies suggest that NTS-accumbens projections are modulated by ECS in the NTS and contribute to the learning of aversion.

Evidence from both rats and mice indicate that increased noradrenergic signaling is required for stress-induced reinstatement to cocaine seeking (Erb et al. 2000; Leri et al. 2002; Mantsch et al. 2010). CB1R antagonism blocked stress-induced but not cocaine-induced reinstatement of cocaine place preference in mice (Vaughn et al. 2012). Furthermore, the combination of nonreinstating doses of a CB1R agonist and an alpha2 adrenergic antagonist produced significant reinstatement of cocaine place preference (Vaughn et al. 2012). These data suggest that ECS and NE are both required for stress-induced relapse to drug seeking and act synergistically to produce the behavioral effect.

It has been argued that stress-induced increases in ECS potentiate stress-induced memory consolidation through increased NE signaling (Campolongo et al. 2009). Although interactions between ECS and noradrenergic signaling in this context have not been demonstrated directly, ECS inhibits GABA signaling in the basolateral amygdala (BLA) during stress (Sumislawski et al. 2011) and inhibition of GABAergic activity in the BLA enhances memory consolidation by increasing NE signaling (Hatfield et al. 1999). These and other observations (Campolongo et al. 2009) are consistent with a model in which stress produces behavioral changes through recruitment of both ECS and NE signaling.

As was discussed above, CB1R agonists increase NE release in the PFC in unstressed rodents. However, Win 55212-2 suppresses stress-induced NE release and increases immobility and decreases climbing in the FST (Reyes et al. 2012). The reduction in immobility and reduced non-swimming active behaviors are consistent with decreased NE release in the PFC (Detke et al. 1995). Stress also reverses the effect of chronic Win 55212-2 to block clonidine-induced increases in PFC excitability (Reyes et al. 2012). These studies indicate that ECS regulation of NE signaling is significantly altered by stress and suggest the intriguing possibility that ECS is a pivot point with regard to the effects of stress on NE signaling.

9.9.2 Sustained CB1R Activation Alters Noradrenergic Signaling

As was described above, serotonergic signaling plays an important role in the regulation of active and passive coping behavioral paradigms employed by rodents in the FST. In a preceding section, data that CB1R agonists exert antidepressant-like effects in the FST through increased 5-HT signaling in rats were presented. Like 5-HT, elevations in NE also produce antidepressant effects in the FST. Acute treatment of rodents with antidepressants that inhibit NE uptake, such as desipramine, reduce immobility and promote the active coping behavior of struggling and climbing rather than swimming in the FST (Detke et al. 1995). Chronic treatment of rats with a synthetic CB1R agonist, HU210, reduces immobility and increases struggling without affecting swimming in the FST, consistent with increased NE signaling (Morrish et al. 2009). In support of this mechanism, the effect of chronic HU210 was attenuated by a ß-adrenergic receptor antagonist and, to a lesser extent, by an alpha-adrenergic receptor antagonist. In light of the data that chronic treatment with Win 55212-2 increases TH expression in the LC and increases CB1R-agonist induced NE release (Page et al. 2007), a possible mechanism is that repeated exposure to CB1R agonists enhances NE synthesis and neurotransmission as a result. An alternative mechanism is that chronic CB1R agonist treatment results in down-regulation of CB1R on NE terminals (Hillard and Bloom 1982; Reyes et al. 2009), resulting in dysinhibition of NE release. The second possibility is supported by data that acute treatment with a CB1R antagonist produces immobility in the FST through a mechanism that requires catecholamines (Haring et al. 2013). If a subset of tonically active CB1R inhibit NE release, acute CB1R antagonism could enhance NE release and thereby increase active coping in FST.

Moranta et al. 2009 investigated the role of enhanced monoaminergic signaling in the aversive effects of withdrawal from chronic cannabinoid agonist treatment. Antagonist-induced withdrawal in rats chronically treated with Win 55212-2 was accompanied by decreased TH activity, likely because of desensitization of alpha2 autoreceptors regulating the synthesis of NE in cortex, hippocampus, and cerebellum. Similar adaptations of alpha2 autoreceptors have been demonstrated in morphine- and ethanol-dependent rats (Esteban et al. 2002; Sastre-Coll et al. 2002) and could contribute to the somatic symptoms of withdrawal. Lesion data indicate that noradrenergic inputs to the BNST are not involved in these effects (Carvalho et al. 2010b).

9.9.3 Sustained Changes in NE Signaling Alter ECS

Chronic treatment of rats with the tricyclic antidepressant (TCA) desipramine (Hill et al. 2006) but not fluoxetine (Hill et al. 2008b) produces significant increases in the density of CB1R in the hypothalamus and hippocampus. Desipramine is an antidepressant whose primary mechanism is inhibition of NE reuptake (Frazer 1997). Chronic desipramine treatment also results in reduced HPA axis activation by stress, evidenced by reductions in both FST-induced c fos expression in the PVN and circulating corticosterone (Hill et al. 2006). Acute treatment with rimonabant before FST completely abolished the effect of chronic desipramine treatment to suppress HPA axis activation. Since decreased ECS in the hypothalamus is associated with HPA axis hyperresponsiveness (Cota et al. 2007; Patel et al. 2004), the authors of this study hypothesized that desipramine-induced, increased CB1R expression in the hypothalamus has the opposite effect, i.e., decreases HPA axis responsivity (Hill et al. 2006). It is interesting that TCAs are more efficacious than SSRIs in reducing HPA axis reactivity in humans (Connor et al. 2000;

Duncan et al. 1996). Perhaps this is because of the differential effects of these two drug classes on ECS in the hypothalamus. The mechanism by which desipramine increases CB1R expression is not known, but could be related to its effect to decrease alpha1 adrenoceptor expression (Subhash et al. 2003), which could result in decreased 2-AG synthesis and up-regulation of CB1R. The effect of desipramine on CB1R expression could oppose the effect of the TCA to increase NE concentrations in the synapse. Up-regulation of CB1R specifically on noradrenergic terminals could result in reduced NE release, which would normalize NE synaptic concentrations. However, this mechanism is speculative and needs to be tested experimentally.

9.9.4 Summary

The biphasic effects of exogenous CB1R agonists on anxiety, place preference and HPA axis activation by stress are well appreciated. High doses of exogenously administered CB1R agonists produce increased stress-induced activation of the HPA axis and induce conditioned place aversion rather than preference; these effects require intact NE signaling to occur. In addition, the effects of stress to induce cocaine reinstatement require both ECS and NE signaling. Recent data showing that CB1R agonist treatment exerts opposite effects on NE concentrations and signaling in the PFC in the absence and presence of stress indicates that ECS regulation of NE signaling is context dependent. Chronic treatment with CB1R agonists decreases expression and/or function of most adrenoceptors in the brain, likely as a result of sustained increases in NE concentrations. On the other hand, chronic treatment with the TCA desipramine causes increased CB1R density in hypothalamus and hippocampus but not in the PFC, changes that contribute to the decreased HPA axis reactivity following chronic desipramine treatment.

9.10 Interactions Between ECS and DA Signaling: Mechanisms

Dopaminergic cell bodies are present in the substantia nigra (SN) and ventral tegmental area (VTA) of the midbrain. Dopaminergic projections from the SN innervate the dorsal striatum via the nigrostriatal circuit while those from the VTA terminate in ventral striatum (nucleus accumbens) and PFC and form the mesocorticolimbic dopaminergic circuit. The amygdala also receives dopaminergic innervation from the ventral mesencephalon via a mesoamygdaloid projection (Fallon et al. 1978; Ungerstedt 1971). The tuberoinfundibular dopaminergic pathway projects from the hypothalamus to the pituitary and is relevant for the effects of DA ligands on endocrine function. Considerable experimental evidence demonstrates that ECS and dopaminergic signaling influence each other in a complex and bidirectional manner. I will focus here on the mechanisms of interaction in the mesocorticolimbic dopaminergic circuit; recent excellent reviews summarize the interactions of ECS and DA in the nigrostriatal circuit (El Khoury et al. 2012; Fitzgerald et al. 2012; Lovinger and Mathur 2012).

Dopaminergic afferents from the VTA play a central role in motivation, reward-related behaviors and cognition (Schultz 2002), thus the mesocorticolimbic DA circuit is often considered the "reward circuit". The mesocorticolimbic dopaminergic circuit provides the brain with information about the emotional valence of all sensory stimulation, not just rewarding stimuli (Laviolette and Grace 2006). While the circuit underlies the rewarding effects of drugs, deficits are thought to underlie the compromised ability of those with schizophrenia or addictions to accurately assign emotional significance to sensory input. There are considerable data to support the contentions that cannabinoid-induced effects on the mesocorticolimbic circuit could contribute to several of the consequences of *Cannabis* intoxication in humans, including its rewarding effects (Danovitch and Gorelick 2012); impairment of working memory (Volk and Lewis 2010) and, at high doses, psychosis (Solowij and Michie 2007). In addition, considerable evidence demonstrates that loss of CB1R function reduces the rewarding effects of most abused substances and natural rewards (Serrano and Parsons 2011), suggesting that the regulation of the mesocorticolimbic circuit by ECS has broad implications for abuse and addiction. Indeed, recent experience with the use of the CB1R antagonist, rimonabant, in humans suggests that ECS is vital for the maintenance of hedonia, particularly in individuals prone to depression (Nathan et al. 2011).

Although the density is low relative to other brain regions, CB1R are present in the VTA on presynaptic terminals of both glutamatergic and GABAergic neurons (Matyas et al. 2008). Excitatory, afferent inputs to VTA dopaminergic neurons from the PFC and pontine tegmental nuclei and GABAergic projections from the accumbens have been shown to express CB1R (Marsicano and Lutz 1999; Matsuda et al. 1993). The axon terminals containing CB1R impinge on VTA neurons expressing TH, evidence that they are dopaminergic (Fitzgerald et al. 2012; Matyas et al. 2008). The 2-AG synthesizing enzyme, diacylglycerol lipase (DGL) alpha is detected in regions adjacent to the postsynaptic areas targeted by CB1R-containing axons, which provides further support for the hypothesis that ECS regulates synaptic activity at these synapses (Matyas et al. 2008). Although CB1R expression in TH positive neurons in the VTA has been detected using light microscopy (Wenger et al. 2003), electron microscopy studies indicate that very few TH-positive neurons coexpress CB1R (Fitzgerald et al. 2012).

Electrophysiological studies of the synapse between GABAergic interneurons and presumptive dopaminergic neurons in VTA brain slices demonstrate that CB1R agonists inhibit GABA release at a subset of synapses (Riegel and Lupica 2004; Szabo et al. 2002). CB1R-mediated inhibition of excitatory inputs to VTA dopaminergic neurons has also been described (Melis et al. 2004b; Riegel and Lupica 2004). Increased postsynaptic activity induces ECS at both GABAergic and glutamatergic synapses onto DA neurons in the VTA (Riegel and Lupica 2004). ECS in the VTA requires both depolarization and mGluR1 activation and is blocked by inhibitors of DGL and PLC activity, indicating that 2-AG is the likely eCB involved

(Melis et al. 2004a). Morphological data that DGL alpha is in close proximity to CB1R axon terminals (Matyas et al. 2008) support this hypothesis. In vivo evidence that stimulation of the PFC recruits ECS in the VTA (Melis et al. 2004a) is consistent with the hypothesis that ECS in the VTA is triggered by glutamate. Thus, the role of ECS in tuning VTA output likely depends on context: under control conditions, ECS could maintain VTA output through inhibition of GABA release while during times of stress or high emotional tone when the PFC is active, ECS reduces the responsiveness of the VTA through inhibition of glutamate release.

CB1R-dependent DSE is significantly reduced by the D2 dopamine receptor (D2R) antagonist, eticlopride, suggesting that this form of eCB-mediated plasticity is augmented by depolarization-induced release of DA (Melis et al. 2004b). In addition, chronic cocaine treatment recruits CB1R-mediated long-term depression of inhibition (I-LTD) in the VTA through a mechanism that requires group 1 mGluR and D2R (Pan et al. 2008a, b). The combination of increased D2R and CB1R activation produces a long-term suppression of GABA release through inhibition of protein kinase A (Pan et al. 2008a). This effect of chronic cocaine is particularly interesting as some hypothesize that overactive reward circuits could underlie addictive behaviors following chronic drug exposure (Vanderschuren and Kalivas 2000).

Exogenous administration of CB1R agonists produces intense burst firing of VTA dopaminergic neurons (Diana et al. 1998; French 1997; French et al. 1997; Gessa et al. 1998; Wu and French 2000) and a concomitant increase in DA release in the accumbens (Cheer et al. 2004; Chen et al. 1990b; Tanda et al. 1997) and the PFC (Chen et al. 1990a). These pharmacological effects are consistent with cannabinoid-mediated dysinhibiton of GABAergic interneurons within the VTA. Support for this mechanism comes from a recent study of the roles of CB1R on GABA and glutamatergic terminals in the VTA on voluntary wheel running (Dubreucq et al. 2012), a rewarding activity in mice that alters neuronal activity of dopaminergic neurons in the VTA (Novak et al. 2012). Conditional deletion of CB1R selectively from GABAergic neurons in brain decreases wheel running performance in mice, which is consistent with tonic and permissive effect of CB1R activation to increase DA release in the striatum which promotes reward and/or causes increased locomotor activity (Dubreucq et al. 2012).

It is possible that exogenous cannabinoids act in extra-VTA regions of the circuit to increase VTA-ventral striatum outflow. For example, a c fos study from our laboratory suggests that activation of excitatory, noradrenergic afferents from the LC contributes to the effects of exogenous CB1R agonists to activate VTA (Patel and Hillard 2003). As is discussed further below, CB1R activation in the PFC also increases the firing rate of PFC neurons projecting to the VTA (Pistis et al. 2001), providing an additional, indirect mechanism for increased VTA firing.

In sum, available evidence indicates ECS can significantly alter the discharge pattern of dopaminergic neurons in the VTA. It has been hypothesized that ECS in the VTA, through regulation of glutamate and GABA release, could fine-tune phasic versus tonic DA release in the mesocorticolimbic circuit (Melis et al. 2004a). As a result, ECS functions as a modulator of the mesocorticolimbic circuit.

A primary projection of VTA dopaminergic neurons is the shell of the nucleus accumbens (NAc). This region of the accumbens also receives glutamatergic input from the cortex which converges on many of the same medium spiny neurons as the dopaminergic projections from the VTA (Sesack and Pickel 1990). Glutamatergic and dopaminergic inputs to the accumbens interact in a bidirectional manner and together regulate activity of the medium spiny neurons.

CB1R are abundantly expressed in the accumbens (Herkenham 1992) on both GABAergic terminals that are likely collaterals of medium spiny neurons and on glutamatergic afferents from cortex (Pickel et al. 2004; Pickel et al. 2006). CB1R are not present on dopaminergic projections (Fitzgerald et al. 2012; Julian et al. 2003). As expected from their localization, CB1R activation suppresses evoked release of GABA (Centonze et al. 2007; Hoffman and Lupica 2001) and glutamate (Pistis et al. 2002b; Robbe et al. 2001) in the accumbens. Stimulation of prelimbic cortical afferents to the accumbens using naturally occurring frequencies induces CB1R-dependent long-term depression (LTD) of excitatory synaptic transmission (Robbe et al. 2002) and LTD is absent in rats chronically treated with THC (Hoffman et al. 2003). OBX reduces glutamatergic inputs to the ventral striatum and also reduces ECS in this brain region—an effect that could contribute to the reduced responsivity to novel stimuli observed in animals with OBX (Eisenstein et al. 2010). In contrast to the dorsal striatum, D2R activation is not required for CB1R-mediated LTD in the amygdala (Robbe et al. 2002). Thus, current evidence indicates that the role of presynaptic ECS in the accumbens is to regulate glutamatergic input, which likely also has an indirect effect on the dopaminergic input to medium spiny neurons.

CB1R are also expressed by striatal neurons (Ferre et al. 2009; Kearn et al. 2005; Kofalvi et al. 2005; Pickel et al. 2006), suggesting that CB1R have postsynaptic effects in this brain region. Interestingly, CB1R can be colocalized with D2R in striatal neurons (Pickel et al. 2006) and recent electrophysiological data suggests that postsynaptic CB1R could enhance DR function, particularly when dopamine concentrations are low (Seif et al. 2011). In addition, there is evidence that D2R and CB1R can form heterodimers (Kearn et al. 2005) and that dual activation of D2R and CB1R results in a stimulatory effect on cAMP, although both CB1R and D2R signal individually through Gi/o family G proteins (Glass and Felder 1997; Jarrahian et al. 2004; Kearn et al. 2005). Thus, ECS and dopaminergic signaling have the ability to influence each other at the level of the receptor.

The mesocortical DA system modulates activity of pyramidal neurons of the PFC. Early studies found a predominant effect of DA to inhibit pyramidal neuron firing through enhanced release of GABA from inhibitory interneurons (Gellman and Aghajanian 1993; Pistis et al. 2001) likely through D1R activation (Arnsten 2011). Since CB1R are present on GABAergic terminals in the PFC and inhibit GABA release (Hill et al. 2011b), the ECS is in a position to oppose the inhibitory effect of dopamine in this circuit. Indeed, systemic CB1R agonist treatment reverses the inhibitory effects of VTA activation on PFC activity in rodents (Pistis et al. 2001), however, whether this occurs through CB1R activation in the PFC or in the VTA is not clear.

Synergistic effects between ECS and DA have also been demonstrated in PFC. In particular, CB1R and D2R are colocalized on GABAergic terminals and simultaneous activation of both triggers CB1R-mediated LTD of GABA release (I-LTD) (Chiu et al. 2010). Further evidence in this study argues against D2R-mediated mobilization of eCBs, rather supports synergism between signaling cascades activated by CB1R and D2R. As was discussed above for the synergistic effects of DA and ECS in the VTA (Pan et al. 2008a), it is possible that CB1R and D2R activation in the PFC suppresses PKA below a threshold that allows for induction of I-LTD.

Biochemical data suggest bidirectional interactions between ECS and dopaminergic signaling at the level of ligand availability; however, none of the available data are specific to the mesocorticolimbic DA system. For example, acute treatment of mice with systemic methylphenidate, GBR 12909 and D1R agonist individually reduce AEA content in the forebrain (Patel et al. 2003). Similarly, mice that do not express the dopamine transporter (DAT) have significantly lower striatal AEA concentrations (Tzavara et al. 2006). On the other hand, forebrain and striatal AEA concentrations are increased by D2R activation (Centonze et al. 2004; Giuffrida et al. 1999; Patel et al. 2003). CB1R activation increases DA release (Pistis et al. 2002a; Tanda et al. 1997) and genetic deletion of the CB1R results in D2R overexpression, which is consistent with decreased DA tone in the absence of ECS (Houchi et al. 2005).

In summary, the interactions between dopaminergic and CB1R signaling are complex and occur in the VTA, accumbens and PFC. There is little evidence in rodent models that ECS regulates the release of DA, although the situation could be different in humans since CB1R are present on DA terminals in human neocortex and inhibit DA release from human synaptosomes (Steffens et al. 2004). There is evidence from morphological and electrophysiological studies that ECS regulation of glutamate release indirectly alters responses to DA in the mesocorticolimbic circuit. Intriguingly, there is also evidence that postsynaptic CB1R are important in this circuit and participate in a bidirectional modulation of postreceptor signaling with D2R. These interactions clearly contribute to the role of ECS in regulation of reward and could also underlie the propsychotic effects of *Cannabis* use in humans.

9.11 Interactions Between ECS and DA Signaling: Stress Context

The VTA is activated by stress and DA concentrations are increased in the mPFC, accumbens and dorsal striatum in response to restraint stress (Deutch et al. 1991). Considerable preclinical and clinical data demonstrate that DA, through both D1R and D2R receptors, is one of the most important neuromodulators of fear and anxiety (LeDoux 2000). ECS is also activated by stress, and CB1R signaling regulates DA neuronal activation in the mesencephalon; release of DA in projection sites; and can affect dopamine receptor signaling (previous section). Therefore, it seems quite probable that ECS modulates and/or contributes to the dopaminergic role in stress responding. However, there is not much experimental evidence that tests this proposition directly.

9.11.1 Dysregulation of Stress Reactivity by Chronic Exposure to Drugs of Abuse

Dysregulation of stress responses is a major, long-term adverse consequence of drug addiction and it plays a primary role in relapse to drug use following abstinence (Sidhpura and Parsons 2011). As is outlined wonderfully in a review by Sidhpura and Parsons (2011), the characteristics of ECS, particularly its role in both reward and stress responses, support the hypothesis that chronic drug use causes dysregulated stress responsivity as a result of alterations in ECS-mediated synaptic plasticity. As was described above, ECS is activated by glucocorticoids and long-term drug use results in sensitization of responses to HPA axis activation (Koob and Kreek 2007), which suggests the hypothesis that chronic drug exposure could alter glucocorticoid-ECS signaling. To my knowledge, this hypothesis has not been tested.

9.11.2 Anxiety

While systemic administration of CB1R agonists produces a biphasic effect on behavior in the elevated plus maze (Patel and Hillard 2006), activation of ECS through administration of indirect agonists primarily exerts anxiolytic effects in this assay particularly when stress is high (Patel and Hillard 2006; Sciolino et al. 2011). Similarly, blockade of ECS with CB1R antagonists results in anxiogenic like effects (Patel and Hillard 2006), implicating ECS as a feedback system that limits the expression of anxiety under stressful circumstances. Acute exposure to psychostimulants produces increased anxiety (Lodge and Grace 2011), an effect that is reversed by prior CB1R activation. In particular, exposure of male mice to a single, high dose of psychostimulant results in an anxiogenic phenotype in the elevated plus maze that lasts for at least 5 days (Hayase et al. 2005). Treatment of the mice 1 h prior with several CB1R agonists reverses the anxiogenic effects, suggesting that suppression of ECS could contribute to the anxiogenic effects of the psychostimulants. Previous studies have demonstrated that methylphenidate in particular decreases forebrain tissue contents of both AEA and 2-AG (Patel et al. 2003), it is possible that a long-lasting suppression of ECS tone contributes to the sustained anxiogenic effects of psychostimulant exposure. On the other hand, the ability of amphetamine to induce long-term depression in the lateral amygdala is dependent upon CB1R and not DA receptors (Huang et al. 2003). Thus, prior exposure to CB1R agonists could abrogate the effect of the psychostimulant to induce plasticity in the amygdala.

There is substantial evidence that exposure to most drugs of abuse affects synaptic plasticity in many brain regions and that in many cases, these changes are subsequent to disruption of ECS-mediated plasticity, particularly at glutamatergic synapses (Wolf et al. 2004). The bed nucleus of the stria terminalis (BNST) is a region of the extended amygdala that is stress-responsive and a key relay between limbic cognitive centers and reward, stress, and anxiety nuclei (McElligott and Winder 2009). ECS-mediated LTD occurs in the BNST and is reduced following multiple exposures of mice to cocaine (Grueter et al. 2006). Since the BNST projects to the VTA (Dong and Swanson 2006), changes in synaptic plasticity in the BNST will alter the mesocorticolimbic circuit and alter motivated behavior (Sidhpura and Parsons 2011).

On the other hand, there is considerable evidence that hyperactive ECS could contribute to the anxiogenic effects of withdrawal from addictive drugs. Withdrawal from chronic exposure to drugs and alcohol induces anxiety that is mediated by increases in amygdalar CRF concentrations and can be blocked by antagonism of CRF receptors in the central nucleus of the amygdala (Koob and Le Moal 2008). The anxiogenic effects of alcohol withdrawal are reduced in CB1R null mice (Racz et al. 2003) and are inhibited by CB1R antagonists (Rubio et al. 2008) as are the anxiogenic effects of withdrawal from diazepam and cocaine (Kupferschmidt et al. 2012; Onaivi 2008). These results suggest that, as a stressor, withdrawal has increased ECS and that this contributes to the anxiogenic effects. There is a high degree of co-localization between mRNA for the CB1R and message for both CRF and CRFR1 in stress-regulated brain regions (Hermann and Lutz 2005) and recent data demonstrate that the effects of i.c.v. CRF to induce anxiogenic responses in the elevated plus maze are blocked by CB1R antagonism (Kupferschmidt et al. 2012). CRF1 receptors are expressed in both glutamatergic projection neurons and VTA dopaminergic neurons and exert a bi-directional influence on anxiety behaviors (Refojo et al. 2011). ECS modulates these neurons as well, suggesting that they could be the substrate for interactions between CRF and ECS signaling. Withdrawal from alcohol in rats is associated with decreased VTA DA concentrations, which is reversed by rimonabant (Rubio et al. 2008).

9.11.3 Emotional Learning

An important role for the mesolimbic dopamine circuit is to associate emotional significance to sensory information, called "emotional learning" (Laviolette and Grace 2006). Most animal studies of emotional learning utilize fear conditioning approaches in which a neutral stimulus is paired with a painful stimulus, usually a mild foot shock. Emotional learning is measured as the ability of either the conditioned stimulus (cue) or placing the animal back into the training environment (context) to provoke a fear response (usually freezing). Emotional learning is dependent upon the DA activity in mesocorticolimbic circuit and an intact amygdala. A recent study indicates that ECS in D1R expressing neurons is involved in emotional learning (Terzian et al. 2011). In this study, mice with a selective deletion of the CB1R in neurons expressing the D1R were exposed to foot shock with an auditory conditioned stimulus. Both cue and context-induced freezing were significantly increased on the first day after conditioning in the knock out compared to wild type mice, suggesting that ECS opposes the effect of D1R activation to evoke emotional learning. Earlier studies found that CB1R agonists also functionally oppose the effects of D1R agonists on motor behaviors (Martin et al. 2008), leading to the hypothesis that CB1R coexpressed in D1R containing neurons opposes the effects of D1R activation. Infusion of AEA into the accumbens before conditioning decreased freezing behavior to the context 24 h later (Pedroza-Llinas et al. 2013). Taken together, these data suggest that ECS in the accumbens functions to dampen the magnitude of emotional learning plasticity perhaps through activation of signaling cascades that oppose those of the D1R.

9.11.4 Chronic Stress and Striatal EC

Exposure to inescapable stress alters synaptic plasticity in many brain regions. In particular, social defeat stress suppresses CB1R-mediated inhibition of GABA release in the striatum (Rossi et al. 2008). It is likely that the mechanism involves decreased CB1R expression or function since exogenous agonists were ineffective at inhibition of GABA release as well. CB1R regulation of glutamate release was unaffected in the striatum. Importantly, the effect of chronic stress to suppress CB1R signaling was reversed when stressed mice were given access to running wheels, sucrose, or cocaine (Rossi et al. 2008). These data suggest that reduced ECS is the result of the anhedonic effects of stress and that reduced ECS contributes to the reward deficit of chronic stress. These data are consistent with biochemical data that dopamine increases striatal CB1R expression (Centonze et al. 2004) and that chronic treatment of unstressed rats with the MAO inhibitor, tranylcypromine increases CB1R density in PFC and hippocampus (Hill et al. 2008b). Taken together, these data indicate a positive relationship between DA signaling and CB1R density. Recent data DA-regulation of brain-derived neurotrophic factor (BDNF) could alter CB1R expression through a mechanism that involves changes in cholesterol metabolism that alter CB1R distribution to lipid rafts (De Chiara et al. 2010).

9.11.5 Summary

The interactions between ECS and dopaminergic systems are complex, bidirectional, and affected by stress. With regard to regulation of anxiety, both the anxiogenic and anxiolytic effects of CB1R activation come into play in the dopaminergic circuit. While activation of CB1R opposes the anxiogenic effects of acute psychostimulant exposure, blockade of CB1R ameliorates the anxiogenic effects of withdrawal. Since DA signaling is also highly dysregulated during withdrawal, it is possible that the state of DA signaling is an important determinant of ECS. Around 10 years ago, Sachin Patel and I suggested, based on the dose-response relationship between CB1R agonist and c fos expression in the VTA, that effects of cannabinoids to activate specific subregions of the VTA could contribute to both the reward-promoting and stress-promoting effects of CB1R agonists (Patel and Hillard 2003). Activation of dopaminergic neurons within the parabrachial pigmented and paranigral regions of the VTA exhibits a steep and bell-shaped dose response relationship that we hypothesized matched well the rewarding effects of cannabinoids while activation in the caudal linear region exhibited a more classical relationship with agonist dose and could subserve the aversive and anxiogenic effects of CB1R activation. Until better tools are developed or applied to study subregional roles of ECS, it will be impossible to completely understand the interactions of these two very important systems.

9.12 Concluding Remarks

A recent genetic association study provides strong evidence that CB1R function plays an important role in stress adaptation in humans (Agrawal et al. 2012). This study examined a synonymous polymorphism in exon 4 of the gene encoding the CB1R in two different cohorts of individuals exposed to childhood physical abuse. In both samples, individuals reporting physical abuse in childhood were significantly more likely to report anhedonia. However, in both cohorts, those with one or more copies of the minor allele at rs1049353 were nearly 50 % less likely to have anhedonia as an adult. Thus, this study suggests that CB1R genotype can buffer the effects of childhood physical abuse on major depression. This study, along with other accumulating data that differences in genes associated with ECS influence the likelihood for substance use and mood disorders, suggest that ECS function contributes to human psychiatric illness (Hillard et al. 2012). Thus, enhanced understanding of the many roles of ECS in the regulation of mood, emotion, and reward is an important goal.

The studies described in this review indicate that we are just beginning to understand the intersection of ECS with monoaminergic systems that are also vital for mood, reward, and processing of emotions. However, there is much work to be done. Although we are fortunate to have excellent morphological maps of the subregional and sometimes even subcellular distribution of the CB1R, our understanding of the roles of individual pools of CB1R in the responses to systemic agonist and antagonist treatment is primitive in most cases, which, along with the complexity of the monoamine systems, results in a clouded picture at this stage of our knowledge. However, the development of selective knock out mouse models, and increased use of local injections along with the promise of other more precise techniques, such as optogenetics and live brain imaging, will ultimately improve our focus and allow for the development and use of ECS-targeted therapeutic approaches in a rationale manner.

References

Aghajanian GK, Cedarbaum JM, Wang RY (1977) Evidence for norepinephrine-mediated collateral inhibition of locus coeruleus neurons. Brain Res 136:570–577

Agrawal A, Nelson EC, Littlefield AK, Bucholz KK, Degenhardt L, Henders AK, Madden PA, Martin NG, Montgomery GW, Pergadia ML, Sher KJ, Heath AC, Lynskey MT (2012) Cannabinoid receptor genotype moderation of the effects of childhood physical abuse on anhedonia and depression. Arch Gen Psychiatry 69:732–740

Arnsten AF (2011) Catecholamine influences on dorsolateral prefrontal cortical networks. Biol Psychiatry 69:e89–99

Aso E, Renoir T, Mengod G, Ledent C, Hamon M, Maldonado R, Lanfumey L, Valverde O (2009) Lack of CB(1) receptor activity impairs serotonergic negative feedback. J Neurochem 109:935–944

Aston-Jones G, Cohen JD (2005) An integrative theory of locus coeruleus-norepinephrine function: adaptive gain and optimal performance. Annu Rev Neurosci 28:403–450

Bambico FR, Katz N, Debonnel G, Gobbi G (2007) Cannabinoids elicit antidepressant-like behavior and activate serotonergic neurons through the medial prefrontal cortex. J Neurosci 27:11700–11711

Bambico FR, Hattan PR, Garant JP, Gobbi G (2012) Effect of delta-9-tetrahydrocannabinol on behavioral despair and on pre- and postsynaptic serotonergic transmission. Prog Neuropsycho pharmacol Biol Psychiatry 38:88–96

Bambico FR, Cassano T, Dominguez-Lopez S, Katz N, Walker CD, Piomelli D, Gobbi G (2010) Genetic deletion of fatty acid amide hydrolase alters emotional behavior and serotonergic transmission in the dorsal raphe, prefrontal cortex, and hippocampus. Neuropsychopharmacology 35:2083–2100

Barbon A, Orlandi C, La Via L, Caracciolo L, Tardito D, Musazzi L, Mallei A, Gennarelli M, Racagni G, Popoli M, Barlati S (2011) Antidepressant treatments change 5-HT2C receptor mRNA expression in rat prefrontal/frontal cortex and hippocampus. NeuropsychoBiology 63:160–168

Barnes NM, Sharp T (1999) A review of central 5-HT receptors and their function. Neuropharmacology 38:1083–1152

Beltramo M, Stella N, Calignano A, Lin SY, Makriyannis A, Piomelli D (1997) Functional role of high-affinity anandamide transport, as revealed by selective inhibition. Science 277:1094–1097

Bergamaschi MM, Queiroz RH, Chagas MH, Oliveira DC de, De Martinis BS, Kapczinski F, Quevedo J, Roesler R, Schroder N, Nardi AE, Martin-Santos R, Hallak JE, Zuardi AW, Crippa JA (2011) Cannabidiol reduces the anxiety induced by simulated public speaking in treatment-naive social phobia patients. Neuropsychopharmacology 36:1219–1226

Berridge CW, Waterhouse BD (2003) The locus coeruleus-noradrenergic system: modulation of behavioral state and state-dependent cognitive processes. Brain Res Brain Res Rev 42:33–84

Best AR, Regehr WG (2008) Serotonin evokes endocannabinoid release and retrogradely suppresses excitatory synapses. J Neurosci 28:6508–6515

Bland ST, Hargrave D, Pepin JL, Amat J, Watkins LR, Maier SF (2003) Stressor controllability modulates stress-induced dopamine and serotonin efflux and morphine-induced serotonin efflux in the medial prefrontal cortex. Neuropsychopharmacology 28:1589–1596

Braida D, Limonta V, Malabarba L, Zani A, Sala M (2007) 5-HT1A receptors are involved in the anxiolytic effect of Delta9-tetrahydrocannabinol and AM 404, the anandamide transport inhibitor, in Sprague-Dawley rats. Eur J Pharmacol 555:156–163

Brozoski DT, Dean C, Hopp FA, Hillard CJ, Seagard JL (2009) Differential endocannabinoid regulation of baroreflex-evoked sympathoinhibition in normotensive versus hypertensive rats. Auton Neurosci 150(1–2):82–93

Campolongo P, Roozendaal B, Trezza V, Hauer D, Schelling G, McGaugh JL, Cuomo V (2009) Endocannabinoids in the rat basolateral amygdala enhance memory consolidation and enable glucocorticoid modulation of memory. Proc Natl Acad Sci U S A 106:4888–4893

Carvalho AF, Van Bockstaele EJ (2011) Direct intra-accumbal infusion of a beta-adrenergic receptor antagonist abolishes WIN 55,212-2-induced aversion. Neurosci Lett 500:82–85
Carvalho AF, Van Bockstaele EJ (2012) Cannabinoid modulation of noradrenergic circuits: implications for psychiatric disorders. Prog Neuropsychopharmacol Biol Psychiatry 38:59–67
Carvalho AF, Mackie K, Van Bockstaele EJ (2010a) Cannabinoid modulation of limbic forebrain noradrenergic circuitry. Eur J Neurosci 31:286–301
Carvalho AF, Reyes AR, Sterling RC, Unterwald E, Van Bockstaele EJ (2010b) Contribution of limbic norepinephrine to cannabinoid-induced aversion. Psychopharmacology (Berl) 211:479–491
Cassano T, Gaetani S, Macheda T, Laconca L, Romano A, Morgese MG, Cimmino CS, Chiarotti F, Bambico FR, Gobbi G, Cuomo V, Piomelli D (2011) Evaluation of the emotional phenotype and serotonergic neurotransmission of fatty acid amide hydrolase-deficient mice. Psychopharmacology (Berl) 214:465–476
Cassens G, Roffman M, Kuruc A, Orsulak PJ, Schildkraut JJ (1980) Alterations in brain norepinephrine metabolism induced by environmental stimuli previously paired with inescapable shock. Science 209:1138–1140
Celada P, Puig MV, Casanovas JM, Guillazo G, Artigas F (2001) Control of dorsal raphe serotonergic neurons by the medial prefrontal cortex: involvement of serotonin-1A, GABA(A), and glutamate receptors. J Neurosci 21:9917–9929
Centonze D, Battista N, Rossi S, Mercuri NB, Finazzi-Agro A, Bernardi G, Calabresi P, Maccarrone M (2004) A critical interaction between dopamine D2 receptors and endocannabinoids mediates the effects of cocaine on striatal gabaergic transmission. Neuropsychopharmacology 29:1488–1497
Centonze D, Rossi S, De Chiara V, Prosperetti C, Battista N, Bernardi G, Mercuri NB, Usiello A, Maccarrone M (2007) Chronic cocaine sensitizes striatal GABAergic synapses to the stimulation of cannabinoid CB1 receptors. Eur J Neurosci 25:1631–1640
Cheer JF, Wassum KM, Heien ML, Phillips PE, Wightman RM (2004) Cannabinoids enhance subsecond dopamine release in the nucleus accumbens of awake rats. J Neurosci 24:4393–4400
Chen J, Paredes W, Lowinson JH, Gardner EL (1990a) Delta 9-tetrahydrocannabinol enhances presynaptic dopamine efflux in medial prefrontal cortex. Eur J Pharmacol 190:259–262
Chen JP, Paredes W, Li J, Smith D, Lowinson J, Gardner EL (1990b) Delta 9-tetrahydrocannabinol produces naloxone-blockable enhancement of presynaptic basal dopamine efflux in nucleus accumbens of conscious, freely-moving rats as measured by intracerebral microdialysis. Psychopharmacology (Berl) 102:156–162
Chen CY, Bonham AC, Dean C, Hopp FA, Hillard CJ, Seagard JL (2010) Retrograde release of endocannabinoids inhibits presynaptic GABA release to second-order baroreceptive neurons in NTS. Auton Neurosci 158:44–50
Chiu CQ, Puente N, Grandes P, Castillo PE (2010) Dopaminergic modulation of endocannabinoid-mediated plasticity at GABAergic synapses in the prefrontal cortex. J Neurosci 30:7236–7248
Connor TJ, Kelliher P, Shen Y, Harkin A, Kelly JP, Leonard BE (2000) Effect of subchronic antidepressant treatments on behavioral, neurochemical, and endocrine changes in the forced-swim test. Pharmacol Biochem Behav 65:591–597
Cota D, Steiner MA, Marsicano G, Cervino C, Herman JP, Grubler Y, Stalla J, Pasquali R, Lutz B, Stalla GK, Pagotto U (2007) Requirement of cannabinoid receptor type 1 for the basal modulation of hypothalamic-pituitary-adrenal axis function. Endocrinology 148:1574–1581
Cravatt BF, Giang DK, Mayfield SP, Boger DL, Lerner RA, Gilula NB (1996) Molecular characterization of an enzyme that degrades neuromodulatory fatty-acid amides. Nature 384:83–87
Cryan JF, Valentino RJ, Lucki I (2005) Assessing substrates underlying the behavioral effects of antidepressants using the modified rat forced swimming test. Neurosci Biobehav Rev 29:547–569
Dahlstrom A, Fuxe K (1964) Localization of monoamines in the lower brain stem. Experientia 20:398–399
Danovitch I, Gorelick DA (2012) State of the art treatments for cannabis dependence. Psychiatr Clin North Am 35:309–326

Darmani NA, Janoyan JJ, Kumar N, Crim JL (2003) Behaviorally active doses of the CB1 receptor antagonist SR 141716A increase brain serotonin and dopamine levels and turnover. Pharmacol Biochem Behav 75:777–787

De Chiara V, Angelucci F, Rossi S, Musella A, Cavasinni F, Cantarella C, Mataluni G, Sacchetti L, Napolitano F, Castelli M, Caltagirone C, Bernardi G, Maccarrone M, Usiello A, Centonze D (2010) Brain-derived neurotrophic factor controls cannabinoid CB1 receptor function in the striatum. J Neurosci 30:8127–8137

Dean C (2011) Endocannabinoid modulation of sympathetic and cardiovascular responses to acute stress in the periaqueductal gray of the rat. Am J Physiol Regul Integr Comp Physiol 300:R771–779

Delfs JM, Zhu Y, Druhan JP, Aston-Jones GS (1998) Origin of noradrenergic afferents to the shell subregion of the nucleus accumbens: anterograde and retrograde tract-tracing studies in the rat. Brain Res 806:127–140

Derbenev AV, Stuart TC, Smith BN (2004) Cannabinoids suppress synaptic input to neurones of the rat dorsal motor nucleus of the vagus nerve. J Physiol 559:923–938

Detke MJ, Rickels M, Lucki I (1995) Active behaviors in the rat forced swimming test differentially produced by serotonergic and noradrenergic antidepressants. Psychopharmacology 121:66–72

Deutch AY, Lee MC, Gillham MH, Cameron DA, Goldstein M, Iadarola MJ (1991) Stress selectively increases fos protein in dopamine neurons innervating the prefrontal cortex. Cereb Cortex 1:273–292

Devane WA, Hanus L, Breuer A, Pertwee RG, Stevenson LA, Griffin G, Gibson D, Mandelbaum A, Etinger A, Mechoulam R (1992) Isolation and structure of a brain constituent that binds to the cannabinoid receptor. Science 258:1946–1949

Devlin MG, Christopoulos A (2002) Modulation of cannabinoid agonist binding by 5-HT in the rat cerebellum. J Neurochem 80:1095–1102

Di S, Malcher-Lopes R, Halmos KC, Tasker JG (2003) Nongenomic glucocorticoid inhibition via endocannabinoid release in the hypothalamus: a fast feedback mechanism. J Neurosci 23:4850–4857

Di S, Malcher-Lopes R, Marcheselli VL, Bazan NG, Tasker JG (2005) Rapid glucocorticoid-mediated endocannabinoid release and opposing regulation of glutamate and GABA inputs to hypothalamic magnocellular neurons. Endocrinology 146:4292–4301

Diana M, Melis M, Gessa GL (1998) Increase in meso-prefrontal dopaminergic activity after stimulation of CB1 receptors by cannabinoids. Eur J Neurosci 10:2825–2830

Diorio D, Viau V, Meaney MJ (1993) The role of the medial prefrontal cortex (cingulate gyrus) in the regulation of hypothalamic-pituitary-adrenal responses to stress. J Neurosci 13:3839–3847

Dong HW, Swanson LW (2006) Projections from bed nuclei of the stria terminalis, dorsomedial nucleus: implications for cerebral hemisphere integration of neuroendocrine, autonomic, and drinking responses. J Comp Neurol 494:75–107

Dubreucq S, Durand A, Matias I, Benard G, Richard E, Soria-Gomez E, Glangetas C, Groc L, Wadleigh A, Massa F, Bartsch D, Marsicano G, Georges F, Chaouloff F (2012) Ventral tegmental area cannabinoid type-1 receptors control voluntary exercise performance. Biol Psychiatry 73(9):895–903

Duncan GE, Knapp DJ, Johnson KB, Breese GR (1996) Functional classification of antidepressants based on antagonism of swim stress-induced fos-like immunoreactivity. J Pharmacol Exp Ther 277:1076–1089

Egashira N, Mishima K, Katsurabayashi S, Yoshitake T, Matsumoto Y, Ishida J, Yamaguchi M, Iwasaki K, Fujiwara M (2002) Involvement of 5-hydroxytryptamine neuronal system in Delta(9)-tetrahydrocannabinol-induced impairment of spatial memory. Eur J Pharmacol 445:221–229

Egashira N, Matsuda T, Koushi E, Higashihara F, Mishima K, Chidori S, Hasebe N, Iwasaki K, Nishimura R, Oishi R, Fujiwara M (2008) Delta(9)-tetrahydrocannabinol prolongs the immobility time in the mouse forced swim test: involvement of cannabinoid CB(1) receptor and serotonergic system. Eur J Pharmacol 589:117–121

Eisenstein SA, Clapper JR, Holmes PV, Piomelli D, Hohmann AG (2010) A role for 2-arachidonoylglycerol and endocannabinoid signaling in the locomotor response to novelty induced by olfactory bulbectomy. Pharmacol Res 61:419–429

El Khoury MA, Gorgievski V, Moutsimilli L, Giros B, Tzavara ET (2012) Interactions between the cannabinoid and dopaminergic systems: evidence from animal studies. Prog Neuropsychopharmacol Biol Psychiatry 38:36–50

Erb S, Hitchcott PK, Rajabi H, Mueller D, Shaham Y, Stewart J (2000) Alpha-2 adrenergic receptor agonists block stress-induced reinstatement of cocaine seeking. Neuropsychopharmacology 23:138–150

Esteban S, Moranta D, Sastre-Coll A, Miralles A, Garcia-Sevilla JA (2002) Withdrawal from chronic ethanol increases the sensitivity of presynaptic 5-HT(1A) receptors modulating serotonin and dopamine synthesis in rat brain in vivo. Neurosci Lett 326:121–124

Evanson NK, Tasker JG, Hill MN, Hillard CJ, Herman JP (2010) Fast feedback inhibition of the HPA axis by glucocorticoids is mediated by endocannabinoid signaling. Endocrinology 151:4811–4819

Fallon JH, Koziell DA, Moore RY (1978) Catecholamine innervation of the basal forebrain. II. Amygdala, suprarhinal cortex and entorhinal cortex. J Comp Neurol 180:509–532

Ferre S, Goldberg SR, Lluis C, Franco R (2009) Looking for the role of cannabinoid receptor heteromers in striatal function. Neuropharmacology 56(Suppl 1):226–234

Fitzgerald ML, Shobin E, Pickel VM (2012) Cannabinoid modulation of the dopaminergic circuitry: implications for limbic and striatal output. Prog Neuropsychopharmacol Biol Psychiatry 38:21–29

Frazer A (1997) Pharmacology of antidepressants. J Clin Psychopharmacol 17(Suppl 1):2S–18S

French ED (1997) Delta9-Tetrahydrocannabinol excites rat VTA dopamine neurons through activation of cannabinoid CB1 but not opioid receptors. Neurosci Lett 226:159–162

French ED, Dillon K, Wu X (1997) Cannabinoids excite dopamine neurons in the ventral tegmentum and substantia nigra. Neuroreport 8:649–652

Freund TF, Katona I, Piomelli D (2003) Role of endogenous cannabinoids in synaptic signaling. Physiol Rev 83:1017–1066

Gardiner SM, March JE, Kemp PA, Bennett T (2005) Involvement of CB1-receptors and beta-adrenoceptors in the regional hemodynamic responses to lipopolysaccharide infusion in conscious rats. Am J Physiol Heart Circ Physiol 288:H2280–2288

Gellman RL, Aghajanian GK (1993) Pyramidal cells in piriform cortex receive a convergence of inputs from monoamine activated GABAergic interneurons. Brain Res 600:63–73

Gessa GL, Melis M, Muntoni AL, Diana M (1998) Cannabinoids activate mesolimbic dopamine neurons by an action on cannabinoid CB1 receptors. Eur J Pharmacol 341:39–44

Giuffrida A, Parsons LH, Kerr TM, Rodriguez de Fonseca F, Navarro M, Piomelli D (1999) Dopamine activation of endogenous cannabinoid signaling in dorsal striatum. Nat Neurosci 2:358–363

Glass M, Felder CC (1997) Concurrent stimulation of cannabinoid CB1 and dopamine D2 receptors augments cAMP accumulation in striatal neurons: evidence for a Gs linkage to the CB1 receptor. J Neurosci 17:5327–5333

Gobbi G, Bambico FR, Mangieri R, Bortolato M, Campolongo P, Solinas M, Cassano T, Morgese MG, Debonnel G, Duranti A, Tontini A, Tarzia G, Mor M, Trezza V, Goldberg SR, Cuomo V, Piomelli D (2005) Antidepressant-like activity and modulation of brain monoaminergic transmission by blockade of anandamide hydrolysis. Proc Natl Acad Sci U S A 102:18620–18625

Gobshtis N, Ben-Shabat S, Fride E (2007) Antidepressant-induced undesirable weight gain: prevention with rimonabant without interference with behavioral effectiveness. Eur J Pharmacol 554:155–163

Greenwood BN, Foley TE, Day HE, Campisi J, Hammack SH, Campeau S, Maier SF, Fleshner M (2003) Freewheel running prevents learned helplessness/behavioral depression: role of dorsal raphe serotonergic neurons. J Neurosci 23:2889–2898

Griebel G, Rodgers RJ, Perrault G, Sanger DJ (1997) Risk assessment behaviour: evaluation of utility in the study of 5-HT-related drugs in the rat elevated plus-maze test. Pharmacol Biochem Behav 57:817–827

Griebel G, Stemmelin J, Scatton B (2005) Effects of the cannabinoid CB1 receptor antagonist rimonabant in models of emotional reactivity in rodents. Biol Psychiatry 57:261–267

Grueter BA, Gosnell HB, Olsen CM, Schramm-Sapyta NL, Nekrasova T, Landreth GE, Winder DG (2006) Extracellular-signal regulated kinase 1-dependent metabotropic glutamate receptor 5-induced long-term depression in the bed nucleus of the stria terminalis is disrupted by cocaine administration. J Neurosci 26:3210–3219

Haj-Dahmane S, Shen RY (2005) The wake-promoting peptide orexin-B inhibits glutamatergic transmission to dorsal raphe nucleus serotonin neurons through retrograde endocannabinoid signaling. J Neurosci 25:896–905

Haj-Dahmane S, Shen RY (2009) Endocannabinoids suppress excitatory synaptic transmission to dorsal raphe serotonin neurons through the activation of presynaptic CB1 receptors. J Pharmacol Exp Ther 331:186–196

Haj-Dahmane S, Shen RY (2011) Modulation of the serotonin system by endocannabinoid signaling. Neuropharmacology 61:414–420

Haring M, Marsicano G, Lutz B, Monory K (2007) Identification of the cannabinoid receptor type 1 in serotonergic cells of raphe nuclei in mice. Neuroscience 146:1212–1219

Haring M, Grieb M, Monory K, Lutz B, Moreira FA (2013) Cannabinoid CB(1) receptor in the modulation of stress coping behavior in mice: the role of serotonin and different forebrain neuronal subpopulations. Neuropharmacology 65:83–89

Haroon E, Raison CL, Miller AH (2012) Psychoneuroimmunology meets neuropsychopharmacology: translational implications of the impact of inflammation on behavior. Neuropsychopharmacology 37:137–162

Hatfield T, Spanis C, McGaugh JL (1999) Response of amygdalar norepinephrine to footshock and GABAergic drugs using in vivo microdialysis and HPLC. Brain Res 835:340–345

Hayase T, Yamamoto Y, Yamamoto K (2005) Persistent anxiogenic effects of a single or repeated doses of cocaine and methamphetamine: interactions with endogenous cannabinoid receptor ligands. Behav Pharmacol 16:395–404

Herkenham M (1992) Cannabinoid receptor localization in brain: relationship to motor and reward systems. Ann N Y Acad Sci 654:19–32

Herkenham M, Groen BG, Lynn AB, De Costa BR, Richfield EK (1991a) Neuronal localization of cannabinoid receptors and second messengers in mutant mouse cerebellum. Brain Res 552:301–310

Herkenham M, Lynn AB, Johnson MR, Melvin LS, Costa BR de, Rice KC (1991b) Characterization and localization of cannabinoid receptors in rat brain: a quantitative in vitro autoradiographic study. J Neurosci 11:563–583

Hermann H, Lutz B (2005) Coexpression of the cannabinoid receptor type 1 with the corticotropin-releasing hormone receptor type 1 in distinct regions of the adult mouse forebrain. Neurosci Lett 375:13–18

Hermann H, Marsicano G, Lutz B (2002) Coexpression of the cannabinoid receptor type 1 with dopamine and serotonin receptors in distinct neuronal subpopulations of the adult mouse forebrain. Neuroscience 109:451–460

Hesketh SA, Brennan AK, Jessop DS, Finn DP (2008) Effects of chronic treatment with citalopram on cannabinoid and opioid receptor-mediated G-protein coupling in discrete rat brain regions. Psychopharmacology (Berl) 198:29–36

Hill MN, Gorzalka BB (2005) Pharmacological enhancement of cannabinoid CB(1) receptor activity elicits an antidepressant-like response in the rat forced swim test. Eur Neuropsychopharmacol 15:593–599

Hill MN, Gorzalka BB (2009) The endocannabinoid system and the treatment of mood and anxiety disorders. CNS Neurol Disord Drug Targets 8:451–458

Hill MN, McEwen BS (2009) Endocannabinoids: the silent partner of glucocorticoids in the synapse. Proc Natl Acad Sci U S A 106:4579–4580

Hill MN, Patel S, Carrier EJ, Rademacher DJ, Ormerod BK, Hillard CJ, Gorzalka BB (2005) Downregulation of endocannabinoid signaling in the hippocampus following chronic unpredictable stress. Neuropsychopharmacology 30:508–515

Hill MN, Ho WS, Sinopoli KJ, Viau V, Hillard CJ, Gorzalka BB (2006) Involvement of the endocannabinoid system in the ability of long-term tricyclic antidepressant treatment to suppress stress-induced activation of the hypothalamic-pituitary-adrenal axis. Neuropsychopharmacology 31:2591–2599

Hill MN, Carrier EJ, McLaughlin RJ, Morrish AC, Meier SE, Hillard CJ, Gorzalka BB (2008a) Regional alterations in the endocannabinoid system in an animal model of depression: effects of concurrent antidepressant treatment. J Neurochem 106:2322–2336

Hill MN, Ho WS, Hillard CJ, Gorzalka BB (2008b) Differential effects of the antidepressants tranylcypromine and fluoxetine on limbic cannabinoid receptor binding and endocannabinoid contents. J Neural Transm 115:1673–1679

Hill MN, Hillard CJ, Bambico FR, Patel S, Gorzalka BB, Gobbi G (2009a) The therapeutic potential of the endocannabinoid system for the development of a novel class of antidepressants. Trends Pharmacol Sci 30:484–493

Hill MN, McLaughlin RJ, Morrish AC, Viau V, Floresco SB, Hillard CJ, Gorzalka BB (2009b) Suppression of amygdalar endocannabinoid signaling by stress contributes to activation of the hypothalamic-pituitary-adrenal axis. Neuropsychopharmacology 34:2733–2745

Hill MN, Karatsoreos IN, Hillard CJ, McEwen BS (2010a) Rapid elevations in limbic endocannabinoid content by glucocorticoid hormones in vivo. Psychoneuroendocrinology 35:1333–1338

Hill MN, McLaughlin RJ, Bingham B, Shrestha L, Lee TT, Gray JM, Hillard CJ, Gorzalka BB, Viau V (2010b) Endogenous cannabinoid signaling is essential for stress adaptation. Proc Natl Acad Sci U S A 107:9406–9411

Hill MN, McLaughlin RJ, Pan B, Fitzgerald ML, Roberts CJ, Lee TT, Karatsoreos IN, Mackie K, Viau V, Pickel VM, McEwen BS, Liu QS, Gorzalka BB, Hillard CJ (2011a) Recruitment of prefrontal cortical endocannabinoid signaling by glucocorticoids contributes to termination of the stress response. J Neurosci 31:10506–10515

Hill MN, McLaughlin RJ, Pan B, Fitzgerald ML, Roberts CJ, Lee TT, Karatsoreos IN, Mackie K, Viau V, Pickel VM, McEwen BS, Liu QS, Gorzalka BB, Hillard CJ (2011b) Recruitment of prefrontal cortical endocannabinoid signaling by glucocorticoids contributes to termination of the stress response. J Neurosci 31:10506–10515

Hillard CJ, Bloom AS (1982) Delta 9-Tetrahydrocannabinol-induced changes in beta-adrenergic receptor binding in mouse cerebral cortex. Brain Res 235:370–377

Hillard CJ, Weinlander KM, Stuhr KL (2012) Contributions of endocannabinoid signaling to psychiatric disorders in humans: genetic and biochemical evidence. Neuroscience 204:207–229

Ho WV, Hill MN, Miller GE, Gorzalka BB, Hillard CJ (2012) Serum contents of endocannabinoids are correlated with blood pressure in depressed women. Lipids Health Dis 11:32

Hoffman AF, Lupica CR (2001) Direct actions of cannabinoids on synaptic transmission in the nucleus accumbens: a comparison with opioids. J Neurophysiol 85:72–83

Hoffman AF, Oz M, Caulder T, Lupica CR (2003) Functional tolerance and blockade of long-term depression at synapses in the nucleus accumbens after chronic cannabinoid exposure. J Neurosci 23:4815–4820

Hohmann AG, Suplita RL, Bolton NM, Neely MH, Fegley D, Mangieri R, Krey JF, Walker JM, Holmes PV, Crystal JD, Duranti A, Tontini A, Mor M, Tarzia G, Piomelli D (2005) An endocannabinoid mechanism for stress-induced analgesia. Nature 435:1108–1112

Houchi H, Babovic D, Pierrefiche O, Ledent C, Daoust M, Naassila M (2005) CB1 receptor knockout mice display reduced ethanol-induced conditioned place preference and increased striatal dopamine D2 receptors. Neuropsychopharmacology 30:339–349

Huang YC, Wang SJ, Chiou LC, Gean PW (2003) Mediation of amphetamine-induced long-term depression of synaptic transmission by CB1 cannabinoid receptors in the rat amygdala. J Neurosci 23:10311–10320

Hungund BL, Vinod KY, Kassir SA, Basavarajappa BS, Yalamanchili R, Cooper TB, Mann JJ, Arango V (2004) Upregulation of CB1 receptors and agonist-stimulated [35S]GTPgammaS binding in the prefrontal cortex of depressed suicide victims. Mol Psychiatry 9:184–190

Insel PA, Hammond HK (1993) Beta-adrenergic receptors in heart failure. J Clin Invest 92:2564

Ishac EJ, Jiang L, Lake KD, Varga K, Abood ME, Kunos G (1996) Inhibition of exocytotic noradrenaline release by presynaptic cannabinoid CB1 receptors on peripheral sympathetic nerves. Br J Pharmacol 118:2023–2028

Jarrahian A, Watts VJ, Barker EL (2004) D2 dopamine receptors modulate Galpha-subunit coupling of the CB1 cannabinoid receptor. J Pharmacol Exp Ther 308:880–886

Jentsch JD, Andrusiak E, Tran A, Bowers MB Jr, Roth RH (1997) Delta 9-tetrahydrocannabinol increases prefrontal cortical catecholaminergic utilization and impairs spatial working memory in the rat: blockade of dopaminergic effects with HA966. Neuropsychopharmacology 16:426–432

Julian MD, Martin AB, Cuellar B, Rodriguez De Fonseca F, Navarro M, Moratalla R, Garcia-Segura LM (2003) Neuroanatomical relationship between type 1 cannabinoid receptors and dopaminergic systems in the rat basal ganglia. Neuroscience 119:309–318

Kamprath K, Romo-Parra H, Haring M, Gaburro S, Doengi M, Lutz B, Pape HC (2011) Short-term adaptation of conditioned fear responses through endocannabinoid signaling in the central amygdala. Neuropsychopharmacology 36:652–663

Kano M, Ohno-Shosaku T, Hashimotodani Y, Uchigashima M, Watanabe M (2009) Endocannabinoid-mediated control of synaptic transmission. Physiol Rev 89.309–380

Kearn CS, Greenberg MJ, DiCamelli R, Kurzawa K, Hillard CJ (1999) Relationships between ligand affinities for the cerebellar cannabinoid receptor CB1 and the induction of GDP/GTP exchange. J Neurochem 72:2379–2387

Kearn CS, Blake-Palmer K, Daniel E, Mackie K, Glass M (2005) Concurrent stimulation of cannabinoid CB1 and dopamine D2 receptors enhances heterodimer formation: a mechanism for receptor cross-talk? Mol Pharmacol 67:1697–1704

Kim J, Isokawa M, Ledent C, Alger BE (2002) Activation of muscarinic acetylcholine receptors enhances the release of endogenous cannabinoids in the hippocampus. J Neurosci 22:10182–10191

Kirilly E, Hunyady L, Bagdy G (2012) Opposing local effects of endocannabinoids on the activity of noradrenergic neurons and release of noradrenaline: relevance for their role in depression and in the actions of CB(1) receptor antagonists. J Neural Transm 120(1):177–186

Klaassen T, Riedel WJ, Praag HM van, Menheere PP, Griez E (2002) Neuroendocrine response to meta-chlorophenylpiperazine and ipsapirone in relation to anxiety and aggression. Psychiatry Res 113:29–40

Kofalvi A, Rodrigues RJ, Ledent C, Mackie K, Vizi ES, Cunha RA, Sperlagh B (2005) Involvement of cannabinoid receptors in the regulation of neurotransmitter release in the rodent striatum: a combined immunochemical and pharmacological analysis. J Neurosci 25:2874–2884

Koob G, Kreek MJ (2007) Stress, dysregulation of drug reward pathways, and the transition to drug dependence. Am J Psychiatry 164:1149–1159

Koob GF, Le Moal M (2008) Review. Neurobiological mechanisms for opponent motivational processes in addiction. Philos Trans R Soc Lond B Biol Sci 363:3113–3123

Kupferschmidt DA, Newman AE, Boonstra R, Erb S (2012) Antagonism of cannabinoid 1 receptors reverses the anxiety-like behavior induced by central injections of corticotropin-releasing factor and cocaine withdrawal. Neuroscience 204:125–133

Lanfumey L, Mongeau R, Cohen-Salmon C, Hamon M (2008) Corticosteroid-serotonin interactions in the neurobiological mechanisms of stress-related disorders. Neurosci Biobehav Rev 32:1174–1184

Lau T, Schloss P (2008) The cannabinoid CB(1) receptor is expressed on serotonergic and dopaminergic neurons. Eur J Pharmacol 578:137–141

Laviolette SR, Grace AA (2006) The roles of cannabinoid and dopamine receptor systems in neural emotional learning circuits: implications for schizophrenia and addiction. Cell Mol Life Sci 63:1597–1613

LeDoux JE (2000) Emotion circuits in the brain. Annu Rev Neurosci 23:155–184
Leri F, Flores J, Rodaros D, Stewart J (2002) Blockade of stress-induced but not cocaine-induced reinstatement by infusion of noradrenergic antagonists into the bed nucleus of the stria terminalis or the central nucleus of the amygdala. J Neurosci 22:5713–5718
Lodge DJ, Grace AA (2011) Developmental pathology, dopamine, stress and schizophrenia. Int J Dev Neurosci 29:207–213
Lovinger DM, Mathur BN (2012) Endocannabinoids in striatal plasticity. Parkinsonism Relat Disord 18(Suppl 1):S132–S134
Lowry CA (2002) Functional subsets of serotonergic neurones: implications for control of the hypothalamic-pituitary-adrenal axis. J Neuroendocrinol 14:911–923
Mackie K, Hille B (1992) Cannabinoids inhibit N-type calcium channels in neuroblastoma-glioma cells. Proc Natl Acad Sci 89:3825–3829
Mailleux P, Vanderhaeghen J-J (1992) Distribution of neuronal cannabinoid receptor in the adult rat brain: a comparative receptor binding autoradiography and in situ hybridization histochemistry. Neuroscience 48:655–668
Mallet PE, Beninger RJ (1998) Delta9-tetrahydrocannabinol, but not the endogenous cannabinoid receptor ligand anandamide, produces conditioned place avoidance. Life Sci 62:2431–2439
Malone DT, Taylor DA (1998) Modulation of delta9-tetrahydrocannabinol-induced hypothermia by fluoxetine in the rat. Br J Pharmacol 124:1419–1424
Malone DT, Taylor DA (2001) Involvement of somatodendritic 5-HT(1A) receptors in Delta(9)-tetrahydrocannabinol-induced hypothermia in the rat. Pharmacol Biochem Behav 69:595–601
Mantsch JR, Weyer A, Vranjkovic O, Beyer CE, Baker DA, Caretta H (2010) Involvement of noradrenergic neurotransmission in the stress- but not cocaine-induced reinstatement of extinguished cocaine-induced conditioned place preference in mice: role for beta-2 adrenergic receptors. Neuropsychopharmacology 35:2165–2178
Marco EM, Perez-Alvarez L, Borcel E, Rubio M, Guaza C, Ambrosio E, File SE, Viveros MP (2004) Involvement of 5-HT1A receptors in behavioural effects of the cannabinoid receptor agonist CP 55,940 in male rats. Behav Pharmacol 15:21–27
Marsicano G, Lutz B (1999) Expression of the cannabinoid receptor CB1 in distinct neuronal subpopulations in the adult mouse forebrain. Eur J Neurosci 11:4213–4225
Martin AB, Fernandez-Espejo E, Ferrer B, Gorriti MA, Bilbao A, Navarro M, Rodriguez de Fonseca F, Moratalla R (2008) Expression and function of CB(1) receptor in the rat striatum: localization and effects on D(1) and D(2) dopamine receptor-mediated motor behaviors. Neuropsychopharmacology 33:1667–1679
Mato S, Aso E, Castro E, Martin M, Valverde O, Maldonado R, Pazos A (2007) CB1 knockout mice display impaired functionality of 5-HT1A and 5-HT2A/C receptors. J Neurochem 103:2111–2120
Mato S, Vidal R, Castro E, Diaz A, Pazos A, Valdizan EM (2010) Long-term fluoxetine treatment modulates cannabinoid type 1 receptor-mediated inhibition of adenylyl cyclase in the rat prefrontal cortex through 5-hydroxytryptamine 1A receptor-dependent mechanisms. Mol Pharmacol 77:424–434
Matsuda LA, Lolait SJ, Brownstein MJ, Young AC, Bonner TI (1990) Structure of a cannabinoid receptor and functional expression of the cloned cDNA. Nature 346:561–564
Matsuda LA, Bonner TI, Lolait SJ (1993) Localization of cannabinoid receptor mRNA in rat brain. J Comp Neurol 327:535–550
Matyas F, Urban GM, Watanabe M, Mackie K, Zimmer A, Freund TF, Katona I (2008) Identification of the sites of 2-arachidonoylglycerol synthesis and action imply retrograde endocannabinoid signaling at both GABAergic and glutamatergic synapses in the ventral tegmental area. Neuropharmacology 54:95–107
McElligott ZA, Winder DG (2009) Modulation of glutamatergic synaptic transmission in the bed nucleus of the stria terminalis. Prog Neuropsychopharmacol Biol Psychiatry 33:1329–1335
McEwen BS (2008) Central effects of stress hormones in health and disease: understanding the protective and damaging effects of stress and stress mediators. Eur J Pharmacol 583:174–185

McGregor IS, Issakidis CN, Prior G (1996) Aversive effects of the synthetic cannabinoid CP 55,940 in rats. Pharmacol Biochem Behav 53:657–664

McLaughlin RJ, Hill MN, Morrish AC, Gorzalka BB (2007) Local enhancement of cannabinoid CB1 receptor signalling in the dorsal hippocampus elicits an antidepressant-like effect. Behav Pharmacol 18:431–438

McLaughlin RJ, Hill MN, Gorzalka BB (2009) Monoaminergic neurotransmission contributes to cannabinoid-induced activation of the hypothalamic-pituitary-adrenal axis. Eur J Pharmacol 624(1–3):71–76

McLaughlin RJ, Hill MN, Bambico FR, Stuhr KL, Gobbi G, Hillard CJ, Gorzalka BB (2012) Prefrontal cortical anandamide signaling coordinates coping responses to stress through a serotonergic pathway. Eur Neuropsychopharmacol 22:664–671

Mechoulam R, Ben-Shabat S, Hanus L, Ligumsky M, Kaminski NE, Schatz AR, Gopher A, Almog S, Martin BR, Compton DR et al (1995) Identification of an endogenous 2-monoglyceride, present in canine gut, that binds to cannabinoid receptors. Biochem Pharmacol 50:83–90

Melis M, Perra S, Muntoni AL, Pillolla G, Lutz B, Marsicano G, Di Marzo V, Gessa GL, Pistis M (2004a) Prefrontal cortex stimulation induces 2-arachidonoyl-glycerol-mediated suppression of excitation in dopamine neurons. J Neurosci 24:10707–10715

Melis M, Pistis M, Perra S, Muntoni AL, Pillolla G, Gessa GL (2004b) Endocannabinoids mediate presynaptic inhibition of glutamatergic transmission in rat ventral tegmental area dopamine neurons through activation of CB1 receptors. J Neurosci 24:53–62

Mendiguren A, Pineda J (2004) Cannabinoids enhance N methyl-d-aspartate-induced excitation of locus coeruleus neurons by CB(1) receptors in rat brain slices. Neurosci Lett 363:1–5

Mendiguren A, Pineda J (2006) Systemic effect of cannabinoids on the spontaneous firing rate of locus coeruleus neurons in rats. Eur J Pharmacol 534:83–88

Mendiguren A, Pineda J (2009) Effect of the CB(1) receptor antagonists rimonabant and AM251 on the firing rate of dorsal raphe nucleus neurons in rat brain slices. Br J Pharmacol 158:1579–1587

Merroun I, Errami M, Hoddah H, Urbano G, Porres JM, Aranda P, Llopis J, Lopez-Jurado M (2009) Influence of intracerebroventricular or intraperitoneal administration of cannabinoid receptor agonist (WIN 55,212-2) and inverse agonist (AM 251) on the regulation of food intake and hypothalamic serotonin levels. Br J Nutr 101:1569–1578

Moranta D, Esteban S, Garcia-Sevilla JA (2004) Differential effects of acute cannabinoid drug treatment, mediated by CB(1) receptors, on the in vivo activity of tyrosine and tryptophan hydroxylase in the rat brain. Naunyn Schmiedebergs Arch Pharmacol 369:516–524

Moranta D, Esteban S, Garcia-Sevilla JA (2009) Chronic treatment and withdrawal of the cannabinoid agonist WIN 55,212-2 modulate the sensitivity of presynaptic receptors involved in the regulation of monoamine syntheses in rat brain. Naunyn Schmiedebergs Arch Pharmacol 379:61–72

Morrish AC, Hill MN, Riebe CJ, Gorzalka BB (2009) Protracted cannabinoid administration elicits antidepressant behavioral responses in rats: role of gender and noradrenergic transmission. Physiol Behav 98:118–124

Munro S, Thomas KL, Abu-Shaar M (1993) Molecular characterization of a peripheral receptor for cannabinoids. Nature 365:61–65

Muntoni AL, Pillolla G, Melis M, Perra S, Gessa GL, Pistis M (2006) Cannabinoids modulate spontaneous neuronal activity and evoked inhibition of locus coeruleus noradrenergic neurons. Eur J Neurosci 23:2385–2394

Murphy LL, Munoz RM, Adrian BA, Villanua MA (1998) Function of cannabinoid receptors in the neuroendocrine regulation of hormone secretion. Neurobiol Dis 5:432–446

Nakazi M, Bauer U, Nickel T, Kathmann M, Schlicker E (2000) Inhibition of serotonin release in the mouse brain via presynaptic cannabinoid CB1 receptors. Naunyn Schmiedebergs Arch Pharmacol 361:19–24

Nathan PJ, O'Neill BV, Napolitano A, Bullmore ET (2011) Neuropsychiatric adverse effects of centrally acting antiobesity drugs. CNS Neurosci Ther 17:490–505

Novak CM, Burghardt PR, Levine JA (2012) The use of a running wheel to measure activity in rodents: relationship to energy balance, general activity, and reward. Neurosci Biobehav Rev 36:1001–1014

Oliva JM, Uriguen L, Perez-Rial S, Manzanares J (2005) Time course of opioid and cannabinoid gene transcription alterations induced by repeated administration with fluoxetine in the rat brain. Neuropharmacology 49:618–626

Onaivi ES (2008) An endocannabinoid hypothesis of drug reward and drug addiction. Ann N Y Acad Sci 1139:412–421

Oropeza VC, Page ME, Van Bockstaele EJ (2005) Systemic administration of WIN 55,212-2 increases norepinephrine release in the rat frontal cortex. Brain Res 1046:45–54

Oropeza VC, Mackie K, Van Bockstaele EJ (2007) Cannabinoid receptors are localized to noradrenergic axon terminals in the rat frontal cortex. Brain Res 1127:36–44

Osei-Owusu P, James A, Crane J, Scrogin KE (2005) 5-Hydroxytryptamine 1A receptors in the paraventricular nucleus of the hypothalamus mediate oxytocin and adrenocorticotropin hormone release and some behavioral components of the serotonin syndrome. J Pharmacol Exp Ther 313:1324–1330

Page ME, Oropeza VC, Sparks SE, Qian Y, Menko AS, Van Bockstaele EJ (2007) Repeated cannabinoid administration increases indices of noradrenergic activity in rats. Pharmacol Biochem Behav 86:162–168

Page ME, Oropeza VC, Van Bockstaele EJ (2008) Local administration of a cannabinoid agonist alters norepinephrine efflux in the rat frontal cortex. Neurosci Lett 431:1–5

Pan B, Hillard CJ, Liu QS (2008a) D2 dopamine receptor activation facilitates endocannabinoid-mediated long-term synaptic depression of GABAergic synaptic transmission in midbrain dopamine neurons via cAMP-protein kinase A signaling. J Neurosci 28:14018–14030

Pan B, Hillard CJ, Liu QS (2008b) Endocannabinoid signaling mediates cocaine-induced inhibitory synaptic plasticity in midbrain dopamine neurons. J Neurosci 28:1385–1397

Pandolfo P, Vendruscolo LF, Sordi R, Takahashi RN (2009) Cannabinoid-induced conditioned place preference in the spontaneously hypertensive rat-an animal model of attention deficit hyperactivity disorder. Psychopharmacology (Berl) 205:319–326

Parolaro D, Realini N, Vigano D, Guidali C, Rubino T (2010) The endocannabinoid system and psychiatric disorders. Exp Neurol 224:3–14

Patel S, Hillard CJ (2003) Cannabinoid-induced Fos expression within A10 dopaminergic neurons. Brain Res 963:15–25

Patel S, Hillard CJ (2006) Pharmacological evaluation of cannabinoid receptor ligands in a mouse model of anxiety: further evidence for an anxiolytic role for endogenous cannabinoid signaling. J Pharmacol Exp Ther 318:304–311

Patel S, Hillard CJ (2008) Adaptations in endocannabinoid signaling in response to repeated homotypic stress: a novel mechanism for stress habituation. Eur J Neurosci 27:2821–2829

Patel S, Hillard CJ (2009a) Endocannabinoids as modulators of synaptic signaling. In: Reggio PH (ed) The Cannabinoid Receptors. Humana Press, New York, pp 281–308

Patel S, Hillard CJ (2009b) Role of endocannabinoid signaling in anxiety and depression. In: Alexander DKaS (ed) Behavioral Neurobiology of the Endocannabinoid System. Springer Verlag, Berlin, pp 347–372

Patel S, Rademacher DJ, Hillard CJ (2003) Differential regulation of the endocannabinoids anandamide and 2-arachidonylglycerol within the limbic forebrain by dopamine receptor activity. J Pharmacol Exp Ther 306:880–888

Patel S, Roelke CT, Rademacher DJ, Cullinan WE, Hillard CJ (2004) Endocannabinoid signaling negatively modulates stress-induced activation of the hypothalamic-pituitary-adrenal axis. Endocrinology 145:5431–5438

Patel S, Roelke CT, Rademacher DJ, Hillard CJ (2005) Inhibition of restraint stress-induced neural and behavioural activation by endogenous cannabinoid signalling. Eur J Neurosci 21:1057–1069

Patel S, Kingsley PJ, Mackie K, Marnett LJ, Winder DG (2009) Repeated homotypic stress elevates 2-arachidonoylglycerol levels and enhances short-term endocannabinoid signaling at inhibitory synapses in basolateral amygdala. Neuropsychopharmacology 34:2699–2709

Pattij T, Wiskerke J, Schoffelmeer AN (2008) Cannabinoid modulation of executive functions. Eur J Pharmacol 585:458–463

Pedroza-Llinas R, Mendez-Diaz M, Ruiz-Contreras AE, Prospero-Garcia O (2013) CB1 receptor activation in the nucleus accumbens core impairs contextual fear learning. Behav Brain Res 237:141–147

Pertwee RG, Howlett AC, Abood ME, Alexander SP, Di Marzo V, Elphick MR, Greasley PJ, Hansen HS, Kunos G, Mackie K, Mechoulam R, Ross RA (2010) International Union of Basic and Clinical Pharmacology. LXXIX. Cannabinoid receptors and their ligands: beyond CB_1 and CB_2. Pharmacol Rev 62:588–631

Pfitzer T, Niederhoffer N, Szabo B (2005) Search for an endogenous cannabinoid-mediated effect in the sympathetic nervous system. Naunyn Schmiedebergs Arch Pharmacol 371:9–17

Pickel VM, Chan J, Kash TL, Rodriguez JJ, MacKie K (2004) Compartment-specific localization of cannabinoid 1 (CB1) and mu-opioid receptors in rat nucleus accumbens. Neuroscience 127:101–112

Pickel VM, Chan J, Kearn CS, Mackie K (2006) Targeting dopamine D2 and cannabinoid-1 (CB1) receptors in rat nucleus accumbens. J Comp Neurol 495:299–313

Pistis M, Porcu G, Melis M, Diana M, Gessa GL (2001) Effects of cannabinoids on prefrontal neuronal responses to ventral tegmental area stimulation. Eur J Neurosci 14:96–102

Pistis M, Ferraro L, Pira L, Flore G, Tanganelli S, Gessa GL, Devoto P (2002a) Delta(9)-tetrahydrocannabinol decreases extracellular GABA and increases extracellular glutamate and dopamine levels in the rat prefrontal cortex: an in vivo microdialysis study. Brain Res 948:155–158

Pistis M, Muntoni AL, Pillolla G, Gessa GL (2002b) Cannabinoids inhibit excitatory inputs to neurons in the shell of the nucleus accumbens: an in vivo electrophysiological study. Eur J Neurosci 15:1795–1802

Poddar MK, Dewey WL (1980) Effects of cannabinoids on catecholamine uptake and release in hypothalamic and striatal synaptosomes. J Pharmacol Exp Ther 214:63–67

Racz I, Bilkei-Gorzo A, Toth ZE, Michel K, Palkovits M, Zimmer A (2003) A critical role for the cannabinoid CB1 receptors in alcohol dependence and stress-stimulated ethanol drinking. J Neurosci 23:2453–2458

Rademacher DJ, Meier SE, Shi L, Ho WS, Jarrahian A, Hillard CJ (2008) Effects of acute and repeated restraint stress on endocannabinoid content in the amygdala, ventral striatum, and medial prefrontal cortex in mice. Neuropharmacology 54:108–116

Radley JJ, Sawchenko PE (2011) A common substrate for prefrontal and hippocampal inhibition of the neuroendocrine stress response. J Neurosci 31:9683–9695

Refojo D, Schweizer M, Kuehne C, Ehrenberg S, Thoeringer C, Vogl AM, Dedic N, Schumacher M, Wolff G von, Avrabos C, Touma C, Engblom D, Schutz G, Nave KA, Eder M, Wotjak CT, Sillaber I, Holsboer F, Wurst W, Deussing JM (2011) Glutamatergic and dopaminergic neurons mediate anxiogenic and anxiolytic effects of CRHR1. Science 333:1903–1907

Reyes BA, Rosario JC, Piana PM, Van Bockstaele EJ (2009) Cannabinoid modulation of cortical adrenergic receptors and transporters. J Neurosci Res 87:3671–3678

Reyes BA, Szot P, Sikkema C, Cathel AM, Kirby LG, Van Bockstaele EJ (2012) Stress-induced sensitization of cortical adrenergic receptors following a history of cannabinoid exposure. Exp Neurol 236:327–335

Riegel AC, Lupica CR (2004) Independent presynaptic and postsynaptic mechanisms regulate endocannabinoid signaling at multiple synapses in the ventral tegmental area. J Neurosci 24:11070–11078

Robbe D, Alonso G, Duchamp F, Bockaert J, Manzoni OJ (2001) Localization and mechanisms of action of cannabinoid receptors at the glutamatergic synapses of the mouse nucleus accumbens. J Neurosci 21:109–116

Robbe D, Kopf M, Remaury A, Bockaert J, Manzoni OJ (2002) Endogenous cannabinoids mediate long-term synaptic depression in the nucleus accumbens. Proc Natl Acad Sci U S A 99:8384–8388

Rodriguez-Gaztelumendi A, Rojo ML, Pazos A, Diaz A (2009) Altered CB receptor-signaling in prefrontal cortex from an animal model of depression is reversed by chronic fluoxetine. J Neurochem 108:1423–1433

Rossi S, De Chiara V, Musella A, Kusayanagi H, Mataluni G, Bernardi G, Usiello A, Centonze D (2008) Chronic psychoemotional stress impairs cannabinoid-receptor-mediated control of GABA transmission in the striatum. J Neurosci 28:7284–7292

Rubio M, Fernandez-Ruiz J, Miguel R de, Maestro B, Walker JM, Ramos JA (2008) CB1 receptor blockade reduces the anxiogenic-like response and ameliorates the neurochemical imbalances associated with alcohol withdrawal in rats. Neuropharmacology 54:976–988

Sakamoto Y, Mashiko K, Obata T, Matsumoto H, Hara Y, Kutsukata N, Yamamoto Y (2008) Effectiveness of continuous hemodiafiltration using a polymethylmethacrylate membrane hemofilter after polymyxin B-immobilized fiber column therapy of septic shock. ASAIO J 54:129–132

Sanudo-Pena MC, Tsou K, Delay ER, Hohman AG, Force M, Walker JM (1997) Endogenous cannabinoids as an aversive or counter-rewarding system in the rat. Neurosci Lett 223:125–128

Sapolsky RM, Meaney MJ, McEwen BS (1985) The development of the glucocorticoid receptor system in the rat limbic brain. III. Negative-feedback regulation. Brain Res 350:169–173

Sastre-Coll A, Esteban S, Garcia-Sevilla JA (2002) Supersensitivity of 5-HT1A autoreceptors and alpha2-adrenoceptors regulating monoamine synthesis in the brain of morphine-dependent rats. Naunyn Schmiedebergs Arch Pharmacol 365:210–219

Scavone JL, Mackie K, Van Bockstaele EJ (2010) Characterization of cannabinoid-1 receptors in the locus coeruleus: relationship with mu-opioid receptors. Brain Res 1312:18–31

Schlicker E, Timm J, Zentner J, Gothert M (1997) Cannabinoid CB1 receptor-mediated inhibition of noradrenaline release in the human and guinea-pig hippocampus. Naunyn Schmiedebergs Arch Pharmacol 356:583–589

Schlosburg JE, Blankman JL, Long JZ, Nomura DK, Pan B, Kinsey SG, Nguyen PT, Ramesh D, Booker L, Burston JJ, Thomas EA, Selley DE, Sim-Selley LJ, Liu QS, Lichtman AH, Cravatt BF (2010) Chronic monoacylglycerol lipase blockade causes functional antagonism of the endocannabinoid system. Nat Neurosci 13:1113–1119

Schultz W (2002) Getting formal with dopamine and reward. Neuron 36:241–263

Sciolino NR, Zhou W, Hohmann AG (2011) Enhancement of endocannabinoid signaling with JZL184, an inhibitor of the 2-arachidonoylglycerol hydrolyzing enzyme monoacylglycerol lipase, produces anxiolytic effects under conditions of high environmental aversiveness in rats. Pharmacol Res 64:226–234

Seagard JL, Dean C, Patel S, Rademacher DJ, Hopp FA, Schmeling WT, Hillard CJ (2004) Anandamide content and interaction of endocannabinoid/GABA modulatory effects in the NTS on baroreflex-evoked sympathoinhibition. Am J Physiol Heart Circ Physiol 286:H992–H1000

Seif T, Makriyannis A, Kunos G, Bonci A, Hopf FW (2011) The endocannabinoid 2-arachidonoylglycerol mediates D1 and D2 receptor cooperative enhancement of rat nucleus accumbens core neuron firing. Neuroscience 193:21–33

Serrano A, Parsons LH (2011) Endocannabinoid influence in drug reinforcement, dependence and addiction-related behaviors. Pharmacol Ther 132:215–241

Sesack SR, Pickel VM (1990) In the rat medial nucleus accumbens, hippocampal and catecholaminergic terminals converge on spiny neurons and are in apposition to each other. Brain Res 527:266–279

Shearman LP, Rosko KM, Fleischer R, Wang J, Xu S, Tong XS, Rocha BA (2003) Antidepressant-like and anorectic effects of the cannabinoid CB1 receptor inverse agonist AM251 in mice. Behav Pharmacol 14:573–582

Shimizu T, Lu L, Yokotani K (2010) Possible inhibitory roles of endogenous 2-arachidonoylglycerol during corticotropin-releasing factor-induced activation of central sympatho-adrenomedullary outflow in anesthetized rats. Eur J Pharmacol 641:54–60

Sidhpura N, Parsons LH (2011) Endocannabinoid-mediated synaptic plasticity and addiction-related behavior. Neuropharmacology 61:1070–1087

Sim L, Hampson R, Deadwyler S, Childers S (1996) Effects of chronic treatment with delta-9-tetrahydrocannabinol in cannabinoid-stimulated [^{35}S]GTPgS autoradiography in rat brain. J Neurosci 16:8057–8066

Solowij N, Michie PT (2007) Cannabis and cognitive dysfunction: parallels with endophenotypes of schizophrenia? J Psychiatry Neurosci 32:30–52

Solowij N, Michie PT, Fox AM (1995) Differential impairments of selective attention due to frequency and duration of cannabis use. Biol Psychiatry 37:731–739

Steffens M, Engler C, Zentner J, Feuerstein TJ (2004) Cannabinoid CB1 receptor-mediated modulation of evoked dopamine release and of adenylyl cyclase activity in the human neocortex. Br J Pharmacol 141:1193–1203

Steiner MA, Wanisch K, Monory K, Marsicano G, Borroni E, Bachli H, Holsboer F, Lutz B, Wotjak CT (2008) Impaired cannabinoid receptor type 1 signaling interferes with stress-coping behavior in mice. Pharmacogenomics J 8:196–208

Stelt HM van der, Breuer ME, Olivier B, Westenberg HG (2005) Permanent deficits in serotonergic functioning of olfactory bulbectomized rats: an in vivo microdialysis study. Biol Psychiatry 57:1061–1067

Subhash MN, Nagaraja MR, Sharada S, Vinod KY (2003) Cortical alpha-adrenoceptor downregulation by tricyclic antidepressants in the rat brain. Neurochem Int 43:603–609

Sugiura T, Kondo S, Sukagawa A, Nakane S, Shinoda A, Itoh K, Yamashita A, Waku K (1995) 2-Arachidonoylglycerol: a possible endogenous cannabinoid receptor ligand in brain. Biochem Biophys Res Commun 215:89–97

Sulkowski A, Vachon L, Rich ES Jr (1977) Propranolol effects on acute marihuana intoxication in man. Psychopharmacology 52:47–53

Sumislawski JJ, Ramikie TS, Patel S (2011) Reversible gating of endocannabinoid plasticity in the amygdala by chronic stress: a potential role for monoacylglycerol lipase inhibition in the prevention of stress-induced behavioral adaptation. Neuropsychopharmacology 36:2750–2761

Szabo B, Siemes S, Wallmichrath I (2002) Inhibition of GABAergic neurotransmission in the ventral tegmental area by cannabinoids. Eur J Neurosci 15:2057–2061

Tanda G, Pontieri FE, Di Chiara G (1997) Cannabinoid and heroin activation of mesolimbic dopamine transmission by a common mu1 opioid receptor mechanism. Science 276:2048–2050

Terzian AL, Drago F, Wotjak CT, Micale V (2011) The dopamine and cannabinoid interaction in the modulation of emotions and cognition: assessing the role of cannabinoid CB1 receptor in neurons expressing dopamine D1 receptors. Front Behav Neurosci 5:49

Tournier M, Sorbara F, Gindre C, Swendsen JD, Verdoux H (2003) Cannabis use and anxiety in daily life: a naturalistic investigation in a non-clinical population. Psychiatry Res 118:1–8

Turu G, Varnai P, Gyombolai P, Szidonya L, Offertaler L, Bagdy G, Kunos G, Hunyady L (2009) Paracrine transactivation of the CB1 cannabinoid receptor by AT1 angiotensin and other Gq/11 protein-coupled receptors. J Biol Chem 284:16914–16921

Tzavara ET, Perry KW, Rodriguez DE, Bymaster FP, Nomikos GG (2001) The cannabinoid CB(1) receptor antagonist SR141716A increases norepinephrine outflow in the rat anterior hypothalamus. Eur J Pharmacol 426:R3–4

Tzavara ET, Davis RJ, Perry KW, Li X, Salhoff C, Bymaster FP, Witkin JM, Nomikos GG (2003) The CB1 receptor antagonist SR141716A selectively increases monoaminergic neurotransmission in the medial prefrontal cortex: implications for therapeutic actions. Br J Pharmacol 138:544–553

Tzavara ET, Li DL, Moutsimilli L, Bisogno T, Di Marzo V, Phebus LA, Nomikos GG, Giros B (2006) Endocannabinoids activate transient receptor potential vanilloid 1 receptors to reduce hyperdopaminergia-related hyperactivity: therapeutic implications. Biol Psychiatry 59:508–515

Ungerstedt U (1971) Stereotaxic mapping of the monoamine pathways in the rat brain. Acta Physiol Scand Suppl 367:1–48

Vanderschuren LJ, Kalivas PW (2000) Alterations in dopaminergic and glutamatergic transmission in the induction and expression of behavioral sensitization: a critical review of preclinical studies. Psychopharmacology 151:99–120

Varga K, Wagner JA, Bridgen DT, Kunos G (1998) Platelet- and macrophage-derived endogenous cannabinoids are involved in endotoxin-induced hypotension. Faseb J 12:1035–1044

Varma N, Carlson GC, Ledent C, Alger BE (2001) Metabotropic glutamate receptors drive the endocannabinoid system in hippocampus. J Neurosci 21:RC188

Vaughn LK, Mantsch JR, Vranjkovic O, Stroh G, Lacourt M, Kreutter M, Hillard CJ (2012) Cannabinoid receptor involvement in stress-induced cocaine reinstatement: potential interaction with noradrenergic pathways. Neuroscience 204:117–124

Vinod KY, Arango V, Xie S, Kassir SA, Mann JJ, Cooper TB, Hungund BL (2005) Elevated levels of endocannabinoids and CB1 receptor-mediated G-protein signaling in the prefrontal cortex of alcoholic suicide victims. Biol Psychiatry 57:480–486

Viveros MP, Marco EM, File SE (2005) Endocannabinoid system and stress and anxiety responses. Pharmacol Biochem Behav 81:331–342

Volk DW, Lewis DA (2010) Prefrontal cortical circuits in schizophrenia. Curr Top Behav Neurosci 4:485–508

Wang Y, Liu Y, Ito Y, Hashiguchi T, Kitajima I, Yamakuchi M, Shimizu H, Matsuo S, Imaizumi H, Maruyama I (2001) Simultaneous measurement of anandamide and 2-arachidonoylglycerol by polymyxin b-selective adsorption and subsequent high-performance liquid chromatography analysis: increase in endogenous cannabinoids in the sera of patients with endotoxic shock. Anal Biochem 294:73–82

Wang W, Sun D, Pan B, Roberts CJ, Sun X, Hillard CJ, Liu QS (2010) Deficiency in endocannabinoid signaling in the nucleus accumbens induced by chronic unpredictable stress. Neuropsychopharmacology 35:2249–2261

Wang J, Shen RY, Haj-Dahmane S (2012a) Endocannabinoids mediate the glucocorticoid-induced inhibition of excitatory synaptic transmission to dorsal raphe serotonin neurons. J Physiol 590:5795–5808

Wang M, Hill MN, Zhang L, Gorzalka BB, Hillard CJ, Alger BE (2012b) Acute restraint stress enhances hippocampal endocannabinoid function via glucocorticoid receptor activation. J Psychopharmacol 26:56–70

Weinshenker D, Schroeder JP (2007) There and back again: a tale of norepinephrine and drug addiction. Neuropsychopharmacology 32:1433–1451

Wenger T, Moldrich G, Furst S (2003) Neuromorphological background of cannabis addiction. Brain Res Bull 61:125–128

Wilson RI, Nicoll RA (2001) Endogenous cannabinoids mediate retrograde signalling at hippocampal synapses. Nature 410:588–592

Wolf ME, Sun X, Mangiavacchi S, Chao SZ (2004) Psychomotor stimulants and neuronal plasticity. Neuropharmacology 47(Suppl 1):61–79

Wu X, French ED (2000) Effects of chronic delta9-tetrahydrocannabinol on rat midbrain dopamine neurons: an electrophysiological assessment. Biochem J 346:835–840

Chapter 10
Chronic Effects of Cannabinoid Drugs on Monoaminergic Systems and the Role of Endocannabinoids and Cannabinoid Receptors in Human Brain Disorders

Leyre Urigüen, Susana Esteban, Luís F. Callado, María Álvaro-Bartolomé, J. Javier Meana and Jesús A. García-Sevilla

Abstracts The endocannabinoid system and cannabinoid (CB) receptors participate in the regulation of a variety of psychiatric and neurological disorders through a functional coupling with the monoaminergic systems in the brain. Norepinephrine, serotonin (5-HT) and dopamine systems are modulated via inhibitory CB_1 receptors by direct or indirect effects. The repeated stimulation of CB_1 receptors (and receptor desensitization) can lead to the induction of tolerance on the activity of monoaminergic systems. The chronic administration of CB drugs can also alter the function of presynaptic inhibitory monoamine autoreceptors and heteroreceptors and thus modulate the final effects on these systems. The functional interactions between endocannabinoids, CB receptors, and monoaminergic systems suggest a potential role for CB receptor signaling in the pathophysiology and treatment of various psychiatric and neurological disorders, including drug addiction, which are discussed on evidence from postmortem and living human brain studies.

Abbreviations

AEA	Anandamide
Am	Basolateral amygdala
2-AG	2-Arachidonoylglycerol
CB	Cannabinoid

J. A. García-Sevilla (✉) · S. Esteban · M. Álvaro-Bartolomé
Laboratori de Neurofarmacologia, Institut Universitari d'Investigació en Ciències de la Salut (IUNICS), Universitat de les Illes Balears (UIB), and Redes Temáticas de Investigación Cooperativa en Salud–Red de Trastornos Adictivos (RETICS-RTA), Palma de Mallorca, Spain
e-mail: jesus.garcia-sevilla@uib.es

L. Urigüen · L. F. Callado · J. J. Meana
Departamento de Farmacología e Instituto Biocruces, Universidad del País Vasco/Euskal Herriko Unibertsitatea (UPV/EHU), Centro de Investigación Biomédica en Red de Salud Mental (CIBERSAM), Leioa, Bizkaia, Spain

E. J. Van Bockstaele (ed.), *Endocannabinoid Regulation of Monoamines in Psychiatric and Neurological Disorders*, DOI 10.1007/978-1-4614-7940-6_10,

CC	Cerebral cortex
CNS	Central nervous system
CP55940	(-)-Cis-3-[2-hydroxy-4-(1,1-dimethylheptyl)phenyl]-trans-4-(3-hydroxypropyl) cyclohexanol
CP93129	3-(1,2,5,6-Tetrahydropyrid-4-yl)pyrrolo[3, 2-b]pyrid-5-one
DA	Dopamine
DOPA	3,4-Dihydroxy-phenylalanine
DPAT	(±)-8-Hydroxy-2-(di-n-propylamino)-tetralin
DR	Dorsal raphe
FAAH	Fatty acid amide hydrolase
GABA	γ-Aminobutyric acid
GLU	Glutamate or glutamic acid
GTPγS	Guanosine triphosphate
HC	Hippocampus
HT	Hypothalamus
5-HT	5-Hydroxytryptamine or serotonin
5-HTP	5-Hydroxy-tryptophan
HU210	(6aR)-Trans-3-(1,1-dimethylheptyl)-6a,7,10,10a-tetrahydro-1-hydroxy-6,6-dimethyl-6H-dibenzo[b, d]pyran-9-methanol
LC	Locus coeruleus
LH	Lateral habenula
NAcc	Nucleus accumbens
NAE	*N*-Acylethanolamines
NE	Norepinephrine
OEA	*N*-Oleoylethanolamine
PEA	*N*-Palmitoylethanolamine
PrH	Prepositus hypoglossal nucleus
SD7015	1-(2-Iodophenyl)-4-cyano-5-(4-methoxyphenyl)-N-(piperidin-1-yl)-1H-pyrazole-3-carboxylate
SN	Substantia nigra
SR141617A	Rimonabant
St	Corpus striatum
TH	Tyrosine hydroxylase
THC	Δ^9-Tetrahydrocannabinol
TPH	Tryptophan hydroxylase
URB597	Cyclohexyl carbamic acid 3′-carbamoyl-biphenyl-3-yl ester
VTA	Ventral tegmental area
WIN55212-2	R-(+)-[2,3-dihydro-5-methyl-3-[(morpholinyl)-methyl]pyrrolol-[1,2,3-de]-1,4-benzoxazinyl]-(1-naphthalenyl) methanone.

10.1 Introduction

The endocannabinoids (e.g., anandamide (AEA), 2-arachidonoylglycerol (2-AG)) function in the brain as retrograde lipid signaling messengers (Vaughan and Christie 2005; Mechoulam and Parker 2013) which, similarly to cannabinoid (CB) drugs, mediate their effects through the activation of two inhibitory G protein-coupled receptors termed CB_1 and CB_2 receptors (Howlett et al. 2002; Pertwee et al. 2010). The predominant CB_1 receptor, highly expressed in the central nervous system (CNS), is mainly located on inhibitory γ-aminobutyric acid (GABA) and excitatory (e.g., glutamate) synapses where it regulates the release of the corresponding transmitter (Katona et al. 1999; Schlicker and Kathmann 2001; Hashimotodani et al. 2007). Moreover, numerous nuclei and axon terminals in a variety of brain regions also express CB_1 receptors whose function is to inhibit the release of excitatory and inhibitory neurotransmitters (Alger 2002). The brain regions enriched in CB_1 receptors include the locus coeruleus/norepinephrine (LC/NE) neurons and axon NE terminals (Oropeza et al. 2007; Carvalho et al. 2010; Scavone et al. 2010) and the dorsal raphe/serotonin (DR/5-HT) neurons and 5-HT terminal fields (Hohmann and Herkenham 2000; Häring et al. 2007). CB_1 receptors are also abundant in limbic mood-regulatory dopamine (DA) rich areas (brain reward circuitry) including the ventral tegmental area (VTA), nucleus accumbens (NAcc), and corpus striatum (Herkenham et al. 1991). CB_1 receptors, however, are not located on VTA/DA neurons (Matsuda et al. 1993) but rather on presynaptic glutamatergic and GABAergic neurons in the VTA. The anatomical localizations of CB_1 receptors indicate that the direct or indirect stimulation/blockade of these inhibitory receptors can result in the fine modulation of the activity of monoaminergic systems in specific brain regions. CB_1 receptors display a high level of constitutive activity (Gifford and Ashby 1996), which can exert a tonic control (i.e. ligand-independent activity) on its endocytic cycle (Leterrier et al. 2004) as well as on the function of other receptors (Canals and Milligan 2008). The CB_1 receptor basal tone, however, might also be related to the ongoing production of endocannabinoids (AEA and 2-AG) which would stimulate CB receptors given the appearance of constitutive activity (Howlett et al. 2011). In the CNS, the less abundantly expressed and less well understood CB_2 receptor is mainly associated with the regulation of neuroinflammatory processes (microglia and immune responses) which can be of importance in the pathogenesis of some psychiatric and neurological diseases (Atwood et al. 2012; Onaivi et al. 2012).

The endocannabinoid system and CB_1 receptors participate, in part, in the control of emotional behavior and mood through a functional coupling with monoaminergic systems in the brain (Bambico et al. 2007; Ashton and Moore 2011). These functional interactions have suggested a potential role for CB_1 receptor signaling in the neurobiology of various psychiatric disorders (Hill and Gorzalka 2005a, 2005b; Parolaro et al. 2010; Carvalho and Van Bockstaele 2012; Esteban and García-Sevilla 2012). This chapter summarizes and discusses the chronic effects of CB drugs modulating brain monoamine systems (spontaneous neuronal activity, synthesis and release of neurotransmitters) as well as the activity of presynaptic monoaminergic receptors (autoreceptors and heteroreceptors) that regulate the synthesis and release of classic neurotransmitters. The chapter also deals with the possible relevance of the endocannabinoid system and CB receptors in the pathophysiology and treatment of several psychiatric and neurological disorders, including drug addiction, with a special focus on evidence from postmortem and living human brain studies.

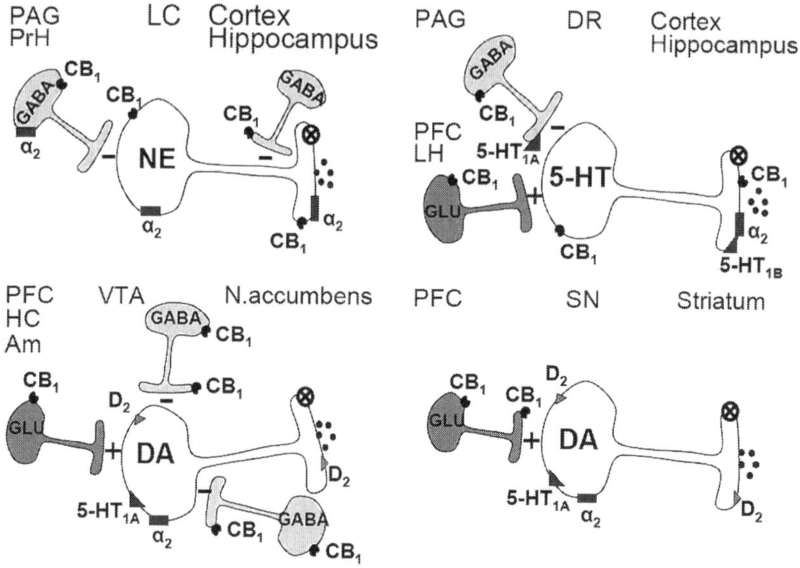

Fig. 10.1 Neuronal structures and neurotransmitters involved in effects of cannabinoid drugs acting at CB_1 receptors on locus coeruleus/norepinephrine (LC/NE) neurons, dorsal raphe/serotonin (DR/5-HT) neurons, ventral tegmental area/dopamine (VTA/DA neurons, and substantia nigra/dopamine (SN/DA) neurons. The most important projections to the LC are GABA local interneurons and GABA afferents from the periaqueductal gray matter (PAG) and the prepositus hypoglossal nucleus (PrH). The relevant neurotransmitter systems that project to the DR are GABA afferents from PAG, and glutamate (GLU) afferents from the medial prefrontal cortex (PFC), and possibly the lateral habenula (LH). The most important projections to the SN are glutamate (GLU) afferents from the medial prefrontal cortex (PFC). The relevant neurotransmitter systems that project to the VTA are glutamatergic (GLU) afferents from the PFC, hippocampus (HC), and basolateral amygdala (Am), as well as GABA inputs from the nucleus accumbens (NAcc) and local GABA interneurons. α_2: inhibitory α_2-adrenoceptor (somatodendritic and terminal NE autoreceptor and heteroreceptor on 5-HT terminals); 5-HT_{1A}: inhibitory somatodendritic autoreceptor; 5-HT_{1B}: inhibitory terminal autoreceptor; D_2: inhibitory somatodendritic and terminal DA autoreceptor. See the main text for specific comments on the chronic effects and interactions of CB_1 drugs regulating monoaminergic systems, including the modulatory role of presynaptic monoaminergic receptors (autoreceptors and heteroreceptors). (Modified from Esteban and García-Sevilla 2012)

10.2 Chronic Effects of Cannabinoid Drugs on Brain Monoaminergic Systems. Induction of Tolerance to the Acute Effects of CB_1 Agonists

Cannabinoid (CB) drugs modify the functioning of monoaminergic systems via inhibitory CB_1 receptors by direct or indirect effects, which depend on receptor localization on monoaminergic neurons themselves and/or inhibitory (GABAergic) and/or excitatory (glutamatergic) regulatory neurons (Fig. 10.1) The acute stimulatory/inhibitory effects of CB drugs on monoaminergic systems have recently been discussed (Esteban and García-Sevilla 2012). In addition, several studies have investigated the chronic effects of CB drugs on brain monoaminergic systems, and some of them have also assessed the possible induction of tachyphylaxis (neurochemical tolerance) to the acute effects of CB_1 receptor agonists (Esteban and García-Sevilla 2012). The long-term regulation of monoaminergic systems by CB drugs can be of importance in the context of the beneficial and deleterious effects of these drugs.

10.2.1 Noradrenergic System

Chronic treatment with URB597 (4 days), a fatty acid amide hydrolase (FAAH) inhibitor, and WIN55,212-2 (8 days), a preferential CB_1 receptor agonist, have been shown to markedly increase the spontaneous firing rate of NE neurons and the expression of tyrosine hydroxylase (TH) in the rat LC (Table 10.1). A longer chronic WIN55,212-2 treatment (20 days) in rats was reported not to alter the firing rate of LC neurons (Table 10.1). Notably, repeated treatment with URB597 (resulting in an increased content of AEA) was not associated with the induction of tolerance to its acute enhancing effect on LC/NE neurons (Table 10.1). Chronic WIN55212-2 (5 days) was also shown to increase the synthesis of DOPA/NE in the hippocampus and cerebellum (lack of tolerance) but not in the cerebral cortex (induction of tolerance) of rats (Table 10.1 and Fig. 10.2). Chronic WIN55,212-2 (8 days) also induced an increase in the release of NE in rat brain cortex with a concomitant upregulation of TH in the LC (Table 10.1).

10.2.2 Serotonergic System

Chronic URB597 (4 days) also induced marked increases in the spontaneous firing rate of rat DR/5-HT neurons (Table 10.1; lack of tolerance). The repeated application (three times) of low and high doses of WIN55,121-2 induced biphasic effects on the firing rate (increases and decreases) of rat DR 5-HT neurons (Table 10.1; apparent lack of tolerance). A prolonged WIN55,212-2 treatment in rats (20 days) did not result in alterations of the basal firing rate of DR neurons (Table 10.1), which could indicate the induction of some degree of tolerance to the acute effect of the agonist. Chronic WIN55,121-2 (5 days) in rats did not significantly alter the synthesis of 5-HTP in the cerebral cortex, hippocampus, and cerebellum (Table 10.1; induction of tolerance) (Table 10.1 and Fig. 10.2). In contrast, chronic WIN55,121-2 (5 days), similarly to the acute agonist treatment, also reduced 5-HTP synthesis in rat striatum (Table 10.1; lack of tolerance) (Table 10.1 and Fig. 10.2).

10.2.3 Dopaminergic System

Chronic Δ^9-tetrahydrocannabinol (THC) treatment (14 days) in rats was also reported to enhance the spontaneous firing rate of SN/DA and VTA/DA neu-

Table 10.1 Effects of chronic treatment with cannabinoid drugs on monoaminergic systems in adult rat brain regions

Cannabinoid drug (dose and duration of treatment)	Brain region and net effect (% basal change)	Induction of tolerance	Reference
Norepinephrine system			
URB597 (0.1 mg/kg, 4 days)	LC, ↑ firing rate (~50 %)	–	Gobbi et al. (2005)
WIN55,212-2 (3 mg/kg, 8 days)	LC, ↑ TH expression (125 %)	NT	Page et al. (2007)
WIN55,212-2 (1 mg/kg, 20 days)	LC, ≈ firing rate	NT	Bambico et al. (2010)
WIN55,212-2 (4–16 mg/kg, 5 days)	HC/CB ↑ DOPA synthesis (30–41 %)	–	Moranta et al. (2009)
WIN55,212-2 (4–16 mg/kg, 5 days)	CC, ≈ DOPA synthesis	+	Moranta et al. (2009)
WIN55,212-2 (3 mg/kg, 8 days)	CC, ↑ NE release (40 %)	NT	Page et al. (2007)
Serotonergic system			
URB597 (0.1 mg/kg, 4 days)	DR, ↑ firing rate (138 %)	–	Gobbi et al. (2005)
WIN55,212-2 (0.1–0.2 mg/kg, 3 times)	DR, ↑ firing rate (65–126 %)	–	Bambico et al. (2007)
WIN55,212-2 (2 mg/kg, 3 times)	DR, ↓ firing rate (64 %)	–	Bambico et al. (2007)
WIN55,212-2 (0.2–1 mg/kg, 20 days)	DR, ≈ firing rate	NT	Bambico et al. (2010)
WIN55,212-2 (4–16 mg/kg, 5 days)	CC/HC/CB, ≈ 5-HTP synthesis	+	Moranta et al. (2009)
WIN55,212-2 (4–16 mg/kg, 5 days)	St, ↓ 5-HTP synthesis (29 %)	–	Moranta et al. (2009)
Dopaminergic system			
THC (5 mg/kg, 14 days)	SN, ↑ firing rate (33 %)	+	Wu and French (2000)
THC (5 mg/kg, 14 days)	VTA, ↑ firing rate (44 %)	–	Wu and French (2000)
HU210 (5 μM, 5 applications)	VTA, ↑ firing rate (400 %)	–	Cheer et al. (2000)
WIN55,212-2 (4–16 mg/kg, 5 days)	St, ↓ DOPA synthesis (25 %)	–	Moranta et al. 2009

Cannabinoid drugs: URB597, an inhibitor of fatty acid amide hydrolase (FAAH); WIN55,212-2, THC (Δ^9 -tetrahydrocannabinol), and HU210, cannabinoid receptor agonists.

Net effect (% basal change): ↑ increase, ↓ decrease, ≈ no significant change.

Pharmacological tolerance: + induction of tolerance or – lack of tolerance after repeated agonist treatment (chronic effect versus acute effect). *NT* not tested

Brain region: *LC* locus coeruleus, *CC* cerebral cortex, *HC* hippocampus, *CB* cerebellum, *DR* dorsal raphe, *SN* substantia nigra, *St* corpus striatum, *VTA* ventral tegmental area, *TH* tyrosine hydroxylase, *DOPA* 3,4-dihydroxy-phenylalanine, *NE* norepinephrine, *5-HTP* 5-hydroxy-tryptophan

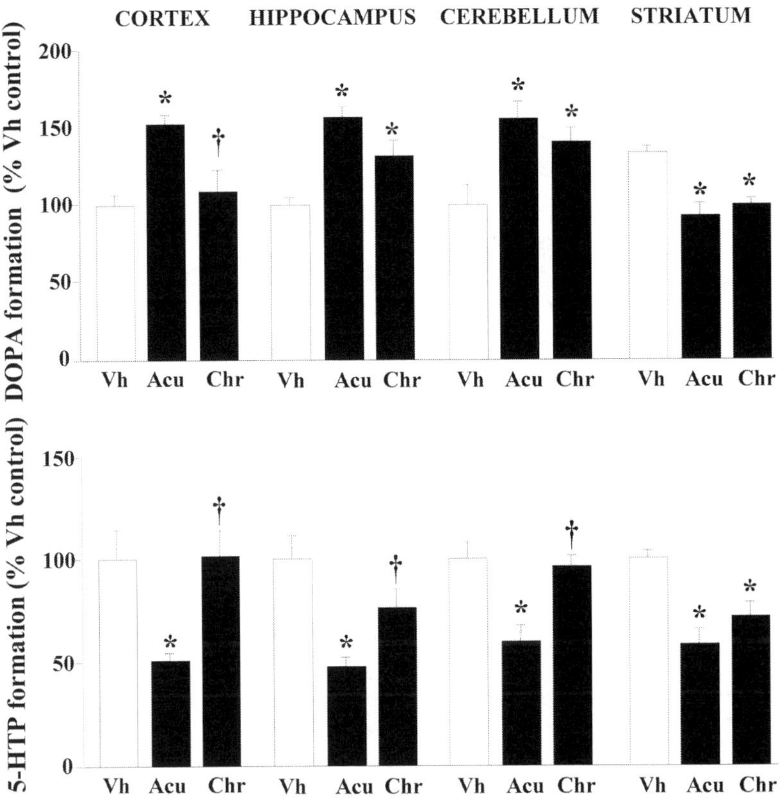

Fig. 10.2 Acute and chronic effects of the cannabinoid receptor agonist WIN 55,212-2 on DOPA and 5-HTP formation in various rat brain regions, expressed as percentages of vehicle-treated animals (Vh control). Groups of rats were treated (i.p.) with drug Vh ($n=10$), acute WIN (Acu, 8 mg/kg, 1 h, $n=6$) and chronic WIN (Chr, 2–8 mg/kg, twice daily for 5 days, $n=6$). * denotes that $P<0.05$ at least when compared with the corresponding vehicle (Vh)-treated group. † denotes that $P<0.05$ at least when compared with the corresponding acute (Acu)- treated group. (Modified from Moranta et al. 2009)

rons (Table 10.1; induction of tolerance in SN and lack of tolerance in VTA). Similarly, the firing rate of VTA/DA neurons was markedly increased after the repeated in vitro application (five times) of HU210, a selective CB_1 receptor agonist (Table 10.2; lack of tolerance). The increase in VTA neuronal activity induced by HU210 was blocked by rimonabant (SR141716A), which by itself was ineffective in altering basal neuronal firing (Cheer et al. 2000). Chronic treatment with WIN55,212-2 (5 days) in rats resulted in a sustained inhibition of DOPA synthesis in striatum (Table 10.1 and Fig. 10.2; lack of tolerance).

These chronic studies in laboratory animals revealed the existence of a complex crosstalk between the endocannabinoid system and monoaminergic neurons in the brain. Notably, chronic CB treatments (FAAH inhibitor and CB_1 receptor agonists) are not associated with the induction of tolerance (neurochemical adaptation) to the acute stimulatory effects of CB drugs on LC/NE, DR/5-HT and VTA/DA neurons (Table 10.1). In contrast, the chronic effects of CB receptor agonists on the synthesis of DOPA and 5-HTP and/or the release of the corresponding neurotransmitter are associated with the induction of tolerance in specific brain regions (Table 10.1 and Fig. 10.2). The process of CB drug tolerance appears to reflect the desensitization of CB_1 receptors after repeated drug exposure, the extent of which being dependent on time exposure, agonist efficacy, and the brain region targeted (Sim-Selley 2003). In this context, recent behavioral studies in rhesus monkeys have shown that CB_1 receptor tolerance/cross-tolerance (after 14 days THC treatment) is greater for low-efficacy agonists (e.g., THC) compared with high-efficacy agonist (e.g., CP55940), which suggested that differences in CB_1 receptor efficacy are relevant in vivo (Hruba et al. 2012). Importantly, the induction of drug tolerance upon CB_1 receptor agonist treatment could alter the direct and/or indirect effects of CB drugs modulating the functionality of monoaminergic systems (Fig. 10.1).

10.3 Modulation of Presynaptic Monoaminergic Receptors After Chronic Cannabinoid Exposure. Autoreceptors and Heteroreceptors

Presynaptic inhibitory receptors (autoreceptors and heteroreceptors) on monoaminergic neurons are involved in the regulation of neuronal (spontaneous firing rate) activity, synthesis, and release of NE, 5-HT, and DA (Esteban et al. 1996; Ichikawa and Meltzer 2000; Starke 2001; Fink and Göthert 2007). Thus, changes in the function of α_2-adrenoceptors and 5-$HT_{1A/1B}$ receptors mediating negative feedback mechanisms in specific neuronal systems (Fig. 10.1) may contribute to the sustained activation of LC/NE, DR/5-HT, SN/DA and VTA/DA neurons induced by chronic CB exposure (Table 10.1). Similarly, the rate-limiting monoamine enzymes TH and tryptophan hydroxylase (TPH) are under the tonic inhibitory control of somatodendritic α_{2A}-autoreceptors and 5-$HT_{1A/1B}$-autoreceptors, which regulate the synthesis of the monoamine precursors DOPA and 5-HTP.

10.3.1 α_2-Adrenoceptors

Chronic treatment of rats with WIN55,212-2 (2-8 mg/kg, 5 days) was associated with the induction of desensitization of somatodendritic and terminal α_{2A}-autoreceptors and α_{2A}-heteroreceptors regulating the synthesis of DOPA and 5-HTP in brain regions enriched in noradrenergic, serotonergic, or dopaminergic nerve terminals (Moranta et al. 2009). Thus, the ability of the α_2-agonist clonidine to decrease the formation of DOPA/NE (α_2-autoreceptor), DOPA/DA (α_2-heteroreceptor), or 5-HTP/5-HT (α_2-heteroreceptor) was markedly reduced or abolished in the cerebral cortex, cerebellum, and striatum of chronic WIN55,212-2 rats (Fig. 10.3). In line with these findings, chronic WIN55,212-2 in rats (3 mg/kg, 7 days) was reported to reduce α_2-adrenoceptor expression in some brain regions (Carvalho et al. 2010). The reduced sensitivity and expression of α_2-adrenoceptors (desensitization of autoreceptors and heteroreceptors) modulating brain monoaminergic systems could be the result of an increased NE release induced by CB_1 receptor agonists (Oropeza et al. 2005; Page et al. 2007), which in turn would explain the downregulation of postsynaptic β-adrenoceptors induced by chronic THC in the brain (Hillard and Bloom 1982).

10.3.2 5-HT$_{1A}$ and 5-HT$_{1B}$ Receptors

Chronic WIN55,212-2 treatment in rats (2–8 mg/kg, 5 days) was also reported to induce supersensitivity of somatodendritic 5-HT$_{1A}$-autoreceptors regulating the synthesis of 5-HTP in the cerebellum and striatum and of 5-HT$_{1A}$-heteroreceptors modulating DOPA/NE and DOPA/DA in these brain regions (Moranta et al. 2009). Thus, a low dose of the selective 5-HT$_{1A}$ receptor agonist 8-OH-DPAT, which was ineffective in the vehicle-treated rat, reduced 5-HTP formation in the cerebellum and striatum of chronic WIN55,212-2 rats (Fig. 10.3). This increased sensitivity of somatodendritic 5-HT$_{1A}$ auto/heteroreceptors could be the result, in part, of a reduced 5-HT release induced by CB drugs (Nakazi et al. 2000). Chronic WIN55,212-2 treatment in rats (2–8 mg/kg, 5 days) also induced supersensitivity of terminal 5-HT$_{1B}$- auto/heteroreceptors regulating the synthesis of DOPA and 5-HTP. Thus, a low dose of the selective 5-HT$_{1B}$ receptor agonist CP93129 reduced DOPA formation (cerebellum) or potentiated the reduction of 5-HTP formation (cerebellum and striatum) in chronic WIN55,212-2 rats (Fig. 10.3).

The changes in presynaptic monoamine receptor function induced by the sustained stimulation of CB_1 receptors (Fig. 10.3) would finally result in less efficient (α_2- auto/heteroreceptors) or more efficient (5-HT$_{1A/B}$-auto/heteroreceptors) feedback autoinhibition leading to alterations in the synthesis/release of NE, 5-HT, and/or DA. These adaptations of presynaptic receptor function (autoreceptors and heteroreceptors) in chronically agonist-treated animals could finally modulate the net effects of chronic CB_1 receptor stimulation (induction or lack of tolerance) on monoaminergic systems in specific brain regions (Fig. 10.1).

10.4 Role of Endocannabinoids and CB Receptors in Human Brain Disorders

Several comprehensive reviews have discussed the potential involvement of the endocannabinoid system and CB receptors in several CNS disorders (most evidence from animal models) with an emphasis on the major psychiatric syndromes major depression and schizophrenia (Bambico et al. 2009; Parolaro et al. 2010; Ashton and

Moore 2011; Gorzalka and Hill 2011; Mechoulam and Parker 2013). Interestingly, the CB_1 receptor deficient mouse has been proposed as a useful model of depression (Valverde and Torrens 2012). Animal models of depression (postulated defective endocannabinoid system), however, have shown paradoxical results concerning the regulation of CB_1 receptors and the effects of antidepressant drugs (Griebel et al. 2005; Hill and Gorzalka 2005b; Bambico et al. 2007; Mato et al. 2010; Gorzalka and Hill 2011). In the CNS, the less abundant CB_2 receptor is mainly associated with the regulation of neuroinflammatory processes which might be of importance in the pathogenesis of neurodegenerative processes such as Alzheimer's disease and Huntington's disease (Fernández-Ruiz et al. 2008). Recently, the CB_2 deficient mouse has been proposed as a model of schizophrenia-like behaviors (Ortega-Alvaro et al. 2011). The participation of endocannabinoids and CB_1 or CB_2 receptors in the pathophysiology and treatment of several psychiatric and neurological disorders is discussed below from data directly obtained in humans.

10.4.1 Basal Serum or Cerebrospinal Fluid (CSF) Concentrations of Endocannabinoids. Effects of Psychotropic Medications

10.4.1.1 Major Depression and Schizophrenia

Little is known on the status of endocannabinoids in the pathogenesis and treatment of major depression. In recent studies, the serum concentrations of AEA and 2-AG, but not *N*-palmitoylethanolamine (PEA) or *N*-oleoylethanolamine (OEA), were reported reduced in depressed women relative to matched controls (Hill et al. 2008, 2009). Conversely, in patients with minor depression, serum AEA was increased whereas 2-AG levels showed a similar but statistically insignificant trend (Hill et al. 2008).

In schizophrenia, four studies of the same research group have reported elevated AEA levels in CSF of patients with schizophrenia (Leweke et al. 1999, 2007; Giuffrida et al. 2004; Koethe et al. 2009). Moreover, CSF AEA contents remained high in patients treated with atypical antipsychotics, but they were similar to controls in patients medicated with typical antipsychotics (Giuffrida et al. 2004). No significant differences in serum AEA levels were found among schizophrenia patients and controls (Leweke et al. 2007, Koethe et al. 2009). The neuronal origin of CSF endocannabinoids remains conjectural and it might reflect an elevation in the peripheral content of these lipid signaling messengers. Thus, blood AEA was increased in patients with acute schizophrenia probably as a consequence of the modified immune response observed during the course of the disease (De Marchi et al. 2003). In fact, patients in initial prodromal states of psychosis with lower levels of AEA in CSF showed a higher risk for transiting to psychosis earlier (Koethe et al. 2009).

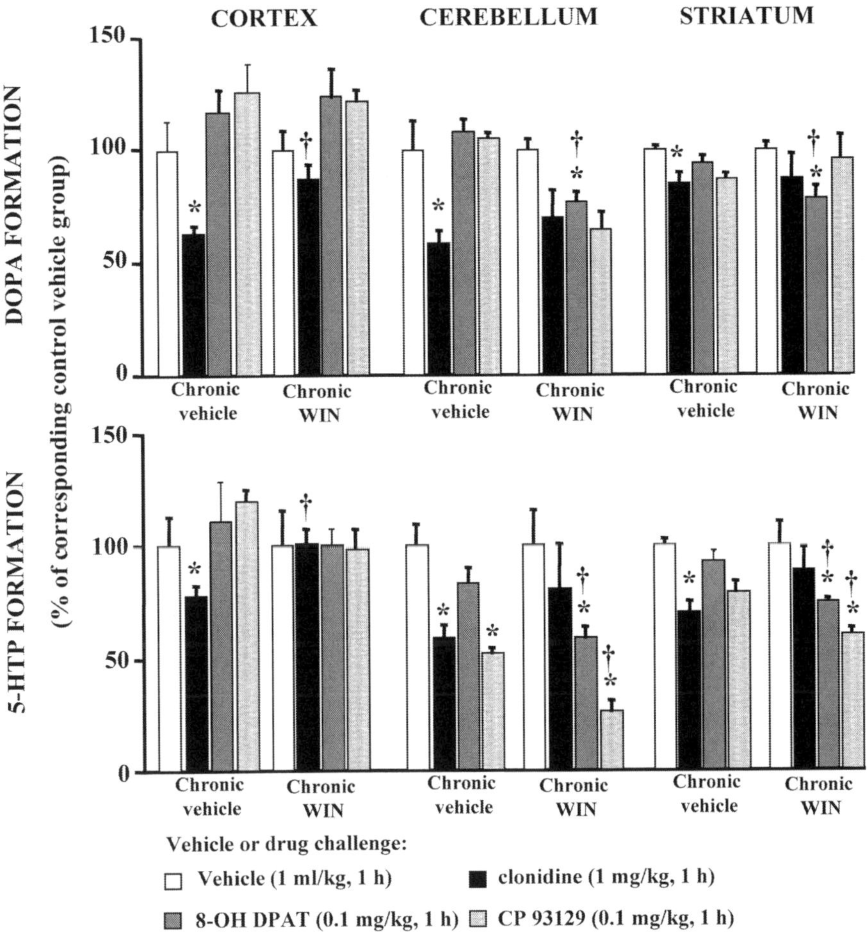

Fig. 10.3 Acute effects of clonidine (α_2-adrenoceptor agonist; 1 mg/kg), 8-OH DPAT (5-HT_{1A} receptor agonist; 0.1 mg/kg), and CP93129 (5-HTP_{1B} receptor agonist; 0.1 mg/kg) on DOPA and 5-HTP formation in various brain regions of chronically vehicle- and WIN55,212-2 (WIN)-treated rats, expressed as percentages of the corresponding control vehicle group. * denotes that $P<0.05$ at least when compared with the corresponding chronic vehicle group. † denotes that $P<0.05$ at least when compared with the corresponding drug challenge in the chronic vehicle group. (Modified from Moranta et al. 2009; data for CP93129, Esteban and García-Sevilla unpublished)

10.4.1.2 Stress and Anxiety

Preclinical studies have revealed the involvement of endocannabinoids in the regulation of stress and anxiety through interactions with monoaminergic systems (e.g., see McLaughlin et al. 2012). However, the clinical evidence is scarce (Mechoulam and Parker 2013). In a recent study, social stress exposure evoked a significant increase of blood 2-AG in women immediately following the stress, and both PEA and OEA blood levels declined during the phase of stress recovery (Hill et al. 2009).

Another study has measured circulating endocannabinoids (AEA, 2-AG, and various *N*-acylethanolamides) in healthy subjects after acute stress (Dlugos et al. 2012). The data indicate that stress increased serum AEA and *N*-acylethanolamides, but not 2-AG, immediately after the stress period. Interestingly, anxiety ratings at baseline were negatively correlated with baseline concentrations of AEA in blood (Dlugos et al. 2012).

10.4.1.3 Parkinson's Disease, Alzheimer's Disease, and Huntington's Disease

Two studies have reported an increased content of AEA in CSF of unmedicated patients with Parkinson's disease (Pisani et al. 2005, 2010). Notably, the CSF AEA levels were at least twofold higher in unmedicated patients compared to control subjects. In medicated patients, AEA levels in CSF were indistinguishable from those measured in controls, regardless of the type of treatment with either levodopa or dopamine agonists (Pisani et al. 2005, 2010).

In Alzheimer's disease, the blood concentrations of AEA and 2-AG were found unaltered when compared with those in matched control subjects (Koppel et al. 2009). In the CSF, the content of 2-AG was similar in patients with Alzheimer's disease and controls, and AEA was not detected in any CSF sample (Koppel et al. 2009). This study also reported a lack of correlation between 2-AG in CSF and any measured domain of cognition (Koppel et al. 2009).

In Huntington's disease, a greater content of AEA in lymphocytes, with reduced activity of the enzyme FAAH, have been reported in patients with this neurodegenerative process. Other peripheral markers of the endocannabinoid system were found unaltered (Battista et al. 2007).

10.4.2 Basal Content of Endocannabinoids and CB Receptors in the Postmortem and Living Human Brains. Effects of Psychotropic Medications

10.4.2.1 Major Depression and Schizophrenia

Several studies have assessed the status of CB_1 receptors in the pathophysiology of major depression and/or suicide in the human brain. Two independent postmortem studies have reported an increased density of CB_1 receptors (agonist radioligand binding sites and receptor protein) and/or a greater CB_1 receptor-mediated G-protein activation (agonist stimulated [^{35}S]GTPγS binding) in the prefrontal cortex of antidepressant-free depressed suicides (Hungund et al. 2004; Valdizán et al. 2011) (Table 10.2). Interestingly, cortical CB_1 receptor-stimulated [^{35}S]GTPγS binding was not altered in antidepressant-treated depressed suicides (Valdizán et al. 2011). In line with these findings, the expression of CB_1 receptor mRNA has been reported to be greater in the prefrontal cortex of depressed patients when compared with matched controls (Choi et al. 2012) (Table 10.2). Other postmortem studies, however, did not find significant differences in CB_1 receptor immunoreactivity in the prefrontal cortex of subjects with major depression (Eggan et al. 2010). Furthermore, the numerical density of cortical CB_1-immunoreactive glial cells was reduced in major depression which could be related to the effects of psychotropic drugs (Koethe et al. 2007) (Table 10.2). The postmortem data (radioligand agonist sites and receptor function) suggest a role for enhanced CB_1 receptor signaling in brains of antidepressant-free depressed suicides. These human postmortem findings, however, conflict with the postulated endocannabinoid deficiency in animal models of depression (Gorzalka and Hill 2011; Valverde and Torrens 2012). It should be noted, however, that the consequences of the reported alterations of the endocannabinoid system in depression (human and animal studies) remain to be clarified: e.g., the CB_1 receptor has both inhibitory and excitatory effects on synaptic transmission in the prefrontal cortex, indicating complex interactions between endocannabinoids and monoamine systems. Interestingly, an increased content of AEA and 2-AG with upregulation of CB_1 receptor density and signaling have been reported in the prefrontal cortex of alcoholic suicides compared with alcoholic nonsuicide subjects (Vinod et al. 2005), which further appears to link sensitization of cortical CB_1 receptors to suicide (Table 10.2).

Several studies have assessed the status of endocannabinoids in the pathogenesis and treatment of schizophrenia. Early studies had shown high AEA content in the CSF of schizophrenia subjects (Leweke et al. 1999) and that cannabis abuse could aggravate existing psychosis (Mathers and Ghodse 1992). Recently, 2-AG and AEA contents have been quantified in postmortem brain regions of subjects with schizophrenia (Muguruza et al. 2012). This study has revealed an opposite pattern for the regulation of endocannabinoids in schizophrenia: 2-AG was increased in cerebellum, hippocampus, and prefrontal cortex, whereas AEA and other *N*-acylethanolamines (dihomo-γ-linolenoylethanolamine, PEA, OEA, and docosahexaenoylethanolamine) were decreased in the same brain regions (Muguruza et al. 2012). Interestingly, antipsychotic medications appeared to reduce the content of endocannabinoids in the prefrontal cortex and hippocampus, but not in cerebellum, when antipsychotic-treated and antipsychotic-free subjects were compared (Muguruza et al. 2012).

On the other hand, several reports have linked schizophrenia with a differential expression of CB_1 receptors in the postmortem human brain. A significant upregulation of CB_1 receptors (autoradiographic density) has been reported in the different brain regions (including the cingulate cortex and dorsolateral prefrontal cortex) of subjects with schizophrenia, irrespective of the treatment given to the patients (Dean et al. 2001; Zavitsanou et al. 2004; Newell et al. 2006; Dalton et al. 2011; Jenko et al. 2012) (Table 10.2). In line with these findings, a neuroimaging (positron emission tomography (PET)) study has reported a generalized increase in CB_1 receptor density in most brain regions of schizophrenia subjects compared to controls, although the increase was significant in the pons only (Wong et al. 2010) (Table 10.2). Interestingly, CB_1 receptor binding in the frontal lobe and middle and posterior cingulate regions significantly correlated with the ratio of the brief psychiatry rating score psychosis to withdrawal score (Wong et al. 2010).

Table 10.2 Basal regulation of brain CB receptors in various psychiatric and neurological disorders

Brain disorder	Brain region and net effect (% basal change)	Reference
Major depression (postmortem)		
CB_1 functional binding	CC, ↑ (45 %)	Hungund et al. (2004)
	CC, ↑ (30 %)	Valdizán et al. (2011)
CB_1 radioligand binding	CC, ↑ (24 %)	Hungund et al. (2004)
CB_1 immunoreactivity	CC, ≈	Eggan et al. (2010)
	CC, ↓	Koethe et al. (2007)
CB_1 mRNA	CC, ↑	Choi et al. (2012)
Schizophrenia (PET)		
CB_1 availability	BS/pons ↑	Wong et al. (2010)
Schizophrenia (postmortem)		
CB_1 immunodensity	CC, ≈ (drug-free subjects)	Urigüen et al. (2009)
	CC, ↓ (29 %) (treated subjects)	Urigüen et al. (2009)
CB_1 radioligand binding	CC, ↑ (23 %)	Dean et al. (2001)
	CC, ↑ (64 %)	Zavitsanou et al. (2004)
	CC, ↑ (25 %)	Newell et al. (2006)
	CC, ↑ (22 %)	Dalton et al. (2011)
	CC, ↑ (20 %)	Jenko et al. (2012)
	STG, ≈	Deng et al. (2007)
CB_1 immunoreactivity	CC, ↓ (12–14 %)	Eggan et al. (2008)
	CC, ↓ (19 %)	Eggan et al. (2010)
	CC, ≈	Koethe et al. (2007)
CB_1 mRNA	CC, ↓ (15 %)	Eggan et al. (2008)
Parkinson (PET)		
CB_1 availability	SN, ↓	Van Laere et al. (2012)
Parkinson (postmortem)		
CB_1 functional binding	CN, ↑ (65 %); P, ↑ (144 %); GP, ↑ (672 %); SN ↑ (53 %)	Lastres-Becker et al. (2001)
CB_1 radioligand binding	P, CN, ≈	Farkas et al. (2012a)
CB_1 mRNA	CN, P, GP, ↓	Hurley et al. (2003)
Alzheimer (postmortem)		
CB_1 radioligand binding	HP, CN, SN, GP, ↓	Westlake et al. (1994)
	FB, BG, ≈	Lee et al. (2010)
	CC, ↑	Farkas et al. (2012b)
CB_1 density	HP, CC, ≈	Benito et al. (2003)
	CC, ↓	Ramirez et al. (2005)
	FB, BG, ≈	Lee et al. (2010)
CB_1 functional binding	CC, ↓	Ramirez et al. (2005)
CB_1 mRNA	HP, CN, SN, GP, ≈	Westlake et al. (1994)
Huntington (PET)		
CB_1 availability	CRB, CB, BS, ↓	Van Laere et al. (2010)

Table 10.2 (continued)

Brain disorder	Brain region and net effect (% basal change)	Reference
Huntington (postmortem)		
CB_1 radioligand binding	SN, ↓	Glass et al. (1993)
	St, GP, ↓	Richfield and Herkenham (1994)
	CN, P, GP, ↓	Glass et al. (2000)
CB_1 immunoreactivity	GP, ↓	Allen et al. (2009)
Alcohol dependence (postmortem)		
CB_1 functional binding	CC, ↑ (34 %)	Vinod et al. (2005)
CB_1 radioligand binding	CC, ↑ (39 %)	Vinod et al. (2005)
CB_1 immunoreactivity	CC, ↑ (67 %)	Vinod et al. (2005)
	Vt, ↓ (26–52 %)	Vinod et al. (2010)
Cannabis dependence (postmortem)		
CB_1 radioligand binding	CC, HP, St, SN, ↓ (25–40 %)	Villares (2007)
CB_1 mRNA	CC, St, SN, ↓	Villares (2007)
CB_1 radioligand binding	CN, P, ↑ (25 %)	Dean et al. (2001)
Cocaine addiction (postmortem)		
CB_1 immunodensity	CC, ↓ (40 %)	Álvaro-Bartolomé and García-Sevilla (2013)
CB_2 immunodensity	CC, ≈	
Opiate addiction (postmortem)		
CB_1 immunodensity	CC, ≈	Álvaro-Bartolomé and García-Sevilla (2013)

Net effect (% basal change): ↑ increase, ↓ decrease, ≈ no significant change
PET positron emission tomography
Brain region: *CC* cerebral cortex, *CB* cerebellum, *CRB* cerebrum, *HP* hippocampus, *SN* substantia nigra, *St* corpus striatum, *Vt* ventral striatum, *CN* caudate nucleus, *P* putamen, *GP* globus pallidus, *FB* forebrain, *BG* basal ganglia, *BS* brain stem/pons, *STG* superior temporal gyrus

In other postmortem studies, however, CB_1 receptor immunodensity was found decreased (with or without changes in CB_1 receptor mRNA) in the prefrontal cortex of antipsychotic-treated subjects with schizophrenia but not in drug-free schizophrenia subjects (Eggan et al. 2008, 2010; Urigüen et al. 2009) (Table 10.2). Other studies did not find alterations in CB_1 receptor density or CB_1 receptor mRNA in the cingulate cortex and superior temporal gyrus of schizophrenia subjects (Deng et al. 2007; Koethe et al. 2007) (Table 10.2). The reported discrepancies between postmortem studies might be related to confounding factors such as the subtype of schizophrenia or the presence of antipsychotic medications. Thus, a recent study has reported increased CB_1 receptor binding in the dorsolateral prefrontal cortex of paranoid schizophrenia subjects when compared with control and nonparanoid schizophrenia subjects (Dalton et al. 2011) (Table 10.2). Mostly, these postmortem studies suggest that the modulation of CB_1 receptor density in the prefrontal cortex seems to be a consequence of antipsychotic treatment and it

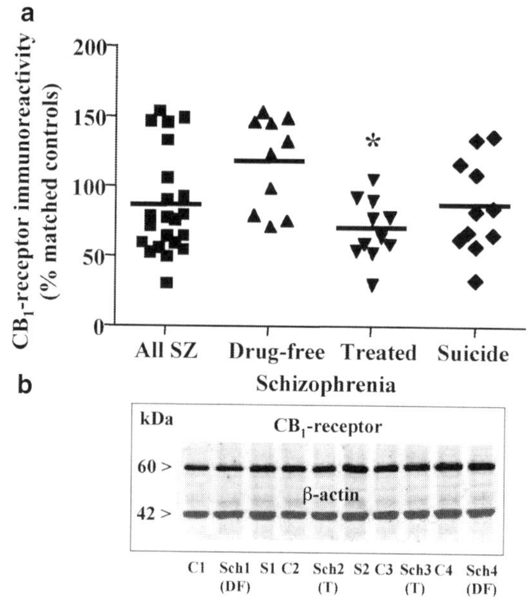

Fig. 10.4 **a** Immunodensity of cannabinoid CB_1 receptor in the prefrontal cortex of drug-free ($n=10$) and antipsychotic-treated suicide schizophrenia subjects ($n=11$) and non-schizophrenia suicide subjects ($n=11$), expressed as a percentage of immunoreactivity in the corresponding matched controls ($*P<0.05$, comparison of antipsychotic-treated and drug-free schizophrenia suicide subjects). **b** Representative immunoblots of CB_1 receptor and β-actin for control subjects (C), drug-free schizophrenia (Sch, DF) and antipsychotic-treated schizophrenia (Sch, T) subjects, and non-schizophrenia suicide subjects (S). The molecular masses (kDa) of target proteins were estimated from referenced standards. (Modified from Urigüen et al. 2009)

represents an adaptative mechanism (Fig. 10.4). Since reductions in markers of GABA neurotransmission have been identified in the prefrontal cortex of subjects with schizophrenia (Lewis et al. 2005), a lower CB_1 receptor density induced by antipsychotic drugs could reduce the endocannabinoid-mediated suppression of GABA release, thus contributing to the normalization of cognitive functions. Consistent with this hypothesis, selective CB_1 receptor antagonists would be beneficial for the treatment of schizophrenia symptoms (Miyamoto et al. 2005). Although rimonabant, the first marketed CB_1 receptor antagonist, was suspended because of the induction of depression and suicide risk in some patients with abdominal obesity and coronary artery disease (Nissen et al. 2008), the identification of high-risk patients for these side effects could be important for the safe use of CB_1 receptor antagonists in various pathologies (Lazary et al. 2011).

Although these findings in the postmortem and living human brains are important, further studies are still needed to substantiate the status of endocannabinoids and CB_1 receptors in the pathogenesis and treatment of major depression and schizophrenia.

10.4.2.2 Parkinson's Disease, Alzheimer's Disease, and Huntington's Disease

In Parkinson's disease the postmortem findings related to CB_1 receptors in the basal ganglia (radioligand binding sites and agonist stimulated [^{35}S]GTPγS binding) are contradictory (Table 10.2). An early study reported an enhanced stimulation of [^{35}S]GTPγS binding by WIN55,212-2 in the caudate nucleus, putamen, lateral globus pallidus, and substantia nigra of subjects with Parkinson's disease (Lastres-Becker et al. 2001). This study also reported an increase in CB_1 receptor binding sites in the same caudate nucleus and putamen samples (Lastres-Becker et al. 2001) (Table 10.2). In contrast, a recent autoradiographic study with the CB_1 receptor inverse agonist [^{125}I]SD7015 demonstrated unchanged CB_1 receptor density in the putamen and nucleus caudatus of subjects with Parkinson's disease (Farkas et al. 2012a) (Table 10.2). Other postmortem studies showed reductions in the expression of CB_1 receptor messenger RNA (mRNA) in the caudate nucleus, anterior dorsal putamen, and external segment of the globus pallidus (Hurley et al. 2003) (Table 10.2). A recent PET study has reported a reduced CB_1 receptor availability in the SN with an increased receptor availability in nigrostriatal, mesolimbic, and mesocortical dopaminergic projection areas (Van Laere et al. 2012) (Table 10.2).

In Alzheimer's disease, compared to normal brains, an early postmortem investigation reported reductions in the density of CB_1 receptors in several brain regions (Westlake et al. 1994). In this study, the specific binding of the agonist [^{3}H]CP55940 was strongly reduced in the hippocampus and caudate nucleus and to a lesser extent in the SN and globus pallidus (Table 10.2). In contrast, the expression of CB_1 receptor mRNA did not differ between Alzheimer's and control brains (Westlake et al. 1994) (Table 10.2). In line with these findings, G-protein coupling and CB_1 receptor protein expression were also shown markedly decreased in the frontal cortex of subjects with Alzheimer's disease (Ramírez et al. 2005) (Table 10.2). In these Alzheimer's brains, moreover, protein nitration was increased, and, more specifically, CB_1 and CB_2 receptor proteins showed enhanced nitration (Ramírez et al. 2005). In contrast, a recent autoradiographic study with [^{125}I]SD7015 has shown upregulation of CB_1 receptors in the prefrontal cortex of subjects with Alzheimer's disease (Farkas et al. 2012b) (Table 10.2). Another immunohistochemical study has reported that CB_1 receptor density was not modified in hippocampus and entorhinal cortex sections from brains of Alzheimer's disease patients (Benito et al. 2003) (Table 10.2). This latter study also showed that FAAH protein and activity as well as CB_2 receptor protein in Alzheimer's disease were selectively overexpressed in glial cells (Benito et al. 2003). Another study has also reported no differences in the immunoreactivity of cannabinoid CB_1 receptors in various areas of the forebrain and basal ganglia of subjects with Alzheimer's disease, a negative finding corroborated with saturation binding assays using the antagonist [^{3}H]SR141716A (rimonabant) (Lee et al. 2010) (Table 10.2).

In Huntington's disease, postmortem quantitative autoradiographic studies with [^{3}H]CP55940 revealed a massive loss of CB_1 receptors in the SN (pars reticulata) of subjects with this neurodegenerative process (Glass et al. 1993). In an independent autoradiographic investigation, the density of CB_1 receptors in striatum and pallidum was also markedly decreased in Huntington's disease (Richfield and Herkenham 1994). Similarly, CB_1 receptor immunoreactivity was markedly reduced in the globus pallidus of subjects with Huntington's disease (Allen et al. 2009) (Table 10.2). These postmortem findings indicating a loss of CB_1 receptors in specific brain regions agree well with the known massive death of GABAergic neurons (enriched in CB_1 receptors) in the neostriatum of subjects with Huntington's disease (DiFiglia 1990). An interesting investigation assessed the distribution and density changes of CB_1 receptors in the basal ganglia in early, intermediate, and advanced neuropathological grades of Huntington's disease (Glass et al. 2000) (Table 10.2). The results showed that the very early stages of the disease were characterized by a major loss of CB_1 receptors in the caudate nucleus, putamen, and globus pallidus externus; the intermediate neuropathological grades were associated with further decreases of CB_1 receptors, and advanced neuropathological grades revealed an almost total loss of CB_1 receptors (Glass et al. 2000). In line with these findings, a PET study has also reported decreases of CB_1 receptor availability in various brain regions (gray matter of cerebrum, cerebellum, and brainstem) in symptomatic patients with Huntington's disease, including the early stages of the disease (Van Laere et al. 2010) (Table 10.2).

10.4.2.3 Alcohol Dependence

A hyperactivity of the endocannabinoid signaling system has been reported in the prefrontal cortex of suicidal alcoholic subjects compared to alcoholic subjects dying of causes other than suicide. These suicidal alcoholic subjects showed a greater CB_1 receptor density and functionality through G protein signaling, as well as higher contents of AEA and 2-AG in brain (Vinod et al. 2005) (Table 10.2). The same group reported decreased CB_1 receptor binding and functionality in the ventral striatum of nonsuicidal alcoholic subjects compared to controls (Vinod et al. 2010) (Table 10.2). However, these parameters were elevated in the suicidal alcoholics when compared to nonsuicidal alcoholic subjects (Vinod et al. 2010). On the other hand, it has been reported that the C allele of the single nucleotide polymorphism (SNP) rs2023239 of the gene that codes for the CB_1 receptor is associated with greater CB_1 receptor binding in postmortem prefrontal cortex, greater alcohol cue-elicited brain activation in the midbrain and prefrontal cortex, greater subjective reward when consuming alcohol, and more positive outcomes after treatment with a medication that targets the mesocorticolimbic neurocircuitry (Hutchison et al. 2008). In regard to the differences between Cloninger type 1 and 2 alcoholics, reduced AEA contents were observed in the NAcc and frontal cortex in type 1 alcoholics (Lehtonen et al. 2010). These findings suggest that endocannabinoids, and mainly AEA, are increased in specific brain regions of impulsive type 2 alcoholics. In contrast, brain AEA content was decreased in anxiety-prone type 1 alcoholics (Lehtonen et al. 2010).

10.4.2.4 Drug Addiction: Cannabis, Cocaine, and Opiates

Chronic CB drug exposure in laboratory animals leads to drug tolerance and dependence, demonstrating that these drugs of abuse possess addictive properties (Hutcheson et al. 1998; Aceto et al. 2001; Lichtman and Martin 2005). The endocannabinoids can also participate, as a modulatory system, in the mechanisms of other drugs of abuse including cocaine and opiates (Maldonado et al. 2006). For example, a complex crosstalk between CB and opioid receptors has been unraveled (e.g., see Fattore et al. 2011; Scavone et al. 2013). A postmortem study has shown that the chronic abuse of marijuana (heavy user subjects) was associated with reduced CB_1 receptor density ([^{3}H]SR141716A antagonist binding) in various regions (NAcc, caudate nucleus, putamen, hippocampus, mesencephalon, and others) of the human brain (Villares 2007) (Table 10.2). Furthermore, the number of CB_1 receptor mRNA–positive neurons was also reduced in various brain regions of heavy cannabis users compared with control brains (Villares 2007) (Table 10.2). In marked contrast, significant increases in the density of CB_1 receptors, using the agonist radioligand [^{3}H]CP55940, have been reported in the caudate-putamen areas from subjects who had been taking cannabis within 5 days of death, which was independent of a diagnosis of schizophrenia (Dean et al. 2001) (Table 10.2). These striking differences may reflect the use of different radioligand (agonist or antagonist receptor sites), the outcomes of long-term cannabis use, the different routes of cannabis intake, or brain regional differences in the effects of THC in humans.

In laboratory animals, chronic treatment with cocaine was shown to decrease the expression of CB_1 receptor mRNA without altering the number of receptor agonist binding sites (^{3}H-CP55940) in rat brain cortex (González et al. 2002). Other studies have shown that chronic cocaine increased AEA content (partly due to FAAH inhibition) and potentiated the effect of the CB_1 receptor agonist HU210 in the rat corpus striatum (Centonze et al. 2004). In human cocaine addiction, however, the immunodensity of CB_1 receptor protein was markedly decreased in the prefrontal cortex of pure cocaine abusers, whereas receptor protein content was not significantly altered in mixed cocaine/opiate addicts and pure opiate (heroin/methadone) addicts (Table 10.2, Fig. 10.5a). In contrast, cortical CB_2 receptor protein in cocaine addicts was similar to that quantified in control subjects (Álvaro-Bartolomé and García-Sevilla, 2013). In mice, acute cocaine exposure increased the activation of mTOR (mammalian target of rapamycin) in brain cortex (Fig. 10.5b). Interestingly, chronic treatment with cocaine was associated with the induction of tolerance to the acute activation of cortical mTOR (Fig. 10.5b). Similarly, the basal activation of mTOR in the prefrontal cortex of long-term cocaine addicts was not significantly altered when compared with that quantified in matched controls (Fig. 10.5b). These postmortem findings strongly suggest that cocaine addiction in humans induces downregulation of CB_1 receptors and dampens the associated mTOR signaling in the prefrontal cortex.

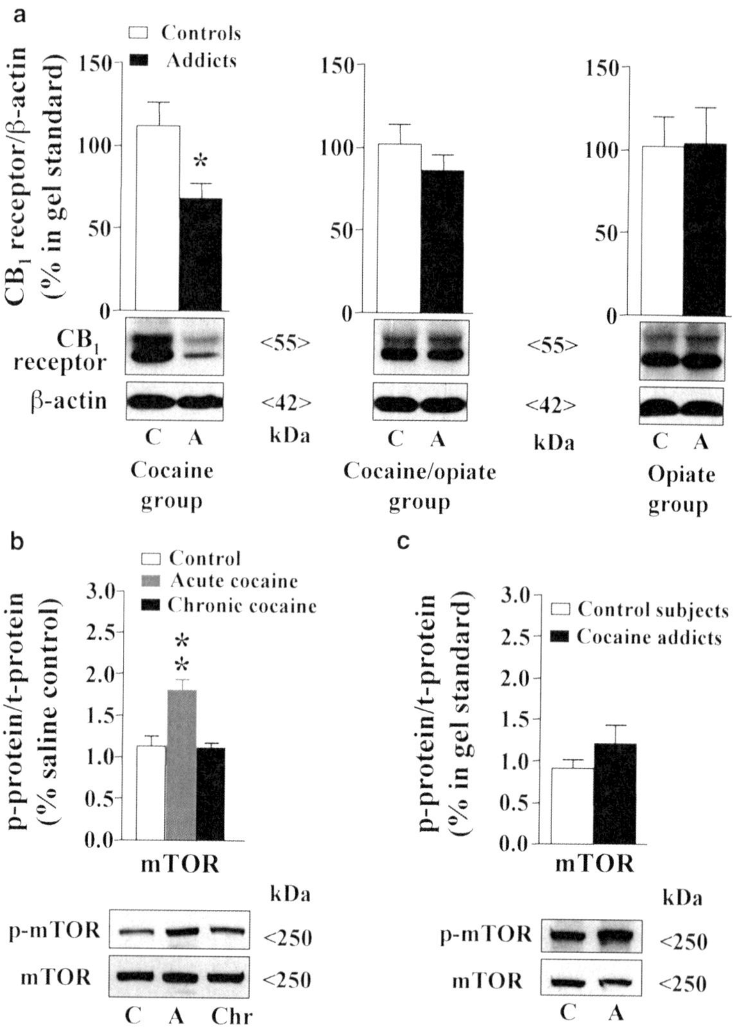

Fig. 10.5 **a** Immunodensity of cannabinoid CB_1 receptor in the prefrontal cortex of control subjects (C, $n=8$–11), pure cocaine addicts (A, $n=9$), mixed cocaine/opiate addicts (A, $n=11$), and pure opiate addicts (C, $n=8$), expressed as mean ± standard error of mean percentages of an in-gel standard (100 %, pool of control samples) (*$P<0.001$ when compared with the corresponding control group, C). **b** Effects of acute (20 mg/kg, i.p. 2 h) and chronic (40 mg/kg, i.p., 7 days) treatments with cocaine on the activation of mammalian target of rapamycin (ratio of phosphorylated mTOR to total mTOR) in mouse brain cortex, expressed as percentages of saline-treated animals (control). **c** Activation of mTOR (ratio of phosphorylated mTOR to total mTOR) in the prefrontal cortex of control subjects ($n=9$) and long-term cocaine addicts ($n=9$), expressed as percentages of an in-gel standard (100 %, pool of control samples). The molecular masses (kDa) of target proteins were estimated from referenced standards. (Modified from Álvaro-Bartolomé and García-Sevilla 2013)

Acknowledgment The studies performed in the authors' laboratories were funded by grants SAF2009-08460 (MINECO/FEDER), Basque Government (IT199-07, IT616-13, SAIOTEK S-PE12UN033) and the University of the Basque Country (UFI 11/35) to JJMM and LFC, and grants SAF2004-03685 and SAF2008-01311 (MINECO/FEDER) and 2007I032 (Plan Nacional sobre Drogas) to JAGS. The studies were also funded by RETICS-RTA (RD12/0028/0011; Instituto de Salud Carlos III, MINECO/FEDER) to JAGS. J.A. García-Sevilla is a member of the Institut d'Estudis Catalans (Barcelona, Catalonia, Spain).

Author Contributions The authors jointly wrote the manuscript. The authors declare no competing financial interests.

References

Aceto MD, Scates SM, Martin BB (2001) Spontaneous and precipitated withdrawal with a synthetic cannabinoid, WIN 55212-2. Eur J Pharmacol 416:75–81

Alger BE (2002) Retrograde signaling in the regulation of synaptic transmission: focus on endocannabinoids. Prog Neurobiol 68:247–286

Allen KL, Waldvogel HJ, Glass M, Faull RL (2009) Cannabinoid CB_1, $GABA_A$ and $GABA_B$ receptor subunit changes in the globus pallidus in Huntington's disease. J Chem Neuroanat 37:266–281

Álvaro-Bartolomé M, García-Sevilla JA (2013) Dysregulation of cannabinoid CB1 receptor and associated signaling networks in brains of cocaine addicts and cocaine-treated rodents. Neuroscience, doi: http://dx.doi.org/10.1016/j.neuroscience.2013.05.035

Ashton CH, Moore PB (2011) Endocannabinoid system dysfunction in mood and related disorders. Acta Psychiatr Scand 124:250–261

Atwood BK, Straiker A, Mackie K (2012) CB_2: therapeutic target-in-waiting. Progr Neuropsychopharmacol Biol Psychiatry 38:16–20

Bambico FR, Duranti A, Tontini A, Tarzia G, Gobbi G (2009) Endocannabinoids in the treatment of mood disorders: evidence from animal models. Curr Pharm Des 15:1623–1646

Bambico FR, Katz N, Debonnel G, Gobbi G (2007) Cannabinoids elicit antidepressant-like behavior and activate serotonergic neurons through the medial prefrontal cortex. J Neurosci 27:11700–11711

Bambico FR, Nguyen NT, Katz N, Gobbi G (2010) Chronic exposure to cannabinoids during adolescence but not during adulthood impairs emotional behavior and monoaminergic neurotransmission. Neurobiol Dis 37:641–655

Battista N, Bari M, Tarditi A, Mariotti C, Bachoud-Levi AC, Zuccato C, Finazzi-Agro A, Genitrini S, Peschanski M, Di Donato S, Cattaneo E, Maccarrone M (2007) Severe deficiency of the fatty acid amide hydrolase (FAAH) activity segregates with the Huntington's disease mutation in peripheral lymphocytes. Neurobiol Dis 27:108–116

Benito C, Nunez E, Tolon RM, Carrier EJ, Rabano A, Hillard CJ, Romero J (2003) Cannabinoid CB2 receptors and fatty acid amide hydrolase are selectively overexpressed in neuritic plaque-associated glia in Alzheimer's disease brains. J Neurosci 23:11136–11141

Canals M, Milligan G (2008) Constitutive activity of the cannabinoid CB1 receptor regulates the function of co-expressed mu opioid receptors. J Biol Chem 283:11424–11434

Carvalho AF, Mackie K, Van Bockstaele EJ (2010) Cannabinoid modulation of limbic forebrain noradrenergic circuitry. Eur J Neurosci 31:286–301

Carvalho AF, Van Bockstaele EJ (2012) Cannabinoid modulation of noradrenergic circuits: implications for psychiatric disorders. Progr Neuropsychopharmacol Biol Psychiatry 38:59–67

Centonze D, Battista N, Rossi S, Mercuri NB, Finazzi-Agrò A, Bernardi G, Calabresi P, Maccarrone M (2004) A critical interaction between dopamine D2 receptors and endocannabinoids mediates the effects of cocaine on striatal gabaergic transmission. Neuropsychopharmacology 29:1488–1497

Cheer JF, Marsden CA, Kendall DA, Mason R (2000) Lack of response suppression follows repeated ventral tegmental cannabinoid administration: an in vitro electrophysiological study. Neuroscience 99:661–667

Choi K, Le T, McGuire J, Xing G, Zhang L, Li H, Parker CC, Johnson LR, Ursano RJ (2012) Expression pattern of the cannabinoid receptor genes in the frontal cortex of mood disorder patients and mice selectively bred for high and low fear. J Psychiatric Res 46:882–889

Dalton VS, Long LE, Weickert CS, Zavitsanou K (2011) Paranoid schizophrenia is characterized by increased CB1 receptor binding in the dorsolateral prefrontal cortex. Neuropsychopharmacol 36:1620–1630

De Marchi N, De Petrocellis L, Orlando P, Daniele F, Fezza F, Di Marzo V (2003) Endocannabinoid signaling in the blood of patients with schizophrenia. Lipids Health Dis 2:5

Dean B, Sundram S, Brabury R, Scarr E, Copolov D (2001) Studies on [^{3}H]CP-55940 binding in the human central nervous system: regional specific changes in density of cannabinoid-1 receptors associated with schizophrenia and cannabis use. Neuroscience 103:9–15

Deng C, Han M, Huang XF (2007) No changes in densities of cannabinoid receptors in the superior temporal gyrus in schizophrenia. Neurosci Bull 23:341–347

DiFiglia M (1990) Excitotoxic injury of the neostriatum: a model for Huntington's disease. Trends Neurosci 13:286–289

Dlugos A, Childs E, Stuhr KL, Hillard CJ, de Wit H (2012) Acute stress increases circulating anandamide and other N-acylethanolamines in healthy humans. Neuropsychopharmacol 37:2416–2427

Eggan SM, Hashimoto T, Lewis DA (2008) Reduced cortical cannabinoid 1 receptor messenger RNA and protein expression in schizophrenia. Arch Gen Psychiatry 65:772–784

Eggan SM, Stoyak SR, Verrico CD, Lewis DA (2010) Cannabinoid CB1 receptor immunoreactivity in the prefrontal cortex: comparison of schizophrenia and major depressive disorder. Neuropsychopharmacology 35:2060–2071

Esteban S, García-Sevilla JA (2012) Effects induced by cannabinoids on monoaminergic systems in the brain and their implications for psychiatric disorders. Progr Neuropsychopharmacol Biol Psychiatry 38:78–87

Esteban S, Lladó J, García-Sevilla JA (1996) α_2-Autoreceptors and α_2-heteroreceptors modulating tyrosine and tryptophan hydroxylase activity in the rat brain in vivo: an investigation into the α_2-adrenoceptor subtypes. Naunyn-Schmiedebergs Arch Pharmacol 353:391–399

Fattore L, Spano MS, Melis V, Fadda P, Fratta W (2011) Differential effect of opioid and cannabinoid receptor blockade on heroin-seeking reinstatement and cannabinoid substitution in heroin-abstinent rats. Br J Pharmacol 163:1550–1562

Farkas S, Nagy K, Jia Z, Harkany T, Palkovits M, Donohou SR, Pike VW, Halldin C, Mathe D, Csiba L, Gulyas B (2012a) The decrease of dopamine D2/D3 receptor densities in the putamen and nucleus caudatus goes parallel with maintained levels of CB1 cannabinoid receptors in Parkinson's disease: a preliminary autoradiographic study with the selective dopamine D2/D3 antagonist [3H]raclopride and the novel CB1 inverse agonist [^{125}I]SD7015. Brain Res Bull 87:504–510

Farkas S, Nagy K, Palkovits M, Kovacs GG, Jia Z, Donohue S, Pike V, Halldin C, Mathe D, Harkany T, Gulyas B, Csiba L (2012b) [125I]SD7015 reveals fine modalities of CB1 cannabinoid receptor density in the prefrontal cortex during progression of Alzheimer's disease. Neurochem Int 60:286–291

Fernández-Ruiz J, Pazos MR, García-Arencibia M, Sagredo O, Ramos JA (2008) Role of CB2 receptors in neuroprotective effects of cannabinoids. Mol Cell Endocrinol 286:S91–S96

Fink KB, Göthert M (2007) 5-HT receptor regulation of neurotransmitter release. Pharmacol Rev 59:360–417

Gifford AN, Ashby CR Jr (1996) Electrically evoked acetylcholine release from hippocampal slices is inhibited by the cannabinoid receptor agonist, WIN 552122–2, and is potentiated by the cannabinoid antagonist, SR 141716A. J Pharmacol Exp Ther 277:1431–1436

Giuffrida A, Leweke FM, Gerth CW, Schreiber D, Koethe D, Faulhaber J, Klosterkötter J, Piomelli D (2004) Cerebrospinal anandamide levels are elevated in acute schizophrenia and are inversely correlated with psychotic symptoms. Neuropsychopharmacology 29:2108–2114

Glass M, Dragunow M, Faull RL (2000) The pattern of neurodegeneration in Huntington's disease: a comparative study of cannabinoid, dopamine, adenosine and GABA-A receptor alterations in the human basal ganglia in Huntington's disease. Neuroscience 97:505–519

Glass M, Faull RL, Dragunow M (1993) Loss of cannabinoid receptors in the substantia nigra in Huntington's disease. Neuroscience 56:523–527

Gobbi G, Bambico FR, Mangieri R, Bortolato M, Campolongo P, Solinas M et al (2005) Antidepressant-like activity and modulation of brain monoaminergic transmission by blockade of anandamide hydrolysis. Proc Natl Acad Sci U S A 102:18620–18625

González S, Fernández-Ruiz J, Sparpaglione V, Parolaro D, Ramos JA (2002) Chronic exposure to morphine, cocaine or ethanol in rats produced different effects in brain cannabinoid CB1 receptor binding and mRNA levels. Drug Alcohol Depend 66:77–84

Gorzalka BB, Hill MN (2011) Putative role of endocannabinoid signaling in the etiology of depression and actions of antidepressants. Progr Neuropsychopharmacol Biol Psychiatry 35:1575–1585

Griebel G, Stemmelin J, Scatton B (2005) Effects of the cannabinoid CB_1 receptor antagonist rimonabant in models of emotional reactivity in rodents. Biol Psychiatry 57:261–267

Häring M, Marsicano G, Lutz B, Monory K (2007) Identification of the cannabinoid receptor type 1 in serotonergic cells of raphe nuclei in mice. Neuroscience 146:1212–1219

Hashimotodani Y, Ohno-Shosaku T, Kano M (2007) Endocannabinoids and synaptic function in the CNS. Neuroscientist 13:127–137

Herkenham M, Lynn AB, Johnson MR, Melvin LS, de Costa BR, Rice KC (1991) Characterization and localization of cannabinoid receptors in rat brain: a quantitative in vitro autoradiographic study. J Neurosci 11:563–583

Hill MN, Gorzalka BB (2005a) Is there a role for the endocannabinoid system in the etiology and treatment of melancholic depression? Behav Pharmacol 16:333–352

Hill MN, Gorzalka BB (2005b) Pharmacological enhancement of cannabinoid CB1 receptor activity elicits an antidepressant-like response in the rat forced swim test. Eur Neuropsychopharmacol 15:593–599

Hill MN, Miller GE, Carrier EJ, Gorzalka BB, Hillard CJ (2009) Circulating endocannabinoids and N-acylethanolamines are differentially regulated in major depression and following exposure to social stress. Psychoneuroendocrinology 34:1257–1262

Hill MN, Miller GE, Ho WSV, Gorzalka BB, Hillard CJ (2008) Serum endocannabinoid content is altered in females with depressive disorders: a preliminary report. Pharmacopsychiatry 41:48–53

Hillard CJ, Bloom AS (1982) Delta 9-tetrahydrocannabinol-induced changes in beta-adrenergic receptor binding in mouse cerebral cortex. Brain Res 235:370–377

Hohmann AG, Herkenham M (2000) Localization of cannabinoid CB(1) receptor mRNA in neuronal subpopulations of rat striatum: a double-label in situ hybridization study. Synapse 37:71–80

Howlett AC, Barth F, Bonner TI, Cabral G, Casellas P, Devane WA et al (2002) International Union of Pharmacology. XXVII. Classification of cannabinoid receptors. Pharmacol Rev 54:161–202

Howlett AC, Reggio PH, Childers SR, Hampson RE, Ulloa NM, Deutsch DG (2011) Endocannabinoid tone versus constitutive activity of cannabinoid receptors. Br J Pharmacol 163:1329–1343

Hruba L, Ginsburg BC, McMahon LR (2012) Apparent inverse relationship between cannabinoid agonist efficacy and tolerance/cross-tolerance produced by Δ^9-tetrahydrocannabinol treatment in rhesus monkeys. J Pharmacol Exp Ther 342:843–849

Hungund BL, Vinod KY, Kassir SA, Basavarajappa BS, Yalamanchili R, Cooper TB et al (2004) Upregulation of CB_1 receptors and agonist-stimulated [^{35}S]GTPγS binding in the prefrontal cortex of depressed suicide victims. Mol Psychiatry 9:184–190

Hurley MJ, Mash DC, Jenner P (2003) Expression of cannabinoid CB1 receptor mRNA in basal ganglia of normal and parkinsonian human brain. J Neural Transm 110:1279–1288

Hutcheson DM, Tzavara ET, Smadja C, Valjent E, Roques BP, Hanoune J, Maldonado R (1998) Behavioral and biochemical evidence for signs of abstinence in mice chronically treated with delta-9-tetrahydrocannabinol. Br J Pharmacol 125:1567–1577

Hutchison KE, Haughey H, Niculescu M, Schacht J, Kaiser A, Stitzel J, Horton WJ, Filbey F (2008) The incentive salience of alcohol: translating the effects of genetic variant in CNR1. Arch Gen Psychiatry 65:841–850

Ichikawa J, Meltzer HY (2000) The effect of serotonin1A receptor agonism on antipsychotic drug-induced dopamine release in rat striatum and nucleus accumbens. Brain Res 858:252–263

Jenko KJ, Hirvonen J, Henter ID, Anderson KB, Zoghbi SS, Hyde TM, Deep-Soboslay A, Innis RB, Kleinman JE (2012) Binding of a tritiated inverse agonist to cannabinoid CB1 receptors is increased in patients with schizophrenia. Schizophr Res 141:185–188

Katona I, Sperlagh B, Sik A, Kafalvi A, Vizi ES, Mackie K et al (1999) Presynaptically located CB1 cannabinoid receptors regulate GABA release from axon terminals of specific hippocampal interneurons. J Neurosci 19:4544–4558

Koethe D, Giuffrida A, Schreiber D, Hellmich M, Schultze-Lutter F, Ruhrmann S, Klosterkötter J, Piomelli D, Leweke FM (2009) Anandamide elevation in cerebrospinal fluid in initial prodromal states of psychosis. Br J Psychiatry 194:371–372

Koethe D, Llenos IC, Dulay JR, Hoyer C, Torrey EF, Leweke FM, Weis S (2007) Expression of CB1 cannabinoid receptor in the anterior cingulate cortex in schizophrenia, bipolar disorder, and major depression. J Neural Transm 114:1055–1063

Koppel J, Bradshaw H, Goldberg TE, Khalili H, Marambaud P, Walker MJ, Pazos M, Gordon ML, Christen E, Davies P (2009) Endocannabinoids in Alzheimer's disease and their impact on normative cognitive performance: a case-control and cohort study. Lipids Health Dis 8:2

Lastres-Becker I, Cebeira M, de Ceballos ML, Zeng BY, Jenner P, Ramos JA, Fernandez-Ruiz JJ (2001) Increased cannabinoid CB1 receptor binding and activation of GTP-binding proteins in the basal ganglia of patients with Parkinson's syndrome and of MPTP-treated marmosets. Eur J Neurosci 14:1827–1832

Lazary J, Juhasz G, Hunyady L, Bagdy G (2011) Personalized medicine can pave the way for the safe use of CB_1 receptor antagonists. Trends Pharmacol Sci 32:270–280

Lee JH, Agacinski G, Williams JH, Wilcock GK, Esiri MM, Francis PT, Wong PT, Chen CP, Lai MK (2010) Intact cannabinoid CB1 receptors in the Alzheimer's disease cortex. Neurochem Int 57:985–989

Lehtonen M, Storvik M, Tupala E, Hyytia P, Tiihonen J, Callaway JC (2010) Endogenous cannabinoids in post-mortem brains of Cloninger type 1 and 2 alcoholics. Eur Neuropsychopharmacol 20:245–252

Leterrier C, Bonnard D, Carrel D, Rossier J, Lenkei Z (2004) Constitutive endocytic cycle of the CB1 cannabinoid receptor. J Biol Chem 279:36013–36021

Leweke FM, Giuffrida A, Koethe D, Schreiber D, Nolden BM, Kranaster L, Neatby MA, Schneider M, Gerth CW, Hellmich M, Klosterkötter J, Piomelli D (2007) Anandamide levels in cerebrospinal fluid of first-episode schizophrenic patients: impact of cannabis use. Schiz Res 94:29–36

Leweke FM, Giuffrida A, Wurster U, Emrich HM, Piomelli D (1999) Elevated endogenous cannabinoids in schizophrenia. Neuroreport 10:1665–1667

Lewis DA, Hashimoto T, Volk DW (2005) Cortical inhibitory neurons and schizophrenia. Nat Rev Neurosci 6:312–324

Lichtman AH, Martin BR (2005) Cannabinoid tolerance and dependence. Handb Exp Pharmacol 168:691–717

Maldonado R, Valverde O, Berrendero F (2006) Involvement of the endocannabinoid system in drug addiction. Trends Neurosci 29:225–232

Mathers FM, Ghodse AH (1992) Cannabis and psychotic illness. Br J Psychiatry 161:648–653

Mato S, Vidal R, Castro E, Díaz A, Pazos A, Valdizán EM (2010) Long-term fluoxetine treatment modulates cannabinoid type 1 receptor-mediated inhibition of adenylyl cyclase in the rat prefrontal cortex through 5-hydroxytryptamine$_{1A}$ receptor-dependent mechanisms. Mol Pharmacol 77:424–434

Matsuda LA, Bonner TI, Lolait SJ (1993) Localization of cannabinoid receptor mRNA in rat brain. J Comp Neurol 327:535–550

McLaughlin RJ, Hill MN, Bambico FR, Stuhr KL, Gobbi G, Hillard CJ, Gorzalka BB (2012) Prefrontal cortical anandamide signaling coordinates coping responses to stress through a serotonergic pathway. Eur. Neuropsychopharamcol 22:664–671

Mechoulam R, Parker LA (2013) The endocannabinoid system and the brain. Annu Rev Psychol 64:6.1–6.27

Miyamoto S, Duncan GE, Mar CE, Lieberman JA (2005) Treatments for schizophrenia: a critical review of pharmacology and mechanisms of action of antipsychotic drugs. Mol Psychiatry 10:79–104

Moranta D, Esteban S, García-Sevilla JA (2009) Chronic treatment and withdrawal of the cannabinoid agonist WIN55,212–2 modulate the sensitivity of presynaptic receptors involved in the regulation of monoamine syntheses in rat brain. Naunyn Schmiedebergs Arch Pharmacol 379:61–72

Morera-Herreras T, Ruiz-Ortega JA, Gómez-Urquijo S, Ugedo L (2008) Involvement of subthalamic nucleus in the stimulatory effect of Δ^9-tetrahydrocannabinol on dopaminergic neurons. Neuroscience 151:817–823

Muguruza C, Lehtonen M, Aaltonen N, Arrieta J, Morentin B, Meana JJ, Callado LF (2012) Altered levels of endocannabinoids in post-mortem human brain of schizophrenic subjects. Int J Neuropsychopharmacol 15:126–127

Nakazi M, Bauer U, Nickel T, Kathmann M, Schlicker E (2000) Inhibition of serotonin release in the mouse brain via presynaptic cannabinoid CB1 receptors. Naunyn-Schmiedebergs Arch Pharmacol 361:19–24

Newell KA, Deng C, Huang XF (2006) Increased cannabinoid receptor density in the posterior cingulate cortex in schizophrenia. Exp Brain Res 172:556–560

Nissen SE, Nicholls SJ, Wolski K, Rodés-Cabau J, Cannon CP, Deanfield JE et al (2008) Effect of rimonabant on progression of atherosclerosis in patients with abdominal obesity and coronary artery disease: the STRADIVARIUS randomized controlled trial. JAMA 299:1547–1560

Onaivi ES, Ishiguro H, Gu S, Liu Q-R (2012) CNS effects of CB2 cannabinoid receptors beyond neuro-immuno-cannabinoid activity. J Psychopharmacol 26:92–103

Oropeza VC, Mackie K, Van Bockstaele EJ (2007) Cannabinoid receptors are localized to noradrenergic axon terminals in the rat frontal cortex. Brain Res 1127:36–44

Oropeza VC, Page ME, Van Bockstaele EJ (2005) Systemic administration of WIN 55,212–2 increases norepinephrine release in the rat frontal cortex. Brain Res 1046:45–54

Ortega-Alvaro A, Aracil-Fernández A, García-Gutiérrez AS, Navarrete F, Manzanares J (2011) Deletion of CB_2 cannabinoid receptor induces schizophrenia-related behaviors in mice. Neuropsychopharmacology 36:1489–1504

Page ME, Oropeza VC, Sparks SE, Qian Y, Menko AS, Van Bockstaele EJ (2007) Repeated cannabinoid administration increases indices of noradrenergic activity in rats. Pharmacol Biochem Behav 86:162–168

Parolaro D, Realini N, Vigano D, Guidali C, Rubino T (2010) The endocannabinoid system and psychiatric disorders. Exp Neurol 224:3–14

Pertwee RG, Howlett AC, Abood ME, Alexander SP, Di Marzo V, Elphick MR et al (2010) International Union of Basic and Clinical Pharmacology. LXXIX. Cannabinoid receptors and their ligands: beyond CB1 and CB2. Pharmacol Rev 62:588–631

Pisani A, Fezza F, Galati S, Battista N, Napolitano S, Finazzi-Agro A, Bernardi G, Brusa L, Pierantozzi M, Stanzione P, Maccarrone M (2005) High endogenous cannabinoid levels in the cerebrospinal fluid of untreated Parkinson's disease patients. Ann Neurol 57:777–779

Pisani V, Moschella V, Bari M, Fezza F, Galati S, Bernardi G, Stanzione P, Pisani A, Maccarrone M (2010) Dynamic changes of anandamide in the cerebrospinal fluid of Parkinson's disease patients. Mov Disord 25:920–924

Ramirez BG, Blazquez C, Gomez del Pulgar T, Guzman M, de Ceballos ML (2005) Prevention of Alzheimer's disease pathology by cannabinoids: neuroprotection mediated by blockade of microglial activation. J Neurosci 25:1904–1913

Richfield EK, Herkenham M (1994) Selective vulnerability in Huntington's disease: preferential loss of cannabinoid receptors in lateral globus pallidus. Ann Neurol 36:577–584

Scavone JL, Mackie K, Van Bockstaele EJ (2010) Characterization of cannabinoid-1 receptors in the locus coeruleus: relationship with mu-opioid receptors. Brain Res 1312:18–31

Scavone JL, Sterling RC, Van Bockstaele EJ (2013) Cannabinoid and opioid interactions: implications for opiate dependence and withdrawal. Neuroscience, http://dx.doi.org/10.1016/j.neuroscience2013.04.034

Schlicker E, Kathmann M (2001) Modulation of transmitter release via presynaptic cannabinoid receptors. Trends Pharmacol Sci 22:565–572

Sim-Selley LJ (2003) Regulation of cannabinoid CB_1 receptors in the central nervous system by chronic cannabinoids. Crit Rev Neurobiol 15:91–119

Starke K (2001) Presynaptic autoreceptors in the third decade: focus on α_2-adrenoceptors. J Neurochem 78:685–693

Urigüen L, García-Fuster J, Callado LF, Morentin B, La Harpe R, Casadó V, Lluis C, Franco R, García-Sevilla JA, Meana JJ (2009) Immunodensity and mRNA expression of A_{2A} adenosine, D_2 dopamine, and CB_1 cannabinoid receptors in postmortem frontal cortex of subjects with schizophrenia: effect of antipsychotic treatment. Psychopharmacol (Berl) 206:313–324

Valdizán EM, Mato S, González-Maeso J, Rodríguez-Puertas R, Meana JJ, Sallés J, et al (2011) Functionality of cannabinoid receptors in the prefrontal cortex of major depression suicide victims: influence of antidepressant treatment at the time of death. XIV Congreso Nacional Sociedad Española de Neurociencia (SENC), Salamanca, September 2011. Abstract P-105

Valverde O, Torrens M (2012) CB_1 receptor-deficient mice as a model for depression. Neuroscience 204:193–206

Van Laere K, Casteels C, Dhollander I, Goffin K, Grachev I, Bormans G, Vandenberghe W (2010) Widespread decrease of type 1 cannabinoid receptor availability in Huntington disease in vivo. J Nucl Med 51:1413–1417

Van Laere K, Casteels C, Lunskens S, Goffin K, Grachev ID, Bormans G, Vandenberghe W (2012) Regional changes in type 1 cannabinoid receptor availability in Parkinson's disease in vivo. Neurobiol aging 33:621–628

Vaughan CW, Christie MJ (2005) Retrograde signaling by endocannabinoids. Handb Exp Pharmacol 168:367–383

Villares J (2007) Chronic use of marijuana decreases cannabinoid receptor binding and mRNA expression in the human brain. Neuroscience 145:323–334

Vinod KY, Arango V, Xie S, Kassir SA, Mann JJ, Cooper TB et al (2005) Elevated levels of endocannabinoids and CB_1 receptor-mediated G-protein signaling in the prefrontal cortex of alcoholic suicide victims. Biol Psychiatry 57:480–486

Vinod KY, Kassir SA, Hungund BL, Cooper TB, Mann JJ, Arango V (2010) Selective alterations of the CB1 receptors and the fatty acid amide hydrolase in the ventral striatum of alcoholics and suicides. J Psychiatr Res 44:591–597

Westlake TM, Howlett AC, Bonner TI, Matsuda LA, Herkenham M (1994) Cannabinoid receptor binding and messenger RNA expression in human brain: an in vitro receptor autoradiography and in situ hybridization histochemistry study of normal aged and Alzheimer's brains. Neuroscience 63:637–652

Wong DF, Kuwabara H, Horti AG, Raymont V, Brasic J, Guevara M, Ye W, Dannals RF, Ravert HT, Nandi A, Rahmin A, Ming JE, Grachev I, Roy C, Cascella N (2010) Quantification of cerebral cannabinoid receptors subtype 1 (CB1) in healthy subjects and schizophrenia by the novel PET radioligand [11C]OMAR. Neuroimage 52:1505–1513

Wu X, French ED (2000) Effects of chronic D^9-tetrahydrocannabinol on rat midbrain dopamine neurons: an electrophysiological assessment. Neuropharmacology 39:391–398

Zavitsanou K, Garrick T, Huang XF (2004) Selective antagonist [^{3}H]SR141716A binding to cannabinoid CB_1 receptors is increased in the anterior cingulate cortex in schizophrenia. Prog Neuropsychopharmacol Biol Psychiatry 28:355–360

Chapter 11
Endocannabinoid Signaling and the Regulation of the Serotonin System

Samir Haj-Dahmane and Roh-Yu Shen

Abstract Endogenous cannabinoids, also called endocannabinoids (eCBs), are lipid signaling molecules in the mammalians' central nervous system (CNS), where they regulate neuronal functions and behaviors by activating cannabinoid receptors. The ubiquitous distribution of eCBs in neuronal populations that are associated with stress responses, such as dorsal raphe nucleus (DRn) serotonin (5-HT) neurons suggests that eCB signaling plays a central role in the regulation of stress-related behaviors. Consistent with this notion, human and animal studies have established that eCB signaling is a key modulator of emotional homeostasis and that a dysfunction of eCB signaling contributes to stress-related psychiatric disorders, including anxiety and depression. This leads to the current view that the eCB signaling could be an excellent target for the development of novel therapeutic intervention for stress-related mood disorders. Over the past few years, extensive research has focused on the functional interaction between eCB signaling and 5-HT systems. As a result, steady progress is made in our understanding of the cellular mechanisms by which eCB signaling regulates the function of 5-HT system. In this chapter, we review the most recent advances in our understanding of the cellular mechanisms by which eCBs modulate the function of the 5-HT system and how stress mediators regulate eCB signaling in the DRn.

Abbreviations

eCBs	endocannabinoids
5-HT	5-hydroxytryptamine (serotonin)
DRn	dorsal raphe nucleus
DAGLs	diacyl-glycerol lipase
COX-2	cyclooxygenase type 2
MGL	monoglyceride lipases
FAAH	fatty acid amid hydrolase
EPSC	excitatory postsynaptic current

S. Haj-Dahmane (✉) · R.-Y. Shen
Research Institute on Addictions, University at Buffalo, State University of New York, 1021 Main Street, Buffalo, NY 14203, USA
e-mail: dahmane@ria.buffalo.edu

E. J. Van Bockstaele (ed.), *Endocannabinoid Regulation of Monoamines in Psychiatric and Neurological Disorders*, DOI 10.1007/978-1-4614-7940-6_11,

WIN 55,212-2	R-(+)-[2,3-dihydro-5-methyl-3-(4-morpholinylmethyl)pyrrolol[1,2,3-de]-1,4-benzoxazin-6-yl]-1-naphthalenylmethanone mesylate
DSE	depolarization-induced suppression of excitation
PPR	paired-pulse ratio
CV	coefficient of variation
LTD	long-term depression
JZL 184	4-[Bis(1,3-benzodioxol-5-yl)hydroxymethyl]-1-piperidinecarboxyl acid 4-nitrophenyl ester
PF 750	N-phenyl-4-(3-quinolinylmethyl)-1-piperidinecarboxamide

11.1 Introduction

The identification of Δ^9-tetrahydrocannabinol (THC) as the main psychoactive ingredient in cannabis (Gaoni and Mechoulam 1964) along with the cloning of cannabinoid type 1 (CB1) and type 2 (CB2) receptors (Matsuda et al. 1990; Munro et al. 1993), and the isolation of their endogenous agonists (endocannabinoids, eCBs) (Devane et al. 1992; Mechoulam et al. 1995; Sugiura et al. 1995) are major cornerstones of modern cannabinoid research. Since these seminal discoveries, extensive work conducted over the last 2 decades has considerably enhanced our understanding of the molecular basis and function of the eCB signaling. Now, it is widely accepted that eCBs are critical modulators of synaptic function in the mammalian brain. Generally, eCBs act as retrograde messengers to induce transient or long lasting depression of neurotransmitter release at both the excitatory and inhibitory synapses (for review see, Kano et al. 2009). By modulating synaptic strength and plasticity throughout the brain, eCBs are involved in the regulation of an amazingly vast array of behavioral and physiological functions such as sleep, feeding behavior, mood, anxiety, pain, and stress (Mechoulam and Parker 2012).

The widespread distribution of CB1 receptors and eCBs in the stress neuronal circuits (Herkenham et al. 1991; Egortová and Elphick 2000; Egortová et al. 2003) suggests that eCB signaling plays a central role in the modulation of stress and emotional homeostasis. Consistent with this prediction, results from behavioral studies have established that eCBs reduce the behavioral and neuroendocrine responses to various stressors (Patel et al. 2004, 2005) and exert anxiolytic-like effects (Marco et al. 2004; Patel and Hillard 2006). In addition, a dysfunction of eCB signaling is implicated in stress-related psychiatric disorders, including anxiety and depression (Hillard et al. 2012). More importantly, the results of these studies suggest that the eCB signaling influences stress and emotional homeostasis, at least in part, by the modulation of the 5-hydroxytryptamine (5-HT) system (Marco et al. 2004; Greibel et al. 2005). Accordingly, numerous studies conducted over the last few years have focused on delineating the mechanisms by which eCB signaling modulates the function of 5-HT system. In this chapter, we review the new advances in eCB-mediated modulation of the 5-HT system with a particular emphasis on the modulation of the activity of dorsal raphe nucleus (DRn) 5-HT neurons. We will also discuss how stress mediators may alter eCB signaling in these neurons.

11.2 Biochemistry of Endocannabinoid System

The eCB system is composed of at least two G-protein-coupled receptors (GPCR), commonly known as CB1 and CB2 receptors (for review see, Howlett 2005), their endogenous agonists (eCBs), and the enzymatic machinery that synthesizes and metabolizes eCBs (for review see, Muccioli 2010). CB1 receptors are expressed at a high density throughout the central nervous system (CNS) (Herkenham et al. 1991; Matsuda et al. 1993; Egertová and Elphick 2000) and at a lower density in peripheral tissues and immune cells (Galiègue et al. 1995). In the CNS, CB1 receptors are predominantly located on both excitatory and inhibitory presynaptic terminals and play a prominent role in controlling neurotransmitter's release (for review see, Katona and Freund 2012). On the other hand, CB2 receptors are predominantly expressed in the peripheral tissues (Munro et al. 1993) and at a very limited level in neurons and microglias in the CNS (Van Sickle et al. 2005).

The two best characterized eCBs are anandamide and 2-arachidonyl glycerol (2-AG). Both eCBs are produced following an elevation of intracellular calcium (Ca^{2+}) concentration (Suguira et al. 1996; Stella et al. 1997). Anandamide is produced from enzymatic hydrolysis of the phospholipid precursor N-arachidonyl-phosphatidylethanolamines (NAPE) which is catalyzed by NAPE specific phospholipase D (NAPE-PLD) (Okamoto et al. 2004). In addition, anandamide can be produced from NAPE by a complex three-step enzymatic catalysis involving α,β-hydrolase-4 (ABHD4), glycerophosphodiesterase-1 (GDE1) (Simon et al. 2008), and lyso-PLD (Sun et al. 2004). On the other hand, 2-AG is synthesized from lipid precursor sn-1-acyl-2-arachidonylglycerols (DAGs), which are mostly produced by the hydrolysis of phosphatidyl-inositols (PIs) via PI specific phospholipase C (PI-PLC) (Stella et al. 1997). DAGs can be directly converted into 2-AG by two calcium-dependent sn-2-selective diacyl-glycerol lipases (DAGLs), DAGL-α, and DAGL-β (Bisogno et al. 2003; Tanimura et al. 2010). Genetic deletion of DAGL-α and DAGL-β leads to 80 and 50 % reduction in brain 2-AG levels, respectively (Tanimura et al. 2010).

The physiological effects of both anandamide and 2-AG are terminated by a two-step process involving their active uptake, by a membrane mechanism that remains to be determined, and subsequent intracellular hydrolysis (Beltramo et al. 1997). Anandamide is degraded by fatty acid amid hydrolase (FAAH) (Cravatt et al. 1996) and 2-AG by monoglyceride lipase (MGL) (Goparaju et al. 1999). Pharmacological inhibition of FAAH and MGL increases the accumulation and efficacy of anandamide and 2-AG, respectively (Long et al. 2009; Pan et al. 2009). In addition, 2-AG can also be metabolized by a serine hydrolase α,β-hydrolase domain 6 (ABHD6) (Marrs et al. 2010). Indeed, an increase in brain 2-AG level has been reported in $ABHD6^{-/-}$ mice (Marrs et al. 2010). In addition to these metabolic pathways, both anandamide and 2-AG can be metabolized into prostaglandin by cyclooxygenase type 2 (COX-2) (Kozak et al. 2000).

11.3 Regulation of the Serotonin System by Endocannabinoid Signaling

The first indirect evidence for a role of eCB signaling in regulating the function of the 5-HT system comes from early studies showing a high level of functional overlap between the eCB and 5-HT systems. Indeed, both eCB and 5-HT systems regulate the sleep–wake cycle (Freemon 1976; Murillo-Rodriguez et al. 2008), body temperature (Englert et al. 1973; Malone and Taylor 2001), feeding behavior (Ward et al. 2008), and stress homeostasis and emotional processes (Williamson and Evans 2000; Lanfumey et al. 2008; Häring et al. 2012). More importantly, observations from these studies indicate that some of the behavioral effects of the eCBs require the participation of the 5-HT system. For instance, local administration of FAAH inhibitor in the DRn, which presumably could increase the level of anandamide, enhances wakefulness (Murillo-Rodriguez et al. 2007). It is also demonstrated that the wake-promoting effect of eCB is associated with an increase in central 5-HT neurotransmission (Murillo-Rodriguez et al. 2011).

Similarly, eCB signaling has been shown to reduce the behavioral and neuroendocrine responses to stress, via at least in part, the recruitment of the 5-HT system. Indeed, results from behavioral studies, using various stress models, have shown that CB1 receptor agonists as well as drugs that increase eCB tone reduce stress response and exert anxiolytic-like effects (Kathuria et al. 2003; Haller et al. 2004; Marco et al. 2004). The anxiolytic-like effect induced by eCB signaling is, in a large part, mediated by the 5-HT system (Marco et al. 2004; Griebel et al. 2005). Taken together, these studies support the existence of a functional cross-talk between eCB and 5-HT systems. They also suggest that eCB-induced modulation of the 5-HT system plays an important role in the regulation of many physiological functions such as wakefulness and stress responses.

Additional support for eCB modulation of the 5-HT system comes from neurochemical studies. An early in vivo neurochemical study showing that systemic administration of Δ^9-THC increases the brain level of 5-HT and its metabolite 5-Hydroxyindoleacetic acid (5-HIAA) in rats (Segawa et al. 1976). Since this early observation, several neurochemical studies using selective CB1 receptor agonists and antagonists have examined the impact of eCB signaling on 5-HT release (Egashira et al. 2002; Tao and Ma 2012). These studies show that systemic administration of the CB1 receptor agonists (i.e., R-(+)-[2,3-dihydro-5-methyl-3-(4-morpholinylmethyl) pyrrolol[1,2,3-de]-1,4-benzoxazin-6-yl]-1-naphthalenylmethanone mesylate (WIN 55,212-2) or 2-[1R,2R,5R)-5-hydroxy-2-(3-hydroxypropyl) cyclohexyl]-5-(2-methyloctan-2-yl)phenol (CP 55950) exert different effects on 5-HT release depending on the brain area studied. In the medial prefrontal cortex (mPFC) and hippocampus, activation of CB1 receptors reduces 5-HT release (Egashira et al. 2002). In contrast, activation of these receptors enhances 5-HT release in the nucleus accumbens (Tao and Ma 2012). The CB1 receptor-induced increase in 5-HT release in the nucleus accumbens appears to be mediated by an indirect effect involving γ-aminobutyric acid (GABA)ergic neurons. In addition, using in vitro brain slice preparation, several studies have shown that activation of CB1 receptors reduce calcium-dependent 5-HT release induced by electrical stimulation in the PFC (Nakazi et al. 2000). In contrast, pharmacological blockade of these receptors enhances basal extracellular levels of 5-HT in the PFC (Tzavara et al. 2003; Aso et al. 2009). Taken together, observations from these neurochemical studies establish that eCBs regulate 5-HT release and central 5-HT neurotransmission via the activation of CB1 receptors.

In addition to the modulation of 5-HT release, studies using CB1 knot-out mice and selective CB1 receptor ligands have shown that eCB signaling regulates central 5-HT neurotransmission by modulating the function and expression of various 5-HT receptors. Genetic deletion of CB1 receptors reduces the function of 5-HT_{1A} receptors in the DRn as measured by the ability of 5-HT_{1A} agonists to inhibit the firing activity of DRn 5-HT neurons (Aso et al. 2009). It also reduces the function of 5-HT_{2A} receptors in the PFC (Mato et al. 2007). In addition, several studies have examined the impact of chronic increase in eCB signaling on the function and expression of 5-HT receptors. The results of these studies show that chronic administration of Δ^9-THC or synthetic CB1 agonist WIN 55,212-2 increases the expression and function of 5-HT_{1A} receptors in the hippocampus (Moranta et al. 2009; Zavitsanou et al. 2010). Taken together, these studies provide compelling evidence that eCB signaling controls central 5-HT neurotransmission by regulating 5-HT release and the function of 5-HT receptors, which in turn mediate many of the eCB-mediated behavioral effects.

11.4 Neurophysiology of Endocannabinoid Signaling in the Dorsal Raphe Nucleus

CB1 receptors are one of the most abundant GPCRs in the mammalian brain (Herkenham et al. 1991). These receptors are expressed throughout the CNS including the DRn (Egortová and Elphick 2000; Matsuda et al. 1993), suggesting that these receptors may play an important role in regulating the function of DRn 5-HT neurons. Consistent with this notion, results from in vivo and in vitro electrophysiological studies have shown that activation of CB1 receptors with eCBs or synthetic agonists regulates the excitability of DRn 5-HT neurons (Gobbi et al. 2005; Haj-Dahmane and Shen 2005; 2009). Indeed, in vivo extracellular recordings from putative DRn 5-HT neurons show that systemic administration of FAAH inhibitors, which presumably increases anandamide levels in the brain, increases the firing activity of DRn 5-HT neurons (Gobbi et al. 2005; Bambico et al. 2007). Administration of a low dose of CB1 receptor agonist WIN 55,212-2 has been shown to increase the firing activity of DRn 5-HT neurons, whereas administration a high dose of WIN 55,212-2 reduces the excitability of these neurons. The excitatory effect reported in these studies is totally blocked following the lesion of the PFC, suggesting that it is signaled by CB1 receptor located in the PFC through an excitatory neuronal circuitry (Bambico et al. 2007).

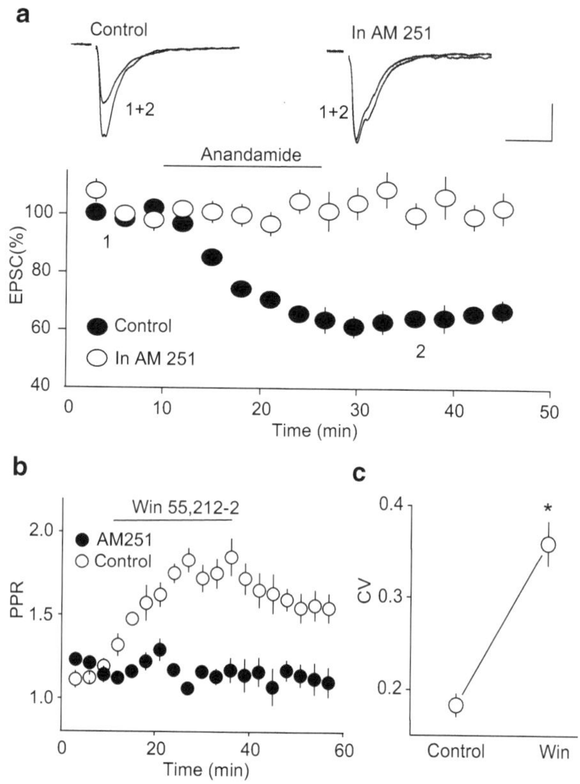

Fig. 11.1 Endocannabinoids (eCBs) suppress glutamate release in the dorsal raphe nucleus (DRn) via activation of cannabinoid type 1 (CB1) receptors. **a** Anandamide inhibits the amplitude of glutamate-mediated excitatory postsynaptic currents (EPSCs) through the activation of CB1 receptors. *Lower* panel is a summary graph of the effect of anandamide (10 μM) on the amplitude of EPSCs in control condition (•) and the presence of the CB1 receptor antagonist AM 251 (3 μM, ○). *Upper* panel depicts superimposed EPSC traces taken at the time points indicated by number in the *lower* panel. Calibration bars; 50 pA, 10 ms. **b** Effect of the CB1 receptor agonist R-(+)-[2,3-dihydro-5-methyl-3-(4-morpholinylmethyl)pyrrolol[1,2,3-de]-1,4-benzoxazin-6-yl]-1-naphthalenylmethanone mesylate (WIN 55,212-2) on the paired-pulse ratio (PPR) in control condition (○) and in the presence of AM 251 (3 μM, •). Note that activation of CB1 receptors increases the PPR. **c** Activation of CB1 receptors enhances the coefficient of variation (CV)

Direct evidence for a role of DRn CB1 receptors in regulating the excitability of 5-HT neurons comes from in vitro intracellular electrophysiological studies (Haj-Dahmane and Shen 2005, 2009). These studies show that administration of the eCB anandamide induces a rapid inhibition of glutamatergic synaptic transmission to DRn 5-HT neurons (Haj-Dahmane and Shen 2005; Fig. 11.1a). As expected for a CB1 receptor-mediated response, the effect of anandamide is readily blocked by the selective CB1 receptor antagonist AM 251 and mimicked by CB1 receptor agonist WIN 55,212-2. Detailed examination of the cellular mechanism underlying this response reveals that activation of CB1 receptors consistently increases the paired-pulse ratio (PPR) (Fig. 11.1b) and coefficient of variation (CV) (Fig. 11.1**c**), which indicate a decrease in the probability of glutamate release (Haj-Dahmane and Shen 2009). Taken together, these studies establish that eCBs reduce the strength of glutamate synapses impinging on DRn 5-HT neurons by inhibiting glutamate release. They also indicate that this effect is signaled by CB1 receptors most likely located presynaptically on glutamatergic inputs to 5-HT neurons. However, additional anatomical studies are required to further determine the precise cellular distribution of CB1 receptors in the DRn.

In addition to the modulation of glutamate synapses, results from a recent study suggests that activation of CB1 receptors may also control GABAergic synaptic transmission to DRn 5-HT neurons (Mendiguren and Pineda 2009). The results of this study show that blockade of CB1 receptors with AM 251 decreases the firing activity of presumably DRn 5-HT neurons, by increasing GABAergic tone in the DRn (Mendiguren and Pineda 2009). However, the cellular mechanisms by which eCBs regulate GABAergic synaptic transmission in the DRn remains to be determined. Taken together, the above studies support the view that eCBs, via activation of CB1 receptors, regulate the excitability of DRn 5-HT neurons primarily by controlling glutamatergic and GABAergic inputs in the DRn. By modulating excitatory and inhibitory synapses, CB1 receptors can exert a bidirectional control on the overall excitability of DRn 5-HT neurons, and hence modulate 5-HT release in their projection areas. The net effect of CB1 receptors on central 5-HT transmission will likely depend on the relative strength of excitatory and inhibitory synapses impinging on individual DRn 5-HT neurons and the density of presynaptic CB1 receptors on these inputs.

11.5 Phasic and Tonic Endocannabinoid Signaling in the Dorsal Raphe Nucleus

Results from previous anatomical studies have shown that DRn express not only CB1 receptors, but also the enzymatic machinery required for eCB production and degradation (Breder et al. 1995; Egertova et al. 2003; Bisogno et al. 2003). These findings provide some indication that DRn 5-HT neurons could synthesize and release eCBs. Consistent with this prediction, neurophysiological studies have shown that DRn 5-HT neurons release eCB in an activity-dependent manner, which mediate retrograde inhibition of synaptic transmission in the DRn (Haj-Dahmane and Shen 2009). This activity-driven phasic eCB release is initiated by membrane depolarization and subsequent increase in intracellular calcium concentration. As illustrated in Fig. 11.2a, a brief strong (−70–0 mV) membrane depolarization of DRn 5-HT neurons elicits a transient suppression of glutamatergic synaptic transmission, commonly known as depolarization-induced suppression of excitation (DSE) (Fig. 11.2a). This form of short-term synaptic plasticity is initiated by a postsynaptic

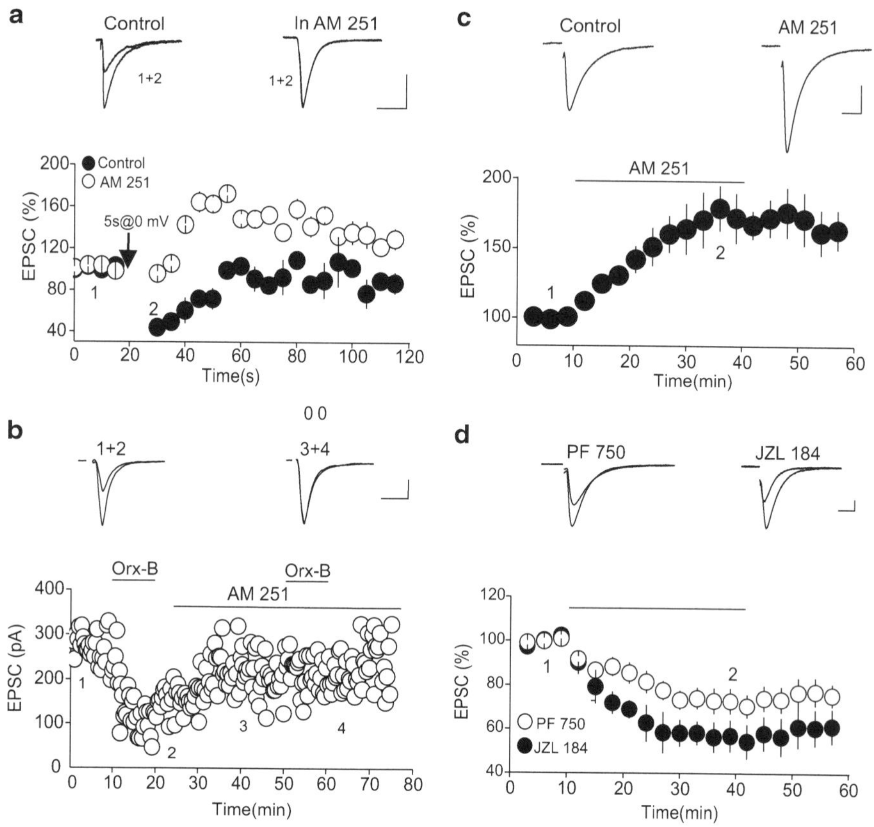

Fig. 11.2 Phasic and tonic mode of endocannabinoid (eCB) release by dorsal raphe nucleus (DRn) 5-hydroxytryptamine (5-HT) neurons. **a** A brief membrane depolarization of 5-HT neurons elicits phasic eCB release in the DRn. *Lower* panel is a summary graph of the eCB-mediated depolarization-induced suppression of excitation (DSE) induced by 5 s membrane depolarization from −70 to 0 mV obtained in control condition (•) and in the presence of the CB1 receptor antagonist AM 251 (○). *Top* graph illustrates sample excitatory postsynaptic current (EPSC) traces taken at time, points indicated by numbers in the *lower* panel. Calibration bars; 50 pA, 10 ms. **b** Activation of postsynaptic orexin receptors (OXR) reduces the strength of glutamate synapses of DRn 5-HT neurons via phasic eCB release. *Lower* graph illustrates the effect of orexin B (Orx-B, 300 nM) on the amplitude of EPSCs in the absence and presence of AM 251 (3 μM). Note that the Orx-B-induced inhibition of EPSC amplitude is blocked by the CB1 receptor antagonist AM 251, indicating that it is mediated by eCB release. *Upper* graph depicts sample EPSC traces taken at the time, points indicated by numbers in the *lower* panel. Calibration bars; 50 pA, 10 ms. **c** Tonic mode of eCB release control the strength of glutamate synapses of DRn 5-HT neurons. *Lower* graph is a summary of the CB1 receptor antagonist AM 251 (3 μM) on the amplitude of EPSCs. *Top* panel illustrates sample EPSC traces taken in control condition (*left* trace) and during bath application of AM 251 (*right* trace). Calibration bars; 50 pA, 10 ms. Note that AM 251 profoundly increases the amplitude of EPSCs indicating the presence of a tonic eCB-mediated inhibition of glutamatergic transmission in the DRn. **d** Both anandamide and 2-arachidonyl glycerol (2-AG) are tonically released by DRn 5-HT neurons. *Lower* graph is a summary of the effect of bath application of the fatty acid amid hydrolase (FAAH) inhibitor N-phenyl-4-(3-quinolinylmethyl)-1-piperidinecarboxamide (PF 750) and the monoglyceride lipase (MGL) inhibitor 4-[Bis(1,3-benzodioxol-5-yl)hydroxymethyl]-1-piperidinecarboxyl acid 4-nitrophenyl ester (JZL 184) on the amplitude of EPSCs. *Upper* panel illustrates sample traces taken, indicated by numbers in the *lower* panel. Calibration bars; 50 pA, 10 ms

increase in intracellular calcium and is caused by a decrease in glutamate release (Haj-Dahmane and Shen 2009). More importantly, like in other brain areas, the DSE is blocked by CB1 receptor antagonist AM 251 (Fig. 11.2a), indicating that it is mediated by retrograde eCB signaling. These neurophysiological results demonstrate that depolarization of DRn 5-HT neuron can induce a phasic eCB release from DRn 5-HT neurons, which in turn inhibit glutamatergic transmission in the DR. However, the eCB species released by this process and mediate the DSE in the DRn remain to be determined.

Phasic eCB release from DRn 5-HT neurons can also be triggered by activation of postsynaptic GPCRs coupled to the canonical $G_{\alpha q/11}$ type G-proteins, such as orexin receptors (OXRs). Activation of OXRs with orexin B (also called hypocretin 2) increases the excitability of DRn 5-HT neurons and strongly inhibits the strength of glutamatergic synapses (Haj-Dahmane and Shen 2005). The orexin-induced inhibition of glutamatergic synaptic transmission is mediated by postsynaptic OXRs and caused by a decrease in glutamate release. More importantly, the inhibition of glutamate release is readily blocked by the CB1 receptor antagonist AM 251, indicating that the effect of orexin is mediated by eCB signaling (Haj-Dahmane and shen 2005; Fig. 11.2b). Examination of the signaling cascade involved in this response reveals that pharmacological inhibition of PLC and diacylglycerol lipase (DAGL) pathway abolishes the orexin-induced depression of glutamatergic synaptic. These observations indicate that 2-AG is most likely the eCB species synthesized and released by 5-HT neurons in response to the activation of OXRs.

In addition to the phasic eCB signaling, tonic eCB release from DRn 5-HT neurons has been reported. This mode of signaling is revealed as an increase in basal synaptic neurotransmission in response to the blockade of CB1 receptors. As illustrated in Fig. 11.2c, administration of the CB1 receptor antagonist AM 251, profoundly enhances the amplitude of glutamate-mediated excitatory postsynaptic currents (EPSCs) by increasing glutamate release. Such results suggest the presence of a constitutive eCB tone which exerts a tonic depression of glutamatergic synaptic transmission to DRn 5-HT neurons. Additional evidence for a constitutive eCB tone in the DRn comes from experiments examining the effects of inhibition of 2-AG and anandamide degradation on eCB levels and basal synaptic transmission. These experiments show that inhibition of MGL and FAAH, enzymes that degrade 2-AG and anandamide, respectively, enhances eCB release and suppresses glutamatergic synaptic transmission to DRn 5-HT neurons. Indeed, application of the MGL inhibitor 4-[Bis(1,3-benzodioxol-5-yl)hydroxymethyl]-1-piperidinecarboxyl acid 4-nitrophenyl ester (JZL 184) strongly inhibits the amplitude EPSCs. This effect is blocked by CB1 receptor antagonist AM 251. Similarly, inhibition of FAAH with N-phenyl-4-(3-quinolinylmethyl)-1-piperidinecarboxamide (PF 750) induces a CB1 receptor mediated depression of EPSC amplitude (Fig. 11.2d). Taken together, these neurophysiological experiments demonstrate the presence of a constitutive eCB tone in DRn, which persistently reduces glutamate-mediated synaptic transmission to 5-HT neurons. In addition, the results of these experiments indicate that both 2-AG and anandamide are the major eCB species synthesized and released by DRn 5-HT neurons.

11.6 Stress Mediators Regulate Tonic and Phasic Endocannabinoid Signaling in the Dorsal Raphe Nucleus

Exposure to stressful stimuli initiates the activation of the arousal (noradrenergic) system and the hypothalamus-pituitary-adrenal axis, which leads to the rapid release of the stress mediators norepinephrine and corticosteroids (Krugers et al. 2012). These stress mediators play a crucial role in helping the organism to adapt and cope with stress by modulating the function of various brain areas including the DRn. Noradrenaline enhances arousal (Stone et al. 2007) and controls the stress response, at least in part, via the activation of α_1 adrenergic receptors (α_1-ARs) located on DRn 5-HT neurons (Morilak et al. 2005). Furthermore, behavioral studies using the learned helplessness model have shown that the activation of α_1-ARs in the DRn is required for fear conditioning and for enhanced escape performance (Grahn et al. 2002). More importantly, disruption of α_1-AR signaling in the DRn reduces the behavioral effects of anxiolytic drugs (O'Leary et al. 2007; Doze et al. 2009).

Given the prominent and convergent role of eCB signaling and 5-HT system in the regulation of stress responses, we have recently examined how noradrenaline may affect eCB signaling in DRn 5-HT neurons. Our results show that, in in vitro brain slices preparation, administration of the α_1-ARs agonist phenylephrine induces a profound and long-lasting depression of glutamatergic synaptic transmission to DRn 5-HT neurons (Fig. 11.3a). This long-term depression (LTD) is mediated by eCB release. In addition, we show that blockade of 2-AG synthesis using the DAGL inhibitor tetrahydrolipstatin (THL), abolishes the α_1-ARs-induced LTD, suggesting that 2-AG is the eCB species that mediates the LTD. Based on these results, we propose that activation of DRn α_1-ARs triggers the release of 2-AG which in turn mediates the α_1-ARs-induced modulation of synaptic transmission and plasticity in DRn 5-HT neurons. The results of this study also suggest that the α_1-ARs-induced eCB release in the DRn could mediate some of the behaviors effect of noradrenaline, including its role in the regulation of stress responses. However, future behavioral studies are needed to thoroughly test this hypothesis.

Corticosteroid hormones regulate the neuroendocrine and behavioral responses to stress by interaction with various neurotransmitter systems such as the 5-HT system. In general, they act through intracellular mineralocorticoid and glucocorticoid receptors, which are transcriptional regulators (Duma et al. 2006). Activation of these receptors regulates gene expression and mediates the slow- and long-lasting effects of stress on neuronal activity (Morsink et al. 2006). However, over the last decade, accumulating evidence indicates that corticosteroid hormones also exert rapid effects on neuronal activity via a nongenomic signaling mechanism (Di et al. 2003; Karst et al. 2005; Wang et al. 2012). Indeed, neurophysiological studies conducted in the hippocampus and paraventricular nucleus (PVN) of the hypothalamus have shown that corticosteroids induce a rapid modulation of synaptic transmission which is mediated by membrane located mineralocorticoid receptors (Karst et al. 2005) and putative GPCRs (Di et al. 2003), respectively. Similarly, recent study in

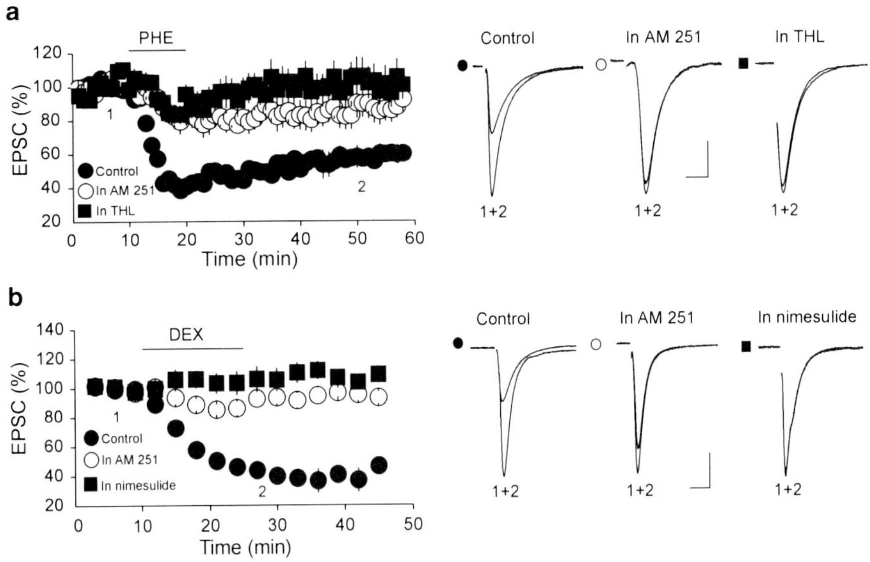

Fig. 11.3 The stress mediators noradrenaline and glucocorticoids control endocannabinoid (eCB) signaling in the dorsal raphe nucleus (DRn). **a** Activation of α_1-adrenergic receptors (α_1-ARs) induces a long-term depression (LTD) of excitatory postsynaptic currents (EPSCs) by increasing phasic eCB release in the DRn. *Left* panel is a summary graph of the α_1-ARs-induced LTD obtained in control condition (•), in the presence of CB1 receptors antagonist AM 251 (○) and in the presence of the diacylglycerol lipase (DAGL) inhibitor tetrahydrolipstatin (THL) (■). *Right* graph illustrates superimposed EPSC traces taken at the time, point indicated by the numbers in the *left* graph. Scale bars; 50 pA, 10 ms. **b** The glucocorticoid dexamethasone (DEX) depresses the amplitude of EPSCs by increasing tonic eCB release via an inhibition of COX-2. *Left* graph illustrates a summary of the DEX-induced inhibition of EPSC amplitude in control condition (•), in the presence of a CB1 receptor antagonist AM 251 (○) and in the presence of the cyclooxygenase type 2 (COX-2) inhibitor nimesulide (■). Note that blocking cannabinoid type 1 (CB1) receptor or inhibiting COX-2 abolishes the effect of DEX. *Right* panel represent sample EPSC trace taken during the experiment as indicated by numbers in the *left* panel. Scale bars; 50 pA, 10 ms

the DRn has shown that glucocorticoids exert a rapid inhibition of glutamatergic synaptic transmission to DRn 5-HT neurons. Detailed examination of the underlying mechanism reveals that it is signaled by putative membrane located GPCRs and mediated by an increase in eCBs tone (Wang et al. 2012). More importantly, this study shows that the glucorticoid-induced increase in eCB tone is not caused by the "de novo" eCB synthesis, but rather by an inhibition of COX-2-dependent eCB degradation. Indeed, as illustrated in Fig. 11.3b, the selective COX-2 inhibitors nimesulide completely abolishes the glucocorticoid-induced inhibition of EPSC amplitude. Such findings indicate that corticosteroid hormones can rapidly control the excitability of 5-HT neurons by increasing tonic eCB signaling in the DRn, which represents an additional mechanism by which stress can control the activity of 5-HT system. In addition, the increase in DRn eCB signaling may be involved in the regulation of behavioral responses to stress.

11.7 Concluding Remarks

As outlined in this chapter, it is well established that eCBs through the activation of CB1 receptors exert an important regulatory control on the 5-HT system. This regulatory control occurs both at the level of DRn 5-HT neuron cell body and their projection areas. In the projection areas, eCBs modulate 5-HT transmission by suppressing 5-HT release and regulating the expression and function of 5-HT receptors (i.e., 5-HT_{1A} and 5-HT_{2A}). At the level of the DRn, eCB signaling controls the excitability of 5-HT neurons primarily by modulating the strength of glutamatergic and GABAergic inputs impinging on DRn 5-HT neurons. One of the most important aspects of eCB modulation of 5-HT system is the discovery that DRn 5-HT can synthesize and release eCBs in an activity-dependent "phasic" mode. In addition, constitutive "tonic" eCBs release from DRn 5-HT neurons has also been reported. Both modes of eCB release play a critical role in retrograde modulation of synaptic transmission to DRn 5-HT neurons. As such, eCB signaling in the DRn represents an additional mechanism that enables 5-HT neurons to fine tune their electrical activity and control central 5-HT transmission.

Studies of eCB-modulation of the 5-HT system have also unraveled important functional interactions between the eCB signaling and various stress mediators such as noradrenaline and stress hormones (i.e., glucocorticoids). They have revealed that eCB release from DRn 5-HT neurons is modulated by noradrenaline and glucocorticoids. Noradrenaline, via the activation of α_1-ARs, enhances phasic eCBs release, whereas, glucocorticoids increase tonic eCB signaling by stimulating membrane receptors and subsequent inhibition of Cox-2-dependent eCB degradation. The increase in eCB signaling in DRn 5-HT neurons induced by the stress mediators (noradrenaline and corticosteroids) represents a potential mechanism by which stress exposure recruits eCB system. The results of these studies also provide additional support to the notion that eCB signaling in the DRn is a key modulator and integrator mediating the homeostatic response to stress (Häring et al. 2012). In addition, these findings suggest that a dysfunction of eCB signaling in 5-HT system could contribute to stress-related mood disorders.

Although considerable progress has been made toward understanding the mechanism of eCB modulation of the 5-HT system and the effects of stress mediators on eCB signaling in the DRn, several other mechanistic questions remain unanswered. For example, do other eCB receptors, such as transient receptor potential vanilloid 1 (TRPV1) and peroxisome proliferators-activated receptors (PPARs) modulate the function of the 5-HT system? What is the precise role of "tonic" eCB release in the DRn? Are 2-AG and anandamide recruited by different types of stress? Finally, what are the effects of current anxiolytic drugs, such as benzodiazepines and selective serotonin re-uptake inhibitors (SSRIs) on eCB signaling in DRn 5-HT neurons? Clearly, future studies are needed to address these outstanding questions. The answers to these questions will certainly enhance our understanding of the precise role of eCB signaling in the modulation of the 5-HT system and in the pathophysiology of mood disorders.

Acknowledgements Work in the Author's lab was supported by research grants from the National Institutes of Health research grant MH 078009 to S. HD and AA12435 to R-Y. S.

References

Aso E, Renoir T, Mengod G, Ledent C, Hamon M, Maldonado R, Lanfumey L, Valverde O (2009) Lack of CB1 receptor activity impairs serotonergic negative feedback. J Neurochem 109:935–944

Bambico FR, Katz N, Debonnel G, Gobbi G (2007) Cannabinoids elicit antidepressant-like behavior and activate serotoninergic neurons through the medial prefrontal cortex. J Neurosci 27:11700–11711

Bisogno T, Howell F, Williams G, Minassi A, Cascio MG, Ligresti A, Matias I, Schiano-Moriello A, Paul P, Williams EJ, Gangadharan U, Hobbs C, Di Marzo V, Diherty P (2003) Cloning of the first sn1-DAG lipases points to the spatial and temporal regulation of endocannabinoid signaling in the brain. J Cell Biol 163:463–468

Beltramo M, Stella N, Calignano A, Lin SY, Makriyannis A, Piomelli D (1997) Functional role of high-affinity anandamide transport, as revealed by selective inhibition. Science 277:1094–1097

Breder CD, Dewitt D, Kraig RP (1995) Characterization of inducible cyclooxygenase in rat brain. J Comp Neurol 355:296–315

Cravatt BF, Giang DK, Mayfield SP, Boger DL, Lerner RA, Gilula NB (1996) Molecular characterization of an enzyme that degrades neuromodulatory fatty-acid amides. Nature 384:83–87

Devane WA, Hanus L, Breuer A, Pertwee RG, Stevenson LA, Griffin G, Gibson D, Mandelbaum A, Etinger A, Mechoulam R (1992) Isolation and structure of brain constituent that binds to cannabinoid receptor. Science 258:1946–1949

Di S, Malcher-Lopes R, Halmos KC, Tasker JG (2003) Nongenomic glucocorticoid inhibition via endocannabinoid release in the hypothalamus: a fast feedback mechanism. J Neurosci 23:4850–4857

Doze VA, Handel EM, Jensen KA, Darsie B, Luger EJ, Haselton JR, Talbot JN (2009) Rorabaugh BR. α_{1A}- and α_{1B}-adrenergic receptors differentially modulate antidepressant-like behavior in the mouse. Brain Res 1285:148–137

Duma D, Jewell C, Cidlowski JA (2006) Multiple glucocorticoid receptor isoforms and mechanisms of post-translational modifications. J Steroid Biochem Mol Biol 102:11–21

Egashira N, Mishima K, Katsurabayashi S, Yoshitake T, Matsumoto Y, Ishida J, Yamaguchi M, Iwasaki K, Fujiwara M (2002) Involvement of 5-hydroxytryptamine neuronal system in Delta (9)-tetrahydrocannabinol-induced impairment of spatial memory. Eur J Pharmacol 445:221–229

Egertová M, Elphick MR (2000) Localization of cannabinoid receptors in the rat brain using antibodies to the intracellular C-terminal tail of CB_1. J Comp Neurol 422:159–171

Egertová M, Cravatt BF, Elphick MR (2003) Comparative analysis of fatty acid amid hydrolase and CB1 cannabinoid receptor expression in the mouse brain: evidence of a widespread role of fatty acid amid hydrolase. Neuroscience 119:481–496

Englert LF, Ho BT, Taylor D (1973) The effects of (-)-Δ^9-tetrahydrocannabinol on reserpine-induced hypothermia in rats. Br J Pharmacol 49:243–252

Freemon FR (1976) Effects of marihuana on sleeping states. JAMA 220:1364–1365

Galiègue S, Mary S, Marchand J, Dussossoy D, Carrière D, Carayon P, Bouaboula M, Shire D, Le Fur G, Casellas P (1995) Expression of central and peripheral cannabinoid receptors in human immune tissues and leukocyte subpopulations. Eur J Biochem 232:54–61

Gaoni Y, Mechoulam R (1964) Isolation, structure and partial synthesis of an active constituent of hashish. J Am Chem Soc 86:1646–1647

Gobbi G, Bambico ER, Mangieri R, Bortolato M, Campolongo P, Solinas M, Cassano M, Morgese MG, Debonnel G, Duranti A, Tontini A, Tarzia G, Mor M, Trezza V, Goldberg SR, Cuomo V,

Piomelli D (2005) Antidepressant-like activity and modulation of brain monoaminergic transmission by blockade of anandamide hydrolysis. Proc Natl Acad Sci USA 102:18620–18625

Goparaju SK, Ueda N, Taniguchi K, Yamamoto S (1999) Enzymes of porcine brain hydrolyzing 2-arachidonoylglycerol an endogenous ligand of cannabinoid receptors. Biochem Pharmacol 57:417–423

Grahn RE, Hammack SE, Will MJ, O'Connor KA, Deak T, Sparks PD, Watkins LR, Maier SF (2002) Blockade of alpha1 adrenoreceptors in the dorsal raphe nucleus prevents enhanced conditioned fear and impaired escape performance following uncontrollable stressor exposure in rats. Behav Brain Res 134:387–392

Griebel G, Stemmelin J, Scatton B (2005) Effects of cannabinoid CB_1 receptor antagonist rimonabant in models of emotional reactivity in rodents. Biol Psychiatry 57:261–267

Haj-Dahmane S, Shen R-Y (2005) The wake-promoting peptide orexin-B inhibits glutamatergic transmission to dorsal raphe nucleus serotonin neurons through retrograde endocannabinoid signaling. J Neurosci 25:896–905

Haj-Dahmane S, Shen R-Y (2009) Endocannabinoids suppress excitatory synaptic transmission to dorsal raphe serotonin neurons through the activation of presynaptic CB_1 receptors. J Pharmacol Exp Ther 331:186–196

Haller J, Varga B, Ledent C, Freund T (2004) CB1 cannabinoid receptors mediate anxiolytic effects: convergent genetic and pharmacological evidence with CB1-specific agents. Behav Pharmacol 15:299–304

Häring M, Guggenhuber S, Lutz B (2012) Neuronal populations mediating the effects of endocannabinoids on stress and emotionality. Neuroscience 204:145–158

Herkenham M, Lynn A, Johnson MR, Melvin LS, Costa BR de, Rice KC (1991) Characterization and localization of cannabinoid receptors in rat brain: a quantitative in vitro autoradiographic study. J Neurosci 11:563–583

Hillard CJ, Weinlander KM, Stuhr KL (2012) Contribution of endocannabinoid signaling to psychiatric disorders in human: genetic and biochemical evidence. Neuroscience 204:207–229

Howlett AC (2005) Cannabinoid receptor signaling. Handb Exp Pharmacol 168:53–79

Kano M, Ohno-Shosaku T, Hashimotodani Y, Uchigashima M, Watanabe M (2009) Endocannabinoid-mediated control of synaptic transmission. Physiol Rev 89:309–380

Karst H, Berger S, Turiault M, Tronche F, Schütz G, Joëls M (2005) Mineralocorticoid receptors are indispensable for nongenomic modulation of hippocampal glutamate transmission by corticosterone. Proc natl Acd Sci USA 102:19204–19207

Kathuria S, Gaetani S, Fegley D, Valino F, Duranti A, Tontini A, Mor M, Tarzia G, La Rana G, Calignano A, Giustino A, Tattoli M, Palmery M, Cuomo V, Piomelli D (2003) Modulation of anxiety through blockade of anandamide hydrolysis. Nat Med 9:76–81

Katona I, Freund TF (2012) Multiple functions of endocannabinoid signaling in the brain. Ann Rev Neurosci 35:529–558

Kozak KR, Rowlinson SW, Marnett LJ (2000) Oxygenation of the endocannabinoid, 2-arachidonylglycerol, to glyceryl prostaglandins by cyclooxygenase-2. J Biol Chem 275:33744–33749

Krugers H, Karst H, Joels M (2012) Interaction between noradrenaline and corticosteroids in the brain: from electrical activity to cognitive performance. Front Cell Neurosci 6:15

Lanfumey L, Mongeau R, Cohen-Salmon C, Hamon M (2008) Corticosteroid-serotonin interactions in the neurobiological mechanisms of stress-related disorders. Neurosci Biobehav Rev 32:1174–1184

Long JZ, Nomura DK, Vann RE, Walentiny DM, Booker L, Jin X, Burston JJ, Sim-Selley LJ, Lichman AH, Wiley JL, Cravatt BF (2009) Dual blockade of FAAH MAGL identifies behavioral processes regulated by endocannabinoid crosstalk in vivo. Proc Natl Acad Sci 106:20270–20275

Malone DT, Taylor DA (2001) Involvement of somatodendritic 5-HT (1A) receptors in Delta (9)-tetrahydrocannabinol-induced hypothermia in the rat. Pharmacol Biochem Behav 69:595–601

Marco EM, Perez-Alverez L, Borcel E, Rubio M, Guaza C, Ambrosio E, File SE, Viveros MP (2004) Involvement of $5\text{-}HT_{1A}$ receptors in behavioral effects of the cannabinoid receptor agonist CP 55, 94 in male rats. Behav Pharmacol 15:21–27

Marrs WR, Blankman JL, Horne EA, Thomazeau A, Lin YH, Coy J, Bodor AL, Muccioli GG, Hu SS-J, Woodruff G, Fung S, Lafourcade M, Alexander JP, Long JZ, Xu C, Möller T, Mackie K, Manzoni OJ, Cravatt BF, Stella N (2010) The serine hydrolase ABHD6 controls the accumulation and efficacy of 2-AG at cannabinoid receptors. Nat Neurosci 13:651–957

Mato S, Aso E, Martin M, Valverde O, Maldonado R, Pazos A (2007) CB1 knockout mice display impaired functionality of 5-HT1A and 5-HT2A/C receptors. J Neurochem 103:2111–2120

Matsuda LA, Bonner TL, Lolait SJ (1993) Localization of cannabinoid receptor mRNA in rat brain. J Comp Neurol 327:535–550

Matsuda LA, Lolait SJ, Brownstein MJ, Young AC, Bonner TI (1990) Structure of cannabinoid receptor and functional expression of the cloned cDNA. Nature 346:561–564

Mechoulam R, Parker LA (2012) The endocannabinoid system and the brain. Annu Rev Psychol 64:6.1–6.27

Mechoulam R, Ben-Shabat S, Hanus L, Ligumsky M, Kaminski NE, Schatz AR, Gopher A, Almog BR, Compton DR, Pertwee RG, Griffin G, Bayewitch M, Barg J, Vogel Z (1995) Identification of endogenous 2-monoglyceride, present in canine gut that binds to cannabinoid receptors. Biochem Pharmacol 50:83–90

Mendiguren A, Pineda J (2009) Effect of the CB1 receptor antagonists rimonabant and AM251 on the firing rate of dorsal raphe nucleus neurons in rat brain slices. Br J Pharmacol 158:1579–1587

Moranta D, Esteban S, Garcia-Sevilla JA (2009) Chronic treatment and withdrawal of the cannabinoid agonist WIN 55,212–2 modulate the sensitivity of presynaptic receptors involved in the regulation of monoamine syntheses in rat brain. Naunyn-Schmiedeberg's Arch Pharmacol 379:61–72

Morilak DA, Barrera G, Echevarria DJ, Garcia AS, Hernandez A, Ma S, Petre CO (2005) Role of brain norepinephrine in the behavioral response to stress. Prog Neuropsychopharmacol Biol Psychiatry 29:1214–1224

Morsink MC, Steenbergen PJ, Vos JB, Karst H, Joëls M, Kloet ER de, Datson NA (2006) Acute activation of hippocampal glucocorticoid receptors results in different waves of gene. J Neuroendocrinol 18:239–252

Muccioli GG (2010) Endocannabinoid biosynthesis and inactivation, from simple to complex. Drug Discov Today 15:474–483

Munro S, Thomas KL, Abu-Shaar M (1993) Molecular characterization of a peripheral receptor for cannabinoids. Nature 365:61–65

Murillo-Rodriguez E, Vázquez E, Millan-Aldaco D, Palomero-Rivero M, Drucker-Colin R (2007) Effects of the fatty acid amide hydrolase inhibitor URB597 on the sleep-wake cycle, c-Fos expression and dopamine levels in the rats. Eur J Pharmacol 562(1–2):82–91

Murillo-Rodriguez E, Millán-Aldaco D, Di Marzo V, Drucker-Colin R (2008) The anandamide membrane transporter inhibitor, VDM-11 modulates sleep and c-Fos expression in the rat brain. Neuroscience 157:1–11

Murillo-Rodriguez E, Palomero-Rivero M, Millan-Aldaco D, Aris-Carrion O, Drucker-Colin R (2011) Administration of URB597, oleoylethanolamide or palmitoylethanolamide increases waking and dopamine in rats. PloS One 6:e20766

Nakazi M, Bauer U, Nickel T, Kathmann M, Schlicker E (2000) Inhibition of serotonin release in the mouse brain via presynaptic CB receptors. Naunyn-Schmiedeberg's Arch Pharmacol 361:19–24

Okamoto Y, Morishita J, Tsuboi K, Tonai T, Useda N (2004) Molecular characterization of phospholipase D generating anandamide and its congeners. J Biol Chem 279:5298–5305

O'Leary OF, Bechtholt AJ, Crowley JJ, Valentino RJ, Lucki I (2007) The role of noradrenergic tone in the dorsal raphe nucleus of the mouse in the acute behavioral effects of antidepressant drugs. Eur Neuropsychopharmacol 17:215–226

Pan B, Wang W, Long JZ, Sun D, Hillard CJ, Cravatt BF, Liu QS (2009) Blockade of 2-arachidonoylglycerol hydrolysis by selective monoacylglycerol lipase inhibitor 4-nitrophenyl 4-(dibenzo[d][1,3]dioxol-5-yl(hydroxyl)methyl)piperidine-1-carboxylate (JZL 184) enhances retrograde endocannabinoid signaling. J Pharmacol Exp Ther 331:591–597

Patel S, Hillard CJ (2006) Pharmacooogical evaluation of cannabinoid receptor ligands in a mouse model of anxiety: further evidence for an anxiolytic role for endogenous cannabinoid signaling. J Pharmacol Exp Ther 318:304–311

Patel S, Roelke CT, Rademacher DJ, Cullinan WE, Hillard CJ (2004) Endocannabinoid signaling negatively modulates stress-induced activation of the hypothalamic-pituitary-adrenal axis. Endocrinology 145:5431–5438

Patel S, Roelke CT, Rademacher DJ, Hillard CJ (2005) Inhibition of restraint stress-induced neuronal and behavioral activation of endogenous cannabinoid signaling. Eur J Neurosci 21:1057–1069

Segawa T, Takeuchi S, Nakano M (1976) Mechanism for the increase of brain 5-hydroxytryptamine and 5-hydroxyindoleacetic acid following Δ^9- tetrahydrocannabinol administration to rats. Japan J Pharmacol 26:377–379

Simon GM, Cravatt BF (2008) Anandamide biosynthesis catalyzed by the phosphodiesterase GDE1 and detection glycerophospho-N-acyl ethanolamine precursors in mousebrain. J Biol Chem 283:9341–9349

Stella N, Schweittzer P, Piomelli D (1997) A second endogenous cannabinoid that modulates long-term potentiation. Nature 388:773–778

Stone EA, Quartermain D, Lin Y, Lehmann MJ (2007) Central α_1-adrenergic system in behavioral activity and depression. Biochem Pharmacol 73:1063–1075

Sugiura T, Kondo S, Sukagawa A, Nakane S, Shinoda A, Itoh K, Yamashita A (1995) Waku K. 2-arachidonoylglycerol: a possible endogenous cannabinoid receptor ligand in brain. Biochem Biophys Res Commun 215:89–97

Sugiura T, Kondo S, Sukagawa A, Tonegawa T, Nakane S, Yamashita A, Waku K (1996) Enzymatic synthesis of anandamide, an endogenous cannabinoid receptor ligand, through N-acylphosphatidylethanolamine pathway in testis: involvement of Ca(2+)-dependent trancyclase and phosphodiesterase activities. Biochem Biophys Res Commun 218:113–117

Sun YX, Tsuboi K, Okamoto Y, Tonai T, Murakami M, Kudo I, Ueda N (2004) Biosynthesis of anandamide and N-palmitoylethanolamine by sequential actions of phospholipase A2 and lysophospholipase D. Biochem J 380:749–756

Tanimura A, Yamazaki A, Hashimotodani Y, Uchigashima M, Kawata S, Abe M, Kita Y, Hashimoto K, Shimizu T, Watanabe M, Sakimura K, Kano M (2010) The endocannabinoid 2-arachidonoylglycerol produced by diacylglycerol lipase α mediates retrograde suppression of synaptic transmission. Neuron 65:320–327

Tao R, Ma Z. (2012) Neuronal circuit in the dorsal raphe nucleus responsible for cannabinoid-mediated increases in 5-Ht efflux in the nucleus accumbens of the rat. ISRN Pharmacol 2012:276902 doi: 10.5402/2012/276902

Tzavara ET, Davis RJ, Perry KW, Li X, Salhoff C, Bymaster FP, Witkin JM, Nomikos GG (2003) The CB1 receptor antagonist SR 141716A selectively increases monoaminergic neurotransmission in the medial prefrontal cortex: implications for therapeutic actions. Br J Pharmacol 138:544–553

Van Sickle MD, Duncan M, Kingsley PJ, Mouihate A, Urbani P, Mackie K, Stella N, Makriyannis A, Piomelli D, Davison JS, Mernett LJ, Di Marzo V, Pittman QJ, Patel KD, Sharkey KA (2005) Identification and functional characterization of brainstem cannabinoid CB_2 receptors. Science 31:329–332

Wang J, Shen R-Y, Haj-Dahmane S (2012) Endocannabinoids mediate the glucocorticoid-induced inhibition of excitatory synaptic transmission to dorsal raphe serotonin neurons. J Physiol 590:5795–5808

Ward SJ, Lefever TW, Jackson C, Tallarida RJ, Walker EA (2008) Effect of cannabinoid 1 receptor antagonist and serotonin 2C receptor agonist alone and in combination on motivation for palatable food: a dose-addition analysis study in mice.

Williamson EM, Evans FJ (2000) Cannabinoids in clinical practice. Drugs 60:1303–1314

Zavitsanou K, Wang H, Dalton VS, Nguyen V (2010) Cannabinoid administration increases 5HT1A receptor binding and mRNA expression in the hippocampus of adult but not adolescent rats. Neuroscience 169:315–324

Chapter 12
Modulation of Serotonin Firing Activity Through CB1 Agonists and FAAH Inhibitors

Gabriella Gobbi

Abstract The psychological feelings produced by cannabis have been described as fatuous euphoria, elation, and talkativeness. Alternatively, cannabis can induce low mood and depression especially after chronic use. Despite these clinical evidences, little was known about the capacity of cannabis to modulate serotonin (5-Hydroxytryptamine, 5-HT), the main neurotransmitter implicated in the regulation of mood and the pathology of mood disorders.

In the past 10 years, our laboratory has attempted to clarify how the cannabinoid type 1 receptor (CB1R) agonists, antagonists, and the fatty acid amide hydrolase (FAAH) inhibitors modulate the firing activity of 5-HT neurons located in the dorsal raphe nucleus.

While the CB1R agonist R-(+)-[2,3-dihydro-5-methyl-3-(4-morpholinylmethyl) pyrrolol[1,2,3-de]-1,4-benzoxazin-6-yl]-1-naphthalenylmethanone mesylate (WIN 55,212-2) produces a bell-shaped curve, increasing 5-HT firing at low doses (0.1–0.3 mg/kg) and decreasing firing at higher doses (>0.3 mg/kg), the FAAH inhibitor [3-(3-carbamoylphenyl)phenyl] N-cyclohexylcarbamate (URB597) produces a sigma-shaped curve, with a plateau at the highest doses tested (0.3 mg/kg). Δ9-Tetrahydrocannabinol (THC) produces a mixed response on 5-HT firing activity with 26% of neurons showing an increase, 33% showing a decrease, and 42% showing no response. However, after 4 days, intraperitoneal (i.p.) injections of THC (1 mg/kg) produced a significant elevation of firing. The increase in firing following WIN 55,212-2 and THC was prevented by the CB1R antagonist rimonabant. Finally, both WIN 55,212-2 and THC evoked a robust decrease in 5-HT firing after long-term administration in adolescence.

These data show that CB1R agonists and FAAH inhibitors interact with the 5-HT system and that this effect may be related to emotional behaviors.

G. Gobbi (✉)
Neurobiological Psychiatry Unit, Department of Psychiatry, McGill University, 1033, Av des Pins Ouest (room 220), Montreal, QC H3A1A1, Canada
e-mail: gabriella.gobbi@mcgill.ca

E. J. Van Bockstaele (ed.), *Endocannabinoid Regulation of Monoamines in Psychiatric and Neurological Disorders*, DOI 10.1007/978-1-4614-7940-6_12,

12.1 Introduction

For many centuries, cannabis *sativa* has been known to possess psychotropic effects linked to the regulation of mood. The English cleric Robert Burton in "The Anatomy of Melancholy", published in 1621, considered cannabis as a possible treatment for depression. It was not until 1845 that this claim was put to test in Western society when the French psychiatrist Jacques-Joseph Moreau de Tours, having come back to Paris from a journey to India, tried its psychotropic effects on his depressive patients. His initiative, however, yielded mixed results.

Certainly, the most important effects of cannabinoids are related to elevation of mood: subjective effects include a sensation of "high", anxiety relief, and the feeling of being "relaxed" and in a good mood (Iversen 2003; Earleywine 2005). The experience is highly variable, depending on the dose of drug, the environment, and the experience and expectations of the drug user. The "high" produced by cannabis is a complex experience characterized by quickening of mental associations and sharpened sense of humor, sometimes described as a state of "fatuous euphoria". The user feels relaxed and calm, talkative, in a dreamlike state, and disconnected from the real world (Iversen 2003). Other psychophysiological effects include pain relief, time distortion, impairments of perception and short-term memory, amotivation/laziness, and paranoia.

Despite this evidence, very little was known about the mechanism(s) that underlie such cannabis-induced "high mood", but it seemed very likely that it was a cannabinoid type 1 (CB1)-mediated mechanism since the cannabinoid type 1 receptor (CB1R) antagonist rimonabant blocks this effect (Huestis et al. 2001).

More surprisingly, very little has been published regarding the link between the CB1R and serotonin (5-Hydroxytryptamine, 5-HT), the main neurotransmitter implicated in the pathogenesis of depression and in the mechanism of action of antidepressants. When my laboratory started this line of research in 2002, only a few articles were published on the link between cannabis and 5-HT. At that time, it was known that CB1Rs were present at the level of the nucleus of the dorsal raphe (DRN) (Matsuda et al. 1993; Tsou et al. 1998; Moldrich and Wenger 2000), the major source of 5-HT neurons in the brain, while fatty acid amide hydrolase (FAAH), the enzyme responsible for the degradation of the endocannabinoid anandamide (AEA), was shown to be present in DRN oligodendrocytes (Egertova et al. 2003). These anatomical data were consistent with an electrophysiolgical study reporting that postsynaptic orexin receptors can modulate glutamatergic synaptic transmission to DRN 5-HT neurons through retrograde endocannabinoid signaling (Haj-Dahmane and Shen 2005). It was known that 5-HT_{1B}, 5-HT_{2A} (Devlin and Christopoulos 2002), and 5-HT_3 receptor subtypes were coexpressed with CB1Rs (Hermann et al. 2002) in γ-aminobutyric acid (GABA) interneurons (Morales et al. 2004), as well as the inhibitory functional interaction between CB1 and 5-HT_{2A} receptors (Kimura et al. 1998; Darmani 2001) and between CB1 and 5-HT_3 receptors (Fan 1995; Barann et al. 2002). Moreover, it was reported that somatodendritic 5-HT_{1A} receptors are involved in Δ^9-tetrahydrocannabinol- (THC) induced hypothermia (Malone and Taylor 2001), suggesting an interaction between the two systems at the hypothalamic level. At this time, Gorzalka and colleagues showed that chronic treatment with the CB1R agonist HU-210 alters the pharmacological responses to 5-HT_{2A} and 5-HT_{1A} receptor agonists (Hill et al. 2006).

Two in vitro studies also demonstrated that CB1R agonists inhibited 5-HT reuptake into rat brain synaptosomes (Banerjee et al. 1975; Johnson et al. 1976), and others demonstrated that they suppress 5-HT release from cortical slices (Nakazi et al. 2000).

However, whether CB1R agonists and FAAH inhibitors could directly modulate the firing rate of 5-HT neurons was unknown. For these reasons, my laboratory started to better dissect the possible involvement of these compounds in modulating 5-HT firing activity. The activity of a putative drug on 5-HT firing rate is indeed essential to predict its potential effect on mood modulation (see next paragraph).

12.2 In Vivo Serotonin (5-HT) Electrical Activity Recording: Pharmacological Meaning

In vivo electrophysiology allows us to study changes in the spontaneous firing activity of 5-HT neurons in the DRN after acute and chronic treatments with psychotropic drugs, such as antidepressants or drugs of abuse, and the responsiveness of pre- and postsynaptic 5-HT receptors (Blier and de Montigny 1994; Bambico and Gobbi 2008). However, the ability of drug treatments to increase 5-HT neurotransmission depends on their particular mechanisms of action. For example, selective serotonin reuptake inhibitors (SSRIs) increase synaptic 5-HT availability by blocking the serotonin transporter (5-HTT). Acutely, this decreases the firing activity of 5-HT neurons located in the DRN, due to the increased agonism on 5-HT_{1A} autoreceptors; but after chronic treatment, a desensitization of 5-HT_{1A} autoreceptors occurs, restoring 5-HT firing activity in the DRN to basal levels (Artigas 1993; Blier and de Montigny 1994). The α_2-adrenoceptor antagonist mirtazapine, whose direct mechanism of action is on norepinephrine (NE) activity, can also increase 5-HT firing activity in DRN neurons (Haddjeri et al. 1997). This is achieved via local antagonism of α_2-adrenoceptors in NE neurons within the locus coeruleus (LC), which subsequently increases DRN 5-HT firing by increasing NE-mediated activation of α_1-adrenoceptors located in the DRN (Freedman and Aghajanian 1984; Haddjeri et al. 2004). Neurokinin 1 (NK1) antagonists also increase 5-HT firing activity through the activation of NE neurons of the LC (Gobbi et al. 2007). In addition to the effects on 5-HT firing activity in the DRN, long-term antidepressant treatment results in increased tonic activation of forebrain postsynaptic 5-HT_{1A} receptors (Haddjeri et al. 1998). The activation of 5-HT firing activity, even if through a direct or indirect mechanism on the DRN, has been correlated to a modulation of "mood" tonus in animal models, and the enhanced 5-HT neurotransmission may be considered a common indicator of antidepressant efficacy (see Bambico and Gobbi 2008 for a review of this literature).

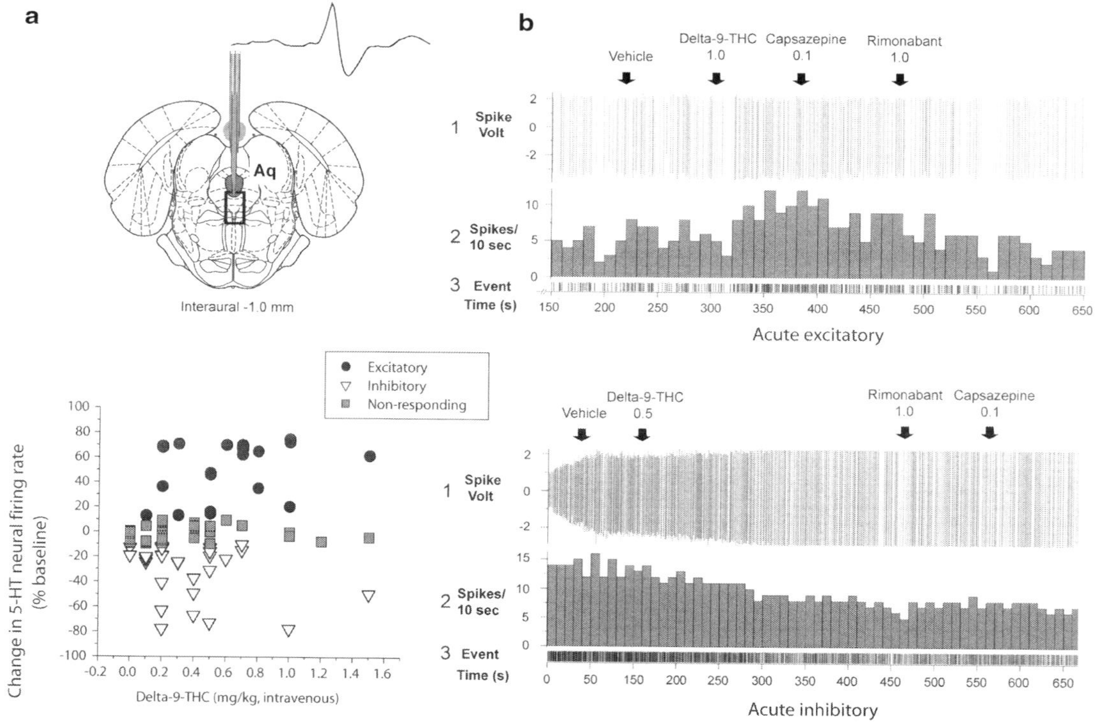

Fig. 12.1 Acute intravenous (i.v.) administrations of variable doses of Δ^9-tetrahydrocannabinol (THC) elicited a complex response profile from dorsal raphe nucleus (DRN) 5-hydroxytryptamine (5-HT) neurons, with neurons exhibiting either excitatory or inhibitory responses, as well as inert ones. *Inset* shows a plate from Paxinos and Watson (2007) illustrating a brain coronal section through the DRN. Additional illustrations indicate the area (*boxed*) from where putative 5-HT neurons were recorded. An exemplary spike waveform of a putative 5-HT neuron is shown *above* (**a**). (*top* panel) A representative 5-HT neural firing rate histogram showing that a THC-evoked excitatory response was reversed by the CB1R antagonist rimonabant, but not by the transient receptor potential cation channel subfamily V member 1 ($TRPV_1$) receptor antagonist capsazepine. (*bottom* panel) A representative 5-HT neural firing rate histogram showing that a THC-evoked inhibitory response was reversed neither by rimonabant nor capsazepine. Numbers above *arrows* indicate the i.v. dose administered (**b**). (With permission from: Prog Neuropsychopharmacol Biol Psychiatry. 2012 Jul 2;38(1):88–96)

Recordings of 5-HT neurons in the DRN are usually restricted to the midline regions, where about half of the 5-HT neurons are found. Although a substantial number of non-5-HT neurons exist in the DRN, especially GABAergic neurons, their distribution differs from that of 5-HT neurons and seems to be segregated to the lateral wings or the ventrolateral subdivision of the DRN (Descarries et al. 1982; Calizo et al. 2011). Aghajanian et al. first described, in 1968, the electrophysiological characteristics of DRN neurons in rats. 5-HT neurons are characterized by a slow firing pattern with a positive-negative, biphasic action potential (see Fig. 12.1a), discharging between 0.5–4 Hz. Their action potential is usually biphasic or triphasic spikes, with an initial positive, rising segment of long duration ranging from 0.7 to 2 ms (the total duration of the action potentials range from 2.7 to 8.7 ms). In addition, a subpopulation of 5-HT neurons fired short trains (burst) of two and occasionally three or four spikes with a mean burst interspike interval (ISI) of approximately 6 ms (Domínguez-López et al. 2011). Further identification of 5-HT neurons can be confirmed by systemically or microiontophoretically injecting lysergic acid diethylamide (LSD), 5-HT or 5-HT_{1A} agonists that selectively suppress the firing activity of 5-HT neurons in the DRN (Haigler and Aghajanian 1974; Sprouse and Aghajanian 1987). Juxtacellular labeling techniques provide additional methods to better identify 5-HT neurons after recording (Allers and Sharp 2003; Hajos et al. 2007). 5-HT neurons may discharge in single firing or burst firing patterns. Burst activity pattern is defined as a train of at least two spikes within a regular low-frequency firing pattern, with the maximal interval signifying burst onset of 20 ms or less and the longest ISI allowed within a burst of 40 ms (Gobbi et al. 2005). Single-firing neurons represent the majority of 5-HT neurons (approximately 70–85 % of recorded cells) (Domínguez-López et al. 2011).

The functional role of burst-firing in 5-HT DRN neurons has been associated with increased 5-HT release in synaptic terminals (Gartside et al. 2000). Our studies have confirmed that high 5-HT firing rates (> 3 Hz) with an increased percentage of burst-firing cells (more than 70 %) is paralleled by an enhancement of 5-HT release (measured with microdialysis) in the hippocampus (Gobbi et al. 2005).

Another parameter that is evaluated in *in vivo* electrophysiology is "neurons per track", providing information about the number of neurons that are spontaneously active. Several pharmacological treatments as well as chronic stress can influence this parameter.

12.3 Δ^9-Tetrahydrocannabinol (THC) and 5-Hydroxytryptamine (5-HT) Firing Activity

The main pharmacologically active cannabinoid principle in cannabis, (−)-trans-Δ^9-tetrahydrocannabinol (THC), likely mediates most of its psychoactive and mood-related effects. Although heavy or high-dose cannabis use has been associated with an elevated risk of mood disorders, anxiety, psychosis, and cognitive impairment, especially among teenagers, it is also used for self-medicating depressive symptoms, suggesting that it could have therapeutic benefits in primary and secondary depression (for review see Bambico and Gobbi 2008). Nabilone (Cesamet®), a synthetic THC derivative, commercialized in Canada, has been reported to increase mood in 38 % of people and to induce euphoria in another 14 %, suggesting a dual effect on mood, dependent on subjects (from *Compendium of Pharmaceuticals and Specialties, Canada*, CPHA 2012).

Similarly in animals, antidepressant-like effects (but also depressogenic effects; see Egashira et al. 2008) have been reported in the forced swim test (FST) (Moreira et al. 2004; El-Alfy et al. 2010; Bambico et al. 2012), and in other behavioral models such as the olfactory bulbectomy model (Elbatsh et al. 2012; Rodriguez-Gaztelumendi et al. 2009) and tail suspension test (TST) in mice (El-Alfy et al. 2010).

We have examined whether THC modulates 5-HT firing activity. The intravenous (i.v.) administration of different doses of THC (0–1.6 mg/kg), yielded a complex response profile. In fact, we identified three different groups of 5-HT neurons on the basis of their response to THC within this dose range: 25.58 % ($n=22$) were excited (≥10 % of baseline), 32.56 % ($n=28$) were inhibited (≥10 % of baseline), and 41.86 % ($n=36$) were nonresponding ($\chi^2=3.5$, $p=0.17$). These neurons, regardless of their response, exhibited identical electrophysiological characteristics and were all recorded from the rostrocaudal midline extent of the DRN. The excitatory responses were mainly produced by doses >0.45 mg/kg and were maximal at 1.0 mg/kg. The inhibitory responses were mainly produced by doses <0.45 mg/kg and were maximal at 0.4 mg/kg. Inert responses were equally distributed between low and high dose ranges. In an attempt to determine which receptor subsystems were responsible for the excitatory and inhibitory responses, the CB1R antagonist rimonabant (1.0 mg/kg) or the transient receptor potential vanniloid 1 ($TRPV_1$) receptor antagonist capsazepine (0.01 mg/kg) were administered i.v. following a THC-induced increase or decrease in 5-HT neural firing activity. We found that the excitatory response was attenuated by rimonabant, but not by capsazepine, suggesting that THC-induced excitations were instigated by CB1R activation. On the other hand, the inhibitory response was only partially reversed by rimonabant in one out of three neurons tested, and was not at all sensitive to capsazepine, indicating a non-CB1 and non-$TRPV_1$ receptor-mediated mechanism (Bambico et al. 2012; Fig. 12.1).

Intraperitoneal (i.p.) administration at a dose, previously shown to evoke the maximal excitatory response from 5-HT neurons (1.0 mg/kg), slightly but non-significantly enhanced the mean spontaneous 5-HT firing rate, even after 1.5 h of electrophysiological recordings. Increasing the dose to 2 and 4 mg/kg similarly yielded non-significant elevations in mean 5-HT firing rates. On the other hand, repeated administration (once daily for 5 days) of 1.0 mg/kg THC (i.p.) increased the mean spontaneous 5-HT firing rate in the DRN after 1.5 h of electrophysiological recordings ($p<0.05$). This increase was blocked by the coapplication of rimonabant (1.0 mg/kg, i.p.) and was therefore mediated by activation of CB1Rs (Fig. 12.2). This discrepancy between the acute vs. chronic i.p. treatment may be due to the pharmacodynamic/pharmacokinetic properties of THC, as well as potential long-term

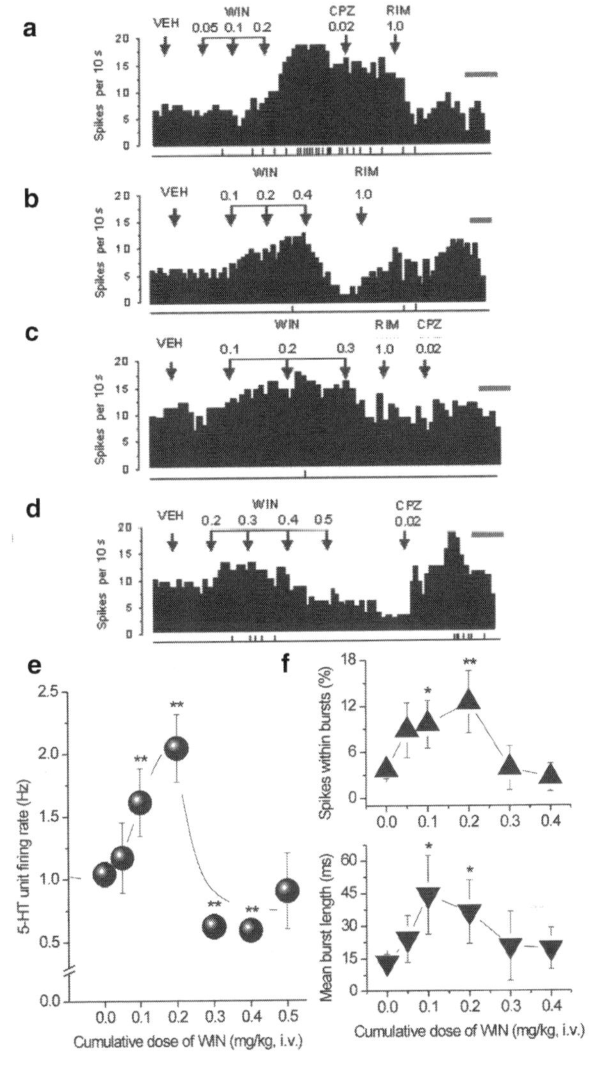

Fig. 12.2 Effect of intravenous (i.v.) administration of cumulative doses of R-(+)-[2,3-dihydro-5-methyl-3-(4-morpholinylmethyl)pyrrolol[1,2,3-de]-1,4-benzoxazin-6-yl]-1-naphthalenylmethanone mesylate (WIN 55,212-2) on the firing activity of DRN 5-Hydroxytryptamine (5-HT) neurons. **a**–**d** Integrated firing rate histograms of 5-HT neurons illustrating that low doses of WIN 55,212-2 (0.1–0.2 mg/kg, i.v.) rapidly increased single-unit firing activity. **a** This effect was reversed by rimonabant (RIM; 1.0 mg/kg, i.v.; $n=4$) but not by capsazepine (CPZ; 0.02 mg/kg, i.v.; $n=4$). **b**–**d** High dose of WIN 55,212-2 (0.30–0.50 mg/kg, i.v.) rapidly decreased single-unit firing activity. This effect was reversed by CPZ (0.02 mg/kg, i.v.) in two of three neurons (**d**) and partially reversed (**b**) or unchanged (**c**) by RIM (1 mg/kg, i.v.) in one and three neurons, respectively. 5-HT neuronal firing rate in each histogram is plotted as spikes per 10 s. Calibration bar on *right* side of each histogram, 1 min. The *vertical lines* depicted below each histogram represent the frequency of neuronal burst activity such that each tick corresponds to a burst discharge event. **e** WIN 55,212-2 (0.05–0.5 mg/kg, i.v.) produced a biphasic response profile in 5-HT single-unit activity. **f** Line graphs showing that cumulative doses of WIN 55,212-2 modulated 5-HT neuronal burst activity measured as percentage of spikes within bursts (Devlin and Christopoulos) and mean burst length (*bottom*). $*p<0.05$ or $**p<0.01$ vs. baseline (vehicle). (With permission from: J Neurosci. 2007 Oct 24;27(43):11700-11)

neuroplastic or synaptic modifications maintained by prolonged CB1R activation (Bambico et al. 2012).

Similar data were found in the FST, a behavioral paradigm commonly used to assess the antidepressive properties of a putative drug (Castagne et al. 2011). Following the FST, we found that single administration of THC (1.0 mg/kg, i.p.) was not sufficient to elicit any difference in either the total duration or frequency of coping behaviors in comparison to control. Conversely, a repeated administration schedule (5 days), elicited an antidepressant-like response, characterized by an increase in swimming and a decrease in immobility, similarly to the SSRI citalopram.

Moreover THC, but not citalopram, also significantly increased climbing duration ($p=0.007$) suggesting that THC may also act through a NE-mediated mechanism (Page et al. 2003). These effects of THC were shown to be blocked by the coapplication of rimonabant (1.0 mg/kg, i.p.) and were therefore mediated by CB1R activation.

Repeated (5 days), but not single, administration of THC (1.0 mg/kg, i.p.), similar to repeated citalopram treatment, increased the excitatory response of hippocampal Cornu Ammonis 3 (CA3) pyramidal neurons to the 5-HT_{1A} receptor antagonist *N*-[2-[4-(2-methoxyphenyl)-1-piperazinyl]ethyl]- *N*-(2-pyridyl)cyclohexanecarboxamide (WAY-100635), suggesting that THC, similar to other classes of antidepressants, increases the tonic activity of postsynaptic 5-HT_{1A} receptors in the hippocampus, but only after long-term treatment (Haddjeri et al. 1998; Besson et al. 2000).

In summary, THC elicits antidepressant-like activity, and this behavioral effect is paralleled by an increase in spontaneous DR 5-HT neural activity and an enhanced tonic activity of 5-HT_{1A} receptors in the dorsal hippocampus, which are neurobiological signatures of antidepressant action.

These electrophysiological experiments revealed that i.v. administration of different doses of THC produces a composite response, with excitation, inhibition, and nonresponse among different populations of 5-HT neurons. Both excitatory and inhibitory responses each assumed a bell-shaped profile. The i.p. injection of THC (1–4 mg/kg) failed to significantly modify the mean discharge rate of DRN 5-HT neurons. This weak response profile associated with acute or single THC treatment may be ascribed to its partial agonist activity at CB1Rs. Recent evidence suggests that THC could act both as a partial and full agonist at CB1Rs, depending on whether the receptor is localized on GABAergic or glutamatergic synapses (Laaris et al. 2010).

Remarkably, repeated THC treatment (1.0 mg/kg) yielded a significant elevation in the mean discharge rate of DRN 5-HT neurons, an effect that was reversed by co-administration with the CB1R antagonist rimonabant, suggesting a CB1-mediated mechanism. This modulation in 5-HT neural activity has been observed in our lab with the CB1R agonist WIN 55,212-2 (Bambico et al. 2007) or the endocannabinoid enhancing FAAH inhibitor [3-(3-carbamoylphenyl)phenyl] N-cyclohexylcarbamate (URB597), demonstrated in vivo (Gobbi et al. 2005; Palazzo et al. 2006) and WIN55,212-2 ex vivo (Mendiguren and Pineda 2009). On the other hand, the CB1R antagonism rimonabant and AM251 depress DR 5-HT neural activity in brain slices

(Mendiguren and Pineda 2009). Additional research on the pharmacological effects of exogenous CB1R agonists and antagonists on monoaminergic transmission and emotional behavior is warranted.

12.4 Cannabinoid Type 1 Agonists and Serotonin Firing Activity

It has been reported that CB1R agonists exhibit antidepressant-like properties in FST models (Hill and Gorzalka 2005, Bambico et al. 2007). Conversely, genetic CB1R deletion (Martin et al. 2002; Aso et al. 2008; for review, Valverde and Torrens 2012) or chronic CB1R antagonism (Beyer et al. 2010) have been shown to produce depressive-like behaviors in animal models. Clinical studies on rimonabant (Acomplia) have likewise ascertained these risks for depression, anxiety, and suicidality (Mitchell and Morris 2007).

In our laboratory, we have assessed the spontaneous single-unit firing activity of DRN 5-HT neurons after cumulative i.v. administration of the CB1R agonist WIN 55,212-2. R-(+)-[2,3-dihydro-5-methyl-3-(4-orpholinylmethyl) pyrrolol[1,2,3-de]-1,4-benzoxazin-6-yl]-1-naphthalenylmethanone mesylate and 5-Hydroxytryptamine Firing Activity. Increasing doses of WIN 55,212-2 (0.05-0.2 mg/kg) evoked a dose-dependent increase in 5-HT unit firing activity, which was half-maximal (ED_{50}) at a dose of 0.1 mg/kg and was not blocked by capsazepine (20 μg/kg, i.v.), but was blocked by rimonabant (1 mg/kg, i.v.) in 100 % of neurons tested.

WIN 55,212-2 treatment also increased burst activity, a pattern that is associated with enhanced 5-HT release in postsynaptic regions (Gartside et al. 2000) as well as antidepressant-like activity (Gobbi et al. 2005). The maximal increase in burst frequency and in the mean number of spikes in a burst from baseline was recorded following WIN 55,212-2 treatment (up to 0.2 mg/kg, i.v.). Among all neurons recorded, 66.67 % of 5-HT neurons responded to increasing dose injections of WIN 55,212-2, while 33.33 % of neurons were nonresponding. All responding and nonresponding neurons showed the same electrophysiological characteristics, inhibited by the $GABA_B$ receptor agonist baclofen and were localized in the DRN, indicating that not all DRN neurons are activated by CB1R stimulation.

Remarkably, cumulative doses of WIN 55,212-2 higher than 0.2 mg/kg i.v. generally produced a decline in neuronal excitation significant at both 0.3 and 0.4 mg/kg and achieved a maximal level 45 % below baseline (vehicle) following 0.4 mg/kg of WIN55,212-2. A waning of stimulatory effects was also observed with different parameters of burst activity: burst frequency, mean number of spikes in a burst, and mean burst length. In two out three neurons tested, the $TRPV_1$ antagonist capsazepine reversed the decrease induced by high doses of WIN 55,212-2, suggesting that $TRPV_1$ receptors, but not CB1Rs, are involved in the 5-HT effects induced by high doses of WIN 55,212-2. Interestingly, neither rimonabant (1 mg/kg, i.v.) alone nor capsazepine (20 μg/kg, i.v.) alone had a significant effect on 5-HT single-unit firing activity, meaning that at low doses these two drugs act as inert antagonists or alternatively, are unable to unmask a putative cannabinoid or vanilloid tonic activity at these doses (Bambico et al. 2007; Fig 12.2)

We found that in the FST, WIN 55,212-2 at the dose of 0.1–0.2 mg/kg administered 23, 5, and 0.75 h before the test, produced a sustained antidepressant-like effect characterized by decreased immobility and increased swimming activity. However, the antidepressant-like activity was not evident when higher doses were injected (Bambico et al. 2007). In order to test whether the antidepressant-like effects of WIN 55,212-2 in the FST were paralleled by enhanced 5-HT neuronal firing activity, we treated animals using the same schedule as used for the FST, but instead performed single unit recordings of DRN 5-HT neurons. In one cohort of animals, rimonabant (1 mg/kg, i.p.) was injected 10 min before the third administration of WIN 55,212-2 (0.2 mg/kg, i.p.).

Increasing doses of WIN 55,212-2 produced a biphasic response on the mean spontaneous firing rate of 5-HT neurons. Specifically, there was a dose-dependent increase in 5-HT firing with lower doses of WIN 55,212-2 and the coadministration of rimonabant prevented this increase. WIN 55,212-2 (0.2 mg/kg) yielded a maximal 126.32 % increase in 5-HT neuronal activity. On the other hand, a high dose of WIN 55,212-2 (2.0 mg/kg) yielded a significant 64 % decrease compared to vehicle. We also calculated the mean number of neurons encountered per electrode descent, which serves as an indirect measure of spontaneously active neurons (Gobbi et al. 2007). In comparison with vehicle injections, there were 28 % more spontaneously active 5-HT neurons encountered after treatment with 0.1 mg/kg WIN 55,212-2 and 33.33 % more active neurons with 0.2 mg/kg WIN 55,212-2 ($p<0.01$), while a high dose of WIN 55,212-2 had 48.8 % fewer active neurons than the control (WIN 2.0 mg/kg $=1.92\pm0.39$, $p<0.01$). The number of spontaneously active neurons in rats treated with a low dose of WIN 55,212-2 (0.2 mg/kg) coapplied with rimonabant (1.0 mg/kg) did not significantly differ from those treated with the vehicle.

We also examined whether stimulation of DRN 5-HT neurons by WIN 55,212-2 is mediated by CB1Rs locally within the DRN, or by CB1Rs located in neurons of the prefrontal cortex (PFC) which project extensively to the DRN. Indeed, DRN 5-HT neurons receive important excitatory inputs from pyramidal (glutamatergic) cells of the PFC (Jankowski and Sesack 2004). In order to answer this question, we performed systematic transections of the PFC–DRN pathway before electrophysiological recordings. Three types of transections were performed: a total bilateral PFC (tPFC) transection, a selective transection of the medial PFC (mPFC; areas transected included the dorsal peduncular, infralimbic, prelimbic Cg3, and cingulate Cg1 cortices), and a transection of the lateral aspect of the PFC (latPFC): mainly the lateral prefrontal/agranular insular, but also the frontal Fr2, Fr1, and Fr3; the ventrolateral and lateral orbital cortices; and some parts of parietal area 1 (modified after Hajos et al. 1999). 5-HT single-unit recordings were conducted 1.5–2 h after each transection.

Following tPFC transection, the i.v. administration of WIN 55,212-2 failed to increase 5-HT single-unit firing activity at otherwise stimulatory doses in intact brains. To pinpoint the specific subregion of the PFC that is critical in mediating the modulation of 5-HT single-unit activity, we compared transection of the mPFC with that of the latPFC. The response of 5-HT single units to the latPFC did not significantly differ from the control, but on the other hand, mPFC transection produced an effect similar to tPFC transection and was significantly different from the control, thus indicating that the medial, but not lateral, subregions of the PFC are responsible for the enhanced 5-HT firing elicited by global CB1R agonism (Bambico et al. 2007). The transection procedure did not significantly modify the basal discharge rate of DRN 5-HT neurons, as was also observed by Hajos et al. (1999).

Since the results obtained from intracerebral WIN 55,212-2 microinfusions with electrophysiology seemed to point to the ventromedial PFC (mPFCv) as a structure that plays an important role in cannabinoid-induced activation of DRN 5-HT neurons, we therefore examined whether local bilateral microinfusion of WIN 55,212-2 into the mPFCv is sufficient to alter antidepressant-like responding in the FST. Both microdoses of WIN 55,212-2 used (1 and 5 μg in 0.5 μl of vehicle), compared with vehicle, produced a reduction of 47.43 and 36.24 %, respectively, in total immobility time, with no significant changes observed in climbing behavior, implying that enhancement in NE transmission may not be as important as enhancement in 5-HT transmission in mediating the antidepressant-like effects of WIN 55,212-2 in the FST. A microdose of rimonabant (1 μg) that by itself did not produce any significant effect in the FST, blocked the effect of 1 μg of WIN 55,212-2 when microinfused 1 min before WIN 55,212-2.

Altogether, these results indicate that the mPFCv plays an instrumental role in mediating the increase in 5-HT firing and the antidepressant-like effects of CB1R agonists (Bambico et al. 2007).

12.5 Fatty Acid Amide Hydrolase Inhibitors (FAAH), Fatty Acid Amide Hydrolase Knockout and Serotonin (5-HT) Firing Activity

Even if THC and WIN 55,212-2 show potent antidepressant-like effect and increase of 5-HT firing activity, their clinical use is limited by several warnings including addiction, tolerance, and their narrow therapeutic window. Selective inhibition of the enzyme FAAH, which catalyzes the intracellular hydrolysis of the AEA, has been proposed to be a useful alternative to the direct CB1R agonists for their capacity to increase endogenous cannabinoid signaling without inducing the typical cannabis side-effects such as dependence and sedation (Gobbi et al. 2005).

One of the first experiments carried out in our laboratory was to test whether the FAAH inhibitor URB597 modulates 5-HT monoaminergic transmission in vivo. We first measured spontaneous activity of 5-HT neurons in the DRN of anesthetized rats. Single injections of URB597 (0.03–0.3 mg/kg, i.v.) evoked a slow increase in 5-HT neuronal firing, which was half-maximal at a dose of $\approx$ 0.06 mg/kg and was blocked by pretreatment with rimonabant (1 mg/kg, i.v.). Interestingly, the increase in firing was not immediate, as observed with the direct CB1R agonist, but occurred after 15–20 min; this delay is compatible with the pharmacodynamics of URB597, which after passing the blood–brain barrier, inhibits FAAH in an irreversible manner, leading to an increase in AEA, which activates CB1Rs (Gobbi et al. 2005; Fig. 12.3).

The increase in 5-HT firing activity was confirmed also after subchronic treatment. Indeed subchronic treatment with URB597 (0.1 mg/kg, i.p., once daily for 4 days) evoked an even stronger response, which was also reversed by rimonabant (1 mg/kg, i.p.). This sustained increase in 5-HT activity following subchronic treatment was also associated with an increase in bursting activity and a sustained 5-HT outflow in the hippocampus, but not in the PFC, as assessed with in vivo microdialysis in awake rats (whereas a single injection of URB597 had no such effect).

Finally, 4-day treatment with URB597 did not affect the responsiveness of 5-HT neurons to local iontophoretic administration of the 5-HT_{1A} agonist 8-OH-DPAT (Fig 12.4e), suggesting that URB597, unlike classical antidepressants (Artigas et al. 1996, Gobbi and Blier 2005), does not produce desensitization of 5-HT_{1A} autoreceptors (Gobbi et al. 2005; Fig. 12.3). Importantly, at the same doses (0.1 and 0.3 mg/kg) and duration (daily for 4 days), URB597 induced antidepressant-like effects in the mouse TST and the rat FST. This effect was more robust after repeated injections (4 days) and was reversed by the preadministration of rimonabant.

We were also able to replicate the electrophysiological and behavioral findings in FAAH knockout (FAAH(−/−)) mice (Bambico et al. 2010), in which we observed a marked increase (+34.68 %) in DRN 5-HT neural firing compared to their littermates that was reversed by rimonabant. This effect was particularly significant in a subset of neurons exhibiting high firing rates (33.15 % mean decrease). FAAH(−/−) mice also showed reduced immobility in the FST and TST, predictive of antidepressant activity, which was attenuated by rimonabant. FAAH(−/−) mice also exhibited an anxiolytic-like profile, with increased duration of open arm visits in the elevated plus maze, and a decrease in thigmotaxis, as well as an increase in exploratory rearing in the open field test (Bambico et al. 2010a). Chronic treatment with some classes of antidepressants, such as SSRIs, that potentiate 5-HT efflux in the PFC can also lead to a downregulation of 5-$HT_{2A/2C}$ receptors in this region (Hollander et al. 1991; Quested et al. 1997; Hill et al. 2009). This downregulation has been associated with the anxiolytic efficacy of these drugs, as antagonism of these receptors elicits identical therapeutic effects (Deakin 1988; Griebel et al. 1997; Adamec et al. 2004) or augments the effects of antidepressants (Marek et al. 2003; Hill et al. 2009), whereas agonism produces panic and anxiety in otherwise healthy humans (Germine et al. 1994). We also assessed the response of the prefrontocortical pyramidal cells to the 5-$HT_{2A/2C}$ receptor agonist (+/−)-1-(2,5-dimethoxy-4-iodophenyl)-2-aminopropane ((+/−)-DOI), using microiontophoresis. In FAAH(−/−) mice, we found a desensitization of prefrontocortcal 5-$HT_{2A/2C}$ receptors, indicating that this neuroplastic change, induced by the genetic deletion of the FAAH enzyme, may be responsible of the low anxiety-like behavior found in these mice (Bambico et al. 2010a).

The delay in therapeutic onset of antidepressants has been attributed to gradual neuroplastic adaptations at the presynaptic and postsynaptic levels that result

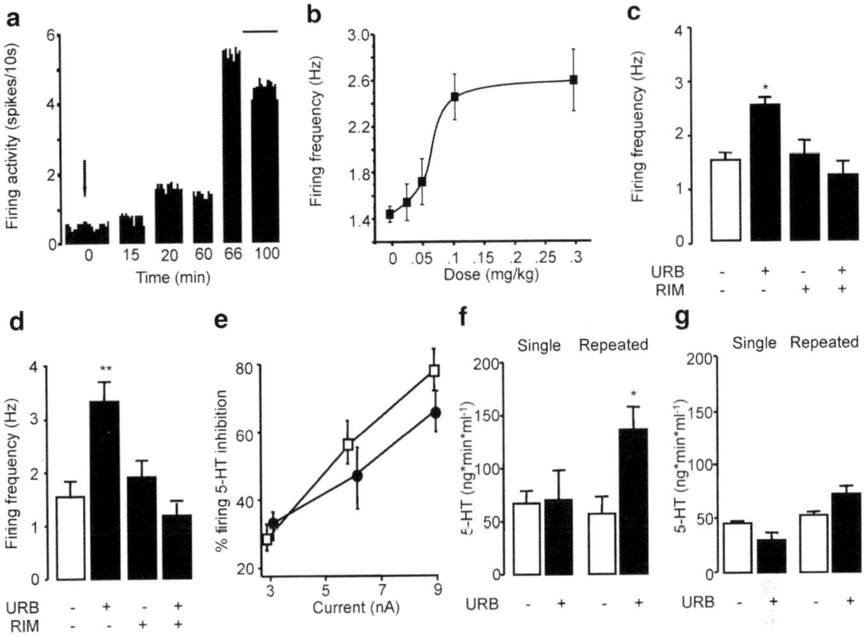

Fig. 12.3 Effects of [3-(3-carbamoylphenyl)phenyl] N-cyclohexylcarbamate (URB597) on 5-Hydroxytryptamine (5-HT) neuron firing in the rat dorsal raphe nucleus (DRN). Integrated firing rate histogram of DRN neurons, illustrating the time-dependent effects of URB597; *arrow* indicates time of URB597 injection (0.1 mg/kg, i.v.; calibration bar: 1 min) (**a**). Dose-dependent effects of URB597 on spontaneous firing rate (**b**). Single administration of rimonabant (*RIM*) (1 mg/kg, i.v.) prevents the effects of single (0.1 mg/kg) (**c**) and repeated (**d**) URB597 injections (0.1 mg/kg, i.p., once daily for 4 days) on 5-HT neuron firing. Repeated URB597 administration does not affect the response of 5-HT neurons to 8-hydroxy-2-(di-n-propylamino)tetralin, expressed as percent inhibition of 5-HT-neuron firing rate. *Open* symbols represent vehicle (**e**). Effects of single or repeated URB597 injections on 5-HT outflow over 3 h in hippocampus (**f**) and prefrontal cortex (PFC) (**g**) of awake rats. *$p<0.05$ vs. vehicle; **$p<0.01$ vs. vehicle. (With permission from: Proc Natl Acad Sci U S A. 2005 Dec 20;102(51):18620–5)

from the progressive augmentation of 5-HT activity. These modifications include desensitization of autoinhibitory 5-HT_{1A} receptors, and sensitization or increased tonic activation of postsynaptic 5-HT_{1A} receptors (Haddjeri et al. 1998; Besson et al. 2000; Szabo and Blier 2001). The hippocampal pyramidal response to the 5-HT_{1A} receptor antagonist, WAY-100635, indicates enhanced tonus on the hippocampal 5-HT_{1A} heteroreceptors, a hallmark of antidepressant-like action. FAAH(−/−) mice, compared to their wild-type littermates, showed an increased tonic activity of 5-HT_{1A} receptors, as tested with administration of WAY-100635 (0.5 mg/kg, i.p.), that potently disinhibited hippocampal pyramidal neural activity in FAAH(−/−) mice, but not wild-type controls. Together, these results suggest that genetic deletion of FAAH enhances anxiolytic-like and antidepressant-like effects, paralleled by augmented 5-HT transmission and postsynaptic 5-HT_{1A} and 5-$HT_{2A/2C}$ receptor function (Bambico et al. 2010a).

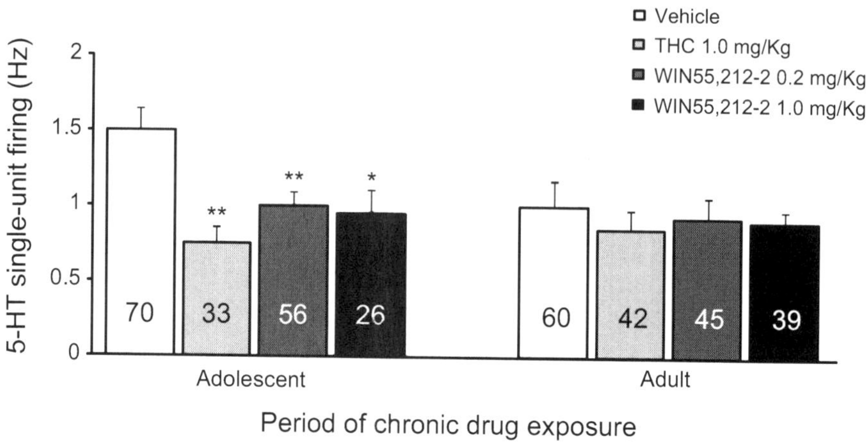

Fig. 12.4 Alteration in serotonergic (5-HT) neurotransmission following adolescent cannabinoid exposure. Chronic daily treatment with R-(+)-[2,3-dihydro-5-methyl-3-(4-morpholinylmethyl) pyrrolol[1,2,3-de]-1,4-benzoxazin-6-yl]-1-naphthalenylmethanone mesylate (WIN 55,212-2) (0.2 or 1.0 mg/kg, intraperitoneal (i.p.)) or Δ^9-tetrahydrocannabinol (THC) (1.0 mg/kg, i.p.) resulted in a significant decrease in dorsal raphe nucleus (DRN) 5-HT spontaneous single-spiking rate when administered during adolescence but not when administered during adulthood. Values at the bottom of each bar denote the number of neurons recorded. $^*p<0.05$; $^{**}p<0.01$ (unpublished results)

12.6 Chronic Administration of Cannabinoid Type 1 Agonists in Adolescence and Adulthood and Serotonin Firing Activity

Cannabis remains the most abused illicit substance by adolescents, with reports of unabated escalation in the last decades in particular during puberty (Schneider 2008). This is particularly alarming since retrospective correlational (Deas 2006) and longitudinal prospective (Wittchen et al. 2007) studies have suggested that its long-term use early in life increases the risk for anxiety, depression, and amotivational syndrome, as well as other neuropsychiatric disorders (for review, Howlett et al. 2004; Bambico and Gobbi 2008; Bambico et al. 2009a), independent of whether the person uses other illicit drugs (Hayatbakhsh et al. 2007). The limited neurobiological data scrutinizing this association are somewhat inconsistent, with reports of both increased (O'Shea et al. 2004; O'Shea et al. 2006) and decreased (Biscaia et al. 2003; Rubino et al. 2008) emotional reactivity across a number of animal models, which have been ascribed to differences in treatment duration and regimen, as well as in drug potency. Many studies have also not compared the impact of adolescent exposure to adult exposure, and the neural mechanisms underlying pathogenesis/pathophysiology remain largely unexplored.

The neurobiological impact of drug use is especially compounded during the critical adolescence period when brain development is punctuated by constant neuroplastic shaping, synaptic reorganization, and extensive neurochemical changes

(Spear 2000) paralleled by a peaking of emotional volatility, anxiety and self-consciousness (Buchanan et al. 1992), disproportionately extensive reckless, novelty- and sensation-seeking and risk-taking behaviors, and partial anhedonia (Spear 2000). Brain imaging studies have ascertained that dynamic neuroanatomical modifications occur throughout adolescence (Giedd 2004). During this stage, corticolimbic CB1R density is at its peak, undergoing gradual pruning thereafter (Belue et al. 1995), which may well relate to the crucial role of endocannabinoids in brain developmental processes, including neurogenic control, neural progenitor proliferation, lineage segregation, and the migration and phenotypic specification of immature neurons (Harkany et al. 2008). Several lines of evidence suggest that prolonged aberrations in CB1R signaling may dramatically alter the density of CB1Rs (Ellgren et al. 2008) in ways that potentially disrupt the development of the monoaminergic systems, hence, influencing mood and anxiety control. In light of this evidence, my laboratory examined the impact of CB1 agonist administration on 5-HT firing activity after long-term exposure during adolescence and adulthood, hypothesizing that the adolescent brain would be influenced by prolonged CB1 agonism in a more detrimental manner compared to mature subjects. Adolescent rats were treated with WIN 55,212-2 (0.1 or 1 mg/kg, daily), THC (1 mg/kg, daily), or vehicle from postnatal day (PND) 30–50, followed by a drug washout period from PND 50–70. At PND 70, these groups were tested with behavioral assays or electrophysiology. Four distinct groups of adult rats were similarly treated for 20 days (from PND 70–90) and tested 20 days later (PND 110).

In the adolescent rats, all drug treatments significantly attenuated spontaneous 5-HT single-spike activity (Bambico et al. 2010b; Fig. 12.4). Further analyses of neural activity revealed a trending, but nonsignificant decline in burst firing activity (decreased number of spikes per burst and burst length, increased burst ISI and decreased ratio (%) of spikes within bursts to the total number of spikes) following exposure to the low dose of WIN 55,212-2. Opposite effects (nonsignificant increase in burst activity) were observed after exposure to the high dose of WIN 55,212-2 as well as THC, which corresponded to positively skewed ISI distributions, further indicating irregular and burst-like neural firing. The cannabinoid-induced decrease in 5-HT neural activity was not due to an increased tone on inhibitory DRN 5-HT_{1A} autoreceptors, since cumulative i.v. administrations of the 5-HT_{1A} receptor antagonist WAY-100635 did not significantly modify firing activity, similar to vehicle-treated controls. Drug exposure during adulthood did not yield significant changes in 5-HT single-spike and burst firing activity (Bambico et al. 2010b; Fig. 12.4)

From a behavioral point of view, chronic adolescent exposure, but not adult exposure, to WIN 55,212-2 (low 0.2 mg/kg and high 1.0 mg/kg) and THC (1 mg/kg) led to depressive-like behavior in the FST and sucrose preference test, while the high dose also induced anxiety-like consequences in the novelty suppressed feeding test.

Together these data suggest that the 5-HT system in adolescents is particularly sensitive to chronic consumption of CB1R agonists leading to a net decrease in electrical activity and ensuing behavioral consequences. Translating these findings to human populations, more studies are needed to understand the reasons for these

Table 12.1 Effects of THC, WIN 55,212-2, and URB597 on 5-HT firing activity

Drug	Acute treatment	Repeated treatment	Long-term use in adolescence followed by washout	Long-term use in adulthood followed by washout
THC	Increase or decrease	Increase	Decrease[a]	No effect
WIN 55,212-2	Increase (low doses) or decrease (high doses)	Increase (low doses) or decrease (high doses)	Decrease	No effect
URB597	Increase	Increase	n.d.	n.d.

THC Δ^9-tetrahydrocannabinol, *WIN 55,212-2* R-(+)-[2,3-dihydro-5-methyl-3-(4-morpholinylmethyl) pyrrolol[1,2,3-de]-1,4-benzoxazin-6-yl]-1-naphthalenylmethanone mesylate, *URB597* [3-(3-carbamoylphenyl)phenyl] N-cyclohexylcarbamate, *n.d.* nondetermined, *5-HT* 5-Hydroxytryptamine
[a] in-house data

plastic changes induced by cannabis derivatives and more efforts are needed in the prevention and treatment of mental health consequences induced by cannabis in these vulnerable populations.

12.7 Conclusion

Extensive electrophysiological and behavioral studies have undoubtedly confirmed the capacity of CB1 agonists, antagonists, and FAAH inhibitors to directly modulate the serotonin neurotransmission (see Table 12.1) evoking changes in mood and emotions. On one hand, these studies have clarified several important issues concerning the acute effects of cannabis consumption and the long-term consequences of cannabis use among adolescents and adults; on the other hand, they have also opened novel avenues in the field of drug discovery in mental health. In summary, these researches have indicated the following translational conclusions. (1) Acute injection of THC may produce differential responses of 5-HT firing activity, even if a sub-chronic treatment with relatively low doses (1 mg/kg) seems to elicit a stable 5-HT increase. The effects of higher doses are still not known. (2) The CB1 agonist WIN 55,212-2 increases 5-HT firing at low doses, but decreases this at higher doses, with a very narrow pharmaceutical (therapeutic) window, meaning that the amount giving an increase in 5-HT (and antidepressant-like effect) and the amount producing a decrease in 5-HT firing is in the range of micrograms. (3) The CB1 antagonist rimonabant blocks the elevation in 5-HT at low doses. (4) The long-term consumption of THC and WIN 55,212-2 during adolescence, but not adulthood decrease 5-HT firing activity even after a washout period. (5) The FAAH inhibitors represent an alternative to the CB1 agonists, increasing 5-HT activity following a sigmoidal curve without the biphasic effects determined by CB1 direct agonists and without dependence warnings. However, more research has to be done in order to validate these novel ligands in clinical studies.

References

Adamec R, Creamer K, Bartoszyk GD, Burton P (2004) Prophylactic and therapeutic effects of acute systemic injections of EMD 281014, a selective serotonin 2A receptor antagonist on anxiety induced by predator stress in rats. Eur J Pharmacol 504:79–96

Aghajanian GK, Foote WE, Sheard MH (1968) Lysergic acid diethylamide: sensitive neuronal units in the midbrain raphe. Science 161:706–708

Allers K, Sharp T (2003) Neurochemical and anatomical identification of fast- and slow -firing neurones in the rat dorsal raphe nucleus using juxtacellular labelling methods in vivo. Neuroscience 122:193–204

Artigas F (1993) 5-HT and Antidepressants: new views from Microdialysis Studies. Trends Pharmacol Sci 14:262

Artigas F, Romero L, Montigny C de, Blier P (1996) Acceleration of the effect of selected antidepressant drugs in major depression by 5-HT1A antagonists. Trends Neurosci 19:378–383

Aso E, Ozaita A, Valdizan EM, Ledent C, Pazos A, Maldonado R, Valverde O (2008) BDNF impairment in the hippocampus is related to enhanced despair behavior in CB1 knockout mice. J Neurochem 105:565–572

Bambico FR, Gobbi G (2008) The cannabinoid CB_1 receptor and the endocannabinoid anandamide: possible antidepressant targets. Expert Opin Ther Targets 12:1347–1366

Bambico FR, Katz N, Debonnel G, Gobbi G (2007) Cannabinoids elicit antidepressant-like behavior and activate serotonergic neurons through the medial prefrontal cortex. J Neurosci 27:11700–11711

Bambico FR, Duranti A, Tontini A, Tarzia G, Gobbi G (2009) Endocannabinoids in the treatment of mood disorders: evidence from animal models. Current Pharm Des 15:1623–1646

Bambico FR, Cassano T, Dominguez-Lopez S, Katz N, Walker CD, Piomelli D, Gobbi G (2010a) Genetic deletion of fatty acid amide hydrolase alters emotional behavior and serotonergic transmission in the dorsal raphe, prefrontal cortex, and hippocampus. Neuropsychopharmacol 35:2083–2100

Bambico FR, Nguyen NT, Katz N, Gobbi G (2010b) Chronic exposure during adolescence but not during adulthood impairs emotional behaviour and monoaminergic transmission. Neurobiol Dis 37:641–655

Bambico FR, Hattan PR, Garant JP, Gobbi G (2012) Effect of delta-9-tetrahydrocannabinol on behavioral despair and on pre- and postsynaptic serotonergic transmission. Progress Neuro-Psychoph 38:88–96

Banerjee SP, Snyder SH, Mechoulam R (1975) Cannabinoids: influence on neurotransmitter uptake in rat brain synaptosomes. J Pharmacol Exp Ther 194:74–81

Barann M, Molderings G, Bruss M, Bonisch H, Urban BW, Gothert M (2002) Direct inhibition by cannabinoids of human 5-HT3A receptors: probable involvement of an allosteric modulatory site. British J Pharmacol 137:589–596

Belue RC, Howlett AC, Westlake TM, Hutchings DE (1995) The ontogeny of cannabinoid receptors in the brain of postnatal and aging rats. Neurotoxicol Teratol 17:25–30

Besson A, Haddjeri N, Blier P, Montigny C de (2000) Effects of the co-administration of mirtazapine and paroxetine on serotonergic neurotransmission in the rat brain. European Neuropsychopharm 10:177–188

Beyer CE, Dwyer JM, Piesla MJ, Platt BJ, Shen R, Rahman Z, Chan K, Manners MT, Samad TA, Kennedy JD, Bingham B, Whiteside GT (2010) Depression-like phenotype following chronic CB1 receptor antagonism. Neurobiol Dis 39:148–155

Biscaia M, Marin S, Fernandez B, Marco EM, Rubio M, Guaza C, Ambrosio E, Viveros MP (2003) Chronic treatment with CP 55,940 during the peri-adolescent period differentially affects the behavioural responses of male and female rats in adulthood. Psychopharmacol 170:301–308

Blier P, Montigny C de (1994) Current Advances and Trends in the Treatment of Depression. Trends Pharmacol Sci 15:220–226

Buchanan CM, Eccles JS, Becker JB (1992) Are adolescents the victims of ragin hormones—evidence for activational effects of hormones on moods and behavior at adolescence. Psychol Bull 111:62–107

Calizo LH, Akanwa A, Ma X, Pan Y, Lemos JC, Craige C, Heemstra LA, Beck SG (2011) Raphe serotonin neurons are not homogenous: electrophysiological, morphological and neurochemical evidence. Neuropharmacol 61:524–543

Castagne V, Moser P, Roux S, Porsolt RD (2011) Rodent models of depression: forced swim and tail suspension behavioral despair tests in rats and mice. Current protocols in neuroscience/ editorial board, Jacqueline N Crawley [et al] Chapter 8:Unit 8 10A

CPHA (2012) CPS: Compendium of Pharmaceuticals and Specialties. Canadian Pharmacists Assoc, Canada

Darmani NA (2001) Cannabinoids of diverse structure inhibit two DOI-induced 5-HT(2A) receptor-mediated behaviors in mice. Pharmacol Biochem Be 68:311–317

Deakin JF (1988) 5HT2 receptors, depression and anxiety. Pharmacol Biochem Be 29:819–820

Deas D (2006) Adolescent substance abuse and psychiatric comorbidities. J Clin Psych 67 (Suppl 7):18–23

Descarries L, Watkins K, Garcia S, Beaudet A (1982) The serotonin neurons in nucleus raphe dorsalis of adult rat: a light and electron microscope radioautographic study. J Comp Neurol 207:239–254

Devlin MG, Christopoulos A (2002) Modulation of cannabinoid agonist binding by 5-HT in the rat cerebellum. J Neurochem 80:1095–1102

Domànguez-López S, Mahar I, Bambico FR, Labonté B, Ochoa-Sánchez R, Leyton M, Gobbi G (2011) Short term effects of melatonin and pinealectomy on serotoner-gic neuronal activity across the light-dark cycle. J Psychopharmacol 26(6):830–844

Earleywine M (2005) Understanding Marijuana: A New Look at the Scientific Evidence. Oxford Univ. Press, USA

Egashira N, Matsuda T, Koushi E, Higashihara F, Mishima K, Chidori S, Hasebe N, Iwasaki K, Nishimura R, Oishi R, Fujiwara M (2008) Delta(9)-tetrahydrocannabinol prolongs the immobility time in the mouse forced swim test: involvement of cannabinoid CB(1) receptor and serotonergic system. Eur J Pharmacol 589(1–3):117–121

Egertova M, Cravatt BF, Elphick MR (2003) Comparative analysis of fatty acid amide hydrolase and cb(1) cannabinoid receptor expression in the mouse brain: evidence of a widespread role for fatty acid amide hydrolase in regulation of endocannabinoid signaling. Neuroscience 119:481–496

El-Alfy AT, Ivey K, Robinson K, Ahmed S, Radwan M, Slade D, Khan I, ElSohly M, Ross S (2010) Antidepressant-like effect of delta9-tetrahydrocannabinol and other cannabinoids isolated from *Cannabis sativa L.* Pharmacol Biochem Be 95:434–442

Elbatsh M, Moklas M, Marsden C, Kendall D (2012) Antidepressant-like effects of Δ9-tetrahydrocannabinol and rimonabant in the olfactory bulbectomised rat model of depression Pharm Bioch Behaviour 102:357–365

Ellgren M, Artmann A, Tkalych O, Gupta A, Hansen HS, Hansen SH, Devi LA, Hurd YL (2008) Dynamic changes of the endogenous cannabinoid and opioid mesocorticolimbic systems during adolescence: THC effects. Eur Neuropsychopharmacol 18:826–834

Fan P (1995) Cannabinoid agonists inhibit the activation of 5-HT3 receptors in rat nodose ganglion neurons. J Neurophysiol 73:907–910

Freedman J, Aghajanian G (1984) Idazoxan (RX 781094) Selectively antagonizes alpha 2-adrenoceptors on rat central neurons. Eur J Pharmacol 105:265–272

Gartside S, Hajos-Korcsok E, Bagdy E, Harsing L, Sharp T, Hajos M (2000) Neurochemical and electrophysiological studies on the functional significance of burst firing in serotonergic neurons. Neurosci 98:295–300

Germine M, Goddard AW, Sholomskas DE, Woods SW, Charney DS, Heninger GR (1994) Response to meta-chlorophenylpiperazine in panic disorder patients and healthy subjects: influence of reduction in intravenous dosage. Psychiatry Res 54:115–133

Giedd JN (2004) Structural magnetic resonance imaging of the adolescent brain. Ann N Y Acad Sci 1021:77–85

Gobbi G, Blier P (2005) Effect of neurokinin-1 receptor antagonists on serotoninergic, noradrenergic and hippocampal neurons: comparison with antidepressant drugs. Peptides 26:1383–1393

Gobbi G, Bambico FR, Mangieri R, Bortolato M, Campolongo P, Solinas M, Cassano T, Morgese MG, Debonnel G, Duranti A, Tontini A, Tarzia G, Mor M, Trezza V, Goldberg SR, Cuomo V, Piomelli D (2005) Antidepressant-like activity and modulation of brain monoaminergic transmission by blockade of anandamide hydrolysis. PNAS 102:18620–18625

Gobbi G, Cassano T, Radja F, Morgese MG, Cuomo V, Santarelli L, Hen R, Blier P (2007) Neurokinin 1 receptor antagonism requires norepinephrine to increase serotonin function. European neuropsychopharmacol 17:328–338

Griebel G, Rodgers RJ, Perrault G, Sanger DJ (1997) Risk assessment behaviour: evaluation of utility in the study of 5-HT-related drugs in the rat elevated plus-maze test. Pharmacol Biochem Be 57:817–827

Haddjeri N, Blier P, Montigny C de (1997) Effects of long-term treatment with the alpha 2-adrenoceptor antagonist mirtazapine on 5-HT neurotransmission. Naunyn Schmiedebergs Arch Pharmacol 355:20–29

Haddjeri N, Blier P, Montigny C de (1998) Long-term antidepressant treatments result in a tonic activation of forebrain 5-HT1A receptors. J Neurosci 18:10150–10156

Haddjeri N, Lavoie N, Blier P (2004) Electrophysiological evidence for the tonic activation of 5-HT1A autoreceptors in the rat dorsal raphe nucleus. Neuropsychopharmacology 29:1800–1806

Haigler HJ, Aghajanian GK (1974) Lysergic acid diethylamide and serotonin: a comparision of effects on serotonergic neurons and neurons receiving a serotonergic input. J Pharmacol Exp Ther 188:688–699

Haj-Dahmane S, Shen RY (2005) The wake-promoting peptide orexin-B inhibits glutamatergic transmission to dorsal raphe nucleus serotonin neurons through retrograde endocannabinoid signaling. J Neurosci 25:896–905

Hajos M, Hajos-Korcsok E, Sharp T (1999) Role of the medial prefrontal cortex in 5-HT1A receptor-induced inhibition of 5-HT neuronal activity in the rat. British J Pharmacol 126:1741–1750

Hajos M, Allers K, Jennings K, Sharp T, Charette G, Sik A, Kocsis B (2007) Neurochemical identification of stereotypic burst-firing neurons in the rat dorsal raphe nucleus using juxtacellular labelling methods. Eur J Neurosci 25:119–126

Harkany T, Keimpema E, Barabas K, Mulder J (2008) Endocannabinoid functions controlling neuronal specification during brain development. Mol Cell Endocrinol 286:84–90

Hayatbakhsh MR, Najman JM, Jamrozik K, Mamun AA, Alati R, Bor W (2007) Cannabis and anxiety and depression in young adults: a large prospective study. J Am Acad Child Adolesc Psychiatry 46:408–417

Hermann H, Marsicano G, Lutz B (2002) Coexpression of the cannabinoid receptor type 1 with dopamine and serotonin receptors in distinct neuronal subpopulations of the adult mouse forebrain. Neuroscience 109:451–460

Hill MN, Gorzalka BB (2005) Pharmacological enhancement of cannabinoid CB1 receptor activity elicits an antidepressant-like response in the rat forced swim test. Eur Neuropsychopharmacol 15:593–599

Hill MN, Sun JC, Tse MT, Gorzalka BB (2006) Altered responsiveness of serotonin receptor subtypes following long-term cannabinoid treatment. Int J Neuropsychopharmacol 9:277–286

Hill MN, Hillard CJ, Bambico FR, Patel S, Gorzalka BB, Gobbi G (2009) The therapeutic potential of the endocannabinoid system for the development of a novel class of antidepressants. Trends Pharmacol Sci 30:484–493

Hollander E, DeCaria C, Gully R, Nitescu A, Suckow RF, Gorman JM, Klein DF, Liebowitz MR (1991) Effects of chronic fluoxetine treatment on behavioral and neuroendocrine responses to meta-chlorophenylpiperazine in obsessive-compulsive disorder. Psychiatry research 36:1–17

Howlett AC, Breivogel CS, Childers SR, Deadwyler SA, Hampson RE, Porrino LJ (2004) Cannabinoid physiology and pharmacology: 30 years of progress. Neuropharmacology 47(Suppl 1):345–358

Huestis MA, Gorelick DA, Heishman SJ, Preston KL, Nelson RA, Moolchan ET, Frank RA (2001) Blockade of effects of smoked marijuana by the CB1-selective cannabinoid receptor antagonist SR141716. Arch Gen Psychiat 58:322–328

Iversen L (2003) Cannabis and the brain. Brain 126:1252–1270

Jankowski MP, Sesack SR (2004) Prefrontal cortical projections to the rat dorsal raphe nucleus: ultrastructural features and associations with serotonin and gamma-aminobutyric acid neurons. J Comp Neurol 468:518–529

Johnson KM, Ho BT, Dewey WL (1976) Effects of delta9-tetrahydrocannabinol on neurotransmitter accumulation and release mechanisms in rat forebrain synaptosomes. Life Sci 19:347–356

Kimura T, Ohta T, Watanabe K, Yoshimura H, Yamamoto I (1998) Anandamide, an endogenous cannabinoid receptor ligand, also interacts with 5-hydroxytryptamine (5-HT) receptor. Biol Pharm Bull 21:224–226

Laaris N, Good CH, Lupica CR (2010) Delta9-tetrahydrocannabinol is a full agonist at CB1 receptors on GABA neuron axon terminals in the hippocampus. Neuropharmacology 59:121–127

Malone DT, Taylor DA (2001) Involvement of somatodendritic 5-HT(1A) receptors in Delta(9)-tetrahydrocannabinol-induced hypothermia in the rat. Pharmacol Biochem Be 69:595–601

Marek GJ, Carpenter LL, McDougle CJ, Price LH (2003) Synergistic action of 5-HT2A antagonists and selective serotonin reuptake inhibitors in neuropsychiatric disorders. Neuropsychopharmacology 28:402–412

Martin M, Ledent C, Parmentier M, Maldonado R, Valverde O (2002) Involvement of CB1 cannabinoid receptors in emotional behaviour. Psychopharmacology 159:379–387

Matsuda LA, Bonner TI, Lolait SJ (1993) Localization of cannabinoid receptor mRNA in rat brain. J Comp Neurol 327:535–550

Mendiguren A, Pineda J (2009) Effect of the CB(1) receptor antagonists rimonabant and AM251 on the firing rate of dorsal raphe nucleus neurons in rat brain slices. British J Pharmacol 158:1579–1587

Mitchell PB, Morris MJ (2007) Depression and anxiety with rimonabant. Lancet 370:1671–1672

Moldrich G, Wenger T (2000) Localization of the CB1 cannabinoid receptor in the rat brain. An immunohistochemical study. Peptides 21:1735–1742

Morales M, Wang SD, Diaz-Ruiz O, Jho DH (2004) Cannabinoid CB1 receptor and serotonin 3 receptor subunit A (5-HT3A) are co-expressed in GABA neurons in the rat telencephalon. J Comp Neurol 468:205–216

Moreira FA, Grieb M, Lutz B (2004) Role of serotonin on the antidepressant-like effect induced by delta-9-tetrahydrocannabinol, but not by rimonabant. In: 18th Symposium of the International Cannabinoid Research Society Avemore, Scotland

Nakazi M, Bauer U, Nickel T, Kathmann M, Schlicker E (2000) Inhibition of serotonin release in the mouse brain via presynaptic cannabinoid CB1 receptors. Naunyn Schmiedebergs Arch Pharmacol 361:19–24

O'Shea M, Singh ME, McGregor IS, Mallet PE (2004) Chronic cannabinoid exposure produces lasting memory impairment and increased anxiety in adolescent but not adult rats. J Psychopharmacol 18:502–508

O'Shea M, McGregor IS, Mallet PE (2006) Repeated cannabinoid exposure during perinatal, adolescent or early adult ages produces similar longlasting deficits in object recognition and reduced social interaction in rats. J Psychopharmacol 20:611–621

Page ME, Brown K, Lucki I (2003) Simultaneous analyses of the neurochemical and behavioral effects of the norepinephrine reuptake inhibitor reboxetine in a rat model of antidepressant action. Psychopharmacol 165:194–201

Palazzo E, Novellis V de, Petrosino S, Marabese I, Vita D, Giordano C, Di Marzo V, Mangoni GS, Rossi F, Maione S (2006) Neuropathic pain and the endocannabinoid system in the dorsal raphe: pharmacological treatment and interactions with the serotonergic system. Eur J Neurosci 24:2011–2020

Paxinos G, Watson C (2007) The rat brain in stereotaxic coordinates. Elsevier, Amsterdam

Quested DJ, Sargent PA, Cowen PJ (1997) SSRI treatment decreases prolactin and hyperthermic responses to mCPP. Psychopharmacol 133:305–308

Rodriguez-Gaztelumendi A, Rojo ML, Pazos A, Diaz A (2009) Altered CB receptor-signaling in prefrontal cortex from an animal model of depression is reversed by chronic fluoxetine. J Neurochem 108:1423–1433

Rubino T, Vigano D, Realini N, Guidali C, Braida D, Capurro V, Castiglioni C, Cherubino F, Romualdi P, Candeletti S, Sala M, Parolaro D (2008) Chronic delta 9-tetrahydrocannabinol during adolescence provokes sex-dependent changes in the emotional profile in adult rats: behavioral and biochemical correlates. Neuropsychopharmacol 33:2760–2771

Schneider M (2008) Puberty as a highly vulnerable developmental period for the consequences of cannabis exposure. Addict Biol 13:253–263

Spear LP (2000) Neurobehavioral changes in adolescence. Curr Dir Psychol Sci 9:111–114

Sprouse J, Aghajanian G (1987) Electrophysiological responses of serotoninergic dorsal raphe neurons to 5-HT1A and 5-HT1B agonists. Synapse 1:3–9

Szabo ST, Blier P (2001) Effects of the selective norepinephrine reuptake inhibitor reboxetine on norepinephrine and serotonin transmission in the rat hippocampus. Neuropsychopharmacol 25:845–857

Tsou K, Brown S, Sanudo-Pena MC, Mackie K, Walker JM (1998) Immunohistochemical distribution of cannabinoid CB1 receptors in the rat central nervous system. Neuroscience 83:393–411

Valverde O, Torrens M (2012) CB1 receptor-deficient mice as a model for depression. Neuroscience 204:193–206

Wittchen HU, Frohlich C, Behrendt S, Gunther A, Rehm J, Zimmermann P, Lieb R, Perkonigg A (2007) Cannabis use and cannabis use disorders and their relationship to mental disorders: a 10-year prospective-longitudinal community study in adolescents. Drug Alcohol Depen 88(Suppl 1):60–70

Chapter 13
Involvement of Serotonergic System in Cannabinoid Analgesia

Ahmet Dogrul, Melik Seyrek, Bulent Yalcin and Ahmet Ulugol

Abstract Plant cannabinoids have been used historically as a therapeutic agent in some folk medicine for the treatment of headache, fibromyalgia, and irritable bowel and related conditions in which serotonergic pathways are considered to play a crucial role in pathogenesis and treatment modalities. Serotonergic system has important modulatory role in acute and chronic pain conditions. The analgesic efficacy of cannabinoids in acute and chronic pain appear to be mediated, at least in part, through the regulation of the serotonergic system. In this chapter, we review the interaction between cannabinoids and serotonergic system in the peripheral, spinal and supraspinal sites with special emphasis on serotonin in central sites by which cannabinoid CB1 receptor activation reinforce descending serotonergic pathways to produce antinociceptive effects.

13.1 Introduction

Cannabis has been used both for recreation and pain management for millennia; however, only after the recent discovery of cannabinoid receptors, together with endocannabinoids, it became apparent how cannabinoids affect pain transmission and/or modulation (Pertwee 2001; Walker and Huang 2002). At present, there are prominent indications substantiating a role for cannabinoids in modulating acute and especially chronic pain states (Pertwee 2001). Δ9-tetrahydrocannabinol and cannabinol, the most important active constituents of plant derived cannabis, synthetic cannabinoids such as WIN 55,212-2 or CP 55,940, and putative endocannabinoids such as anandamide (N-arachidonoylethanolamide, AEA) and 2-arachidonylglycerol (2-AG) have been shown to possess strong antinociceptive and anti-inflammatory properties in different experimental preclinical pain models (Buxbaum 1972; Chester et al. 1973; Dogrul

A. Dogrul (✉) · M. Seyrek
Department of Medical Pharmacology, Gulhane Military Academy of Medicine, Ankara, Turkey
e-mail: dogrula@gata.edu.tr

B. Yalcin
Department of Anatomy, Gulhane Military Academy of Medicine, Ankara, Turkey

A. Ulugol
Department of Medical Pharmacology, Faculty of Medicine, Trakya University, Edirne, Turkey

E. J. Van Bockstaele (ed.), *Endocannabinoid Regulation of Monoamines in Psychiatric and Neurological Disorders*, DOI 10.1007/978-1-4614-7940-6_13,

et al. 2002, 2004; Gunduz et al. 2011; Herzberg et al. 1997; Kwilasz and Negus 2012; Martin et al. 1993; Richardson et al. 1998; Sofia et al. 1973; Ulugol et al. 2004, 2006).

So far, two subtypes of cannabinoid receptors have been identified and cloned: cannabinoid-1 (CB1) and cannabinoid-2 (CB2) receptors (Matsuda et al. 1990; Munro et al. 1993). These receptors are the primary targets of both endogenous and exogenous cannabinoids. CB1 receptors seem to play a pivotal role in the antinociceptive effect of cannabinoids. CB1 receptors are found to be 10 times more abundant than μ-opioid receptors in the brain; they have been demonstrated in the amygdala, basal ganglia, cerebellum, cerebral cortex, hippocampus, and brain areas that play role in descending pain control, such as periaqueductal gray matter, rostral ventromedial medulla, and dorsal horn (Hohmann and Suplita 2006; Kraft 2012; Svizenska et al. 2008). On the other hand, CB2 receptors are found mainly outside the nervous system (Svizenska et al. 2008). CB2 receptors have been indicated to modulate inflammation; they may exist in the central nervous system, but their role is unclear (Mackie and Stella 2006). It is worth mentioning that an antinociceptive role is also proposed for CB2 receptors in some pain models (Malan et al. 2002; Whiteside et al. 2007).

Recent research on pain indicates that targeting the endogenous cannabinoid system seems to be a promising therapeutic approach. The endocannabinoid system is constituted of CB1 and CB2 receptors, endogenous agonists, known as endocannabinoids, that activate these receptors, and the processes accountable for endocannabinoid biosynthesis, release, transport, and degradation (Guindon and Hohmann 2009; Pertwee 2012). Identification of cannabinoid receptors led to the invention of endocannabinoids and the enzymes responsible for their biosynthesis and degradation, and opened up new insights into modulation of pain. Several putative endocannabinoids have been identified, but AEA and 2-AG are the two best characterized. AEA is metabolized to arachidonic acid and ethanolamine, whereas 2-AG is metabolized to arachidonic acid and glycerol, primarily by the enzymes fatty acid amide hydrolase (FAAH) and monoacylglycerol lipase (MAGL), respectively. It has been postulated that inhibition of these catabolic enzymes and augmenting the endocannabinoid tonus may be a promising strategy, which does not accompany significant central nervous system (CNS) side effects associated with exogenous cannabinoids (Di Marzo 2008; Long et al. 2009; Naidu et al. 2010; Pertwee 2012; Schlosburg et al. 2009). FAAH inhibitors have been shown to be effective in preclinical experimental pain models (Caprioli et al. 2012; Schlosburg et al. 2009) and are currently under clinical trials; MAGL inhibitors, on the other hand, are not thoroughly investigated.

It is well known that the noxious stimuli are detected and transduced by the small-diameter afferent fibers as terminating specialized free nerve endings (nociceptors) in tissues and transmitted to the spinal cord, and medullary dorsal horn neurons, in turn, activate ascending pain pathways that carry pain messages to higher brain centers and somatosensorial cortex (Basbaum et al. 2009). It seems that cannabinoids affect nociception through activation of CB1 and CB2 receptors in the peripheral, spinal, and supraspinal sites (Dogrul et al. 2003; Svizenska et al. 2008; Walker and Huang 2002). While analgesic and other pharmacological effects of cannabinoids such as muscle relaxant, antiemetic, and appetite-stimulating appear to be largely mediated by CB1 receptors (Novotna et al. 2011), numerous points of intersection between cannabinoidergic and serotonergic systems are evident in the control of pain. A variety of studies provide evidence that cannabinoids modulate the function of serotonergic systems through CB1 receptor-mediated mechanisms (Bambico et al. 2007; Haj-Dahmane and Shen 2011; Haring et al. 2007). Plant cannabinoids have been used as a therapeutic agent in some folk medicine for headache, fibromyalgia, and abdominal pain in which serotonergic pathways are considered to play a crucial role in their pathogenesis and treatment modalities (Borgelt et al. 2013; Russo 2008). It is obvious that serotonin (5-hydroxytryptamine, 5-HT) in the periphery, spinal cord, and brain has important modulatory roles in acute and chronic pain states. In this chapter, we will describe briefly the site of action of cannabinoids, serotonin and its control on nociception at peripheral and central sites, and descending control of nociception with the focus on serotonergic system, and then report the current evidence of the involvement of serotonin and its receptors in CB1-mediated analgesia.

13.2 Site of Action of Cannabinoids

Cannabinoids produce antinociception predominantly through activity at spinal and supraspinal sites via CB1 receptors (Richardson 2000). Recent evidence also points to a peripheral action for cannabinoids; topical cannabinoid antinociception and its synergy with spinal sites and topical morphine have been proposed (Dogrul et al. 2003; Richardson 2000; Yesilyurt et al. 2003). Accordingly, CB1 receptors are localized in peripheral endings of primary sensory neurons, in dorsal horn and lamina X in the spinal cord, and in brain sites that participate in cannabinoid-induced antinociception (Hohmann et al. 1999; Piomelli et al. 2000). At spinal synapses, CB1 receptors are reported to be present on nerve endings of afferent neurons, on intrinsic spinal neurons, and on terminals of efferent supraspinal neurons (Hohmann et al. 1999; Piomelli et al. 2000). Thus, CB1 receptors seem to mediate both presynaptic and postsynaptic inhibition, by reducing transmitter release from primary afferents and directly inhibiting dorsal horn neurons, respectively.

The spinal cord dorsal horn is a crucial anatomical site with respect to its pivotal role in nociceptive transmission as well as in modulation (Chen et al. 2005). Nociceptive inputs entering dorsal horn of the spinal cord are exposed to descending modulation from several different brain regions (Chen et al. 2005; Millan 2002). The periaqueductal grey region (PAG) and the rostral ventromedial medulla (RVM) have essential roles in the descending modulation of nociception (Millan 2002; Porreca et al. 2002; Vanegas and Schaible 2004). Some glutamatergic neurons project from PAG to RVM (Millan 2002). Cannabinoids have been shown to diminish γ-aminobutyric acid (GABA) release from inhibitory GABAergic interneurons in the PAG via CB1 receptors (Finn et al. 2003; Vaughan et al. 2000). Consequently, it is likely that disinhibition of output neurons leads to activation of descending inhibitory pain pathways. On the other hand, RVM is a critical relay site through which descending inhibitory and facilitatory bulbospinal projections may diminish or augment spinal nociceptive transmission (Bee and Dickenson 2007; Dogrul et al. 2009). Opioids modulate on- and off-cells in the RVM (Heinricher et al. 1994); similarly, cannabinoids act in the same fashion (Meng et al. 1998). Thus, cannabinoid-induced increase in off-cell activity and/or reduction in on-cell activity seem to weaken spinal nociceptive transmission by activating descending inhibition and/or to enhance spinal nociceptive transmission by activating descending facilitation, respectively. Presynaptic inhibition of GABAergic and glutamatergic transmission is likely to be involved in these effects both in the PAG and RVM (Vaughan et al. 1999, 2004). In total, cannabinoids seem to exert analgesic effects through activation of supraspinal, spinal, and peripheral CB1 receptors (Dogrul et al. 2003; Guindon and Hohmann 2009; Hohmann and Suplita 2006).

Although many animal studies demonstrate that cannabinoids are effective analgesics in acute and chronic pain states, analgesic efficacy of cannabinoids in acute pain are not approved by randomized placebo-controlled clinical trials in humans; in case of some chronic pain conditions such as multiple sclerosis-related pain and HIV-associated neuropathic pain, a limited effect is observed (Kraft 2012; Borgelt et al. 2013). However, not only efficacy but also safety problems exist over long-term use of cannabinoids, including serious psychotomimetic side effects as well as development of tolerance and physical dependence. Thus, cannabinoids have been identified as potential adjuvant analgesics.

13.3 Serotonin and Pain Modulation

As one of the oldest known important monoamine and signaling molecules, 5-HT in peripheral tissues and nervous system has been suggested to be highly involved in a variety of physiological or behavioral functions including inflammation, allergy, vascular blood flow, gastrointestinal motility, autonomic activity, stress, depression, mood, and appetite (Wei et al. 2012; Loyd et al. 2012). Now, it is becoming clear that 5-HT in the periphery, spinal, and supraspinal sites have important modulatory roles in acute and chronic pain conditions (Bardin 2011; Loyd et al. 2012; Wei et al. 2012).

A vast majority of 5-HT is located in peripheral tissues, predominantly in the gastrointestinal tract, where it plays an important role in gastrointestinal motility and secretion (Sommer 2010; Loyd et al. 2012). In blood circulation, 5-HT is stored in platelets, mast cells, and immune cells and released with other mediators in response to inflammation, tissue injury, and immune insult (Duerschmied et al. 2012; Loyd et al. 2012). In general, endogenous 5-HT plays a proinflammatory and pronociceptive role in the periphery (Loyd et al. 2012; Wei et al. 2012; Bardin 2011). Although only 1–2 % of whole body 5-HT exists in the brain and the majority of 5-HT-containing neurons are located in midline raphe nuclei and adjacent nuclear groups of the brainstem, serotonergic system influences tremendous important brain functions by its wide distribution of ascending and descending fibers in the brain and spinal cord

(Wei et al. 2012; Kandel et al. 2000). 5-HT, together with noradrenaline, are the two main neurotransmitters involved in top–down endogenous pain inhibition by supraspinal brain areas (Benarroch 2008; Bingel and Tracey 2008). Central serotonergic neurons modulate nociception by spinally projecting descending serotonergic fibers largely derived from RVM through the dorsolateral funiculus at the spinal cord level (Sommer 2010; Zhang et al. 2000; Wei et al. 2010).

5-HT affects pain processing and modulation by acting via seven families of 5-HT receptors (5-HT1–5-HT7) (Bardin 2011). Except 5-HT3 receptor, which is a ligand-gated ion channel (permeable to sodium and potassium), all the other 5-HT receptors belong to G-protein-coupled metabotropic receptor family (Wei et al. 2012). While 5-HT1 receptors are linked to inhibitory G-proteins, 5-HT2, 5-HT4, 5-HT6, and 5-HT7 receptors are linked to stimulatory G-proteins (Bardin 2011). Thus, it is not surprising that 5-HT exerts complex, excitatory (hyperalgesic) or inhibitory (analgesic) actions, depending on the subtype of receptors, localization of receptors on the cell type, and site of action, and because of the downstream effect on neuronal or other cells following 5-HT receptor activation is either inhibitory or stimulatory (Millan 2002; Sommer 2010; Bardin 2011). The current understanding seems to be that 5-HT1A, 5-HT2A, 5-HT3, and 5-HT7 receptors are specifically involved in both pain processing and the antinociceptive and antihyperalgesic mechanism of action of some analgesic dugs.

13.4 Control of Cannabinoids on Serotonergic Neuromodulation of Pain in Peripheral Sites

Following tissue injury or inflammation, a variety of endogenous chemical mediators together with 5-HT are released that activate and/or sensitize nociceptors (Loyd et al. 2012). It has been reported that thermal injury or inflammation induces a rapid increase in peripheral tissue endogenous 5-HT levels (Nakajima et al. 2009; Sasaki et al. 2006). In animals experiments, the local administration of 5-HT into peripheral tissues induce inflammation and hyperalgesia in accordance with the increase in excitability of myelinated A-delta fibers and C-fibers as well as dorsal root ganglion neurons (Babenko et al. 2000; Schmelz et al. 2003; Loyd et al. 2012). Besides animal experiments, human studies also support the pronociceptive role for 5-HT at the peripheral level by demonstrating that intradermal or intramuscular injection of 5-HT elicits pain and hyperalgesia in human volunteers (Ernberg et al. 2006; Lischetzki et al. 2001).

The application of 5-HT to nerve roots also leads to nerve damage, inflammation, and pain behavior in rodents, suggesting the important role of 5-HT in the pathogenesis of nerve injury-related pain (Kato et al. 2008; Kobayashi et al. 2011). Moreover, it has been shown that after nerve transection or chronic constriction injury, the 5-HT content in the lesioned nerve is increased (Anden and Olsson 1967; Vogel et al. 2003; Sommer 2010). The demonstration that the development of thermal hyperalgesia was well correlated with the increased 5-HT levels in the injured sciatic nerve using chronic constriction injury paradigm, together with attenuation of thermal hyperalgesia followed with a concomitant decrease in 5-HT content in nervous tissues after deletion of 5-HT transporter by genetic techniques, indicates the peripheral pronociceptive role of 5-HT after nerve injury (Vogel et al. 2003; Loyd et al. 2012). It has been hypothesized that 5-HT, functioning in combination with other proinflammatory mediators, may ectopically excite and sensitize acutely injured afferent nerve fibers and lower the nociceptive thresholds of sensory neurons to other stimuli (Bardin 2011; Sommer 2010).

One possible mechanism for the peripheral pronociceptive role of 5-HT is to alter transient receptor potential of vanilloid type-1 (TRPV1) channel or tetrodotoxin-resistant (TTX-R) sodium channel properties in sensory neurons followed by painful stimuli (Bardin 2011; Loyd et al. 2012; Sommer 2010). The activation of TRPV1 receptor or TTX-R sodium channel by mechanical and thermal stimuli or inflammatory mediators and many other stimuli arising from intra/extracellular environment are crucial in detection, transmission, and regulation of pain (Palazzo et al. 2010). It has been reported that 5-HT significantly enhances TRPV1 receptor activation-mediated calcium influx and the inflammatory peptide, calcitonin gene-related peptide (CGRP), release in sensory and trigeminal neurons (Loyd et al. 2012). Additionally, it has been shown that 5-HT increases the magnitude of TTX-R sodium channel current, produces a hyperpolarizing shift of its activation curve, and increases its rate of activation and inactivation in sensory neurons (Gold et al. 1996; Loyd et al. 2012). Thus, it is possible that 5-HT alters function or expression of a variety of ion channels in peripheral nociceptors to modulate pain. Alternatively, it has been suggested that 5-HT sensitizes nociceptors indirectly by enhancing the release of cytokines and neurotropins during inflammation or noxious insults (Loyd et al. 2012).

Serotonin in the periphery affects nociception via 5-HT1, 5-HT2, 5-HT3, and 5-HT7 receptors, which have been expressed in sensory neurons, and contribute to peripheral nociceptive transmission (Wei et al. 2012). Peripheral injections of 5-HT2A, 5-HT3, and 5-HT7 antagonists have been shown to prevent inflammation-induced and nerve injury-induced hyperalgesia (Bardin 2011; Loyd et al. 2012; Sommer 2010). Moreover, while peripherally injected 5-HT1B/1D and 5-HT1A agonists attenuate nociception, 5-HT2A, 5-HT3, and 5-HT7 receptor agonists have been shown to elicit hyperalgesic effects in different types of animal models of nociception (Bravo-Hernandez et al. 2012; Brenchat et al. 2012; Colpaert 2006; Granados-Soto et al. 2010; Tokunaga et al. 1998).

Targeting peripheral serotonergic system to treat painful symptoms associated with migraine, fibromyalgia, and irritable bowel syndrome has been suggested as a therapeutic option (Loyd et al. 2012). Although cannabinoids have been used symptomatically for relief of migraine, fibromyalgia, and irritable bowel syndrome historically (Russo et al. 2008), the influence of cannabinoid system on peripheral serotonergic system has not been examined well yet. CB1 receptors are expressed in peripheral sites, albeit at a lower level than in the central nervous system (Carley et al. 2002). However, the demonstration of reduction of systemic and local analgesic effects of cannabinoids due to nociceptor-specific loss of CB1 receptors using conditional gene deletion method provides evidence that the role of CB1 receptors expressed on the peripheral endings of nociceptors in cannabinoid analgesia is of great importance (Agarwal et al. 2007). Thus, peripheral CB1 receptors may modify peripheral serotonin receptor signaling in controlling pain. This speculation is supported by the observation showing that Δ9-tetrahydrocannabinol inhibits release of serotonin from platelets induced by plasma from human migraineurs (Volfe et al. 1985) and that arachidonyl-2-chloroethylamide (ACEA), a selective CB1 receptor agonist, reduced whole blood 5-HT levels, which was reversed by the CB1 antagonist pretreatment in rats (Rutkowska and Gliniak 2009). Additionally, anandamide, WIN 55,212-2, and CP-55940 inhibited 5-HT3 receptor-mediated current in rat nodose ganglion neurons, which has been suggested as a possible mechanism for the analgesic and antiemetic effects of cannabinoids at the peripheral level (Fan 1995). Comparable with the pronociceptive role of 5-HT in the periphery, intradermal injections of 5-HT activate C-fibers and exhibit efficacious pruritogenic properties (Ikoma et al. 2003). Similar to pain suppression, the inhibition of intradermally injected 5-HT-mediated pruritic responses with pretreatment of skin with exogenous or endogenous cannabinoids supports the notion that cannabinoids can modulate peripheral serotonergic system to produce analgesic effects. There is very little work in the area of peripheral serotonin and cannabinoid analgesia, and it can be clearly seen that there is a big need for further research.

13.5 Control of Cannabinoids on Serotonergic Neuromodulation of Pain on Central Sites

The perception of pain is known to be highly complex, not only related to the intensity of nociceptive input but also to significant modulation at peripheral, spinal, and supraspinal level through its neuroaxis to cerebral cortex by complex, top–down pain modulatory pathways linked with cognitive and emotional variables (Bingel and Tracey 2008). Well-characterized brain regions in the context of pain modulation include rostral anterior cingulated cortex, amygdala, and hypothalamus, which send projections to PAG that make dense connections with RVM (Bingel and Tracey 2008; Braz and Basbaum 2008; Dogrul et al. 2012). PAG is the key element of top and down descending pain modulatory pathways providing reciprocally interconnected input to frontal cortex, amygdala, hypothalamus, locus coeruleus, dorsal raphe nucleus (DRN), and RVM (Dogrul et al. 2012; Ossipov et al. 2010). Experimental and clinical studies strongly support that the axis of PAG–RVM circuitry project to spinal or medullar dorsal horn via dorsolateral funiculus constitutes the key structure of descending pain modulatory pathways that control and modulate nociception at the spinal level (Ossipov et al. 2010).

RVM is a functionally critical medullary reticular area composed of nucleus raphe magnus, nucleus reticularis gigantocellularis, and nucleus reticularis paragigantocellularis (Dogrul et al. 2012; Géranton et al. 2010). RVM is considered to be the final common relay station for most of the brain regions at the upper site of midbrain and by which descending inhibitory and facilitatory bulbospinal projections originate from to inhibit or enhance spinal nociceptive transmission (Bee and Dickenson 2007). Although majority of 5-HT-containing neurons are located in the DRN in the brain stem, which send some collaterals to spinal cord, majority of serotonergic neurons involved in pain modulation arise from nucleus raphe magnus in RVM and their spinally projecting descending fibers terminate in the spinal dorsal horn (Braz and Basbaum 2008). It has been suggested that the effects of the DRN to spinal cord are mediated by its connection with nucleus raphe magnus (Wang and Nagai 1994). Electrical stimulation of PAG, RVM, or nucleus raphe magnus produces antinociceptive effects accompanied with 5-HT release in the spinal cord (Bardin 2011; Cui et al. 1999; Fields et al. 2006; Hammond and Yaksh 1984; Nichols et al 1989; Rivot et al. 1982; Sorkin et al. 1993; Wei et al. 2010), and spinal administration of nonselective 5-HT receptor antagonist blocks this stimulation-induced antinociception (Aimone et al. 1987; Bardin 2011; Hammond and Yaksh 1984; Jensen and Yaksh 1986; Millan 2002).

Although descending inhibitory control mechanism of pain is well documented and described as a biological protective mechanism during stressful conditions, anatomical, electrophysiological, and pharmacological studies have shown that descending pain modulatory pathways elicit even a facilitatory role in nociceptive sensory processing (Bardin 2011; Millan 2002; Sommer 2010). Descending pain facilitatory pathways also serve as a biological protective function by signaling to restricting activities and focusing on healing painful body area (Ossipov et al. 2010). Electrophysiological studies have identified three different classes of neurons in RVM: on-cells, off-cells, and neutral cells that increase, decrease, and do not change the action potential activity, respectively (Ossipov et al. 2010). Pharmacological and neurochemical studies support the notion that off-cells and on-cells are putative antinociceptive and pronociceptive neuronal cells that drive the descending antinociceptive and pronociceptive influence, respectively, on the spinal cord Fields et al. 1983; Ossipov et al. 2010. Neutral cells seem to be involved in some pathological pain states such as neuropathic pain but are not active in physiological nociceptive processing (Fields et al. 2006). Interestingly, it has been found that neither off-cells nor on-cells was serotonergic, except some neutral cells in the RVM (Braz and Basbaum 2008; Ossipov et al. 2010; Dogrul et al. 2012). Thus, it is thought that serotonergic neurons in the RVM modulate many of the nonserotonergic descending pathways arising from RVM and are the critical integrator of downstream output in pain modulation (Braz and Basbaum 2008; Inyushkin et al. 2010; Morgan et al. 2008; Pertovaara and Almeida 2006; Wei et al. 2010)

The neuronal organization and circuitry of spinal dorsal horn is fundamentally critical not only because the nociceptive information from the skin and underlying tissues first terminates and reaches the central nervous system but also because it is the only place where brain can exert control over pain sensation through a variety of descending pathways (Brown 1982; Todd 2010). Descending serotonergic pathways make connections with primary afferent fibers, projection neurons, and interneurons in the spinal cord area (Millan 2002). The nociceptive responses modulated by descending serotonergic pathways are dependent on the subtype and localization of 5-HT receptors in the spinal cord. All the seven families of 5-HT receptors (5-HT1–5-HT7) are identified in the spinal cord dorsal horn (Hamon and Bourgoin 1999; Jeong et al. 2004). Current understanding appears to be that descending inhibition of pain is mediated primarily by spinal 5-HT1, 5-HT2, and 5-HT7 receptors, whereas spinal 5-HT3 receptors mediate descending facilitation of pain (Dogrul and Seyrek 2006; Dogrul et al. 2009, 2012; Gu et al. 2011; Iwasaki et al. 2013; Seyrek et al. 2010; Wei et al. 2010).

Medicinal cannabinoids are generally administered systemically by oral formulations (Klumpers et al. 2012). As highly lipophilic substances, they readily cross the blood–brain barrier and distribute in the brain, spinal cord, and peripheral tissues (Huestis 2007). The profound decrease in the antinociceptive effects of systemic cannabinoids, subsequent to surgical dorsolateral funiculus lesion or spinal cord transection in the tail-flick test, implicates that supraspinal sites and descending pathways contribute significantly to systemically administered cannabinoid-induced analgesia (Dogrul et al. 2012; Lichtman and Martin 1991; Seyrek et al. 2010).

The CB1 receptor is one of the most commonly expressed and widely distributed G-protein coupled receptor in the central nervous system (Wilson-Poe et al. 2012). CB1 receptors are also present in high density not only in pain transmission and but also in processing and modulation sites, including the region of descending serotonergic pathways such as amygdala, PAG, RVM, nucleus reticularis gigantocellularis pars alpha, DRN, locus coeruleus, spinal cord dorsal horn, and dorsal root ganglia (Guindon and Hohmann 2009; Haring et al. 2007; Hohmann and Suplita 2006; Scavone et al. 2010; Svizenska et al. 2008; Wilson-Poe et al. 2012). Direct support for the contribution of supraspinal sites and descending serotonergic pathways to the analgesic action of cannabinoids was derived from studies in which plant-derived tetrahydrocannabinol (THC) and synthetic cannabinoids are injected intracerebroventricularly or by microinjection into various local brain regions (Dogrul et al. 2012; Guindon and Hohmann 2009; Litchman et al. 1996). Intracerebroventricular administration of mixed CB1 and CB2 agonists, such as THC, WIN 55,212-2, and CP 55,940, and CB1 agonists, such as ACEA and methanandamide, generated dose-dependent analgesic effects in acute, inflammatory, and nerve injury models via CB1 receptors (Dogrul et al. 2012; Garzon et al. 2009; Litchman et al. 1996; Martin et al. 1993; Raffa et al. 1999; Wakley and Craft 2011; Walker and Hohmann 2005). Microinjection of cannabinoids into amygdala, PAG, RVM, nucleus reticularis gigantocellularis pars alpha, DRN, and locus coeruleus also produces antinociceptive effects in a wide array of nociceptive animal models (Litchman et al. 1996; Maione et al. 2011; Manning et al. 2003; Martin et al. 1993, 1998, 1999; Meng and Johansen 2004; Monhemius et al. 2001; Wilson-Poe et al. 2012). The crucial role of descending pathways in systemic cannabinoid analgesia is also supported by the observation that inactivation of RVM by microinjection of GABA-A receptor agonist, muscimol, into it, totally prevents systemic WIN 55,212-2-induced analgesic effect in the tail-flick test (Meng et al. 1998).

Additionally, a variety of studies showed that endocannabinoid system activated as an adaptive response to environmental stress or as an endogenous protective mechanism to painful threat aims to counteract the establishment of pain

(Zogopoulos et al. 2013). It has been reported that anandamide and/or 2-AG levels in PAG, RVM, dorsal raphe (DR), and dorsal root ganglia were increased following nerve injury (Mitrirattanakul et al. 2006; Palazzo et al. 2006; Petrosino et al. 2007). Moreover, conditional or unconditional stress produces analgesic effects via CB1 receptors in association with increased tissue levels of 2-AG in PAG (Hohhman and Suplita 2006; Olango et al. 2012). Electrical stimulation of PAG induces analgesic activity via a CB1-dependent mechanism together with anandamide release in the PAG (Walker and Hohmann 2005). Taken together, the results of these studies implicate the importance of endocannabinoid–CB1 mediated signaling in the descending pain modulatory pathways, as well as the role of PAG–RVM–spinal cord neuroaxis as a protective mechanism in stress-induced analgesia.

Currently, there is accumulating evidence supporting an interaction between cannabinoids and serotonergic system in the CNS in the context of cannabinoid-induced analgesic effect. It is well known that acute and chronic nociceptive stimuli induce enhanced descending serotonergic activity along with 5-HT release in the spinal cord (Millan 2002). CB1 receptors were identified in the serotonergic cells of raphe nuclei, and it has been shown that endocannabinoid and serotonergic systems were activated together in the DR (Haring et al. 2007; Palazzo et al. 2006). Many findings exist to support a modulatory role for CB1 receptor signaling in the modulation of the DR 5-HT system. Systemic administration of WIN 55,212-2 have been reported to enhance DRN 5-HT neuronal activity through a CB1-dependent mechanism (Bambico et al. 2007). Additionally, DR also expresses FAAH and genetic deletion of FAAH results in an enhancement in the spontaneous activity of DR 5-HT neurons (Bambico et al. 2007, 2010). Anandamide and 2-AG content in PAG, RVM, and DRN were increased at different time courses following sciatic nerve chronic constriction injury (Palazzo et al. 2006; Petrosino et al. 2007). The report showing that CB1-selective antagonists, rimonabant and AM251, reduced the firing rate mostly of DRN 5-HT cells in brain slices suggests the existence of a tonic regulation of DRN 5-HT cells by the endocannabinoid system (Mendiguren and Pineda 2009). While chronically systemic injection of WIN 55,212-2 and the FAAH inhibitor, AM 404, elevated DR neuronal firing and increased 5-HT release in sham-operated control animals, they elicited thermal antihyperalgesic effect in sciatic nerve chronic constriction injury model in association with reduction in nerve injury-induced enhanced firing rate and 5-HT release in serotonergic DRN neurons (Palazzo et al. 2006).

Cannabinoids are effective drugs in persistent pain, and a variety of reports point to the plastic changes of the serotonergic system in chronic pain (Wei et al. 2012). Recent studies indicate that there exists a balance between descending inhibition and facilitation in normal conditions, but following long-lasting nociceptive activity, descending facilitatory serotonergic drive is enhanced, which leads to neuronal hyperexcitability as evidenced behaviorally by allodynia and hyperalgesia (Wei et al. 2010). The antiallodynic and antihyperalgesic effects of systemically administered cannabinoids in nerve injury models, together with the increase in the firing activity of DR neurons, provide evidence that CB1-mediated restoration or increase in descending serotonergic activity may be a mechanism of action of cannabinoids in reducing chronic pain-related symptoms (Dogrul et al. 2012).

Considerable evidence demonstrates that descending serotonergic pathways represent one of the major component of endogenous pain inhibitory system, and analgesic efficacy of some clinically important drugs depends on the integrity of serotonergic system in the CNS (Dogrul et al. 2012; Manning et al. 2003; Millan 2002; Ossipov et al 2010). A variety of approaches, including the selective denervation of spinal serotonergic neurons by neurotoxins, neurosurgical dorsolateral funiculus lesion, and intrathecal injection of selective 5-HT receptor subtype antagonists were used to evaluate the contribution of descending serotonergic pathways to the cannabinoid-induced analgesic effect (Seyrek et al. 2010). It has been reported that antinociceptive effects induced by systemically administered WIN 55,212-2 and the selective CB1 agonist ACEA were totally absent in the tail-flick tests following bilateral surgical lesion of dorsolateral funiculus, which is accepted as the main route for descending pain inhibitory pathways (Seyrek et al. 2010). Furthermore, spinal application of 5,7-dihydroxytryptamine (5,7-DHT), which selectively lesions descending serotonergic pathways, disrupted antinociceptive effects of systemically injected WIN 55,212-2 and ACEA in mice (Seyrek et al. 2010). Consistent with this study, Mallet et al. (2008) showed that systemic ACEA-induced antinociceptive effect was totally diminished in rats spinally pretreated with 5,7-DHT in the paw pressure test. Thus, inhibition of antinociceptive efficacy of systemic cannabinoids in spinal 5-HT-depleted animals indicates that systemic cannabinoid-induced analgesia depends on integrity of descending serotonergic pathways.

Moreover, spinal administration of SB-269970, a selective 5-HT7 receptor antagonist, and ketanserin, a selective 5-HT2A receptor antagonist, completely blocked the antinociceptive effects of systemically administered WIN 55,212-2 and ACEA (Seyrek et al. 2010). It is interesting to note that spinal administration of atypical antipsychotic risperidone also completely blocked the analgesic effects of systemic cannabinoids (Seyrek et al. 2010). Although risperidone exhibits competitive antagonistic properties over 5-HT2A and dopamine D2 receptors, it has unique effects on 5-HT7 receptors; it irreversibly binds to and inactivates the 5-HT7 serotonin receptor (Smith et al. 2006; Toohey et al. 2009). These studies reveal that spinal 5-HT7 and 5-HT2A receptors are critical in systemic cannabinoid analgesia. Previous reports indicating that spinal 5-HT7 receptor blockade inhibits systemic morphine analgesia in the same experimental paradigm (Dogrul and Seyrek 2006) support the notion that cannabinoids and opioids activate similar descending modulatory circuits to produce analgesic effects (Desroches and Beaulieu 2010) and that spinal 5-HT7 receptors are involved in both opioid- and cannabinoid-induced analgesic mechanisms. However, in contrast to cannabinoid analgesia, the demonstration suggesting that intrathecal injection of ketanserin did not block systemic morphine-induced analgesic effects (Dogrul and Seyrek 2006) points to a specific role of 5-HT2A receptors in cannabinoid analgesia as compared with morphine. The important roles of spinal 5-HT7 and 5-HT2A receptors in systemic cannabinoid analgesia are consistent with other studies showing that genetic deletion of CB1 receptor or chronic administration of CB1 receptor agonist altered and impaired the function of 5-HT1A and 5-HT2A receptors in same brain regions (Hill et al. 2006; Mato et al. 2007). It is probable that 5-HT1A receptors in these studies are the same as 5-HT7 receptors, as they show common similarities in their pharmacological profile, and it has been suggested that several functions formerly attributed to 5-HT1A receptors may be mediated by the 5-HT7 subtype (Bonaventure et al. 2002; Meuser et al. 2002).

Regarding positive coupling of 5-HT7 and 5-HT2 receptors to adenylate cyclaseand phospholipase C, respectively, resulting of their activation in excitation inneurons, it is unlikely that the activation of spinal 5-HT7 or 5-HT2A receptor by cannabinoids could directly inhibit primary afferents or nociceptive dorsal horn neurons. Hovewer, codistribution of 5-HT7 and 5-HT2A receptor on the GABA-ergic or enkephalinergic interneurons in the dorsal root ganglions and dorsal horn (Brenchat et al. 2012; Millan 2002), it is possible to infer that following systemic canabinoid administration, CB1 mediated activation of 5-HT7 or 5-HT2A receptors localized on spinal inhibitory enkephalinergic or GABAergic interneurons by descending serotonergic pathways may evoke the release of enkephalins or GABA, and could produce antinociception.

13.6 Concluding Remarks

Besides the ethical and legal obstacles, unwanted side effects limit or preclude the use of cannabinoids in acute and chronic pain states. Novel strategies on the utility of cannabinoids will expectantly overcome these difficulties. Among these, using peripherally restricted CB1 agonists or CB2 agonists seems encouraging; the main aim in both is to minimize central side effects. Another promising approach is to combine low doses of analgesic drugs from different pharmacological groups. The goal is to develop additive or synergistic combinations with enhanced pain relief and reduced CNS effects. Finally, probably the most popular strategy in the last few years is modulating the endocannabinoid system. While there are some disappointing findings, drugs targeting synthesis, reuptake, and degradation of endocannabinoids are expected to be used as effective alternative analgesics.

In conclusion, cannabinoids are likely to have any impact on serotonergic modulation of pain not only in peripheral but also in central sites. While the modulatory roles of 5-HT in acute and chronic pain conditions are beyond doubt, there are only very few studies focusing on this cannabinoid action. Further research is required to elucidate the mechanisms underlying the antinociceptive property of cannabinoids. These researches together with the abovementioned strategies will hopefully provide opportunity for their effective use in pain therapy.

References

Agarwal N, Pacher P, Tegeder I, Amaya F, Constantin CE, Brenner GJ, Rubino T, Michalski CW, Marsicano G, Monory K, Mackie K, Marian C, Batkai S, Parolaro D, Fischer MJ, Reeh P, Kunos G, Kress M, Lutz B, Woolf CJ, Kuner R (2007) Cannabinoids mediate analgesia largely via peripheral type 1 cannabinoid receptors in nociceptors. Nat Neurosci 10(7):870–879

Aimone LD, Jones SL, Gebhart GF (1987) Stimulation-produced descending inhibition from the periaqueductal gray and nucleus raphe magnus in the rat: mediation by spinal monoamines but not opioids. Pain 31(1):123–136

Anden NE, Olsson Y (1967) 5-Hydroxytryptamine in normal and sectioned rat sciatic nerve. Acta Pathol Microbiol Scand 70:537–540

Babenko V, Svensson P, Graven-Nielsen T, Drewes AM, Jensen TS, Arendt-Nielsen L (2000) Duration and distribution of experimental muscle hyper algesia in humans following combined infusions of serotonin and bradykinin. Brain Res 853(2):275–281

Bambico FR, Katz N, Debonnel G, Gobbi G (2007) Cannabinoids elicit antidepressant-like behavior and activate serotonergic neurons through the medial prefrontal cortex. J Neurosci 27: 11700–11711

Bambico FR, Cassano T, Dominguez-Lopez S, Katz N, Walker CD, Piomelli D, Gobbi G (2010) Genetic deletion of fatty acid amide hydrolase alters emotional behavior and serotonergic transmission in the dorsal raphe, prefrontal cortex, and hippocampus. Neuropsychopharmacology 35(10):2083–2100

Bardin L (2011) The complex role of serotonin and 5-HT receptors in chronic pain. Behav Pharmacol 22:390–404

Basbaum AI, Bautista DM, Scherrer G, Julius D (2009) Cellular and molecular mechanisms of pain. Cell 139(2):267–284

Bee LA, Dickenson AH (2007) Rostral ventromedial medulla control of spinal sensory processing in normal and pathophysiological states. Neuroscience 147:786–793

Benarroch EE (2008) Descending monoaminergic pain modulation: bidirectional control and clinical relevance. Neurology 71:217–221

Bingel U, Tracey I (2008) Imaging CNS modulation of pain in humans. Physiology (Bethesda) 23:371–380

Bonaventure P, Nepomuceno D, Kwok A, Chai W, Langlois X, Hen R, Stark K, Carruthers N, Lovenberg TW (2002) Reconsideration of 5-hydroxytryptamine (5-HT)7 receptor distribution using [3H]5-carboxamidotryptamine and [3H]8-hydroxy-2-(di-n-propylamino)tetraline: analysis in brain of 5-HT1A knockout and 5-HT1A/1B double-knockout Mice. J Pharmacol Exp Ther 302:240–248

Borgelt LM, Franson KL, Nussbaum AM, Wang GS (2013) The pharmacologic and clinical effects of medical cannabis. Pharmacotherapy 33:195–209

Bravo-Hernández M, Cervantes-Durán C, Pineda-Farias JB, Barragán-Iglesias P, López-Sánchez P, Granados-Soto V (2012) Role of peripheral and spinal 5-HT(3) receptors in development and maintenance of formalin-induced long-term secondary allodynia and hyperalgesia. Pharmacol Biochem Behav 101(2):246–257

Braz JM, Basbaum AI (2008) Genetically expressed transneuronal tracer reveals direct and indirect serotonergic descending control circuits. J Comp Neurol 507:1990–2003

Brenchat A, Zamanillo D, Hamon M, Romero L, Vela JM (2012) Role of peripheral versus spinal 5-HT(7) receptors in the modulation of pain undersensitizing conditions. Eur J Pain 16(1):72–81

Brown AG (1982) The dorsal horn of the spinal cord. Q J Exp Physiol 67(2):193–212

Buxbaum DM (1972) Analgesic activity of delta 9-tetrahydrocannabinol in the rat and mouse. Psychopharmacol Berlin 25:275–280

Caprioli A, Coccurello R, Rapino C, Di Serio S, Di Tommaso M, Vertechy M, Vacca V, Battista N, Pavone F, Maccarrone M, Borsini F (2012) The novel reversible fatty acid amide hydrolase inhibitor ST4070 increases endocannabinoid brain levels and counteracts neuropathic pain in different animal models. J Pharm Exp Ther 342:188–195

Carley DW, Paviovic S, Janelidze M, Radulovacki M (2002) Functional role for cannabinoids in respiratory stability during sleep. Sleep 25(4):391–398

Chen YP, Chen SR, Pan HL (2005) Systemic morphine inhibits dorsal horn projection neurons through spinal cholinergic system independent of descending pathways. J Pharmacol Exp Ther 314:611–617

Chesher GB, Dahl CJ, Everingham M, Jackson DM, Marchant-Williams H, Starmer GA (1973) The effect of cannabinoids on intestinal motility and their antinociceptive effect in mice. Br J Pharmacol 49(4):588–594

Colpaert FC (2006) 5-HT(1A) receptor activation: new molecular and neuroadaptive mechanisms of pain relief. Curr Opin Investig Drugs 7(1):40–47

Cui M, Feng Y, McAdoo DJ, Willis WD (1999) Periaqueductal gray stimulation-induced inhibition of nociceptive dorsal horn neurons in rats is associated with the release of norepinephrine, serotonin, and amino acids. Pharmacol Exp Ther 289(2):868–876

Desroches J, Beaulieu P (2010) Opioids and cannabinoids interactions: involvement in pain management. Curr Drug Targets 11(4):462–473

Di Marzo V (2008) Targeting the endocannabinoid system: to enhance or reduce? Nat Rev Drug Discov 7:438–455

Dogrul A, Seyrek M (2006) Systemic morphine produce antinociception mediated by spinal 5-HT7, but not 5-HT1A and 5-HT2 receptors in the spinal cord. Br J Pharmacol 149(5):498–505

Dogrul A, Gardell LR, Ma S, Ossipov MH, Porreca F, Lai J (2002) Knock-down of spinal CB1 receptors produces abnormal pain and elevates spinal dynorphin content in mice. Pain 100:203–209

Dogrul A, Gul H, Akar A, Yildiz O, Bilgin F, Guzeldemir E (2003) Topical cannabinoid antinociception: synergy with spinal sites. Pain 105:11–16

Dogrul A, Gul H, Yildiz O, Bilgin F, Guzeldemir E (2004) Cannabinoids block tactile allodynia in diabetic mice without attenuation of its antinociceptive effect. Neurosci Lett 368:82–86

Dogrul A, Ossipov MH, Porreca F (2009) Differential mediation of descending pain facilitation and inhibition by spinal 5HT-3 and 5HT-7 receptors. Brain 1280:52–59

Dogrul A, Seyrek M, Yalcin B, Ulugol A (2012) Involvement of descending serotonergic and noradrenergic pathways in CB1 receptor-mediated antinociception. Prog Neuropsychopharmacol Biol Psychiatry 38(1):97–105

Duerschmied D, Suidan GL, Demers M, Herr N, Carbo C, Brill A, Cifuni SM, Mauler M, Cicko S, Bader M, Idzko M, Bode C, Wagner DD (2012) Platelet serotonin promotes the recruitment of neutrophils to sites of acute inflammation in mice. Blood 7 121(6):1008–1015

Ernberg M, Hedenberg-Magnusson B, Kurita H, Kopp S (2006) Effects of local serotonin administration on pain and microcirculation in the human masseter muscle. J Orofac Pain 20(3):241–248

Fan P (1995) Cannabinoid agonists inhibit the activation of 5-HT3 receptors in rat nodose ganglion neurons. J Neurophysiol 73(2):907–910

Fields HL, Basbaum AI, Heinricher MM (2006) Central nervous system mechanisms of pain modulation. In: McMahon S, Koltzenburg M (eds) Wall and Melzack's textbook of pain. Elsevier, London, pp 125–142

Fields HL, Bry J, Hentall I, Zorman G (1983) The activity of neurons in the rostral medulla of the rat during withdrawal from noxious heat. J Neurosci 3(12):2545–2552

Finn DP, Jhaveri MD, Beckett SRG, Roe CH, Kendall DA, Marsden CA, Chapman V (2003) Effects of periaqueductal grey administration of a cannabinoid agonist on nociceptive and aversive responses in rats. Neuropharmacology 45:594–604

Garzón J, la Torre-Madrid E de, Rodràguez-Muñoz M, Vicente-Sánchez A, Sánchez-Blázquez P (2009) Gz mediates the long-lasting desensitization of brain CB1 receptors and is essential for cross-tolerance with morphine. Mol Pain 5:11

Géranton SM, Tochiki KK, Chiu WW, Stuart SA, Hunt SP (2010) Injury induced activation of extracellular signal regulated kinase (ERK) in the rat rostral ventromedial medulla (RVM) is age dependent and requires the lamina I projection pathways. Mol Pain 6:54

Gold MS, Reichling DB, Shuster MJ, Levine JD (1996) Hyperalgesic agents increase a tetrodotoxin-resistant Na + current in nociceptors. Proc Natl Acad Sci U S A 93(3):1108–1112

Granados-Soto V, Argüelles CF, Rocha-González HI, Godànez-Chaparro B, Flores-Murrieta FJ, Villalón CM (2010) The role of peripheral 5-HT1A, 5-HT1B, 5-HT1D, 5-HT1E and 5-HT1F serotonergic receptors in the reduction of nociception in rats. Neuroscience 165(2):561–568

Gu M, Miyoshi K, Dubner R, Guo W, Zou S, Ren K, Noguchi K, Wei F (2011) Spinal 5-HT(3) receptor activation induces behavioral hypersensitivity via a neuronal-glial-neuronal signaling cascade. J Neurosci 31(36):12823–12836

Guindon J, Hohmann AG (2009) The endocannabinoid system and pain. CNS Neurol Disord Drug Targets 8:403–421

Gunduz O, Karadag HC, Ulugol A (2011) Synergistic anti-allodynic effects of nociceptin/orphanin FQ and cannabinoid systems in neuropathic mice. Pharmacol Biochem Behav 99:540–544

Haj-Dahmane S, Shen RY (2011) Modulation of the serotonin system by endocannabinoid signaling. Neuropharmacology 61(3):414–420

Hammond DL, Yaksh TL (1984) Antagonism of stimulation-produced antinociception by intrathecal administration of methysergide or phentolamine. Brain Res 298(2):329–337

Hamon M, Bourgoin S (1999) Serotonin and its receptors in pain control. In: Sawynok J, Cowan A (eds) Novel aspects of pain management: opioids and beyond. Wiley-Liss, Inc., New York, pp 203–228

Haring M, Marsicano G, Lutz B, Monory K (2007) Identification of the cannabinoid receptor type 1 in serotonergic cells of raphe nuclei in mice. Neuroscience 146:1212–1219

Heinricher MM, Morgan MM, Tortorici V, Fields HL (1994) Disinhibition of off-cells and antinociception produced by an opioid action within the rostral ventromedial medulla. Neuroscience 63:279–288

Herzberg U, Eliav E, Bennett GJ, Kopin IJ (1997) The analgesic effect of R(+)-WIN 55,212–2 mesylate, a high affinity cannabinoid agonist, in a rat model of neuropathic pain. Neurosci Lett 221:157–160

Hill MN, Sun JC, Tse MT, Gorzalka BB (2006) Altered responsiveness of serotonin receptor subtypes following long-term cannabinoid treatment. Int J Neuropsychopharmacol 9(3):277–286

Hohmann AG, Suplita RL (2006) Endocannabinoid mechanisms of pain modulation. AAPS J 8:693–708

Hohmann AG, Briley EM, Herkenham M (1999) Pre- and postsynaptic distribution of cannabinoid and mu opioid receptors in rat spinal cord. Brain Res 822:17–25

Huestis MA (2007) Human cannabinoid pharmacokinetics. Chem Biodivers 4(8):1770–1804

Ikoma A, Rukwied R, Ständer S, Steinhoff M, Miyachi Y, Schmelz M (2003) Neurophysiology of pruritus: interaction of itch and pain. Arch Dermatol 139(11):1475–1478

Inyushkin AN, Merkulova NA, Orlova AO, Inyushkina EM (2010) Local GABAergic modulation of the activity of serotoninergic neurons in the nucleus raphe magnus. Neurosci Behav Physiol 40(8):885–893

Iwasaki T, Otsuguro KI, Kobayashi T, Ohta T, Ito S (2013) Endogenously released 5-HT inhibits A and C fiber-evoked synaptic transmission in the rat spinal cord by the facilitation of GABA/glycine and 5-HT release via 5-HT(2A) and 5-HT(3) receptors. Eur J Pharmacol doi:10.1016/j.ejphar.2013.01.058

Jensen TS, Yaksh TL (1986) Examination of spinal monoamine receptors through which brainstem opiate-sensitive systems act in the rat. Brain Res 363:114–127

Jeong CY, Choi JI, Yoon MH (2004) Roles of serotonin receptor subtypes for the antinociception of 5-HT in the spinal cord of rats. Eur J Pharmacol 502(3):205–211

Kandel ER, Schwartz JH, Jessell TM (2000) Principles of Neural Science, 4th edn. McGraw-Hill, New York

Kato K, Kikuchi S, Konno S, Sekiguchi M (2008) Participation of 5-hydroxytryptamine in pain-related behavior induced by nucleus pulposus applied on the nerve root in rats. Spine 33:1330–1336

Klumpers LE, Beumer TL, Hasselt JG van, Lipplaa A, Karger LB, Kleinloog HD, Freijer JI, Kam ML de, Gerven JM van (2012) Novel Δ(9) -tetrahydrocannabinol formulation Namisol® has beneficial pharmacokinetics and promising pharmacodynamic effects. Br J Clin Pharmacol 74(1):42–53

Kobayashi H, Kikuchi S, Konno S, Kato K, Sekiguchi M (2011) Interaction of 5-hydroxytryptamine and tumor necrosis factor-α to pain-related behavior by nucleus pulposus applied on the nerve root in rats. Spine (Phila Pa 1976) 36(3):210–218

Kraft B (2012) Is there any clinically relevant cannabinoid-induced analgesia? Pharmacology 89(5–6):237–246

Kwilasz AJ, Negus SS (2012) Dissociable effects of the cannabinoid receptor agonists Δ9-tetrahydrocannabinol and CP55940 on pain-stimulated versus pain-depressed behavior in rats. J Pharmacol Exp Ther 343(2):389–400

Lichtman AH, Martin BR (1991) Spinal and supraspinal components of cannabinoid-induced antinociception. J Pharmacol Exp Ther 258:517–523

Lichtman AH, Cook SA, Martin BR (1996) Investigation of brain sites mediating cannabinoid-induced antinociception in rats: evidence supporting periaqueductal gray involvement. J Pharmacol Exp Ther 276(2):585–593

Lischetzki G, Rukwied R, Handwerker HO, Schmelz M (2001) Nociceptor activation and protein extravasation induced by inflammatory mediators in human skin. Eur J Pain 5(1):49–57

Long JZ, Nomura DK, Vann RE, Walentiny DM, Booker L, Jin X, Burston JJ, Sim-Selley LJ, Lichtman AH, Wiley JL, Cravatt BF (2009) Dual blockade of FAAH and MAGL identifies behavioral processes regulated by endocannabinoid crosstalk in vivo. Proc Natl Acad Sci U S A 106:20270–20275

Loyd DR, Henry MA, Hargreaves KM (2012) Serotonergic neuromodulation of peripheral nociceptors. Semin Cell Dev Biol 24:51–57

Mackie K, Stella N (2006) Cannabinoid receptors and endocannabinoids: evidence for new players. AAPS J 8:E298–E306

Maione S, Piscitelli F, Gatta L, Vita D, De Petrocellis L, Palazzo E, Novellis V de, Di Marzo V (2011) Non-psychoactive cannabinoids modulate the descending pathway of antinociception in anaesthetized rats through several mechanisms of action. Br J Pharmacol 162(3):584–596

Malan TP Jr, Ibrahim MM, Vanderah TW, Makriyannis A, Porreca F (2002) Inhibition of pain responses by activation of CB2 cannabinoid receptors. Chem Phys Lipids 121:191–200

Mallet C, Daulhac L, Bonnefont J, Ledent C, Etienne M, Chapuy E, Libert F, Eschalier A (2008) Endocannabinoid and serotonergic systems are needed for acetaminophen-induced analgesia. Pain 139(1):190–200

Manning BH, Martin WJ, Meng ID (2003) The rodent amygdala contributes to the production of cannabinoid-induced antinociception. Neuroscience 120(4):1157–1170

Martin WJ, Coffin PO, Attias E, Balinsky M, Tsou K, Walker JM (1999) Anatomical basis for cannabinoid-induced antinociception as revealed by intracerebral microinjections. Brain Res 822(1–2):237–242

Martin WJ, Lai NK, Patrick SL, Tsou K, Walker JM (1993) Antinociceptive actions of cannabinoids following intraventricular administration in rats. Brain Res 629(2):300–304

Martin WJ, Tsou K, Walker JM (1998) Cannabinoid receptor-mediated inhibition of the rat tail-flick reflex after microinjection into the rostral ventromedial medulla. Neurosci Lett 242:33–36

Mato S, Aso E, Castro E, Martàn M, Valverde O, Maldonado R, Pazos A (2007) CB1 knockout mice display impaired functionality of 5-HT1A and 5-HT2A/C receptors. J Neurochem 103:2111–2120

Matsuda LA, Lolait SJ, Brownstein MJ, Young AC, Bonner TI (1990) Structure of a cannabinoid receptor and functional expression of the cloned cDNA. Nature 346:561–564

Mendiguren A, Pineda J (2009) Effect of the CB(1) receptor antagonists rimonabant and AM251 on the firing rate of dorsal raphe nucleus neurons in rat brain slices. Br J Pharmacol 158(6):1579–1587

Meng ID, Johansen JP (2004) Antinociception and modulation of rostral ventromedial medulla neuronal activity by local microinfusion of a cannabinoid receptor agonist. Neuroscience 124(3):685–693

Meng ID, Manning BH, Martin WJ, Fields HL (1998) An analgesia circuit activated by cannabinoids. Nature 395:381–383

Meuser T, Pietruck C, Gabriel A, Xie GX, Lim KJ, Pierce Palmer P (2002) 5-HT7 receptors are involved in mediating 5-HT-induced activation of rat primary afferent neurons. Life Sci 71:2279–2289

Millan MJ (2002) Descending control of pain. Prog Neurobiol 66:355–474

Mitrirattanakul S, Ramakul N, Guerrero AV, Matsuka Y, Ono T, Iwase H, Mackie K, Faull KF, Spigelman I (2006) Site-specific increases in peripheral cannabinoid receptors and their endogenous ligands in a model of neuropathic pain. Pain 126(1–3):102–114

Monhemius R, Azami J, Green DL, Roberts MH (2001) CB1 receptor mediated analgesia from the Nucleus Reticularis Gigantocellularis pars alpha is activated in an animal model of neuropathic pain. Brain Res 908(1):67–74

Morgan MM, Whittier KL, Hegarty DM, Aicher SA (2008) Periaqueductal gray neurons project to spinally projecting GABAergic neurons in the rostral ventromedial medulla. Pain 140:376–386

Munro S, Thomas KL, Abu-Shaar M (1993) Molecular characterization of a peripheral receptor for cannabinoids. Nature 365:61–65

Naidu PS, Kinsey SG, Guo TL, Cravatt BF, Lichtman AH (2010) Regulation of inflammatory pain by inhibition of fatty acid amide hydrolase. J Pharmacol Exp Ther 334:182–190

Nakajima K, Obata H, Ito N, Goto F, Saito S (2009) The nociceptive mechanism of 5-hydroxytryptamine released into the peripheral tissue in acute inflammatory pain in rats. Eur J Pain 13(5):441–447

Nichols DS, Thorn BE, Berntson GG (1989) Opiate and serotonergic mechanisms of stimulation-produced analgesia within the periaqueductal gray. Brain Res Bull 22(4):717–724

Novotna A, Mares J, Ratcliffe S, Novakova I, Vachova M, Zapletalova O, Gasperini C, Pozzilli C, Cefaro L, Comi G, Rossi P, Ambler Z, Stelmasiak Z, Erdmann A, Montalban X, Klimek A, Davies P (2011) A randomized, double-blind, placebo-controlled, parallel-group, enriched-design study of nabiximols* (Sativex(®)), as add-on therapy, in subjects with refractory spasticity caused by multiple sclerosis. Eur J Neurol 18:1122–1131

Olango WM, Roche M, Ford GK, Harhen B, Finn DP (2012) The endocannabinoid system in the rat dorsolateral periaqueductal grey mediates fear-conditioned analgesia and controls fear expression in the presence of nociceptive tone. Br J Pharmacol 165(8):2549–2560

Ossipov MH, Dussor GO, Porreca F (2010) Central modulation of pain. J Clin Invest 120(11):3779–3787

Palazzo E, Novellis V de, Petrosino S, Marabese I, Vita D, Giordano C, Di Marzo V, Mangoni GS, Rossi F, Maione S (2006) Neuropathic pain and the endocannabinoid system in the dorsal raphe: pharmacological treatment and interactions with the serotonergic system. Eur J Neurosci 24(7):2011–2020

Palazzo E, Luongo L, Novellis V de, Berrino L, Rossi F, Maione S (2010) Moving towards supraspinal TRPV1 receptors for chronic pain relief. Mol Pain 6:66

Pertovaara A, Almeida A (2006) Endogenous pain modulation: descending inhibitory systems. In: Cervero F, Jensen TJ (eds) Handbook of clinical neurology—pain, vol. 81 (3rd series), Aminoff MJ, Boller F, Swaab DF (Series eds). Elsevier, London, pp 179–192

Pertwee RG (2001) Cannabinoid receptors and pain. Prog Neurobiol 63:569–611

Pertwee RG (2012) Targeting the endocannabinoid system with cannabinoid receptor agonists: pharmacological strategies and therapeutic possibilities. Phil Trans R Soc 367:3353–3363

Petrosino S, Palazzo E, Novellis V de, Bisogno T, Rossi F, Maione S, Di Marzo V (2007) Changes in spinal and supraspinal endocannabinoid levels in neuropathic rats. Neuropharmacology 52(2):415–422

Piomelli D, Giuffrida A, Calignano A, Rodriguez de Fonseca F (2000) The endocannabinoid system as a target for therapeutic drugs. Trends Pharmacol Sci 21:218–224

Porreca F, Ossipov MH, Gebhart GF (2002) Chronic pain and medullary descending facilitation. Trends Neurosci 25:319–325

Raffa RB, Stone DJ Jr, Hipp SJ (1999) Differential cholera-toxin sensitivity of supraspinal antinociception induced by the cannabinoid agonists delta9-THC, WIN 55,212–2 and anandamide in mice. Neurosci Lett 263(1):29–32

Richardson JD (2000) Cannabinoids modulate pain by multiple mechanisms of action. J Pain 1:2–14

Richardson JD, Kilo S, Hargreaves KM (1998) Cannabinoids reduce hyperalgesia and inflammation via interaction with peripheral CB1 receptors. Pain 75:111–119

Rivot JP, Chiang CY, Besson JM (1982) Increase of serotonin metabolism within the dorsal horn of the spinal cord during nucleus raphe magnus stimulation, as revealed by in vivo electrochemical detection. Brain Res 238(1):117–126

Russo EB (2008) Clinical endocannabinoid deficiency (CECD): can this concept explain therapeutic benefits of cannabis in migraine, fibromyalgia, irritable bowel syndrome and other treatment-resistant conditions? Neuro Endocrinol Lett 29(2):192–200

Rutkowska M, Gliniak H (2009) The influence of ACEA—a selective cannabinoid CB1 receptor agonist on whole blood and platelet-poor plasma serotonin concentrations. Pharmazie 64(9):598–601

Sasaki M, Obata H, Kawahara K, Saito S, Goto F (2006) Peripheral 5-HT2A receptor antagonism attenuates primary thermal hyperalgesia and secondary mechanical allodynia after thermal injury in rats. Pain 122(1–2):130–136

Scavone JL, Mackie K, Van Bockstaele EJ (2010) Characterization of cannabinoid-1 receptors in the locus coeruleus: relationship with mu-opioid receptors. Brain Res 1312:18–31

Schlosburg JE, Kinsey SG, Lichtman AH (2009) Targeting fatty acid amide hydrolase (FAAH) to treat pain and inflammation. AAPS J 11:39–44

Schmelz M, Schmidt R, Weidner C, Hilliges M, Torebjork HE, Handwerker HO (2003) Chemical response pattern of different classes of C-nociceptors to pruritogens and algogens. J Neurophysiol 89(5):2441–2448

Seyrek M, Kahraman S, Deveci MS, Yesilyurt O, Dogrul A (2010) Systemic cannabinoids produce CB1-mediated antinociception by activation of descending serotonergic pathways that act upon spinal 5-HT(7) and 5-HT(2A) receptors. Eur J Pharmacol 649(1–3):183–194

Smith C, Rahman T, Toohey N, Mazurkiewicz J, Herrick-Davis K, Teitler M (2006) Risperidone irreversibly binds to and inactivates the h5-HT7 serotonin receptor. Mol Pharmacol 70(4):1264–1270

Sofia RD, Nalepa SD, Harakal JJ, Vassar HB (1973) Anti-edema and analgesic properties of delta 9-tetrahydrocannabinol (THC). J Pharmacol Exp Ther 186:646–655

Sommer C (2010) Serotonin in pain and pain control. In: Müller CP, Jacobs BL (eds) Handbook of the behavioral neurobiology of serotonin, 1st edn. Academic Press, San Francisco

Sorkin LS, McAdoo DJ, Willis WD (1993) Raphe magnus stimulation-induced antinociception in the cat is associated with release of amino acids as well as serotonin in the lumbar dorsal horn. Brain Res 618(1):95–108

Svizenska I, Dubovy P, Sulcova A (2008) Cannabinoid receptors 1 and 2 (CB1 and CB2), their distribution, ligands and functional involvement in nervous system structure-a short review. Pharmacol Biochem Behav 90:501–511

Todd AJ (2010) Neuronal circuitry for pain processing in the dorsal horn. Nat Rev Neurosci 11(12):823–836

Tokunaga A, Saika M, Senba E (1998) 5-HT2A receptor subtype is involved in the thermal hyperalgesic mechanism of serotonin in the periphery. Pain 76(3):349–355

Toohey N, Klein MT, Knight J, Smith C, Teitler M (2009) Human 5-HT7 receptor-induced inactivation of forskolin-stimulated adenylate cyclase by risperidone, 9-OH-risperidone and other "inactivating antagonists". Mol Pharmacol 76(3):552–559

Ulugol A, Karadag HC, Ipci Y, Tamer M, Dokmeci I (2004) The effect of WIN 55,212–2, a cannabinoid agonist, on tactile allodynia in diabetic rats. Neurosci Lett 371:167–170

Ulugol A, Ozyigit F, Yesilyurt O, Dogrul A (2006) The additive antinociceptive interaction between WIN 55,212–2, a cannabinoid agonist, and ketorolac. Anesth Analg 102:443–447

Vanegas H, Schaible H (2004) Descending control of persistent pain: inhibitory or facilitatory. Brain Res Rev 46:295–309

Vaughan CW, McGregor IS, Christie MJ (1999) Cannabinoid receptor activation inhibits GABAergic neurotransmission in rostral ventromedial medulla neurons in vitro. Br J Pharmacol 127:935–940

Vaughan CW, Connor M, Bagley EE, MacDonald CJ (2000) Actions of cannabinoids on membrane properties and synaptic transmission in rat periaqueductal gray neurons in vitro. Mol Pharmacol 57:288–295

Vogel C, Mössner R, Gerlach M, Heinemann T, Murphy DL, Riederer P, Lesch KP, Sommer C (2003) Absence of thermal hyperalgesia in serotonin transporter-deficient mice. J Neurosci 23(2):708–715

Volfe Z, Dvilansky A, Nathan I (1985) Cannabinoids block release of serotonin from platelets induced by plasma from migraine patients. Int J Clin Pharmacol Res 5:243–246

Wakley AA, Craft RM (2011) Antinociception and sedation following intracerebroventricular administration of Δ^9-tetrahydrocannabinol in female vs. male rats. Behav Brain Res 216(1):200–206

Walker JM, Hohmann AG (2005) Cannabinoid mechanisms of pain suppression. Handb Exp Pharmacol 168:509–554

Walker JM, Huang SM (2002) Cannabinoid analgesia. Pharmacol Therapeut 95:127–135

Wang QP, Nakai Y (1994) The dorsal raphe: an important nucleus in pain modulation. Brain Res Bull 34(6):575–585

Wei F, Dubner R, Zou S, Ren K, Bai G, Wei D, Guo W (2010) Molecular depletion of descending serotonin unmasks its novelfacilitatory role in the development of persistent pain. J Neuroscience 30(25):8624–8636

Wei F, Gu M, Chu YX (2012) New tricks for an old slug: descending serotonergic system in pain. Acta Physiologica Sinica 64(5):520–530

Whiteside GT, Lee GP, Valenzano KJ (2007) The role of the cannabinoid CB2 receptor in pain transmission and therapeutic potential of small molecule CB2 receptor agonists. Curr Med Chem 14:917–936

Wilson-Poe AR, Morgan MM, Aicher SA, Hegarty DM (2012) Distribution of CB1 cannabinoid receptors and their relationship with mu-opioid receptors in the rat periaqueductal gray. Neuroscience 213:191–200

Yesilyurt O, Dogrul A, Gul H, Seyrek M, Kusmez O, Ozkan Y, Yildiz O (2003) Topical cannabinoid enhances topical morphine antinociception. Pain 105:303–308

Zhang YQ, Gao X, Zhang LM, Wu GC (2000) The release of serotonin in rat spinal dorsal horn and periaqueductal gray following carrageenan inflammation. Neuroreport 11:3539–3543

Zogopoulos P, Vasileiou I, Patsouris E, Theocharis SE (2013) The role of endocannabinoids in pain modulation. Fundam Clin Pharmacol 27(1):6

Chapter 14
Cannabinoids, Monoamines, COMT and Schizophrenia: Pathobiological Mechanisms in Psychosis

Colm M. P. O'Tuathaigh, Lieve Desbonnet and John L. Waddington

Abstract Chronic cannabis use is associated with increased risk for developing a psychotic disorder, with risk to develop psychosis highest among individuals who use cannabis during adolescence. The majority of cannabis users, however, do not develop a diagnosable psychiatric disorder. Individuals genetically predisposed to the development of psychosis seem at increased risk to the effects of cannabis. Contemporary models of psychosis posit that genetic predisposition and/or disruption at critical developmental periods is a substrate on which act various biological and psychosocial adversities, resulting in early functional impairments and later emergence of diagnostic symptoms. Recent years has seen the generation of experimental models of psychosis based on the interaction of genetic mutations and environmental factors (e.g. exposure to drugs of abuse). An emerging human and animal literature has shown showing that variation in the genes implicated in dopamine neurotransmission (COMT, AKT1, D2R) moderates the psychotomimetic effects of cannabis exposure. Further studies are required to clarify the molecular underpinnings of dopamine system involvement in cannabis-induced psychosis.

Abbreviations

AKT1	V-akt murine thymoma viral oncogene homolog 1
BDNF	Brain-derived neurotrophic factor
CB1R	Cannabinoid receptor 1
COMT	Catechol-*o*-methyltransferase
CPT	Continuous performance test
CSF	Cerebrospinal fluid
DA	Dopamine
DTI	Diffusion tensor imaging

C. M. P. O'Tuathaigh (✉)
School of Medicine, University College Cork, Cork, Ireland
e-mail: c.otuathaigh@ucc.ie

L. Desbonnet
Biosciences Institute, University College Cork, Cork, Ireland

J. L. Waddington
Molecular & Cellular Therapeutics, Royal College of Surgeons in Ireland, Dublin, Ireland

E. J. Van Bockstaele (ed.), *Endocannabinoid Regulation of Monoamines in Psychiatric and Neurological Disorders*, DOI 10.1007/978-1-4614-7940-6_14,

fMRI	Functional magnetic resonance imaging
GABA	Gamma-aminobutyric acid
G × E	Gene × environment
HPA	Hypothalamic-pituitary-adrenal
KO	Knockout
LI	Latent inhibition
MB-COMT	Membrane-bound catechol-o-methyltransferase
NRG1	Neuregulin-1
PFC	Prefrontal cortex
PET	Positron emission tomography
PND	Postnatal day
PPI	Prepulse inhibition
S-COMT	Soluble catechol-o-methyltransferase
SNP	Single nucleotide polymorphism
Δ^9-THC	Delta-9-tetrahydrocannabinol
VCFS	Velocardiofacial syndrome
WT	Wildtype

14.1 Introduction

It is conservatively estimated that cannabis is a drug that has been used at least once by 75.5 million Europeans, with 31.6% of European adolescents and young adults (15–34 years) reporting ever having used cannabis (EMCDDA 2010). As a widely used illicit drug, having a relationship with risk for psychiatric symptoms in general and psychosis in particular (Murray et al. 2007), it is important to identify and characterise the role of clinically relevant causative or moderating variables. Case-control and longitudinal studies indicate that lifetime cannabis use increases risk for developing a psychotic disorder (see Moore et al. 2007, for a review of the clinical data). Cannabis use has been associated with a number of clinical variables related to psychosis, including higher relapse rates (Linszen et al. 2007), poor treatment outcome and increased severity of symptoms (Grech et al. 2005) and accelerated loss of grey matter volume (Moore et al. 2007; Rais et al. 2008), even after adjusting for potential confounding factors. In a large-scale prospective study conducted in the Netherlands, a positive association was reported between cannabis use and self-reported psychotic symptoms; the risk was greatest among those with greater baseline history of cannabis use, and those with a pre-established vulnerability to psychosis (van Os et al. 2002).

Several authors have questioned the directionality of the relationship between cannabis use and psychosis in longitudinal designs; for example, despite high prevalence of cannabis use in Western countries, only a minority of cannabis users develop subclinical symptoms or a clinical psychotic disorder (van Os et al. 2009; Decoster et al. 2012). This may be explained by potential amplification of cannabis risk when interacting with genetic and other environmental risk factors (van Winkel et al. 2010).

14.2 Neurobiology of Cannabis Use

Cannabinoid CB1R receptors are expressed abundantly throughout the brain, particularly in areas implicated in learning and memory, notably the hippocampus, basal ganglia, cerebellum and prefrontal cortex (PFC; Freund et al. 2003). In the prefrontal cortex, CB1R receptors mediate glutamatergic and GABA release, while cortical CB1R receptors are localised to a subtype of GABAergic interneurons (Eggan and Lewis 2007). It has been postulated that overstimulation of CB1R receptors on GABAergic and glutamatergic terminals modulating activity in dopaminergic projections from the brain stem to the striatum may play a central role in the pathogenesis of cannabis-induced psychosis (Morrison and Murray, 2009). As with most drugs of abuse, CB1R receptor stimulation also causes an increase in extracellular dopamine (DA); cannabinoids stimulate burst firing of midbrain DA neurons and increase DA release in the ventral striatum, which is likely attributable to activation of CB1R receptors on GABAergic interneurons that synapse with DA neurons (Pistis et al. 2002).

14.3 Exogenous and Endogenous Cannabinoids

Cannabis contains greater than 60 cannabinoids, with Δ^9-tetrahydrocannabinol (THC), a partial agonist at the CB1R receptor, thought to be responsible for the principal psychotomimetic effects (Bossong et al. 2012). Cannabidiol is another constituent which is thought to possess antipsychotic properties. Interestingly, a number of studies have shown that cannabis containing a high THC and low cannabidiol concentration is associated with higher risk of a first psychotic episode (Di Forti et al. 2009), and higher levels of positive (i.e., psychotic) but not negative symptoms (Schubart et al. 2011). Zouardi et al. (1982) demonstrated that cannabidiol, co-administered with THC, significantly reduced the psychotomimetic symptoms induced by the latter. Preclinical findings support these findings, suggesting that cannabidiol may in fact antagonise the behavioural effects of THC (McLaren et al. 2008). In summary, these clinical and preclinical data suggest that variation in relative concentration of THC and cannabidiol may moderate the association between cannabis and psychosis.

14.4 Cannabis Use: Relationship with Psychosis Endophenotypes

Schizophrenia is a neurodevelopmental disorder characterised by the presence of positive (hallucinations, delusions, thought disorder), negative (social interactions deficits, avolition, anhedonia), as well as cognitive deficits (Waddington et al. 2012).

Several studies have shown that acute systemic administration of THC induces a transient increase in psychotic-like symptoms in both healthy volunteers (D'Souza et al. 2004, 2008a; Morrison et al. 2009) and patients with schizophrenia (D'Souza et al. 2005). Epidemiological studies have generally supported the link between long-term cannabis use and psychosis (reviewed elsewhere by McLaren et al. 2010 and D'Souza et al. 2009).

Leweke et al. (1999) reported increased concentrations of the endocannabinoid anandamide in cerebrospinal fluid (CSF) of antipsychotic-naive patients with schizophrenia; this was later replicated by the same group (Giuffrida et al. 2004). It was also found that CSF concentrations of anandamide were not increased in patients with affective disorders or dementia. Levels of anandamide in CSF in patients with schizophrenia correlated inversely with the severity of psychotic symptoms (Giuffrida et al. 2004), suggesting that the endocannabinoid system might be upregulated as a protective mechanism in patients with schizophrenia.

14.4.1 Positive Symptoms

A body of evidence has examined acute effects of cannabis or THC on psychosis-relevant symptoms in healthy subjects. In a double-blind study, D'Souza et al. (2004) found that intravenous THC produced positive and negative symptoms, which peaked over the first 80 min after treatment, decreasing to baseline at 4 h. Bhattacharyya et al. (2012) examined acute THC effects on processing of salience, as aberrant salience processes have been linked with presence of positive symptoms such as delusions (Murray et al. 2008; Roiser et al. 2009). They found that during a visual oddball task, THC reduced activation or augmented it in the right PFC during the processing of salient vs. non-salient stimuli, respectively. THC effects in the right caudate were negatively correlated with severity of psychotic symptoms induced and effect on response latency, suggesting that cannabis effects on psychosis may be mediated by influencing the neural substrate of attentional salience processing. Interestingly, cannabidiol administration was associated with the opposite response to THC and enhanced the appropriate response to salient stimuli (Bhattacharyya et al. 2010, 2012). A related study examined THC effects during a reward-based salience processing task; neural responses to behaviourally relevant salience stimuli were attenuated by THC, reflecting increased activation in response to non-salient stimuli and attenuated responsivity to salient stimuli (van Hell et al 2011). It has been hypothesised that DAergic dysfunction might lead to the development of psychotic symptoms by disrupting salience processing (Kapur 2003).

The potential of the THC models to predict antipsychotic efficacy was investigated for the DA D2 receptor (D2R) antagonist haloperidol. Co-administration of a single dose of haloperidol exerted only limited reduction in psychotomimetic symptoms induced by THC (D'Souza et al. 2008b). In a subsequent study, Liem-Moolenaar et al. (2010) reported haloperidol to exert material reduction of psychotomimetic symptoms induced by THC. Kleinloog et al. (2012) demonstrated acute THC to induce a transient psychotomimetic effect; co-administration of the antipsychotic olanzapine reduced this psychotomimetic effect by 33 % overall (50 % in responders) and also reduced the euphoric effects of THC. This work is of limited clinical relevance due to the psychotomimetic effects being transient and small.

14.4.2 Cognitive Dysfunction

Although the extent to which cognitive dysfunction in schizophrenia represents a generalised deficit or an impairment of more restricted scope remains contentious, there is some agreement that patients with the disorder exhibit deficits in attention, memory, reasoning and processing speed; these deficits frequently predate the emergence of clinical symptoms, relate closely to functional outcomes (e.g., relationship success, employment, treatment adherence) and are resistant to currently available antipsychotic treatments (Green et al. 2004; Pelletier et al. 2005; Keefe and Harvey 2012; Seeman 2011; Waddington et al. 2012; O'Tuathaigh et al. 2012b).

A previous meta-analysis indicated that learning and memory retrieval impairments were the only robust cognitive deficits seen in cannabis users (Grant et al. 2003). In a pharmacological functional magnetic resonance imaging (fMRI) study conducted in healthy subjects, acute THC disrupted working memory in the Sternberg intem-recognition test at a lower working memory load, test at a lower working memory load; this disruption was associated with differential activation in relevant brain areas, including the dorsolateral PFC, inferior temporal gyrus and cerebellum (Bossong et al. 2012). fMRI studies have indicated impairments in learning and memory following cannabinoids to be mediated medial temporal, striatal, midbrain and PFC function (Bhattacharyya et al. 2009, 2012; Bossong et al. 2011). Attentional deficits have also been reported following acute THC administration and in chronic cannabis users (Solowij and Michie 2007).

When discussing effects of cannabis on cognition, it is important to distinguish between the short-term effects of ongoing cannabis use vs. the consequences of previous lifetime history of cannabis use; in brief, many studies have documented a short-term negative effect but a positive long-term effect on functioning. In terms of cognition, acute THC has been shown to impair attention and memory in schizophrenia patients and their unaffected siblings relative to healthy controls (D'Souza et al. 2005; Henquet et al. 2006).

Systematic reviews and meta-analyses indicate improved cognitive functioning in cannabis-using relative to non-cannabis-using patients (Loberg and Hugdahl 2009; Rabin et al. 2011; Yucel et al. 2012); furthermore, better cognitive functioning has been reported in cannabis-using patients relative to non-using patients on executive function tasks, visual memory, processing speed, global cognition and working memory (Coulston et al. 2007; Potvin et al. 2008). It has been suggested that the cognition-improving properties of THC may be due to stimulation of PFC neurotransmission (Potvin et al. 2008; Cohen et al. 2008) or, alternatively, that psychotic patients with lifetime cannabis use may constitute a better functioning group of patients from the outset (Schnell et al. 2009; de la Serna et al. 2010). In a recent study (Meijer et al. 2012) conducted in 956 psychotic patients, 953 unaffected siblings, and 554 control subjects, current cannabis use was associated with poorest performance on immediate verbal learning, processing speed and working memory; this association did not differ across the three groups or in terms of frequency of cannabis use during the previous year. In contrast, lifetime cannabis use was not associated with worse cognitive functioning, supporting the hypothesis that cannabis-using patients might constitute a patient subgroup that is less vulnerable to such deficits than those who had not used cannabis on a long-term basis. Indeed, lifetime cannabis users perform better on tests of social cognition and acquired knowledge; this may be attributable to lower genetic vulnerability and premorbid functioning rather than the drug itself (Meijer et al. 2012). Other authors have suggested that poor premorbid functioning in schizophrenia is associated with increased vulnerability for adverse effects of cannabis on cognition (D'Souza et al. 2005; Ringen et al. 2013).

14.4.3 Anatomical Phenotypes Associated with Psychosis

Given that schizophrenia is associated with brain structural abnormalities and cannabis is a risk factor for schizophrenia, might cannabis use contribute to development of anatomical abnormalities and contribute to transition from at-risk state to overt illness (Welch et al. 2012)? Yucel et al. (2008) studied cannabis users who had taken greater than five joints daily for more than 10 years and compared them with non-users. Heavy users displayed reduced hippocampal and amygdale volume; additionally, left hemisphere hippocampal volume was inversely correlated with cumulative exposure to cannabis and expression of sub-threshold psychotic symptoms.

In an MRI study, Welch et al. (2012) showed an association between cannabis use and grey matter loss in currently well individuals at familial risk of developing schizophrenia. Shape analysis in cannabis-using and non-using patients with schizophrenia indicated hippocampal abnormalities in each group relative to controls, with the most prominent changes found in those patients using cannabis (Solowij et al. 2013).

A number of imaging studies have shown region-specific abnormalities of white matter in brains of cannabis users relative to non-using controls (Wilson et al. 2000; Matochik et al. 2005); however, negative findings for both white matter (Block et al. 2000; Gruber and Yurgulen-Todd 2005), overall brain volume and cortical grey matter (Wilson et al. 2000) have been reported. A recent study used diffusion tensor imaging (DTI) to examine white matter tracts in heavy cannabis users vs. non-using controls and reported an increase in diffusivity in cannabis users in the region of the corpus callosum where white matter passes between the prefrontal lobes. Other studies have shown evidence for bilateral hippocampal and amygdalar volume reductions in adults with long-term cannabis use, with left hippocampal volume inversely related to length of cannabis exposure (Yucel et al. 2008).

In a PET study, Stokes et al. (2012) looked at striatal DA D2/D3R availability in patients with a history of cannabis exposure, and found no relationship with history or frequency of lifetime cannabis use. As indicated in other reviews, these data do not exclude the possibility of readaptation of DA or D2/D3R levels. Other studies have noted reduced cortical DA D2R binding following acute THC challenge, with decreased binding most pronounced in individuals homozygous for the catechol-*O*-methyltransferase (COMT) Val allele (Stokes et al. 2011). There is inconclusive evidence from human studies regarding THC modulation of striatal DA release, with some studies showing a modest increase (Bossong et al. 2009) and others showing no difference (Stokes et al. 2009). In a recent study of 11 male volunteers treated intravenously with either THC or placebo, THC did not result in a significant increase in DA release when administered at a dose capable of inducing psychotic symptoms (Bossong and Niesink 2010).

14.4.4 Cannabis Use and Psychosis: Moderating Variables

In a series of studies conducted by Morgan and colleagues (Morgan and Curran 2008; Morgan et al. 2012), it was shown that individuals who had smoked cannabis rich in cannabidiol as well as THC exhibited fewer psychotic-like symptoms symptoms than those who smoked cannabis with a low cannabidiol concentration. In addition, while higher THC concentrations were correlated with dimished prose recall and source memory performance, recognition memory was improved in those consuming hair-confirmed, cannabidiol-rich cannabis. Interestingly, there were greater psychotic-like symptoms in those smoking high THC/low cannabidiol in recreational rather than daily users, suggesting that those more vulnerable to the psychotomimetic effects of cannabis may use the drug less frequently and may prefer to smoke lower THC strains.

14.5 Cannabis and Liability to Psychosis: Focus on Adolescence

Cannabis use is largely concentrated among young people (15–34 years), with highest prevalence being reported among 15–24 year olds (EMCDDA 2010). Epidemiological studies have shown that risk to develop psychosis is highest among individuals who use cannabis during adolescence (Fergusson et al. 2003; Arseneault et al. 2002, 2004; McGrath et al. 2010). In a large scale study examining the relationship between adolescent cannabis use and psychotic symptoms in 3,500 19-year-olds in Greece, adolescent cannabis use was associated with both positive and negative symptoms of schizophrenia. Despite controlling for confounding variables, a study in Spain found that young persons presenting with first episode of psychosis (mean age 15.5) had a higher rate of positive symptoms and less negative symptoms if they were cannabis users rather than non-users (Baeza et al. 2009). Prospective cohort studies have shown increased risk for psychosis among those who use cannabis in young adulthood, with heavy users more likely to develop psychosis during a 21 year follow-up period (McGrath et al. 2010). Schubart et al. (2011) underlined that early (under 12 years of age) and heavy cannabis use were each materially and independently associated with increased risk for psychiatric hospitalisation. These authors suggest several possible explanations for these data patterns: increased tendency of young people with psychotic experiences to commence cannabis use; greater cumulative exposure to cannabis of early users; increased vulnerability to THC during crucial developmental windows.

Adolescence is a critical period in brain development, with considerable maturation occurring in limbic structures such as the hippocampus and also in PFC, which undergoes synaptic pruning, myelination and receptor development during this period (Spear 2000; Andesen and Teacher 2008). Some studies have reported long-term white matter changes in adolescent cannabis-using adults in prefrontal fibre bundles of the corpus callosum (Arnone et al. 2008) and changes in fronto-parietal circuitry (Bava et al. 2009). However, other studies have reported no changes in the integrity of white matter (Delisi et al. 2006) or the hippocampus (Medina et al. 2007) relative to age-matched, cannabis-naive subjects.

14.5.1 Adolescent Cannabis Use and Psychosis: Evidence from Animal Studies

Use of experimental models can help us to test hypotheses regarding the mechanistic role of different neurotransmitter systems in these effects. Complex processes are likely to have an equally complex pathophysiology, with associated difficulties in conducting studies in human subjects that are able to isolate individual factors and quantify their clinical impact. In this context, animal models are crucial for understanding the involvement of putative risk factors in psychiatric disorders and for identifying pathophysiological mechanisms, disease biomarkers and, ultimately, novel and effective therapies for these disorders.

Animal models of positive symptoms, reviewed in detail elsewhere (Van den Buuse 2010; Kirby et al. 2010), have often relied on indirect DA-linked measures, such as novelty- and psychostimulant-induced hyperactivity, or information processing paradigms, such as prepulse inhibition (PPI) or latent inhibition (LI), two measures of sensorimotor gating and learned inattention processes that are disrupted in schizophrenia (Moser et al. 2000; Barak and Weiner 2011). Adolescent exposure to the synthetic cannabinoid Win 55,212 in male rats produced changes across several schizophrenia-related endophenotypes, including PPI deficits, object recognition memory and a deterioration in progressive ratio instrumental performance in adulthood (Schneider and Koch 2003); these deficits were accompanied by abnormal basal neuronal activation across several brain regions (Wegener and Koch 2009). Recently, Gleason et al. (2012) administered the synthetic cannabinoid Win 55,212 during adolescence or adulthood to C57BL6 mice and evaluated long-term effects on fear conditioning, PPI, exploratory activity and social interaction. They reported long-lasting deficits in PPI and contextual fear conditioning in adolescent-treated mice, with no changes in social interaction and exploratory activity; these deficits were accompanied by normal CB1R receptor expression but reduced mGluR5 protein expression in the hippocampus. Other studies have reported conflicting data concerning changes in hippocampal CB1R receptor expression levels following adolescent THC treatment (Ellgren et al. 2007; Rubino et al. 2008).

A survey of the animal literature shows that the link between adolescent cannabinoid exposure and impaired cognition is inconsistent and appears to be dependent upon a number of moderating variables, including cannabinoid administered, treatment schedule and specific cognitive paradigm employed (Realini et al. 2009). Adolescent exposure to the CB1R receptor agonist CP 55,940 in rats produced deficits in working memory (O'Shea et al. 2004, 2006), with similar deficits observed in adulthood in a sex-specific manner. No effect of THC on spatial learning in the Morris water maze was observed (Cha et al. 2006, 2007). Administration of THC to rats during early adolescence [postnatal day (PND)22–40] relative to late adolescence (PND41-60) impaired reversal learning as measured in the active place avoidance task (Harte and Dow-Edwards 2010). A recent study found that rats administered Win 55,212 over PND45-60 exhibited long-lasting effects in a hippocampal-dependent cognitive task, while there was only transient (<30 day) impairment in a more PFC-dependent task (Abush and Akirav 2012).

Sex-specific (female only) deficits in working memory, as measured in the radial arm maze, was observed in adolescent THC-treated mice, and these were accompanied by synaptic impairment in PFC (Rubino et al. 2009a). Deficits in recognition memory were reported in THC-treated adolescent but not adult rats (Quinn et al. 2008; Renard et al. 2012), accompanied by proteomic alterations in the hippocampus related to degenerative and oxidative processes (Quinn et al. 2008). Other studies have reported structural changes in the hippocampus following adolescent THC treatment, including impairment in structural and functional plasticity of both neurons and glia in this region, alongside a reduction in dendrite length and complexity/number of dendritic spines in the dentate gyrus (Rubino et al. 2009b).

In relation to social functioning and other behavioural features related to negative symptomatology, chronic CP 55,940 administration during adolescence impaired social interaction in both male and female rats, but only in males when administered during adulthood (O'Shea et al. 2004, 2006). Other studies have shown that while chronic THC exposure during adolescence had no effect on social approach behaviour in mice (O'Tuathaigh et al. 2010), disruption of sociability was found following treatment with the synthetic cannabinoid Win 55,212 during adolescence (O'Tuathaigh et al. 2012a). In relation to social cognition, THC or synthetic cannabinoid administration has been shown to disrupt social novelty preference (O'Tuathaigh et al. 2010, 2012a). Comparable deficits in social recognition has been reported in rats which were administered Win 55,212 during adolescence (Leweke and Schneider 2011). Chronic THC treatment during adolescence (PND35–45) produced no change in anxiety-related behaviours but, rather, a depressive-like profile in female rats, including increased 'behavioural despair' in the forced swim test and anhedonia as indexed by decreased sucrose consumption (Rubino et al. 2008).

14.5.2 Neurobiological Mechanisms Underlying the Relationship Between Adolescent Cannabis Exposure and Psychosis

During adolescence, endocannabinoid levels and cannabinoid receptors increase (Schneider 2008), indicating that the endocannabinoid system undergoes considerable development over this period (Spear 2009). Also, GABAergic neurons in PFC have CB1R receptors that, when activated, result in a decrease in extracellular GABA release (Egerton et al. 2006). It has been suggested that exogenous activation of CB1R receptors during adolescence may alter the balance of GABAergic inhibitory inputs to pyramidal neurons in PFC (Eggan et al. 2010) that could result in impaired cognitive function in schizophrenia. Both human and rodent studies have noted complementary changes in the expression of CB1R receptors in corticolimbic regions from adolescence into adulthood (Belue et al. 1995; Mato et al. 2003). It is thought that CB1R receptor levels decrease throughout adolescence, although these ontogenetic changes might be quite spatially- and temporally-specific (Malone et al. 2010). Adolescent cannabinoid exposure has been shown to produce a long-lasting decrease in CB1R receptor expression and/or G protein coupling in specific brain regions (Rubino et al. 2008). It has been proposed that cannabis disturbs the protective action of the endocannabinoid system during adolescence and that this is associated with disrupted GABAergic and glutamatergic transmission, especially in PFC, which is in turn associated with abnormal functional (e.g., DAergic hyperfunction) and structural (e.g., wiring defect) changes (Bossong and Niesink 2010).

Psychosis is hypothesised to result from aberrant reward prediction and abnormal attribution of salience caused by disrupted DA neurotransmission (Kapur 2004). It has been hypothesised that repeated cannabis use during adolescence results in sensitisation of the endogenous mesocorticolimbic DA system, such that cannabis use during adolescence results in a worse outcome relative to otherwise comparable use during adulthood (Stefanis et al. 2004). Studies in humans and rodents show that core elements in the DAergic system, such as synthesis and degrading enzymes, DA and DA receptor levels, increase during adolescence (Seeman et al. 1987; Pitts et al. 1990).

Important changes in the level and timing of release of hormones in the hypothalamic-pituitary-adrenal (HPA) axis also occur during adolescence, particularly in response to stress (Spear 2009); both of these systems have been shown to interact with the endocannabinoid system (Freund et al. 2003; Malone et al. 2010). In contrast, other studies have noted changes in 5-HT1A receptor density and messenger RNA (mRNA) expression in the hippocampal CA1 region and dentate gyrus of adult but not adolescent cannabinoid-treated rats (Zavitsanou et al. 2010).

14.6 Genetic Modulation of the Cannabis: Psychosis Association

It should be noted that the majority of cannabis users do not experience or develop a psychotic disorder or a diagnosable psychiatric disorder later in life. Several authors have postulated that cannabis may represent a risk factor in a diathesis-stress model of schizophrenia, whereby risk for developing psychosis is higher in a subset of genetically-vulnerable people (Pelayo-Terán et al. 2012). One variant on this gene [G] × environment [E] model posits that genetic predisposition and/or genetic disruptions at critical developmental periods is a substrate on which various biological and psychosocial adversities act. These can accumulate across each of the developmental periods to result in early functional impairments that are followed by the emergence of diagnostic symptoms over young adulthood.

While hypothesising a reason for interindividual variability in susceptibility to cannabis-induced psychosis, it should be noted that there is limited evidence for genetic effects. Recent years have also seen a movement towards the generation of animal models of psychosis based on the interaction of genetic mutations and well-characterised environmental factors (Gray and Hannan 2007; Ayhan et al. 2009; Desbonnet et al. 2012). Recent studies from our laboratory and others have investigated how candidate gene mutations may modify the consequences of environmental insults that can be delivered in a manner similar to those which may accumulate throughout the disease process.

14.7 COMT, Cannabis, and Schizophrenia

14.7.1 COMT and Schizophrenia Susceptibility

The COMT enzyme is expressed in pyramidal neurons of PFC and hippocampus, and plays a specific role in the catabolism of cortical DA but not noradrenaline (Papaleo et al. 2008). COMT encodes two transcripts from two promoters in humans (membrane bound, MB-COMT; soluble, S-COMT). MB-COMT displays a tenfold greater affinity for DA and noradrenaline than S-COMT, suggesting that MB-COMT is customised for metabolism of catecholamines, including DA, at the physiological levels found in brain (Roth and Pfefferbaum 1992; Tunbridge et al. 2006). The COMT gene is located in chromosome 22q11.2.2, which has repeatedly been linked with schizophrenia (Karayiorgou et al. 2010). COMT is implicated in degradation of DA in the synapse, and plays a specific role in regulating DA-dependent, PFC-mediated cognition. A functional polymorphism of the COMT gene involving the allelic substitution of valine (Val) for methionine (Met) at the 108/158 locus results in a fourfold shift in COMT enzymatic activity, with clinical and preclinical studies indicating COMT genotype to influence cognition, both in normal subjects and patients with schizophrenia (Tunbridge et al. 2006). The enzymatic activity of the Val allele is ~40% higher than that of the Met allele in post-mortem human PFC tissue (Chen et al. 2004). COMT Val108/158Met allelic variation has been associated with differential performance on tasks measuring PFC–mediated cognition: individuals homozygous for the Met allele display increased PFC DA levels and the highest performance in these tasks; Val/Met individuals are intermediate; individuals homozygous for the Val allele display reduced PFC DA levels and the lowest performance (Tunbridge et al. 2006).

Although several association studies have implicated the involvement of the COMT Val108/158Met variant (also referred to as rs4680) with increased risk for schizophrenia (Egan et al. 2001; Shifman et al. 2002; Hoenicka et al. 2010), others have reported no association between this variant and risk for schizophrenia or its clinical symptomatology (Tee et al. 2011; Tovilla-Zarate et al. 2013). However, recent meta-analyses do not support the relationship between this single nucleotide polymorphism (SNP) and schizophrenia susceptibility (Munafo et al. 2005; Okochi et al. 2009). These contradictory findings may be attributable to different confounding factors such as population stratification, clinical heterogeneity or gender differences in COMT expression (Jiang et al. 2003; Dempster et al. 2006; Hoenicka et al. 2010).

A recent meta-analysis of 30,000 samples derived from 51 studies suggested that heterozygosity at the Val108/158Met SNP may represent a protective factor against schizophrenia (Costas et al. 2011). These authors argued that both too high and too low levels of DA signalling may be risk factors for schizophrenia and propose that risk liability may reflect the balance between DA D1R and D2R activation in PFC. Speculating on the underlying receptor mechanisms, several authors have proposed that Met homozygosity is associated with excess DA levels and preferential activation of extrasynaptic DA D1 receptors in the PFC, while Val carriers release reduced amounts of DA, favouring activation of DA D2 receptors at the synaptic cleft (Bilder et al. 2004; Durstewitz and Seamans 2008).

14.7.2 22q11.2-Deletion Syndrome: Relationship with COMT Gene Variation

22q11.2-deletion syndrome, also referred to as velocardiofacial syndrome (VCFS), is a genetic disorder caused by a microdeletion on chromosome 22 and is associated with multiple congenital malformations and several neuropsychiatric disorders (Murphy et al. 1999). It is associated with a 25–30-fold increase in risk for psychosis (Murphy et al. 1999; Murphy and Owen 2001; Prasad et al. 2008). Despite sharing many phenotypic features, 22q11.2-deletion patients do not show the same verbal memory deficits observed in patients with schizophrenia (Lajiness-O'Neill et al. 2005; Kraviriti et al. 2010). Previous studies have shown that 22q11.2 deletion patients with homozygosity at the COMT Met allele are at additional risk for psychiatric disorders (Gothelf et al. 2007, 2008).

The phenotype of mutant mice carrying a multigene deletion across the 22q11.2 region includes impaired PPI, neuronal migratory defects and disruption of cortical neurogenesis, with haploinsufficiency of genes located in this region implicated in these phenotypic deficits (Stark et al. 2008; Meechan et al. 2009). A recent study demonstrated that selective overexpression of COMT, via virus-mediated reintroduction, to PFC in mice with hemizygous deletion of 16 genes in the 22q11.2 region, reversed a number of defects in NMDA receptor (NMDAR) signalling as well as GABA release and expression of GABA-related genes in PFC (Kimoto et al. 2012).

14.7.3 COMT Genotype and Schizophrenia Symptoms

Studies have generally shown little evidence for a relationship between COMT gene variation and severity or type of positive symptoms in patients with schizophrenia (Bertolino et al. 2004; Wonodi et al. 2006). However, it was shown that among patients with schizophrenia, carriers of the Val genotype were characterised by a higher severity of psychotic symptoms than carriers of other genotypes (Molero et al. 2007). In a more recent study, Collip et al. (2011) reported that Met carriers showed increased psychotic and affective reactivity to stress when compared to Val/Met or Val carriers; this was observed only in patients with schizophrenia and not in control subjects.

Studies in mutant mice with COMT knockout (KO) have demonstrated changes in psychostimulant sensitivity but no changes in either cognition, PPI or exploratory activity in males (Gogos et al. 1998, Huotari et al. 2002; Haasio et al. 2003; Babovic et al. 2007). Mice lacking the soluble isoform of COMT displayed higher accumbens DA levels in both sexes but male S-COMT mutants showed lower PFC DA concentrations than WT mice. They also displayed enhanced acoustic startle without any change in PPI or any other measures of sensorimotor gating (Tammimäki et al. 2010).

There is limited evidence to suggest an association between the COMT Val108/Met158 variant and overall severity of negative symptoms in patients with schizophrenia (Wang et al. 2009). In a recent study, patients with schizophrenia who were Met carriers showed olfactory function impairment that was independent of medication status (Kamath et al. 2012). Olfactory deficits have been linked with negative symptoms in patients with schizophrenia (Ishizuka et al. 2010). Other studies have also reported poorer social cognition in COMT Val carriers (Uçok et al. 2010).

Sociability and social novelty preference are unaffected in both heterozygous and homozygous COMT KO (Babovic et al. 2008), but heterozygous COMT mutants showed increased aggression in a resident intruder assay (Gogos et al. 1998). Mice lacking soluble COMT also demonstrated impairment in social functioning, as indexed by increased nonaggressive social dominance behaviour (Tammimäki et al. 2010). COMT Val108/Met158 variation has long been associated with variation in executive cognition/working memory among both normal subjects and patients with schizophrenia (Egan et al. 2001; Tunbridge et al. 2006). DAergic activity in PFC plays a significant role in modulation of higher order cognitive processes, including both executive function and working memory (Seamans and Yang 2004; Mier et al. 2010). Howcver, an extensive meta-analysis indicated only a modest relationship between COMT genotype and executive function in normal subjects and minimal relationship in schizophrenia patients (Barnett et al. 2008) A more recent study demonstrated that the benefit of the Met allele in a visuospatial working memory task only emerged after 10 years of age, indicating that COMT genotype effects on cognition are not static during development (Dumontheil et al. 2011).

COMT gene variation has also been studied in relation to other forms of cognition, notably measures of sensory gating, using auditory evoked potential markers P50 and P300, and early information processing such as smooth pursuit eye movements, which are disturbed in schizophrenia (Lu et al. 2007; Kang et al. 2010; Leitman et al. 2010). However, other studies have shown no association between COMT gene variation and P50 gating deficits in schizophrenia (Shaikh et al. 2011).

COMT transgenic mice (Val-tg) with overexpression of a human COMT-Val polymorphism have been described (Papaleo et al. 2008, 2012). These mutants exhibit deficits in attentional behaviour, working memory and recognition memory. Amphetamine disrupted recognition memory in wildtypes but ameliorated recognition memory in COMT-Val transgenics, providing further support of an inverted-U relationship between the extent of DAergic transmission and cognitive function (Papaleo et al. 2008). COMT Val-tg mice also showed selective impairments in their ability to shift an attentional-set (Papaleo et al. 2008). In addition, COMT Val-tg mice demonstrated working memory deficits, requiring greater time to acquire a discrete paired-trial alternation T-maze task. On the other hand, COMT KO mice show improved working memory, as indexed by: (a) faster acquisition of the discrete paired-trial alternation T-maze task and higher number of correct responses at different intra-trials delays (Papaleo et al. 2008); (b) improved spontaneous alternation performance in the Y-maze (Babovic et al. 2008); and (c) improved delayed alternation in the Y-maze (O'Tuathaigh et al. 2010). Overall, these mutant data demonstrate that chronic increase or decrease in COMT enzymatic activity markedly impacts cognition, particularly working memory and attentional performance.

14.7.4 COMT Modulation of Psychotomimetic Effects of Cannabis

14.7.4.1 Clinical Studies

Caspi et al. (2005) first reported the COMT Val158Met polymorphism to moderate the risk of developing schizophreniform disorder at 26 years of age in subjects who used cannabis in adolescence. Specifically, adolescent cannabis use was associated with increased psychotic symptoms and hallucinatory experiences in adulthood among high-activity Val/Val carriers (odds ratio (OR) 10.9) and Val/Met carriers (OR 2.5) but not in low-activity Met/Met carriers. The same association was still observed even after adjusting for illicit drug use in adulthood, and the same variants were not associated with either risk for psychosis or cannabis use itself. However, several additional studies have failed to replicate this association (Costas et al. 2011; Van Winckel et al. 2011). Alongside the original study by Caspi et al. (2005), several reports have, however, confirmed an association between COMT variation, cannabis consumption and psychosis (Henquet et al. 2006; Pelayo-Terán et al. 2010; Costas et al. 2011), although conflicting data indicate an interaction with either the low activity Met (Zammit et al. 2007; Costas et al. 2011) or high activity Val allele (Caspi et al. 2005).

In addition to symptom-based phenotypes, a small number of studies have focused on age at onset as an outcome measure. Estrada et al. (2011), in a sample of 80 patients with a psychotic disorder and 77 patients with a non-psychotic disorder, found an interaction between COMT Val/Met genotype and cannabis use for age at onset, but only for patients with a psychotic disorder; specifically, Val/Val carriers showed earlier age at onset than Met/Met carriers.

A subsequent study by Zammit et al. (2011) supported the link between cannabis use and onset of psychotic symptoms but failed to find any evidence for association between cannabis-induced psychosis and either the COMT Val/Met polymorphism or any other COMT variants. These authors concluded that even if the relative risk for developing psychosis does differ slightly across COMT genotypes, it is questionable whether such an effect size would advance our understanding of etiological mechanisms or inform potential strategies for prevention or intervention. An alternative perspective might be that the direction of COMT enzyme activity might differentially contribute to the psychotomimetic effects of cannabis exposure depending on the symptom(s) or endophenotype(s) under examination.

14.7.4.2 Data from Genetic Models

Our laboratory has shown that in male COMT KO mice, genotype exerted a specific modulation of responsivity to chronic administration of THC or the synthetic cannabinoid WIN 55,212 during adolescence, but not during adulthood, in terms of phenotypes relevant to positive symptoms (hyperactivity, disrupted sensorimotor gating), negative symptoms (impaired social functioning) and cognitive dysfunction (disrupted working memory); these data support a putative COMT × cannabis exposure interaction over this particular stage of development in expression of the psychosis phenotype (O'Tuathaigh et al. 2010, 2012a). In these mice, we also showed COMT genotype × THC treatment interactions for several morphological indices of endocananbinoid, DA and GABAergic function (Behan et al. 2012). Specifically, COMT deletion is associated with: (a) release of an effect of adolescent THC treatment to reduce ventral tegmental DAergic cell size; (b) increased CBR1 intensity in PFC, reduced CBR1 intensity in hippocampus, with a shift in effect of adolescent THC treatment in hippocampus and PFC from reduction to increase in CBR1 intensity; and (c) increased GABAergic cell size in PFC and hippocampus, with a shift in effect of adolescent THC treatment in PFC and hippocampus from increase to decrease in GABAergic cell size. Adolescent THC treatment also reduced dopaminergic cell density. In particular, the GABAergic changes observed in this G × E mouse model and their relationship to schizophrenia indicate further vulnerability to, and consequent dysregulation of, neurotransmitter systems by developmental insults such as THC treatment during adolescence.

14.8 AKT1 and Cannabis-Induced Psychosis

AKT1 is a protein kinase implicated in processes downstream of the DA D2R. Cannabinoids can activate the AKT1 pathway by acting on CB1R and CB2 receptors. An interaction was reported between cannabis use and an AKT1 polymorphism (rs2494732) on reaction time and accuracy in the continuous performance test (CPT) (Van Winkel et al. 2011). Given that CPT performance has been linked with PFC DAergic function, these results support the general contention that AKT1 modulation of PFC DAergic function may modulate cannabis effects on psychosis. Specifically, it has been proposed that variants in the AKT1 gene could be involved in cannabis-induced psychosis via a mechanism of cannabinoid-regulated AKT1/GSK-3 signalling downstream of the DA D2 receptor (Casadio et al. 2011).

In a recent study, Bhattacharyya et al. (2012) demonstrated that the acute psychotomimetic effects of THC were moderated by variation in two genes implicated in DA neurotransmission. In a study by Van Winkel et al. (2011) in participants with minimal lifetime cannabis use, individuals carrying risk variants of either the DA transporter (9DAT 3'UTR VNTR) or the AKT1 (rs130233) polymorphisms show increased sensitivity to the psychotic effects of THC. Additionally, risk was greatest in subjects carrying both variants. These differences were accompanied by changes in the neural patterns elicited by THC in the striatum and midbrain.

14.9 Additional Genes Implicated in Cannabis-Induced Psychosis

Another study looked at brain-derived neurotrophic factor (BDNF), a neurotrophin implicated in development of mesolimbic DA neurons (Altar et al. 1997). A Val to Met substitution at codon 66 (rs6265) of the BDNF gene results in less efficient intracellular trafficking and decreased activity-dependent BDNF secretion (Egan et al. 2003). In a recent study, Decoster et al. (2011) showed that cannabis use predicted earlier age at onset of psychosis in male patients independently of genotype, while in female patients cannabis use was only associated with age of onset in BDNF Met carriers.

Ho et al. (2011) recently examined the impact of cannabis use/dependency and 12 SNPs in the CB1R receptor gene on white matter volume and performance in a neurocognitive test battery in 235 schizophrenia patients. A specific association was found between the rs12720071 variant and cannabis use on white matter volume and problem-solving skills.

One of the strongest candidate genes for schizophrenia reported so far is neuregulin-1 (NRG1; Harrison and Law 2006; Stefansson et al. 2002; Tosato et al. 2005). There is limited data to indicate increased sensitivity of male (but not female) transmembrane TM-domain NRG1 mutant mice to the neurobehavioural effects of THC; these include PPI enhancement, hypolocomotion and anxiogenic effects, as well as a selective increase in neuronal activation (Boucher et al. 2007a, b; Long et al. 2010). A similar change was observed in relation to cannabinoid responsivity in adult NRG1 mutants to the locomotor and anxiogenic effects of the synthetic cannabinoid CP 55,940 (Boucher et al. 2011). A recent study examining the effects of acute and chronic THC during adolescence on several psychosis-relevant endophenotypes, both during treatment and 48 h after 48 h, found genotype-independent effects of THC on exploratory activity, but no effect of treatment or NRG1 genotype on PPI. This study also revealed differential NRG1 modulation of adolescent THC effects on CB1R and 5-HT2A receptor binding in the substantia nigra and insular cortex (decreased in TM-NRG1 mutants, increased in wildtypes), while the opposite pattern was seen for NMDAR levels. TM-NRG1 mutants were also resistant to THC-induced suppression of investigative social behaviours. Chronic cannabidiol administered during adulthood did not alter PPI, locomotor hyperactivity or 5-HT-2A receptor binding in the substantia nigra, but did selectively enhance social interaction in TM-NRG1 mutants (Long et al. 2012). Chronic cannabidiol, in a dose-dependent fashion, also selectively increased $GABA_A$ receptor binding in the granular retrosplenial cortex in NRG1 mutants. These data would suggest that NRG1 variation might alter sensitivity to the neurobehavioural effects of THC, particularly under conditions of stress, and may suggest new clinical hypotheses regarding G×E interactions in psychosis. These results are heuristic but have yet to be studied in clinical populations.

14.10 Future Research

Insel (2010) has suggested that our understanding of schizophrenia will be improved not only by knowledge on the genomics of psychosis but also by identification of environmental factors and mapping of any crucial epigenetic modifications. In relation to preclinical data examining biological factors modulating induction of cannabis-induced psychosis, many of the studies in animal models can describe environmental or G×E interactions at a descriptive level; until supplemented by subsequent molecular, cellular and physiological studies, they cannot identify the neuronal basis of such interactions. The lack of agreement with respect to the effects of adolescent THC on behaviours related to psychosis and their molecular and cellular correlates may be attributable to variable use across studies of incremental or irregular dosing regimen (Quinn et al. 2008).

Taken together, the clinical and preclinical genetic data provide convergent evidence for the notion of an interaction between cannabis and individual genetic vulnerability, with a focus on genes encoding proteins implicated in DA signalling. In relation to clinical genetic studies implicating COMT in risk for psychosis, a number of key issues have been identified in this field, principally the lack of consistent replication of clinical G×E associations, due to gene selection, underpowered samples, 'pseudoreplications' and continued debate about what constitutes a true replication, as well as heterogeneity in research design and outcome measures. Several authors (e.g., Casadio et al. 2011) have suggested the COMT×cannabis interaction as an example of current problems in the field, since the initial report by Caspi et al. (2005) has yet to be replicated with the same outcome-measure, same genotype, same direction of association or same definition of environmental exposure. Authors have emphasised on the necessity for adequately powered studies and increased density of genetic markers (Duncan et al. 2011; Decoster et al. 2012).

Recent clinical evidence suggest that genetic modulation of DA D2-AKT1-COMT related prefrontal-subcortical circuits could, at least in part, influence cognitive dysfunction in psychosis and its treatment (Tan et al. 2012). COMT Val158Met variation influenced PFC control of both parietal processing in maintenance of working memory and striatal processing in manipulation of working memory. DA D2R and AKT1 polymorphisms implicated in DA D2R signalling influenced only the prefrontal-striatal network associated with manipulation. In the context of antipsychotic drugs, the DA D2R and AKT1 polymorphisms altered dose–response effects of antipsychotic drugs on cognition in schizophrenia. In an earlier study, Caldú et al. (2007) reported that during a response inhibition task carried out by healthy subjects, greater PFC activation was observed in carriers of the DA transporter DAT-9-allele or in COMT Met carriers, as compared to DAT 10/10 carriers or COMT Val carriers. In a recent study by Stelzel et al. (2009), COMT Met carriers outperformed Val carriers in a working memory task, but only under conditions where DA D2R density could be regarded as high—a D2R/ANKK-1-Taq-Li polymorphism related to DA D2 receptor density. This adds to a growing body of knowledge looking at combined influence of COMT and DA D2R polymorphisms on cognitive performance in schizophrenia (Reuter et al. 2005, 2006). Given the growing evidence for additive and multiplicative epistatic interactive effects of DA-linked genes on expression of cognitive deficits in schizophrenia, future studies should focus on genes regulating various DA signalling processes on cannabis-induced psychosis.

As indicated in previous sections, a large number of experimental studies of acute and chronic THC effects in both humans and animals have reported THC-induced changes in DA activation, some of which are normalised by antipsychotic drugs acting at the DA D2R. Improved understanding of the specificity of the role of DA in THC effects would be improved by a more comprehensive investigation of other antipsychotic drugs (e.g., quetiapine, which has low affinity for DA receptors, and aripirazole, a partial D2R agonist), as well as the influence of moderating environmental and genetic background factors.

Acknowledgment The authors' studies are supported by a Science Foundation Ireland Principal Investigator grant (07/IN.1/B960).

References

Abush H, Akirav I (2012) Short- and long-term cognitive effects of chronic cannabinoids administration in late-adolescence rats. PLoS One 7(2):e31731

Altar CA, Cai N, Bliven T et al (1997) Anterograde transport of brain-derived neurotrophic factor and its role in the brain. Nature 389(6653):856–860

Arnone D, Barrick TR, Chengappa S (2008) Corpus callosum damage in heavy marijuana use: preliminary evidence from diffusion tensor tractography and tract-based spatial statistics. Neuroimage 41(3):1067–74

Andersen SL, Teicher MH (2008) Stress, sensitive periods and maturational events in adolescent depression. Trends Neurosci 31(4):183–91

Arseneault L, Cannon M, Poulton R et al (2002) Cannabis use in adolescence and risk for adult psychosis: longitudinal prospective study. BMJ 325(7374):1212–1213

Arseneault L, Cannon M, Witton J et al (2004) Causal association between cannabis and psychosis: examination of the evidence. Br J Psychiatry 184:110–117

Ayhan Y, Sawa A, Ross CA et al (2009) Animal models of gene-environment interactions in schizophrenia. Behav Brain Res 204(2):274–81

Babovic D, O'Tuathaigh CM, O'Sullivan GJ et al (2007) Exploratory and habituation phenotype of heterozygous and homozygous COMT knockout mice. Behav Brain Res 183(2):236–239

Babovic D, O'Tuathaigh CM, O'Connor AM et al (2008) Phenotypic characterization of cognition and social behaviour in mice with heterozygous versus homozygous deletion of catechol-O-methyltransferase. Neuroscience 155(4):1021–1029

Baeza I, Graell M, Moreno D et al (2009) Cannabis use in children and adolescents with first episode psychosis: influence on psychopathology and short-term outcome (CAFEPS study). Schizophr Res 113(2–3):129–137

Barak S, Weiner I (2011) The M_1/M_4 preferring agonist xanomeline reverses amphetamine-, MK801- and scopolamine-induced abnormalities of latent inhibition: putative efficacy against positive, negative and cognitive symptoms in schizophrenia. Int J Neuropsychopharmacol 14(9):1233–46

Barnett JH, Scoriels L, Munafò MR (2008) Meta-analysis of the cognitive effects of the catechol-O-methyltransferase gene Val158/108Met polymorphism. Biol Psychiatry 64(2):137–144

Bava S, Frank LR, McQueeny T et al (2009) Altered white matter microstructure in adolescent substance users. Psychiatry Res 173(3):228–237

Behan AT, Hryniewiecka M, O'Tuathaigh CM et al (2012) Chronic adolescent exposure to delta-9-tetrahydrocannabinol in COMT mutant mice: impact on indices of dopaminergic, endocannabinoid and GABAergic pathways. Neuropsychopharmacology 37(7):1773–1783

Belue RC, Howlett AC, Westlake TM et al (1995) The ontogeny of cannabinoid receptors in the brain of postnatal and aging rats. Neurotoxicol Teratol 17(1):25–30

Bertolino A, Caforio G, Blasi G et al (2004) Interaction of COMT (Val(108/158)Met) genotype and olanzapine treatment on prefrontal cortical function in patients with schizophrenia. Am J Psychiatry 161(10):1798–1805

Bhattacharyya S, Crippa JA, Martin-Santos R et al (2009) Imaging the neural effects of cannabinoids: current status and future opportunities for psychopharmacology. Curr Pharm Des 15(22):2603–2614

Bhattacharyya S, Morrison PD, Fusar-Poli P (2010) Opposite effects of delta-9-tetrahydrocannabinol and cannabidiol on human brain function and psychopathology. Neuropsychopharmacology 35(3):764–774

Bhattacharyya S, Crippa JA, Allen P (2012) Induction of psychosis by Δ9-tetrahydrocannabinol reflects modulation of prefrontal and striatal function during attentional salience processing. Arch Gen Psychiatry 69(1):27–36

Bilder RM, Volavka J, Lachman HM et al (2004) The catechol-O-methyltransferase polymorphism: relations to the tonic-phasic dopamine hypothesis and neuropsychiatric phenotypes. Neuropsychopharmacology 29(11):1943–1961

Block RI, O'Leary DS, Ehrhardt JC (2000) Effects of frequent marijuana use on brain tissue volume and composition. Neuroreport 11(3):491–6

Bossong MG, van Berckel BN, Boellaard R et al (2009) Delta 9-tetrahydrocannabinol induces dopamine release in the human striatum. Neuropsychopharmacology 34(3):759–66

Bossong MG, Niesink RJ (2010) Adolescent brain maturation, the endogenous cannabinoid system and the neurobiology of cannabis-induced schizophrenia. Prog Neurobiol 92(3):370–385

Bossong MG, Jansma JM, van Hell HH (2012) Effects of δ9-tetrahydrocannabinol on human working memory function. Biol Psychiatry 71(8):693–699

Boucher AA, Hunt GE, Karl T et al (2007a) Heterozygous neuregulin 1 mice display greater baseline and Delta(9)-tetrahydrocannabinol-induced c-Fos expression. Neuroscience 149(4):861–870

Boucher AA, Arnold JC, Duffy L et al (2007b) Heterozygous neuregulin 1 mice are more sensitive to the behavioural effects of Delta9-tetrahydrocannabinol. Psychopharmacology 192(3):325–336

Boucher AA, Hunt GE, Micheau J et al (2011) The schizophrenia susceptibility gene neuregulin 1 modulates tolerance to the effects of cannabinoids. Int J Neuropsychopharmacology 14(5):631–643

Caldú X, Vendrell P, Bartrés-Faz D et al (2007) Impact of the COMT Val108/158 Met and DAT genotypes on prefrontal function in healthy subjects. Neuroimage 37(4):1437–1444

Casadio P, Fernandes C, Murray RM et al (2011) Cannabis use in young people: the risk for schizophrenia. Neurosci Biobehav Rev 35(8):1779–1787

Caspi A, Moffitt TE, Cannon M et al (2005) Moderation of the effect of adolescent-onset cannabis use on adult psychosis by a functional polymorphism in the catechol-O-methyltransferase gene: longitudinal evidence of a gene X environment interaction. Biol Psychiatry 57(10):1117–1127

Cha YM, White AM, Kuhn CM et al (2006) Differential effects of delta9-THC on learning in adolescent and adult rats. Pharmacol Biochem Behav 83(3):448–55

Cha YM, Jones KH, Kuhn CM et al (2007) Sex differences in the effects of delta9-tetrahydrocannabinol on spatial learning in adolescent and adult rats. Behav Pharmacol 18(5–6):563–569

Chen J, Lipska BK, Halim N et al (2004) Functional analysis of genetic variation in catechol-O-methyltransferase (COMT): effects on mRNA, protein, and enzyme activity in postmortem human brain. Am J Hum Genet 75(5):807–821

Cohen M, Solowij N, Carr V (2008) Cannabis, cannabinoids and schizophrenia: integration of the evidence. Aust N Z J Psychiatry 42(5):357–368

Collip D, van Winkel R, Peerbooms O et al (2011) COMT Val158Met-stress interaction in psychosis: role of background psychosis risk. CNS Neurosci Ther 17(6):612–619

Costas J, Sanjuán J, Ramos-Ràos R et al (2011) Interaction between COMT haplotypes and cannabis in schizophrenia: a case-only study in two samples from Spain. Schizophr Res 127(1–3):22–27

Coulston CM, Perdices M, Tennant CC (2007) The neuropsychological correlates of cannabis use in schizophrenia: lifetime abuse/dependence, frequency of use, and recency of use. Schizophr Res 96(1–3):169–184

Decoster J, van Os J, Kenis G et al (2011) Age at onset of psychotic disorder: cannabis, BDNF Val66Met, and sex-specific models of gene-environment interaction. Am J Med Genet B Neuropsychiatr Genet 156B(3):363–9

Decoster J, van Os J, Myin-Germeys I et al (2012) Genetic variation underlying psychosis-inducing effects of cannabis: critical review and future directions. Curr Pharm Des 18(32):5015–5023

de la Serna E, Mayoral M, Baeza I (2010) Cognitive functioning in children and adolescents in their first episode of psychosis: differences between previous cannabis users and nonusers. J Nerv Ment Dis 198(2):159–62

Delisi LE, Bertisch HC, Szulc KU (2006) A preliminary DTI study showing no brain structural change associated with adolescent cannabis use. Harm Reduct J 3:17

Dempster EL, Mill J, Craig IW et al (2006) The quantification of COMT mRNA in post mortem cerebellum tissue: diagnosis, genotype, methylation and expression. BMC Med Genet 7:10

Desbonnet L, O'Tuathaigh C, Clarke G et al (2012) Phenotypic effects of repeated psychosocial stress during adolescence in mice mutant for the schizophrenia risk gene neuregulin-1: a putative model of gene × environment interaction. Brain Behav Immun 26(4):660–71

Dumontheil I, Roggeman C, Ziermans T et al (2011) Influence of the COMT genotype on working memory and brain activity changes during development. Biol Psychiatry 70(3):222–229

Duncan LE, Keller MC (2011) A critical review of the first 10 years of candidate gene-by-environment interaction research in psychiatry. Am J Psychiatry 168(10):1041–9

Durstewitz D, Seamans JK (2008) The dual-state theory of prefrontal cortex dopamine function with relevance to catechol-o-methyltransferase genotypes and schizophrenia. Biol Psychiatry 64(9):739–749

Di Forti M, Morgan C, Dazzan P et al (2009) High-potency cannabis and the risk of psychosis. J Psychiatry 195(6):488–491

D'Souza DC, Perry E, MacDougall L et al (2004) The psychotomimetic effects of intravenous delta-9-tetrahydrocannabinol in healthy individuals: implications for psychosis. Neuropsychopharmacology 29(8):1558–1572

D'Souza DC, Abi-Saab WM, Madonick S et al (2005) Delta-9-tetrahydrocannabinol effects in schizophrenia: implications for cognition, psychosis, and addiction. Biol Psychiatry 57(6):594–608

D'Souza DC, Ranganathan M, Braley G et al (2008a) Blunted psychotomimetic and amnestic effects of delta-9-tetrahydrocannabinol in frequent users of cannabis. Neuropsychopharmacology 33(10):2505–2516

D'Souza DC, Braley G, Blaise R et al (2008b) Effects of haloperidol on the behavioural, subjective, cognitive, motor, and neuroendocrine effects of Delta-9-tetrahydrocannabinol in humans. Psychopharmacology 198(4):587–603

D'Souza DC, Sewell RA, Ranganathan M (2009) Cannabis and psychosis/schizophrenia: human studies. Eur Arch Psychiatry Clin Neurosci 259(7):413–431

Egan MF, Goldberg TE, Kolachana BS et al (2001) Effect of COMT Val108/158 Met genotype on frontal lobe function and risk for schizophrenia. Proc Natl Acad Sci U S A 98(12):6917–6922

Egan MF, Kojima M, Callicott JH et al (2003) The BDNF val66met polymorphism affects activity-dependent secretion of BDNF and human memory and hippocampal function. Cell 112(2):257–269

Egerton A, Allison C, Brett RR et al (2006) Cannabinoids and prefrontal cortical function: insights from preclinical studies. Neurosci Biobehav Rev 30(5):680–695

Eggan SM, Lewis DA (2007) Immunocytochemical distribution of the cannabinoid CB1R receptor in the primate neocortex: a regional and laminar analysis. Cereb Cortex 17(1):175–191

Eggan SM, Mizoguchi Y, Stoyak SR et al (2010) Development of cannabinoid 1 receptor protein and messenger RNA in monkey dorsolateral prefrontal cortex. Cereb Cortex 20(5):1164–1174

Ellgren M, Spano SM, Hurd YL (2007) Adolescent cannabis exposure alters opiate intake and opioid limbic neuronal populations in adult rats. Neuropsychopharmacology 32(3):607–615

Estrada G, Fatjó-Vilas M, Muñoz MJ et al (2011) Cannabis use and age at onset of psychosis: further evidence of interaction with COMT Val158Met polymorphism. Acta Psychiatr Scand 123(6):485–492

European Monitoring Centre for Drugs and Drug Addiction (EMCDDA) (2010) Statistical Bulletin. http://www.emcdda.europa.eu/stats10

Fergusson DM, Horwood LJ, Swain-Campbell NR (2003) Cannabis dependence and psychotic symptoms in young people. Psychol Med 33(1):15–21

Freund TF, Katona I, Piomelli D (2003) Role of endogenous cannabinoids in synaptic signalling. Physiol Rev 83(3):1017–1066

Giuffrida A, Leweke FM, Gerth CW (2004) Cerebrospinal anandamide levels are elevated in acute schizophrenia and are inversely correlated with psychotic symptoms. Neuropsychopharmacology 29(11):2108–2114

Gleason KA, Birnbaum SG, Shukla A et al (2012) Susceptibility of the adolescent brain to cannabinoids: long-term hippocampal effects and relevance to schizophrenia. Transl Psychiatry 2:e199

Gogos JA, Morgan M, Luine V (1998) Catechol-O-methyltransferase-deficient mice exhibit sexually dimorphic changes in catecholamine levels and behaviour. Proc Natl Acad Sci U S A 95(17):9991–9996

Gothelf D, Michaelovsky E, Frisch A et al (2007) Association of the low-activity COMT 158Met allele with ADHD and OCD in subjects with velocardiofacial syndrome. Int J Neuropsychopharmacol 10(3):301–308

Gothelf D, Schaer M, Eliez S (2008) Genes, brain development and psychiatric phenotypes in velo-cardio-facial syndrome. Dev Disabil Res Rev 14(1):59–68

Grant I, Gonzalez R, Carey CL et al (2003) Non-acute (residual) neurocognitive effects of cannabis use: a meta-analytic study. J Int Neuropsychol Soc 9(5):679–689

Gray L, Hannan AJ (2007) Dissecting cause and effect in the pathogenesis of psychiatric disorders: genes, environment and behaviour. Curr Mol Med 7(5):470–478

Grech A, Van Os J, Jones Pbet al (2005) Cannabis use and outcome of recent onset psychosis. Eur Psychiatry 20(4):349–353

Green AI, Tohen MF, Hamer RM et al (2004) First episode schizophrenia-related psychosis and substance use disorders: acute response to olanzapine and haloperidol. Schizophr Res 66(2–3):125–135

Gruber SA, Yurgelun-Todd DA (2005) Neuroimaging of marijuana smokers during inhibitory processing: a pilot investigation. Brain Res Cogn Brain Res 23(1):107–118

Haasio K, Huotari M, Nissinen E et al (2003) Tissue histopathology, clinical chemistry and behaviour of adult Comt-gene-disrupted mice. J Appl Toxicol 23(4):213–219

Harte LC, Dow-Edwards D (2010) Sexually dimorphic alterations in locomotion and reversal learning after adolescent tetrahydrocannabinol exposure in the rat. Neurotoxicol Teratol 32(5):515–524

Harrison PJ, Law AJ (2006) Neuregulin 1 and schizophrenia: genetics, gene expression, and neurobiology. Biol Psychiatry 60(2):132–140

Henquet C, Rosa A, Krabbendam L (2006) An experimental study of catechol-o-methyltransferase Val158Met moderation of delta-9-tetrahydrocannabinol-induced effects on psychosis and cognition. Neuropsychopharmacology 31(12):2748–2757

Ho BC, Wassink TH, Ziebell S et al (2011) Cannabinoid receptor 1 gene polymorphisms and marijuana misuse interactions on white matter and cognitive deficits in schizophrenia. Schizophr Res 128(1–3):66–75

Hoenicka J, Garrido E, Martànez I et al (2010) Gender-specific COMT Val158Met polymorphism association in Spanish schizophrenic patients. Am J Med Genet B Neuropsychiatr Genet 153B(1):79–85

Huotari M, Santha M, Lucas LR et al (2002) Effect of dopamine uptake inhibition on brain catecholamine levels and locomotion in catechol-O-methyltransferase-disrupted mice. J Pharmacol Exp Ther 303(3):1309–1316

Insel TR (2010) Rethinking schizophrenia. Nature 468(7321):187–193

Ishizuka K, Tajinda K, Colantuoni C (2010) Negative symptoms of schizophrenia correlate with impairment on the University of Pennsylvania smell identification test. Neurosci Res 66(1):106–110

Jiang H, Xie T, Ramsden DB et al (2003) Human catechol-O-methyltransferase down-regulation by estradiol. Neuropharmacology 45(7):1011–1018

Kamath V, Moberg PJ, Gur RE et al (2012) Effects of the val(158)met catechol-O-methyltransferase gene polymorphism on olfactory processing in schizophrenia. Behav Neurosci 126(1):209–215

Kang C, Xu X, Liu H et al (2010) Association study of catechol-O-methyltransferase (COMT) gene Val158Met polymorphism with auditory P300 in Chinese Han patients with schizophrenia. Psychiatry Res 180(2–3):153–155

Kapur S (2003) Psychosis as a state of aberrant salience: a framework linking biology, phenomenology, and pharmacology in schizophrenia. Am J Psychiatry 160(1):13–23

Kapur S (2004) How antipsychotics become anti-"psychotic"–from dopamine to salience to psychosis. Trends Pharmacol Sci 25(8):402–406

Karayiorgou M, Simon TJ, Gogos JA (2010) 22q11.2.2 microdeletions: linking DNA structural variation to brain dysfunction and schizophrenia. Nat Rev Neurosci 11(6):402–416

Keefe RS, Harvey PD (2012) Cognitive impairment in schizophrenia. Handb Exp Pharmacol 213:11–37

Kimoto S, Muraki K, Toritsuka M et al (2012) Selective overexpression of Comt in prefrontal cortex rescues schizophrenia-like phenotypes in a mouse model of 22q11.2 deletion syndrome. Transl Psychiatry 2:e146

Kirby BP, Waddington JL, O'Tuathaigh CM (2010) Advancing a functional genomics for schizophrenia: psychopathological and cognitive phenotypes in mutants with gene disruption. Brain Res Bull 83(3–4):162–176

Kleinloog D, Liem-Moolenaar M, Jacobs G et al (2012) Does olanzapine inhibit the psychomimetic effects of Δ^9-tetrahydrocannabinol? J Psychopharmacol 26(10):1307–1316

Kravariti E, Jacobson C, Morris R et al (2010) Memory in intellectually matched groups of young participants with 22q11.2.2 deletion syndrome and those with schizophrenia. Res Dev Disabil 31(3):864–868

Lajiness-O'Neill RR, Beaulieu I, Titus JB et al (2005) Memory and learning in children with 22q11.2.2 deletion syndrome: evidence for ventral and dorsal stream disruption? Child Neuropsychol 11(1):55–71

Leitman DI, Laukka P, Juslin PN et al (2010) Getting the cue: sensory contributions to auditory emotion recognition impairments in schizophrenia. Schizophr Bull 36(3):545–556

Leweke FM, Giuffrida A, Wurster U et al (1999) Elevated endogenous cannabinoids in schizophrenia. Neuroreport 10(8):1665–1669

Leweke FM, Schneider M (2011) Chronic pubertal cannabinoid treatment as a behavioural model for aspects of schizophrenia: effects of the atypical antipsychotic quetiapine. Int J Neuropsychopharmacol 14(1):43–51

Liem-Moolenaar M, te Beek ET, de Kam ML et al (2010) Central nervous system effects of haloperidol on THC in healthy male volunteers. J Psychopharmacol 24(11):1697–1708

Linszen D, van Amelsvoort T (2007) Cannabis and psychosis: an update on course and biological plausible mechanisms. Curr Opin Psychiatry 20(2):116–120

Løberg EM, Hugdahl K (2009) Cannabis use and cognition in schizophrenia. Front Hum Neurosci 3:53

Long LE, Chesworth R, Arnold JC et al (2010) A follow-up study: acute behavioural effects of Delta(9)-THC in female heterozygous neuregulin 1 transmembrane domain mutant mice. Psychopharmacology 211(3):277–289

Long LE, Chesworth R, Huang XF et al (2012) Distinct neurobehavioural effects of cannabidiol in transmembrane domain neuregulin 1 mutant mice. PLoS One 7(4):e34129

Lu BY, Martin KE, Edgar JC et al (2007) Effect of catechol O-methyltransferase val(158)met polymorphism on the p50 gating endophenotype in schizophrenia. Biol Psychiatry 62(7):822–825

Malone DT, Hill MN, Rubino T (2010) Adolescent cannabis use and psychosis: epidemiology and neurodevelopmental models. Br J Pharmacol 160(3):511–522

Mato S, Del Olmo E, Pazos A (2003) Ontogenetic development of cannabinoid receptor expression and signal transduction functionality in the human brain. Eur J Neurosci 17(9):1747–1754

McGrath J, Welham J, Scott J (2010) Association between cannabis use and psychosis-related outcomes using sibling pair analysis in a cohort of young adults. Arch Gen Psychiatry 67(5):440–447

McLaren J, Swift W, Dillon P et al (2008) Cannabis potency and contamination: a review of the literature. Addiction 103(7):1100–1109

McLaren JA, Silins E, Hutchinson D et al (2010) Assessing evidence for a causal link between cannabis and psychosis: a review of cohort studies. Int J Drug Policy 21(1):10–19

Medina KL, Nagel BJ, Park A et al (2007) Depressive symptoms in adolescents: associations with white matter volume and marijuana use. J Child Psychol Psychiatry 48(6):592–600

Meechan DW, Tucker ES, Maynard TM et al (2009) Diminished dosage of 22q11.2 genes disrupts neurogenesis and cortical development in a mouse model of 22q11.2 deletion/DiGeorge syndrome. Proc Natl Acad Sci U S A 106(38):16434–16445

Meijer JH, Dekker N, Koeter MW, et al (2012) Cannabis and cognitive performance in psychosis: a cross-sectional study in patients with non-affective psychotic illness and their unaffected siblings. Psychol Med 42(4):705–716

Mier D, Kirsch P, Meyer-Lindenberg A (2010) Neural substrates of pleiotropic action of genetic variation in COMT: a meta-analysis. Mol Psychiatry 15(9):918–927

Munafò MR, Bowes L, Clark TG et al (2005) Lack of association of the COMT (Val158/108 Met) gene and schizophrenia: a meta-analysis of case-control studies. Mol Psychiatry 10(8):765–770

Matochik JA, Eldreth DA, Cadet JL et al (2005) Altered brain tissue composition in heavy marijuana users. Drug Alcohol Depend 77(1):23–30

Molero P, Ortuño F, Zalacain M et al (2007) Clinical involvement of catechol-O-methyltransferase polymorphisms in schizophrenia spectrum disorders: influence on the severity of psychotic symptoms and on the response to neuroleptic treatment. Pharmacogenomics J 7(6):418–426

Moore TH, Zammit S, Lingford-Hughes A, Barnes TR, Jones PB, Burke M, Lewis G (2007) Cannabis use and risk of psychotic or affective mental health outcomes: A systematic review. Lancet 370(9584):319–328

Morgan CJ, Curran HV (2008) Effects of cannabidiol on schizophrenia-like symptoms in people who use cannabis. Br J Psychiatry 192(4):306–7

Morgan CJ, Duffin S, Hunt S et al (2012) Neurocognitive function and schizophrenia-proneness in individuals dependent on ketamine, on high potency cannabis ('skunk') or on cocaine. Pharmacopsychiatry 45(7):269–74

Morrison PD, Murray RM (2009) From real-world events to psychosis: the emerging neuropharmacology of delusions. Schizophr Bull 35(4):668–674

Morrison PD, Zois V, McKeown DA et al (2009) The acute effects of synthetic intravenous Delta9-tetrahydrocannabinol on psychosis, mood and cognitive functioning. Psychol Med 39(10):1607–1616

Moser PC, Hitchcock JM, Lister S et al (2000) The pharmacology of latent inhibition as an animal model of schizophrenia. Brain Res Brain Res Rev 33(2–3):275–307

Murphy KC, Jones LA, Owen MJ (1999) High rates of schizophrenia in adults with velo-cardio-facial syndrome. Arch Gen Psychiatry 56(10):940–945

Murphy KC, Owen MJ (2001) Velo-cardio-facial syndrome: a model for understanding the genetics and pathogenesis of schizophrenia. Br J Psychiatry 179:397–402

Murray RM, Morrison PD, Henquet C et al (2007) Cannabis, the mind and society: the harsh realities. Nat Rev Neurosci 8(11):885–895

Murray RM, Lappin J, Di Forti M (2008) Schizophrenia: from developmental deviance to dopamine dysregulation. Eur Neuropsychopharmacol 18:129–134.

Okochi T, Ikeda M, Kishi T et al (2009) Meta-analysis of association between genetic variants in COMT and schizophrenia: an update. Schizophr Res 110(1–3):140–148

O'Shea M, Singh ME, McGregor IS et al (2004) Chronic cannabinoid exposure produces lasting memory impairment and increased anxiety in adolescent but not adult rats. J Psychopharmacol 18(4):502–508

O'Shea M, McGregor IS, Mallet PE (2006) Repeated cannabinoid exposure during perinatal, adolescent or early adult ages produces similar longlasting deficits in object recognition and reduced social interaction in rats. J Psychopharmacol 20(5):611–621

O'Tuathaigh CM, Hryniewiecka M, Behan A et al (2010) Chronic adolescent exposure to Δ-9-tetrahydrocannabinol in COMT mutant mice: impact on psychosis-related and other phenotypes. Neuropsychopharmacology 35(11):2262–2273

O'Tuathaigh CM, Clarke G, Walsh J et al (2012a) Genetic vs. pharmacological inactivation of COMT influences cannabinoid-induced expression of schizophrenia-related phenotypes. Int J Neuropsychopharmacol 15(9):1331–1342

O'Tuathaigh CM, Desbonnet L, Waddington JL (2012b) Mutant mouse models in evaluating novel approaches to antipsychotic treatment. Handb Exp Pharmacol 213:113–45

Papaleo F, Crawley JN, Song J et al (2008) Genetic dissection of the role of catechol-O-methyltransferase in cognition and stress reactivity in mice. J Neurosci 28(35):8709–8723

Papaleo F, Lipska BK, Weinberger DR (2012) Mouse models of genetic effects on cognition: relevance to schizophrenia. Neuropharmacology 62(3):1204–1220

Pelayo-Terán JM, Pérez-Iglesias R, Mata I et al (2010) Catechol-o-methyltransferase (COMT) Val158Met variations and cannabis use in first-episode non-affective psychosis: clinical-onset implications. Psychiatry Res 179(3):291–296

Pelayo-Terán JM, Suárez-Pinilla P, Chadi N et al (2012) Gene-environment interactions underlying the effect of cannabis in first episode psychosis. Curr Pharm Des 18(32):5024–5035

Pelletier M, Achim AM, Montoya A et al (2005) Cognitive and clinical moderators of recognition memory in schizophrenia: a meta-analysis. Schizophr Res 74(2–3):233–252

Pistis M, Muntoni AL, Pillolla G et al (2002) Cannabinoids inhibit excitatory inputs to neurons in the shell of the nucleus accumbens: an in vivo electrophysiological study. Eur J Neurosci 15(11):1795–1802

Pitts DK, Freeman AS, Chiodo LA (1990) Dopamine neuron ontogeny: electrophysiological studies. Synapse 6(4):309–320

Potvin S, Joyal CC, Pelletier J et al (2008) Contradictory cognitive capacities among substance-abusing patients with schizophrenia: a meta-analysis. Schizophr Res 100(1–3):242–251

Prasad SE, Howley S, Murphy KC (2008) Candidate genes and the behavioural phenotype in 22q11.2.2 deletion syndrome. Dev Disabil Res Rev 14(1):26–34

Quinn HR, Matsumoto I, Callaghan PD et al (2008) Adolescent rats find repeated Delta(9)-THC less aversive than adult rats but display greater residual cognitive deficits and changes in hippocampal protein expression following exposure. Neuropsychopharmacology 33(5):1113–1126

Rabin RA, Zakzanis KK, George TP (2011) The effects of cannabis use on neurocognition in schizophrenia: a meta-analysis. Schizophr Res 128(1–3):111–116

Rais M, Cahn W, Van Haren N et al (2008) Excessive brain volume loss over time in cannabis-using first-episode schizophrenia patients. Am J Psychiatry 165(4):490–496

Realini N, Rubino T, Parolaro D (2009) Neurobiological alterations at adult age triggered by adolescent exposure to cannabinoids. Pharmacol Res 60(2):132–138

Renard D, Taieb G, Gras-Combe G et al (2012) Cannabis-related myocardial infarction and cardioembolic stroke. J Stroke Cerebrovasc Dis 21(1):82–83

Reuter M, Peters K, Schroeter K et al (2005) The influence of the dopaminergic system on cognitive functioning: a molecular genetic approach. Behav Brain Res 164(1):93–99

Reuter M, Schmitz A, Corr P et al (2006) Molecular genetics support Gray's personality theory: the interaction of COMT and DRD2 polymorphisms predicts the behavioural approach system. Int J Neuropsychopharmacol 9(2):155–166

Ringen PA, Melle I, Berg AO (2013) Cannabis use and premorbid functioning as predictors of poorer neurocognition in schizophrenia spectrum disorder. Schizophr Res 143(1):84–9

Roiser JP, Stephan KE, den Ouden HE et al (2009) Do patients with schizophrenia exhibit aberrant salience? Psychol Med 39(2):199–209

Roth WT, Pfefferbaum A (1992) Abnormalities of the left temporal lobe in schizophrenia. N Engl J Med 327(23):1689

Rubino T, Guidali C, Vigano D (2008) CB1R receptor stimulation in specific brain areas differently modulate anxiety-related behaviour. Neuropharmacology 54(1):151–160

Rubino T, Realini N, Braida D, Alberio T, Capurro V, Vigano D (2009a) The depressive phenotype induced in adult female rats by adolescent exposure toTHC isassociated with cognitive impairment and altered neuroplasticity in the prefrontal cortex. Neurotox Res 15(4):291–302

Rubino T, Realini N, Braida D, Guidi S, Capurro V, Vigano D (2009b) Changes in hippocampal morphology and neuroplasticity induced by adolescent THC treatment are associated with cognitive impairment inadulthood. Hippocampus 19(8):763–772

Rubino T, Realini N, Braida D (2012) The depressive phenotype induced in adult female rats by adolescent exposure to THC is associated with cognitive impairment and altered neuroplasticity in the prefrontal cortex. Neurotox Res 15(4):291–302

Schnell T, Koethe D, Daumann J et al (2009) The role of cannabis in cognitive functioning of patients with schizophrenia. Psychopharmacology 205(1):45–52

Schneider M, Koch M (2003) Chronic pubertal, but not adult chronic cannabinoid treatment impairs sensorimotor gating, recognition memory, and the performance in a progressive ratio task in adult rats. Neuropsychopharmacology 28(10):1760–1769

Schneider M (2008) Puberty as a highly vulnerable developmental period for the consequences of cannabis exposure. Addict Biol 13(2):253–263

Schubart CD, Sommer IE, van Gastel WA (2011) Cannabis with high cannabidiol content is associated with fewer psychotic experiences. Schizophr Res 130(1–3):216–221

Seamans JK, Yang CR (2004) The principal features and mechanisms of dopamine modulation in the prefrontal cortex. Prog Neurobiol 74(1):1–58

Seeman P, Bzowej NH, Guan HC et al (1987) Human brain dopamine receptors in children and aging adults. Synapse 1(5):399–404

Seeman P (2011) Schizophrenia diagnosis and treatment. CNS Neurosci Ther 17(2):81–82

Shaikh M, Hall MH, Schulze K et al (2011) Do COMT, BDNF and NRG1 polymorphisms influence P50 sensory gating in psychosis? Psychol Med 41(2):263–276

Shifman S, Bronstein M, Sternfeld M et al (2002) A highly significant association between a COMT haplotype and schizophrenia. Am J Hum Genet 71(6):1296–1302

Solowij N, Michie PT (2007) Cannabis and cognitive dysfunction: parallels with endophenotypes of schizophrenia? J Psychiatry Neurosci 32(1):30–52

Solowij N, Walterfang M, Lubman DI et al (2013) Alteration to hippocampal shape in cannabis users with and without schizophrenia. Schizophr Res 143(1):179–84

Spear LP (2000) The adolescent brain and age-related behavioural manifestations. Neurosci Biobehav Rev 24(4):417–463

Spear LP (2009) Heightened stress responsivity and emotional reactivity during pubertal maturation: implications for psychopathology. Dev Psychopathol 21(1):87–97

Stark KL, Xu B, Bagchi A et al (2008) Altered brain microRNA biogenesis contributes to phenotypic deficits in a 22q11.2-deletion mouse model. Nat Genet 40(6):751–760

Stefanis NC, Delespaul P, Henquet C et al (2004) Early adolescent cannabis exposure and positive and negative dimensions of psychosis. Addiction 99(10):1333–1341

Stefansson H, Sigurdsson E, Steinthorsdottir V et al (2002) Neuregulin 1 and susceptibility to schizophrenia. Am J Hum Genet 71(4):877–892

Stelzel C, Basten U, Montag C et al (2009) Effects of dopamine-related gene-gene interactions on working memory component processes. Eur J Neurosci 29(5):1056–1063

Stokes PR, Mehta MA, Curran HV et al (2009) Can recreational doses of THC produce significant dopamine release in the human striatum? Neuroimage 48(1):186–90

Stokes PR, Rhodes RA, Grasby PM et al (2011) The effects of the COMT Val108/158Met polymorphism on BOLD activation during working memory, planning, and response inhibition: a role for the posterior cingulate cortex? Neuropsychopharmacology 36(4):763–71

Stokes PR, Egerton A, Watson B (2012) History of cannabis use is not associated with alterations in striatal dopamine D2/D3 receptor availability. J Psychopharmacol 26(1):144–9

Tammimäki A, Käenmäki M, Kambur O et al (2010) Effect of S-COMT deficiency on behaviour and extracellular brain dopamine concentrations in mice. Psychopharmacology 211(4):389–401

Tan HY, Chen AG, Kolachana B et al (2012) Effective connectivity of AKT1-mediated dopaminergic working memory networks and pharmacogenetics of anti-dopaminergic treatment. Brain 135(Pt 5):1436–1445

Tee SF, Tang PY, Loh HC (2011) No evidence for association between DRD3 and COMT with schizophrenia in a Malay population. Genet Mol Res 10(3):1850–1855

Tosato S, Dazzan P, Collier D (2005) Association between the neuregulin 1 gene and schizophrenia: a systematic review. Schizophr Bull 31(3):613–617

Tovilla-Zárate C, Medellàn BC, Fresán A et al (2013) No association between catechol-o-methyltransferase Val108/158Met polymorphism and schizophrenia or its clinical symptomatology in a Mexican population. Mol Biol Rep 40(2):2053–2058

Tunbridge EM, Weinberger DR, Harrison PJ (2006) A novel protein isoform of catechol O-methyltransferase (COMT): brain expression analysis in schizophrenia and bipolar disorder and effect of Val158Met genotype. Mol Psychiatry 11(2):116–117
Uçok A, Oztürk M, Duman Z et al (2010) COMT Val 158 Met polymorphism is related with interpersonal problem solving in schizophrenia. Eur Psychiatry 25(6):320–322
van den Buuse M (2010) Modelling the positive symptoms of schizophrenia in genetically modified mice: pharmacology and methodology aspects. Schizophr Bull 36(2):246–70
van Hell HH, Jager G, Bossong MG (2011) Involvement of the endocannabinoid system in reward processing in the human brain. Psychopharmacology 219(4):981–990
van Os J, Bak M, Hanssen M et al (2002) Cannabis use and psychosis: a longitudinal population-based study. Am J Epidemiol 156(4):319–327
van Os J, Rutten BP, Poulton R (2009) Gene-environment interactions in schizophrenia: review of epidemiological findings and future directions. Schizophr Bull 34(6):1066–82
van Winkel R, Genetic Risk and Outcome of Psychosis (GROUP) Investigators (2011) Family-based analysis of genetic variation underlying psychosis-inducing effects of cannabis: sibling analysis and proband follow-up. Arch Gen Psychiatry 68(2):148–157
Waddington JL, Hennessy RJ, O'Tuathaigh CMP et al (2012) Schizophrenia and the lifetime trajectory of psychotic illness: developmental neuroscience and pathobiology. In: Patterson PH (ed) The origins of schizophrenia. Columbia University Press, New York
Wang Y, Hu Y, Fang Y et al (2009) Evidence of epistasis between the catechol-O-methyltransferase and aldehyde dehydrogenase 3B1 genes in paranoid schizophrenia. Biol Psychiatry 65(12):1048–1054
Wegener N, Koch M (2009) Behavioural disturbances and altered Fos protein expression in adult rats after chronic pubertal cannabinoid treatment. Brain Res 1253:81–91
Welch KA, Moorhead TW, McIntosh AM (2012) Tensor-based morphometry of cannabis use on brain structure in individuals at elevated genetic risk of schizophrenia. Psychol Med 29:1–10
Wilson W, Mathew R, Turkington T et al (2000) Brain morphological changes and early marijuana use: a magnetic resonance and positron emission tomography study. J Addict Dis 19(1):1–22
Wonodi I, Mitchell BD, Stine OC et al (2006) Lack of association between COMT gene and deficit/nondeficit schizophrenia. Behav Brain Funct 2:42
Yücel M, Solowij N, Respondek C et al (2008) Regional brain abnormalities associated with long-term heavy cannabis use. Arch Gen Psychiatry 65(6):694–701
Yücel M, Bora E, Lubman DI et al (2012) The impact of cannabis use on cognitive functioning in patients with schizophrenia: a meta-analysis of existing findings and new data in a first-episode sample. Schizophr Bull 38(2):316–330
Zammit S, Spurlock G, Williams H et al (2007) Genotype effects of CHRNA7, CNR1 and COMT in schizophrenia: interactions with tobacco and cannabis use. Br J Psychiatry 191:402–407
Zammit S, Owen MJ, Evans J et al (2011) Cannabis, COMT and psychotic experiences. Br J Psychiatry 199(5):380–385
Zavitsanou K, Wang H, Dalton VS et al (2010) Cannabinoid administration increases 5HT1A receptor binding and mRNA expression in the hippocampus of adult but not adolescent rats. Neuroscience 169(1):315–324
Zuardi AW, Shirakawa I, Finkelfarb E et al (1982) Action of cannabidiol on the anxiety and other effects produced by delta 9-THC in normal subjects. Psychopharmacology 76(3):245–250

Index

E. J. Van Bockstaele (ed.), *Endocannabinoid Regulation of Monoamines in Psychiatric and Neurological Disorders*, DOI 10.1007/978-1-4614-7940-6,

N

About the Editor

Elisabeth Van Bockstaele is a Professor in the Department of Neuroscience and Farber Institute for Neurosciences at Thomas Jefferson University in Philadelphia, PA. She obtained her PhD from New York University and conducted her postdoctoral work at Cornell University Medical College in New York City. She served as Vice-Chair of Research in the Department of Neurological Surgery and is the current founding Director of the Graduate Program in Neuroscience in the College of Biomedical and Graduate Studies at Thomas Jefferson University. She has served on multiple grant review panels including serving as Chair of the Neuroimmunology, Neuroendocrinology and Behavior Study Section at the National Institutes of Health and she served as Chair of the Membership and Chapters Committee of the Society for Neuroscience (SfN) and as a Member of the Committee on Neuroscience Departments and Programs and Professional Development Committee at SfN. She was also a fellow in the Executive Leadership in Academic Medicine program at Drexel University College of Medicine. Dr. Van Bockstaele has devoted her scientific career to understanding the role of norepinephrine in stress-related illness, particularly as it relates to psychiatric disorders. Her research has primarily focused on preclinical studies examining the cellular adaptations of noradrenergic circuits to drugs of abuse (opiates, cannabinoids and psychostimulants) but more recently has expanded to include clinical investigations. Ongoing research efforts are aimed at understanding stress-related psychiatric disorders, novel opiate detoxification approaches for counteracting norepinephrine over-activity following withdrawal from opiates as well as elucidating the impact of stress on vulnerability to substance abuse. Her laboratory is one of the only research groups studying interactions between the endocannabinoid and noradrenergic systems and her group employs state-of-the-art high-resolution neuroanatomical approaches with subcellular precision to understand the nature of state dependent interactions of this integrative system for the treatment of stress-induced anxiety disorders.

E. J. Van Bockstaele (ed.), *Endocannabinoid Regulation of Monoamines in Psychiatric and Neurological Disorders*, DOI 10.1007/978-1-4614-7940-6,

Printed by Publishers' Graphics LLC